Mechanics of Flow-Induced Sound and Vibration, Volume 1

Mechanics of Flow-Induced Sound and Vibration, Volume 1

General Concepts and Elementary Sources

Second Edition

William K. Blake

ELSEVIER

ACADEMIC PRESS
An imprint of Elsevier

Academic Press is an imprint of Elsevier
125 London Wall, London EC2Y 5AS, United Kingdom
525 B Street, Suite 1800, San Diego, CA 92101-4495, United States
50 Hampshire Street, 5th Floor, Cambridge, MA 02139, United States
The Boulevard, Langford Lane, Kidlington, Oxford OX5 1GB, United Kingdom

Notices
Knowledge and best practice in this field are constantly changing. As new research and experience broaden our understanding, changes in research methods, professional practices, or medical treatment may become necessary.

Practitioners and researchers must always rely on their own experience and knowledge in evaluating and using any information, methods, compounds, or experiments described herein. In using such information or methods they should be mindful of their own safety and the safety of others, including parties for whom they have a professional responsibility.

To the fullest extent of the law, neither the Publisher nor the authors, contributors, or editors, assume any liability for any injury and/or damage to persons or property as a matter of products liability, negligence or otherwise, or from any use or operation of any methods, products, instructions, or ideas contained in the material herein.

British Library Cataloguing-in-Publication Data
A catalogue record for this book is available from the British Library

Library of Congress Cataloging-in-Publication Data
A catalog record for this book is available from the Library of Congress

ISBN: 978-0-12-809273-6

Publisher: Joe Hayton
Acquisition Editor: Brian Guerin
Editorial Project Manager: Edward Payne
Production Project Manager: Kiruthika Govindaraju
Cover Designer: Greg Harris

Typeset by MPS Limited, Chennai, India

Dedication

To my wife, Donna

Contents

5. Fundamentals of Flow-Induced Vibration and Noise

6. Introduction to Bubble Dynamics and Cavitation

Preface to the Second Edition

It has been 31 years since the publication of the first edition of this book and I believe that the foundations and fundamentals of the combined subject of aero-hydro acoustics were well established at the time of the first edition. However, in the time since then while there have been developments in those fundamentals there has also been an extensive growth in applications and methods of applications. This growth has been made possible by the development of computational tools, personal computers, data acquisition hardware and software, and sensors. These were not available at the time of Edition 1. In fact personal tools such as, *Matlab, Mathematica, Mathcad*, and *Labview*, now widely used in academic and commercial applications were not available to the reader either. The science of aero-hydro acoustic phenomena has really benefitted from the use of simultaneously-collected multichannel sensor arrays as well. Finally the range of applications has grown under the combined pulls of consumer awareness and intolerance of noise and vibration, public legislation requiring noise control, and military needs.

Computational tools have made possible both direct numerical simulations for research and detailed design engineering applications. I have attempted to selectively extend the coverage of Edition 1 into these new growth areas while at the same time maintaining the structure and philosophy of the book and not substantially increasing its size. In some areas, the newly developed numerical technologies have made it possible to conduct "numerical experiments" that parallel and complement physical experiments, thereby leveraging the capabilities of both. I have used some of these in the areas of jet noise, boundary layer noise, and rotor noise as examples to address the application of numerical techniques. I have avoided going into numerical methods, however, since there are now numerous books on the techniques of computational fluid mechanics, large eddy simulations, and finite element methods making it duplicative to address these techniques, themselves.

The formalisms developed here are suitable for evaluation on a personal computer, but closed-form asymptotic solutions are also given for immediate interpretation for understanding trends in data. The book is written principally as a reference work, although it may be used as a teaching aid. The reader will always find theoretical results supported by step-by-step derivations that identify any assumptions made. For as many sources of sound as possible, each chapter is illustrated with comparisons of leading-order formulas, measured data, and results of numerical simulations.

In writing the first Edition 1 provided a comprehensive list of references in each focus area. Each of these I read and integrated into the text. This was intended in Edition 2, but I soon faced the reality that the number of papers published in any area is now too large to treat in this manner. One journal has a search engine that provides the user with a year-by-year distribution of papers published in a selected area. The annual publication rate in one area increased in that journal by a factor of 10 beginning in 1999–2000. Accordingly in this edition, the list of references has been expanded, but admittedly less exhaustively than in the first.

As noted previously the presentation philosophy and organization of the first edition has been maintained in this second edition with fundamentals central to Volume 1 and more complex geometry and fluid-structure interaction the subjects of Volume 2. Considering Volume 1, an area of addition and change is in Chapter 3, Shear Layer Instabilities, Flow Tones, and Jet Noise, where the discussion of turbulence statistics and jet noise have been changed and expanded; this required an additional section in Chapter 2, Theory of Sound and Its Generation by Flow, on the effects of source convection and the Doppler effect. Chapter 4, Dipole Sound From Cylinders, and Chapter 5, Fundamentals of Flow-Induced Vibration and Noise, have been updated to meet the needs of the other chapters for which they provide fundamentals. Chapter 6, Introduction to Bubble Dynamics and Cavitation, has been revised to present the latest views on bubble dynamics, cavitation inception, and acoustic transmission in bubbly media. Regarding Volume 2, we have changed chapter numbering, but not the chapter subjects. Accordingly, Chapter 1 of Volume 2 now addresses the phenomena related to hull pressure fluctuations on ships due to extensive propeller cavitation and recent measurements of cavitation noise from full scale ships. Chapters 2 and 3 of Volume 2 have been extensively re-worked. The section on the use of sensors and arrays has been moved from Chapter 2 to Chapter 3 of Volume 2; Chapter 2 of Volume 2 now deals exclusively with the science of boundary layer pressure and Chapter 3 of Volume 2 deals with response of sensors, sensor arrays, and elastic structures. Together these chapters now present the modern views of turbulent boundary layer wall pressure fluctuations at low wave number, radiated sound, rough wall boundary layers, and the effects of steps and gaps on sound. Chapter 4 of Volume 2, presents a comprehensive treatment of flow-excitation and radiated sound from elastic cylinders, both ducts and shells. This coverage recognizes the capability of obtaining modal solutions on personal computers. Chapters 5 and 6 of Volume 2 have also been revised, although less extensively so. Turbulence ingestion noise was not well understood when Edition 1 was written; Edition 2 provides an expanded treatment for lifting surfaces and propeller fans.

Chapter 6 of Volume 2 provides more examples of comparisons between theory and measurement than were possible for Edition 2.

A work of this scope could not have been possible, except for the continued collaboration, benefit, and support of a large number of professionals in the field and with whom I have had the pleasure of working; unfortunately many of whom are no longer active. Of these my late mentors, Patrick Leehey, Maurice Sevik, Gideon Maidanik, George Chertock, and Murry Strasberg were particularly close. In their place is a host of contemporary friends and collaborators with whom I have both held discussions and published research that has contributed to the development of the many concepts presented herein. Among these are Hafiz Atassi, David Feit, Stewart Glegg, Marvin Goldstein, Jason Anderson, Rudolph Martinez, John Muench, Ki Han Kim, Robert Minnitti, Denis Lynch, John Wojno, Joseph Katz, Theodore Farabee, Lawrence Maga, Irek Zawadzki, Jonathan Gershfeld, Matthew Craun, William Devenport, Meng, Wang, Douglas Noll, Peter Chang, Yu Tai Lee, Thomas Mueller, Scott Morris, Yaoi Guan, and William Bonness. I am especially grateful to Christine Kuhn who has provided a thoughtful and thorough critique of parts of the work. Thanks are also due to Kiruthika Govindaraju and the Elsevier editorial team.

Finally the main debts are owed to my wife Donna, who has endured yet another writing of this book with enduring gifts of love, support, and patience, and to our daughters Kristen and Helen; all of whom enthusiastically supported this revision.

Preface to the First Edition

Flow-induced vibration and sound occur in many engineering applications, yet it is one of the least well known of all the engineering sciences. This subject area is also one of the most diverse, incorporating many other narrower disciplines: fluid mechanics, structural dynamics, vibration, acoustics, and statistics. Paradoxically it is also this diverse nature that causes this subject to be widely regarded as one reserved for experts and specialists. The main purpose of this book, therefore, is to classify and examine each of the leading sources of vibration and sound induced by various types of fluid motion and unifies the disciplines essential to describing each source.

This book treats a broad selection of flow sources that are widely encountered in many applications of subsonic flow engineering and provides combined physical and mathematical analyses for each of these sources. The sources considered include jet noise, flow-induced tones and self-excited vibration, dipole sound from rigid and flexible acoustically compact surfaces, random vibration of flow-excited plates and cylindrical shells, cavitation noise, acoustic transmission characteristics and sound radiation from bubbly liquids, splash noise, throttling and ventilation system noises, lifting surface flow noise and vibration, and tonal and broadband sounds from rotating machinery. The formalisms developed are suitable for computer modeling, but closed-form asymptotic solutions are emphasized. Many features of the book have evolved, in part, from the author's own requirements for integrating the fundamentals of the subject with the many practicalities of the design of quiet vibration-free machinery.

To achieve the objective of the book to unify the subject, the second chapter provides comprehensive analytical developments of the classical theories of aeroacoustics and hydroacoustics. These developments begin with the equations of motion, progress through derivations of various forms of the wave equation, and end with the setting down of the formalism of integral solutions that are valid for sources near boundaries. The formal treatment is then broadened and applied to various practical source types throughout the remainder of the book. An important feature of the treatment of real sources is the random nature of the exciting flows in both space and time. Thus statistical methods are introduced in these chapters to describe the sound and vibration generation process in such cases. In summary the book treats the essentials of how flow disturbances generate sound in the absence of local surfaces, how flows of practical importance excite bodies into vibration, and then how these excited surfaces radiate sound.

Once a mathematical description of the flow-induced surface motion exists, it is a straightforward matter for design engineers to extend the modeling of this book to address other problems such as flow-induced stress and fatigue in structures. In every case presented the derived relationships in this book are tested against whatever empirical data were made available to the author, from either laboratory or field test results, in order to examine the limitations to the theory. The results are also examined to elucidate effective methods for sound and vibration control by considering both the nature of the flow as well as the classical noise control methods. The results of the book may thus also be used to give insights into how entire processes may be designed for fundamentally quiet operation.

The book is written principally as a reference work, although it may be used as a teaching aid. The reader will always find reasonably sophisticated results supported by step-by-step derivations that clearly identify any assumptions made. Each chapter is illustrated with comparisons of leading formulas and measured data. The reference lists, though not meant to be exhaustive, are extensive and are intended to support all phases of the book with up-to-date background and additional information. Because the physical sources of sound and vibration are developed from fundamental principles, readers who are also well versed in machine design or in any of the related engineering sciences should be able to apply the principles in this book in their work. An attempt has been made to use mathematical notation that is standard in other fields of engineering.

The first six chapters (the contents of Volume I) have been written with emphasis on the elements of fluid mechanics, vibration, and acoustics. These chapters deal with the more fundamental sources of flow noise. Thus this volume might fit into a curriculum that offers courses in applied mathematics, acoustics, vibration, and strength of materials and lacks a relatively generalized course in the physical principles of vibration and sound abatement. Volume II, on the other hand, deals with more advanced and practical subject areas. Both volumes could serve as reference books for graduate courses in vibration, noise control, acoustics, and process design engineering. Draft versions of parts of the book have been used by the author in a graduate course in special topics in acoustics at the Catholic University of America and in short courses.

Due to the interdisciplinary nature of the subject of flow-induced vibration and sound as treated in this book, it is unlikely that the average reader will be equally well versed in all the component disciplines: applied mathematics, fluid mechanics, vibrations, strength of materials, acoustics, and statistical methods. Accordingly readers of the book should be accomplished in senior-level applied mathematics as well as in strength of materials and in at least one of the remaining disciplines listed. An attempt has been made to provide at least a cursory review of certain concepts where it is felt that prior training might be lacking. Readers lacking familiarity in any of the areas will find references to currently available representative texts. An attempt has been made to consolidate the various mathematical developments so that readers who do not seek familiarity with analytical details may focus on the physical properties of the sources. The illustrations will in these cases often provide those readers with insights concerning the parametric dependencies of the various sources.

The author is indebted to his colleagues at the David Taylor Naval Ship Research and Development Center, in academia, and in industry for continuing interest in the project. Special thanks go to Professor Patrick Leehey of the Massachusetts Institute of Technology who provided me with both instruction and inspiration and to Dr. Maurice Sevik who provided encouragement as the work progressed. The book has benefited from conversations with and information provided by A. Powell, J. T. C. Shen, G. Maidanik, G. Franz, M. Strasberg, F. C. DeMetz, W. T. Reader, S. Blazek, A. Paladino, T. Brooks, L. J. Maga, R. Schlinker, J. E. Ffowcs Williams, I. Ver, A. Fagerlund, and G. Reethoff. From time to time, I imposed on a variety of experts to review selected chapters; gratitude is extended to M. Casarella, D. Crighton, M. S. Howe, R. E. A. Arndt, R. Armstrong, F. B. Peterson, A. Kilcullen, D. Feit, M. C. Junger, F. E. Geib, R. Henderson, R. A. Cumming, W. B. Morgan, and R. E. Biancardi. Thanks are also due to C. Knisely, D. Paladino, and J. Gershfeld who read all or part of the manuscript and located many of the inconsistencies and errors.

Finally the main debts are owed to my wife Donna, who initially suggested the project and whose enduring gifts of love, support, and patience made possible its completion, and to our daughters Kristen and Helen for their cheerfulness as they virtually grew up with the book around them.

List of Symbols

AR	Aspect ratio
A_p	Area of a panel, or hydrofoil
B	Number of blades in a rotor or propeller
b	Gap opening (Chapter 3)
C	Blade chord
C_D, C_L, C_f, C_p	Drag, lift, friction, and pressure coefficients, respectively
c	Wave speed, subscripted: 0, acoustic; b, flexural bending; g, group (Chapter 5), gas (Chapters 6 and 7); L, bar; l longitudinal; m, membrane (Chapter 5), mixture (Chapters 3, 5, and 6)
D	Steady drag
D	Diameter (jet; propeller, rotor in Chapters 3, 7, 12)
d	Cylinder diameter, cross section
$E(x)$	Expected value of $x(=\overline{x})$
f	Frequency
$F_i(t)$	Force in i direction
F_i'', F_i'''	Force per unit area, volume
F_r	Froude number
G(x, y), G(x, y, ω)	Green's functions. Subscripted m for monopole, d_i for dipole oriented along i axis.
$H_n(\xi)$	Cylindrical Hankel function, nth order
h	Thickness of plate, or of trailing edge, hydrofoil, propeller blade
h_m	Maximum thickness of an airfoil section
I	Acoustic intensity
J	Propeller advance coefficient
$J_n(\xi)$	Bessel's function, first kind, nth order
K	Cavitation index $(P_\infty - P_v)/q_\infty$
k, k_i	Wave number; i, ith direction; k_{13}, in the 1, 3 plane
k_g	Geometric roughness height
k_n, k_{mn}	Wave numbers of nth or m, n modes
k_p	Plate bending wave number, $k_p = \omega/c_b$
k_s	Equivalent hydrodynamic sand roughness height
k_T, k	Thrust and torque coefficients for propellers and rotors, Eqs. (12-20) and (12-21).
k_0	Acoustic wave number ω/c_0
L	Steady lift
L, L'	Unsteady lift and lift per unit span, Chapter 12, usually subscripted
L, L_3	Length across the stream, span
L_i	Geometric length in ith direction

l_c, l_f	Spanwise correlation length, eddy formation length
l_0	Length scale pertaining to fluid motion without specification
M, M_c, M_T, M_∞	Mach numbers: convection (c), tip (T), free stream (∞)
M	Mass
m_m, m_{mn}	Fluid added mass per unit area for *m* or *mn* vibration mode
M_s	Structural plating mass per unit area
N	Number of bubbles per unit fluid volume
$n(k)$, $n(\omega)$	Mode number densities
n, n_i	Unit normal vector
n	Shaft speed, revolutions per second
$n(R)$	Bubble distribution density number of bubbles per fluid volume per radius increment
$\mathbb{P}$, $\mathbb{P}(\omega, \Delta\omega)$	Power, total and in bandwidth $\Delta\omega$, respectively
$\mathbb{P}_{rad}$	Radiated sound power
P	Average pressure
P_i	Rotor pitch
P_∞	Upstream pressure
p	Fluctuating pressure; occasionally subscripted for clarity: a, acoustic; b, boundary layer, h, hydrodynamic
L	Torque
q	Rate of mass injection per unit volume
q_∞, q_T	Dynamic pressures based on U_∞ and U_T
R_L	Reynolds number based on any given length scale L, $= U_\infty L/v$
R	Radius; used in Chapters 7 and 8 for general bubble radius and in Chapter 12 for propeller radius coordinate
R_b	Bubble radius
R_{ij}	Normalized correlation function of velocity fluctuations u_i and u_j
R_{pp}	Normalized correlation function of pressure
$\hat{R}$	Nonnormalized correlation function Section 2.6.2
R_T, R_H	Fan tip and hub radii
r, r_i	Correlation point separation, the distinction from r is clear in the text
r	Acoustic range, occasionally subscripted to clarify special source point-field identification
S	Strouhal number $f_s l_0/U$ where l_0 and U depend on the shedding body
S_e, S_{2d}	One- and two-dimensional Sear's functions
S_{mn}(k)	Modal spectrum function
$S_p(r, \omega)$	Spectrum function used in Chapter 6 defined in Section 6.4.1
T	Averaging time
T, $T(t)$	Thrust, steady and unsteady
T_{ij}	Lighthill's stress tensor Eq. (2-47)
t	Time
U	Average velocity, subscripted: a, advance, c, convection; s, shedding $\left(= U_\infty \sqrt{1 - C_{pb}}\right)$; T, tip, τ, hydrodynamic friction $(= \sqrt{\tau_w/\rho_0})$; ∞, free stream
u, u_i	Fluctuating velocities
V	Stator vane number in Chapter 12
v	Volume fluctuation

$v(t)$	Transverse velocity of vibrating plate, beam, hydrofoil
W_e	Weber number, Chapter 7
x, x_i	Acoustic field point coordinate
y, y_i	Acoustic source point coordinate
y_i	Cross-wake shear layer thickness at point of maximum streamwise velocity fluctuation in wake, Figs, 11-1 and 11-18
α	Complex wave number, used in stability analyses and as dummy variable
α_s	Stagger angle
β	Volumetric concentration (Chapters 3 and 7), fluid loading factor $\rho_0 c_0/\rho_p h\omega$ (Chapters 1, 5, 9, and 11), hydrodynamic pitch angle (Chapter 12)
Γ, Γ_0	Vortex circulation (0), root mean square vortex strength in Chapter 11
y	Adiabatic gas constant (Chapter 6), rotor blade pitch angle (Chapter 12)
δ	Boundary layer or shear layer thickness, also $\delta(0.99)$ and $\delta(0.995)$
$\delta(x)$	Either of two delta functions, see p. xx
δ^*	Boundary (shear) layer displacement thickness
η_i, η_p	Powering efficiencies; i, ideal; p, propeller
η_T, η_{rad}, η_m, η_v, η_h	Loss factors: T, total; rad, radiation; m, mechanical; v, viscous; h, hydrodynamic
θ	Angular coordinate
θ_τ	Integral time scale of turbulence
θ_m	Moving-axis time scale
κ	von Karman constant (Chapter 8), radius of gyration of vibrating plate $h/\sqrt{12}$,, beam, hydrofoil (Chapters 9, 10, and 11)
κ, κ_{13}	Dummy wave number variables
Λ	Integral correlation length; for spatial separations in i direction Λ_i
λ	Wavelength (also turbulent microscale in Chapter 11)
μ	Viscosity
μ_p	Poisson's ratio, used interchangeably with μ when distinction with viscosity is clear
$\pi(\omega)$	Power spectral density
ρ	Density; ρ_0 average fluid; ρ_g, gas; ρ_m, mixture; ρ_p, plate material
σ_{mn}	Radiation efficiency of mn mode, also σ_{rad}
τ	Time delay, correlation
τ_w	Wall shear
τ_{ij}	Viscous shear stress
$\mathbf{\Phi}_{pp}$ (k, ω)	Wave number, frequency spectrum of pressures
$\mathbf{\Phi}_{vv}$ (ω)	Auto-spectral density of $v(t)$; subscripted: p for $p(t)$; i for $u_i(t)$, f for $F(t)$
$\mathbf{\Phi}_{vv}$ (y, ω)	Auto-spectral density of $v(t)$ with dependence on location y emphasized; other subscripts as previously
ϕ	Angular coordinate
ϕ(y), $\phi(y_i)$	Potential functions
$\phi_i(k_j)$	Wave number spectrum (normalized) of velocity fluctuation u_i
ϕ_{ij} (r, ω)	Cross-spectral density (normalized) between $u_i(\mathbf{y}, t)$ and $u_j(\mathbf{y} + \mathbf{r}, t)$
$\phi_m(\omega - U_c \cdot k)$	Moving-axis spectrum
Ψ_{mn} (y), Ψ_m (y)	Mode shape functions

$\psi(y)$	Stream function
Ω	Shaft rate
ω	Circular frequency
ω, ω_i	Vorticity vector, component in *i* direction
ω_c	Coincidence frequency
ω_{co}	Cutoff frequency of an acoustic duct mode
ω_R	Circular cylinder ring frequency

Chapter 1

Introductory Concepts

1.1 OCCURRENCES OF NOISE INDUCED BY FLOW

Sound may be emitted whenever a relative motion exists in fluids or between a fluid and a surface. Examples for which flow-induced noise has been a subject of concern are industrial jets and valves, automobiles, airplanes, helicopters, wood-cutting machinery, ventilation fans, marine propellers, and household rotary lawn mowers. In these applications the common physical processes that are responsible for noise generation include turbulent fluid motions, vibration of structures, acoustics, and aerodynamics of wings and bodies. In this book we will be focusing on this broad range of topics, restricting ourselves to essentially isothermal mechanisms. We will be studying the subject of flow-induced vibration and sound for a number of purposes in an integrated way that will dwell on the unique matching of spatio-temporal scales that takes place. This is a practical engineering topic for several reasons. First, understanding of the mechanisms of sound generation from fluid–structure interactions can result only from an appreciation of the parameters of fluid dynamics and structural dynamics that promote that coupling. Second, the parameters of various forms of interaction must be known in order to effect productive design changes for noise control. Third, the mechanics of the generation of unsteady flow by various classes of fluid motions must be understood in order to alter fundamentally noise production by sources of disturbance at fluid–fluid and at fluid–body interfaces.

The study of fluid-dynamic vibration and sound is both empirical and analytical. It is analytical in that the formulation of vibration and sound variables in terms of parameters of the fluid-dynamical, structural-dynamical, and acoustical processes is determined from the laws of motion. For certain simple canonical problems in which the interactions are particularly simple, we can derive analytical formulations quite precisely. In most practical cases, however, a number of numerical coefficients, of the flow especially, must be determined by measurement. This may be done by scale model testing, observation of sound or vibration on a given configuration as a function of operating speed, and measurements in laboratory experiments designed to emphasize a particular disturbance mechanism suspected of dominating the performance of the original sound or vibration problem. In an increasingly

Mechanics of Flow-Induced Sound and Vibration, Volume 1.
DOI: http://dx.doi.org/10.1016/B978-0-12-809273-6.00001-4

large number of cases flow-generated sound and forces may be simulated numerically. Typical sources of noise from fluidic machines involve a time-varying system of forces being applied at one or more components of the machine; sound is radiated from fluid reactions to these forces as well as from forced vibration of structures in contact with the flow. The study of noise control in all applications is best approached by examining ways of minimizing the exciting forces, the vibratory response of the machine to those forces, or the efficiency of radiation from the structure by surrounding or cladding the structure with acoustically absorbent material. The control of most flow-induced noise sources follows this common theme; however, the mechanism of force generation is often complex, requiring an understanding of the generation of unsteady flows in specific instances. The subject of fluid-dynamic noise is therefore truly interdisciplinary. It involves the simultaneous study of fluid mechanics, vibration, and acoustics. To understand how structures are excited by flow, one must ascertain what qualities of the flow–structure interaction promote the transference of energy from the fluid flow to the structure, then from the structure into sound in the fluid. In the absence of a structure, the transfer of energy is from one type of fluid motion, hydrodynamic or aerodynamic, to another type, sound. Therefore a thorough study of flow-induced sound and vibration must be concerned with the generation and control of turbulence and unsteady fluid motions, structural dynamics and vibration, and acoustics and sound propagation.

To design a quiet and energy–efficient fluid-borne structure, it is also important to know what the features of both rigid and flexible bodies are that make them both receptive to excitation by flow and generators of sound. Effectively, such bodies are transducers that alter the form of energy from hydrodynamic (aerodynamic) power to acoustic power. In the case of flexible bodies, the mathematical description of flow-induced structural response as an element of this process may depend on whether the unsteady fluid mechanics are representable as a system of broadband random local exciters (as with turbulent boundary layer excitation) or as a system of local forces that are nearly periodic in time (as with Aeolian tones and certain trailing-edge flows) with spatial disturbance scales that are coincident with the spatial response scales in the structure. Therefore the subject of flow-induced vibration must also consider the multi-mode vibration of structures with consideration being given to whether the vibrating boundaries are small or large compared to an acoustic wavelength. Here, again, the spatial scales of acoustic coincidence must also be known to describe the qualities of acoustic radiation from the moving surfaces properly.

The subject of flow-induced sound must also include the acoustic transmission characteristics of two-phase media. This is because in many hydrodynamic flows the acoustic medium is bubbly, and the sound source is bubble or cavitation dynamics. Such flows are important in the noise of throttling of liquid flows and of marine propellers.

1.2 FLUID–BODY INTERACTIONS FOR SOUND PRODUCTION

In nearly all problems of flow-generated noise, the energy source for sound production is some form of flow unsteadiness. This unsteadiness needs not always to be turbulent, or random, as there are numerous cases of tonal sounds (whistles, cavity tones, singing propellers, and turbine blades) that involve sinusoidal disturbances in the fluid. Most other cases of flow-induced sound and vibration, at low velocity (or Mach number) especially, involve a restricted region of turbulence that is either free of solid boundaries (jets) or in contact with a body. An essential ingredient of flow unsteadiness that determines the efficiency of noise production in noncavitating and bubble-fluids is its *vorticity. Vortices*, or *eddies*, are locally rotating, or spiraling, fluid motions. If we imagine that the fluid particles at the center of rotation of a vortex become a small frozen agglomerate, then these frozen particles will rotate at an angular velocity that is exactly one half of the vorticity of the vortex. The axis of rotation of each frozen agglomerate is tangent to a line that connects the centers of rotation of the agglomerates that are entrapped in the core of the vortex. This line is coincident with the core, and it is regarded as the *vortex line*, or the *vortex filament*. Vortex motions are generally associated with regions of flow discontinuity that occur at interfaces between fluids and solids in relative motion or between parallel-moving fluids of differing velocity or density. In turbulent flow, the vorticity is responsible for regions of relatively intense fluid activity and mixing. In Chapter 2, Theory of Sound and its Generation, by Flow, the theory of vortex-generated sound in low-speed (low Mach number) flow will be eloquently developed, but for now it may be stated simply that sound is produced whenever vortex lines are stretched or accelerated relative to the acoustic medium. Analogously, whenever vortex lines are stretched or accelerated relative to a body in the flow, forces are exerted on the body–fluid interface. Classically, this is exactly the mechanism of lift production by wings.

Fig. 1.1 illustrates one example, showing parameters of noise generation from a lifting surface passing through unsteady flow. This example is typical of the sound sources discussed in this book. The example also typifies parameters of flow excitation on a blade element of a fan as it rotates through a nonuniform inflow. The flow unsteadiness consists of free stream turbulence of average vortex size Λ_T, boundary layer turbulence generated on the body and of length scale Λ_f, flow separation near the trailing edge, and additional disturbances in the wake that have an average size Λ_s. All these boundary disturbances, buffeting, boundary layer, and near-wake, generate surface pressures p_h on the body. Pressures induced by the separated flow at the trailing edge and by the wake could be tonal depending on the regularity of the vortices created at the trailing edge and swept downstream by the flow. In response to all these surface pressures, the body may vibrate with a wavelength λ_p.

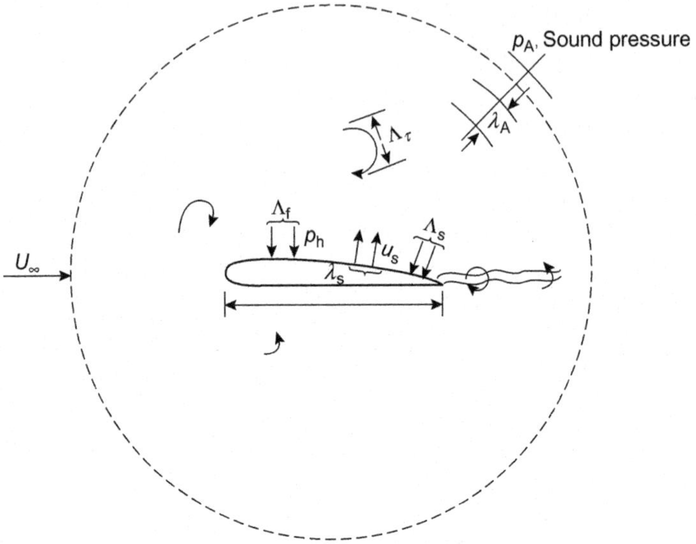

FIGURE 1.1 Illustration of a body subjected to a disturbed flow of scale Λ_T; body vibration u_s of wave length Λ_s resulting from surface pressure p_h of length scale Λ_f.

Sound pressures are therefore radiated from the turbulence itself, the distribution of forces on the surface, and the motion of the body. The net sound pressure from all such sources is dependent on the amplitude and phase of each contributor and these features will be dependent on both the frequency of oscillation through the relative impedances which are characteristic of the motions.

Unless cavitation occurs, the acoustic effects of the interaction of a body with a flowing medium are therefore threefold. First, the flow-body interaction may result in net forces at the body applied to the fluid which radiate sound as dipoles; second, the body alters the sound field radiated by the turbulent sources by acting as a scattering or diffracting surface; third, the body may alter the flow itself by creating additional flow disturbances in the form of vorticity and associated body-flow interaction forces. The case shown in Fig. 1.1 is a good example of this. Without the body present, the sound would be generated by the mixing of the incident turbulent stream. Generally, at low subsonic speeds, this noise will be quite minimal. In interacting with the incident turbulence, the leading edge of the body creates a scattered pressure. This interaction also causes an aerodynamic surface pressure that is regarded by aerodynamicists as the aerodynamic load on the surface. However, this aerodynamic load is just the near field of the scattered pressures caused by the edge–turbulence interaction. All remaining noise sources constitute additional sounds created as a result of new aerodynamic disturbances being generated as the flow passes along the body. It should be borne in mind that these additional disturbances are generally the result of the fluid being viscous,

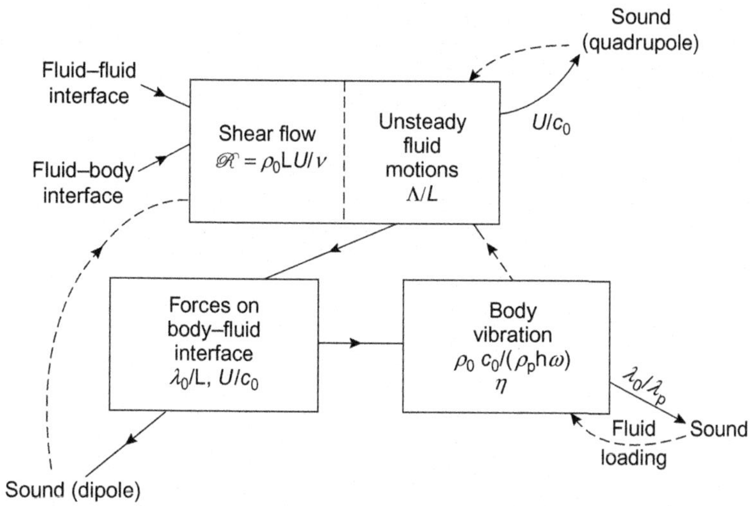

FIGURE 1.2 Idealized functional diagram of fluid–body interactions producing relative motion and sound. Dotted lines show expected feedback paths. Λ represents a spatial scale of turbulence and other symbols are in Table 1.1.

although thickness noise of high-speed lifting surfaces is caused by potential flow (see Chapter 12: Noise from Rotating Machinery).

These interactions are generalized somewhat in Fig. 1.2. The physics of sound production from a body–flow interaction is depicted as a system of interacting elements, although similar interactions occur in free jets. Solid-line arrows denote principal cause–effect interactions, and dotted-line arrows denote possible feedback routes between elements of the system. Unsteady fluid motions and turbulence are generated in all manner of fluid machinery. In one class of acoustic sources, disturbances are generated whenever there is an interface between two liquids at different mean velocities, as in the flow-over cavities, the flow of jets (see Chapter 3: Shear Layer Instabilities, Flow Tones, and Jet Noise), or flow in the wakes of bluff bodies (see Chapter 4: Dipole Sound from Cylinders). These flows are naturally unstable, often producing tones, and they are generally susceptible to acoustic reinforcement. A second class of acoustic sources is produced by turbulent flows adjacent to solid surfaces. Such wall or boundary layers are generally turbulent and excite flexural vibration in the adjacent structures. These sources do not fundamentally require flow instability. Such wall layers may also radiate strongly when the surface is discontinuous, as with a trailing edge (see Chapter 11: Noncavitating Lifting Sections). Noise may also be the result of a surface being buffeted by upstream turbulence or by disturbances generated by inlet guide vanes (see Chapter 11: Noncavitating Lifting Sections and Chapter 12: Noise from Rotating Machinery). Fluctuating velocities produced from free shear layers in the absence of bodies (e.g., the free jet) may radiate directly as *quadrupole* sound, but in the presence of bodies (e.g., the Aeolian type), shear layers may induce surface stresses that also radiate. If the body is rigid and small compared to an acoustic wavelength, these surface stresses are applied locally in equal magnitude and opposite direction to the fluid, resulting in classical *dipole* sound. This is contributed to by additional sound from adjacent body surface vibration, which is also excited by these same surface forces (If the surface is perfectly plane, rigid, and of infinite extent in the plane of the flow, these surface stresses will not radiate dipole sound; see Section 2.4.4).

Feedback to the flow is made possible by each of two major paths. First, the body motion may provide a direct excitation of fluid disturbances in the shear flow; second, the fluid particle motions associated with sound produced may be transmitted back to the region where aerodynamic disturbances are initiated. This fluid-path feedback may also be hydrodynamic at low speeds. Examples of structural feedback are singing of turbine blades, Aeolian tones, and certain cavity tones. Examples of fluid-path feedback are jet tones, edge tones, flow-induced excitation of Helmholtz resonators, and some other cavity tones, and some trailing-edge tones. A common feature of nearly all these feedback phenomena is the creation of tonal or nearly tonal velocity fluctuations. Frequently this feature is accompanied with enhanced response only over a restricted range of flow velocity.

The interactions depicted in Fig. 1.2 also apply in a sense to cavitation and bubble noise, but in this application the body vibration pertains to bubble wall breathing modes; radiation from forces on the fluid–body interface is generally negligible.

1.3 DIMENSIONAL ANALYSIS OF SOUND GENERATION

As noted above, this subject is one that relies heavily on the knowledge obtained empirically. Even without an in-depth understanding of the physical processes involved, experimental evaluation and parametric scaling from one situation to another may be done by considering the general nature of the interactions. These considerations, known as similitude, are based on the way forces are applied and media respond. By maintaining *geometrical* similitude, the locations and general orientations of forces will be maintained. *Dynamical* similitude requires that the relationships (both magnitude and phase) among forces and motions remain fixed. For fluid–structure interaction, this is a tall order!

The practical control or prediction of sound and vibration relies heavily on collections of empirical data, carefully and systematically accumulated and reduced according to theory or hypothesis. These data form the basis of engineering prediction formulas. It is therefore important to develop an appreciation of laws and limitations of fluid mechanics, structural dynamics, and acoustics. Table 1.1 gives a summary of the important dimensionless groups that govern many of the similitude considerations dictated by these laws.

TABLE 1.1 Dimensionless Ratios Appropriate to Requirements of Similitude

Similitude	Parameter[a]	Application (Common Name)
(Inertial forces)/(Viscous forces)	$R = \rho UL/\mu$	All flows (Reynolds number)
Geometry	Body size, shape	All flows
Gross surface finish	$(k_g/L)_1 = (k_g/L)_2$	All flows
(Inertial stress)/(Compressive stress)	$M = U/c$	All flows (Mach number)
Material (fluid) compressibility, ratio	$(\rho c^2)_1/(\rho c^2)_2$	Fluid [1]–fluid [2] and fluid [1]–structure [2] interactions (compressibility)
(Fluid (acoustic) impedance)/(Structural mass impedance)	$\rho c/\rho_p h\omega$	Fluid–structure interactions (fluid-loading factor)
(Energy dissipated)/(Kinetic energy)	η	Fluid–structure interactions, hydroelastic coupling, vibration (loss factor)
(Pressure stresses)/(Inertial stresses)	$K = (P_\infty - P_v)/\frac{1}{2}\rho U^2$	Cavitation (number)
(Inertial forces)/(Gravitational forces)	$F_r = U/\sqrt{gL}$	Buoyancy and hydrostatic effects (Froude number)
(Gas inertial stresses)/(Surface tension)	$W_e = \rho U_g^2 L/S$	Gas jet disintegration in liquids, bubble splitting (Weber number)

[a] *ρ is mass density; U, mean velocity; c, phase speed of sound or vibration; h, plate thickness; L, body dimension; k_g, geometric roughness height; η, loss factor; P, static pressure; S, surface tension; g, gravitational constant; $\mu/\rho = \nu$, $\omega/2\pi$ is frequency; $\lambda = 2\pi c/\omega$ is wave length. Subscript p denotes plate; g, gas; 1, medium 1; 2, medium 2.*

Further introductory discussion of the importance of many of these factors is given in the chapters indicated below. Table 1.1 organizes and extends the similitude quantities that govern the interfaces shown in Fig. 1.2.

First, the generation of sound and vibration by fluid motion involves the reactions (i.e., strains) of fluids and solids to stresses imposed by time-varying flow. The first feature of flow-induced disturbances to be recognized is that the flow is generally usefully regarded as a mean plus a fluctuating part. That is, the local velocity at a point may be regarded as superposition of an average value and an instantaneous fluctuating value. Thus the velocity at a point in the fluid may be regarded as a sum

$$\overline{U} = \overline{U} + u(t)$$

where $\overline{U}$ is the average value and u is the unsteady value, which depends on both time and location in the flow. In dynamically similar flows, say in a model-to-full-size comparison, the ratio

$$u(t)/\overline{U}$$

is a constant, regardless of the value of $\overline{U}$, i.e., the distribution of velocity fluctuations scales on the mean velocity. For this constancy to be maintained throughout the flow, the balance of the various types of stresses that act on fluid particles must also be maintained. Generally these are combinations of inertial and viscous stresses and a measure of the ratio of the inertial to viscous stresses in the flow is the Reynolds number

$$R = \overline{U}L/\nu$$

where ν is the kinematic viscosity of the fluid and L is a geometric length of the body causing or disturbing the flow. Typically, chosen dimensions are diameters of jets, cylinders, and pipes; or chords or thicknesses of wings and propeller blades. For a given geometry or flow, the nature of the disturbances that occur (e.g., whether the disturbances are periodic or turbulent) is greatly dependent on the Reynolds number. In all flows there is a critical value, R_{crit}, above which the flow is turbulent and below which the flow is laminar, periodic, or at least well ordered. It is often the case that when $R < R_{\text{crit}}$ fairly significant changes occur in the flow structure for modest variation in R. However, when $R > R_{\text{crit}}$ in the turbulent regimes of the flow, fairly modest changes or differences in R do not always carry significant changes in fluid stress relationships. In general, turbulent flows are dominated by inertial stresses. It is important to recognize that the specific behavior of the flow dynamics with respect to R will be unique to the flow type, as will be discussed in the chapters which follow.

The above notion of similitude is important because the exciting stresses, denoted here by $|p|$, that produce sound or vibration in a given type of flow will be in direct proportion as

$$|p| \sim \frac{1}{2}\rho_0\overline{U}^2 = q$$

where ρ_0 is the mass density of the fluid and q is the dynamic pressure. The proportionality may hold as long as the fluctuating velocity and the mean velocity are also proportional. Since the sound- and vibration-producing stresses are proportional to q, this quantity will appear frequently in all the chapters as a metric of the intensity of the magnitude of the excitation. The dynamic pressure of the flow therefore also serves as a scaling or reference pressure for radiated sound; and it can be used in combination with other factors as a scale for the magnitude of flow-induced vibration. Dynamical similitude also implies that the hydrodynamic disturbance length Λ retains a proportionality to the body or flow size L; i.e., $\Lambda \propto L$.

A measure of the matching of fluid inertial motions of aerodynamics or hydrodynamics and the particle velocities related to the propagation of sound is the Mach number, $M = \overline{U}/c_o$. In a sense, this factor expresses the ratio of the hydrodynamic velocity to the acoustic particle velocity. Thus these two parameters, the Reynolds number and the Mach number, express the relative importance of inertial, viscous, and compressive stresses in the fluid. Fluid dynamics and acoustic similitude therefore ideally will require, in addition to similar geometries, equal values of Reynolds and Mach numbers. When structures are rigid, flow-induced sound arises only from momentum and entropy exchanges in the flow as it interacts hydrodynamically with the body. In Chapter 3, Shear Layer Instabilities, Flow Tones, and Jet Noise, and Chapter 4, Dipole Sound from Cylinders, these interactions will be examined in detail to determine what features of flow–body interactions facilitate the creation of different types of flow disturbances. The acoustical (scattering) effect of the presence of the body will be examined in an elementary sense in Chapter 4, Dipole Sound from Cylinders, for the case of a cylinder. Chapter 11, Noncavitating Lifting Sections, and Chapter 12: Noise from Rotating Machinery, will cover the more complex hydrodynamic and acoustic interactions of extended structures (i.e., large in terms of an acoustic wavelength) such as wings, rotor blades, fans, and propellers. The discussion of mechanisms of sound radiation will be rooted in an understanding of the underlying fluid mechanics of unsteady flows. This will be done in Chapter 3, Shear Layer Instabilities, Flow Tones, and Jet Noise, for free shear layers and for jets; in Chapter 4, Dipole Sound from Cylinders, and Chapter 10, Sound Radiation from Pipe and Duct Systems, for unsteady wakes of bluff bodies, throttles, and gratings; in Chapter 8, Essentials of Turbulent Wall-Pressure Fluctuations, for turbulent wall layers (boundary layers); and in Chapter 11, Noncavitating Lifting Sections, and Chapter 12, Noise from Rotating Machinery, for the wide variety of flow disturbances generated by lifting surfaces, compressors, and propellers.

In fluid–structure interactions the vibration in the structure depends on the wave-bearing qualities of the structure as they relate to the unsteady hydrodynamic and acoustic qualities of the fluid. We shall see in Chapter 5, Fundamentals of Flow-Induced Vibration and Noise, that there are four minimum important factors that control both the vibration response of the structure to an imposed excitation and the resulting acoustic radiation. These are the damping in the structure, the structure mass, the speed of vibration wave propagation in the structure relative to the acoustic wave propagation on the fluid, and the wavelength of the structural vibration compared with an eddy size Λ of the exciting-fluid stresses. The dimensionless ratios that are important descriptions of the relative magnitudes of the structural–acoustic matching factors are shown in Fig. 1.2 and Table 1.1. These are the fluid-loading factor, the ratio of structural (flexural wave) velocity to the acoustic wave velocity, and the loss (damping) factor of the structure. Chapter 5, Fundamentals of Flow-Induced Vibration and Noise, is devoted to a derivation of all these coupling parameters, and it will give an introductory example of their use in the case of sound radiation from a flow-excited one-dimensional structure—a cylinder. Subsequent coverage in Chapter 9, Response of Arrays and Structures to Turbulent Wall Flow and Random Sound, Chapter 10, Sound Radiation from Pipe and Duct Systems, will examine vibration and noise control of extended two-dimensional structures. Similitude in this area will imply other forms of Mach number that attached to the compressive and shear stresses in the structure.

The remaining three dimensionless groupings shown in the table are applicable for two-phase fluids (liquid and gas), and these will be introduced in Chapter 6, Introduction to Bubble Dynamics and Cavitation. The Froude number, for example, is a measure of the ratio of inertial to gravitational forces on a fluid particle. As we shall see in Chapter 7, Hydrodynamically Induced Cavitation and Bubble Noise, in the formation of gas bubbles in liquids, combinations of both Froude and Weber numbers express relative inertial, surface tension, and gravitational stresses. The cavitation number indicates the relative susceptibility of a moving fluid to form vapor cavities because of local rarefaction zones. As discussed in Chapter 6, Introduction to Bubble Dynamics and Cavitation, and Chapter 7, Hydrodynamically Induced Cavitation and Bubble Noise, this is the primary dimensionless number that governs the occurrence of cavitation noise.

1.4 SIGNAL ANALYSIS TOOLS OF VIBRATION AND SOUND

The subject area of this book relies heavily on signal analysis and the analysis of random data. Accordingly, one must appreciate the fundamentals of statistical signal processing theory. In this section we shall give a summary of the practical results of signal analysis theory in order to provide the reader with a working familiarity to understand the terms and the practical limitations and meanings of quantities. It is not necessary for the reader to have a full background in this subject to understand the material of this book's text fully. A number of references are available to the reader who wishes a more extensive treatment of the subjects of this section. From a classical perspective Beranek [1] gives discussions of transducers, signal filtering, instrumentation, and measurement principles. Readers interested in signal analysis per se are referred to books by Bendat and Piersol [2], Davenport and Root [3], Lee [4], Newland [5], Papoulis and Pillai [6], and by Pierce [7] which cover a variety of related topics.

1.4.1 A Simple Example of an Acoustic Radiator

To illustrate how all these ideas come together, we consider a single physical process as an application of the statistical methods discussed. The example selected is of the simplest acoustic radiator, the pulsating sphere. The theoretical derivation of the sound pressure that results from the expansion mode of a pulsating sphere in an unbounded compressible medium will be derived in the next chapter, but the result is simple enough to comprehend readily and give physical significance to the mathematical manipulations of this section. The source is a spherical shell that vibrates at small amplitude uniformly over its entire surface as shown by the concentric dashed lines in Fig. 1.3. The average radius of the sphere is a. At a distance r away from the center of the sphere the pressure is $p(r, t)$. The acceleration of the surface of the sphere is $A(t)$ in the radial direction, i.e., normal to the surface of the sphere.

Although this is a simple example, many practical sound sources behave in an analogous manner, e.g., underwater sound projectors and the component of splash noise induced by vibration of entrained bubbles.

We assume now that the wall acceleration is a transient, i.e., is a pulse occurring at time t_0, as illustrated in Fig. 1.3. The pressure pulse travels at speed c_0 through the fluid to the point r and beyond. The pressure at r is also a pulse that has the same time shape as the acceleration, but it occurs a delayed time $(r-a)/c_0$. This time duration is required for the pulse to transit from $r=a$ to $r=r$. The formula for the pressure at r and time t is (This formula is a linear acoustics result that applies when the wall velocity is substantially less than the speed of sound.)

$$p(r,t) = \rho_0\left(\frac{a}{r}\right)aA\left(t-\frac{r-a}{c_0}\right) \tag{1.1}$$

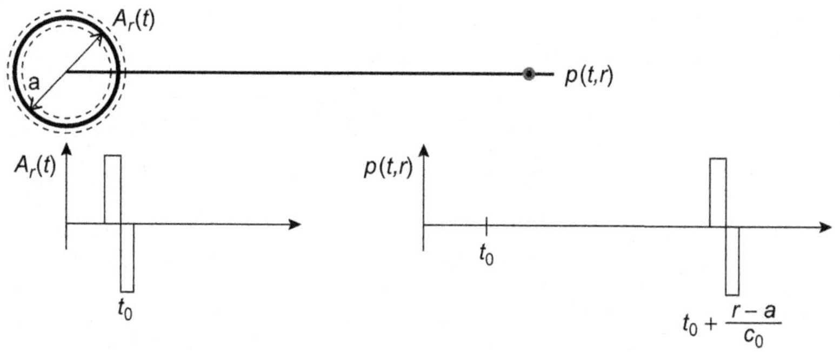

FIGURE 1.3 Acceleration and pressure pulses are shown as examples of physically related signals that are delayed in time.

where ρ_0 is the density of the fluid. In this formula the pressure at t is caused by an acceleration event at the earlier time $t_p = (r-a)/c_0$. This example contains many features of all acoustic sources: the wall acceleration that produces the sound, the sound pressure, and a transit time that provides a delay between these two quantities. Surface acceleration and fluid pressures are the two most commonly reported acoustic and vibration variables.

1.4.2 Fundamentals of Correlation Analysis

1.4.2.1 A Simple Example of Classical Correlation Analysis

Suppose that it was our intention to verify experimentally the dependence on the propagation time that appears in Eq. (1.1). This would, in fact, be a means of measuring the sound speed c_0. Suppose, further that the acceleration pulse has equal positive and negative values as shown in Fig. 1.4 so that the integral over time is identically zero; i.e., the average acceleration is

$$\overline{A} = \frac{1}{T_\mathrm{p}}\int_{-\infty}^{\infty} A(t)dt = \frac{1}{T_\mathrm{p}}\int_{t_0-T_\mathrm{p}/2}^{t_0+T_\mathrm{p}/2} A(t)dt = 0 \tag{1.2}$$

The pulse only exists for a duration T_p so that the integral overall time really reduces to a definite integral over only the duration time. Accordingly, the average sound pressure is also zero; i.e.,

$$\overline{p(r)} = \frac{1}{T_\mathrm{p}}\int_{t_0-T_\mathrm{p}/2}^{t_0+T_\mathrm{p}/2} p(r,t)dt \tag{1.3}$$

The integrals over time of the acceleration squared or of the pressure squared are nonzero; i.e.,

$$\begin{aligned}\overline{A^2} &= \frac{1}{T_\mathrm{p}}\int_{t_0-T_\mathrm{p}/2}^{t_0+T_\mathrm{p}/2} A^2(t)dt \\ &= A_\mathrm{m}^2\end{aligned} \tag{1.4}$$

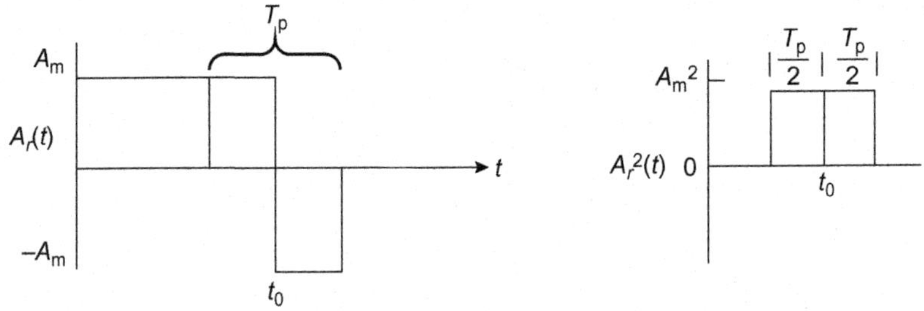

FIGURE 1.4 Details of the wave form of the acceleration pulse shown in Fig. 1.3. The pressure pulse has the same shape.

Similarly,

$$\overline{p^2}(r) = P_\mathrm{m}^2 \tag{1.5}$$

and by Eq. (1.1)

$$P_\mathrm{m}^2 = \rho_0^2(a/r)^2 a^2 A_\mathrm{m}^2 \tag{1.6}$$

These quantities are the mean-square acceleration and pressure, whose magnitudes in this problem are related mathematically by the front terms in Eq. (1.1).

By this example we see that it is often meaningless to speak of an average acceleration or an average pressure, for these two averages may be identically zero. It is, however, most appropriate to speak of the mean-square quantities $\overline{A^2}$ and $\overline{p^2}(r)$ because these are nonzero and easily measurable. We shall see in succeeding chapters that these quantities relate to acoustic energy or power and are thus useful descriptors for energy balances in acoustic systems where the time average of pressure, velocity, or acceleration is generally zero. For signals having zero mean, the mean square of the signal is also called the statistical variance.

A function of considerable importance in the analysis of signals is autocorrelation. This function is defined for such transient signals as this as

$$R_{AA}(\tau) = \int_{-\infty}^{\infty} A(t)A(t+\tau)dt \tag{1.7}$$

and as we shall see later, it contains all the information regarding the frequency content of the signals. The delay time τ may be set to vary continuously; for $\tau = 0$ the autocorrelation function is identical to the mean square times the pulse length, i.e.,

$$R_{AA}(\tau = 0) = \overline{A^2}T_\mathrm{p} \tag{1.8}$$

At other values of time delay as illustrated in Fig. 1.5 for this example, the autocorrelation function reflects the general behavior of the pulse. It is symmetrical about $\tau = 0$, and it is zero for $|\tau| > T_\mathrm{p}$. Since the time pulses of both the acceleration and the pressure are identical in shape, with magnitudes A_m^2 and P_m^2, respectively, the autocorrelation functions are also identical, with magnitudes of A_m^2 and P_m^2, respectively. The more enlightening function in the current context is the cross-correlation function between the acceleration and pressure, which is defined as

$$R_{Ap}(\tau) = \int_{-\infty}^{\infty} A(t)p(r, t+\tau)dt \tag{1.9}$$

This function will show both the general qualities of the pulse and the transmission time from the source to the field point r. We see this from Eq. (1.1), where we have

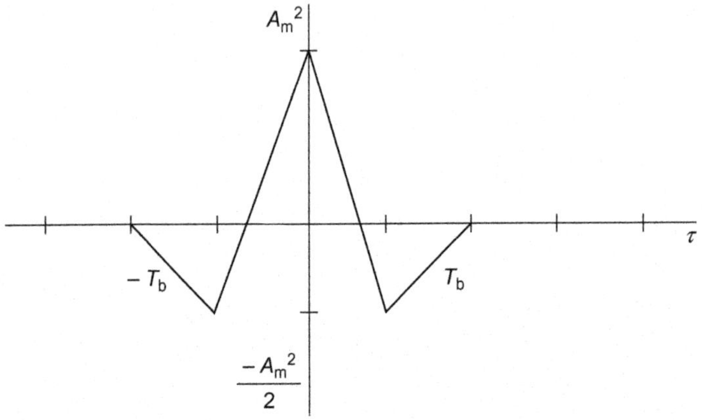

FIGURE 1.5 Autocorrelation function of the acceleration pulse shown in Fig. 1.4.

$$R_{Ap}(\tau) = \rho_0\left(\frac{a}{r}\right)a\int_{-\infty}^{\infty}A(t)A\left(t-\frac{r-a}{c_0}+\tau\right)dt$$

or

$$R_{Ap}(\tau) = \rho_0\left(\frac{a}{r}\right)aR_{AA}\left(\tau-\frac{r-a}{c_0}\right) \tag{1.10}$$

Thus the cross-correlation $R_{Ap}(\tau)$ will have the same shape as the autocorrelation function of the acceleration, but it will be centered at the time delay $\tau_m = (r-a)/c_0$. The cross-correlation function therefore also contains the propagation characteristics of the problems, and from it one could determine the speed $c_0 = (r-a)/\tau_m$. This determination is, in fact, a common use of cross-correlation analysis, particularly in fluid dynamics. In that application the speed indicated by the cross-correlation of two flow properties (e.g., turbulent velocity fluctuations) measured at two appropriately selected locations will yield a "convection velocity" that is defined $U_c = r/\tau_m$, where r is the separation of sensors.

The time function described in this section is a single pulse, existing only once for a finite duration of time. Such signals are called "aperiodic" signals, and to this class of signals belongs all transients. Other signals occurring in practice are more persistent in time and may be regarded as steady state. Of these there are also two extreme types, periodic and random. Random signals are those for which the value from one time to the next has a certain probability of occurrence but no specific predictable value; periodic signals have a definite frequency of occurrence from one time to another. Correlation properties of each will be discussed below.

1.4.2.2 Correlation Functions of Steady-State Signals

1.4.2.2.1 Periodic Signals

The most common periodic signals are the sine and cosine functions, say

$$v(t) = v_0 \cos[(2\pi t/T_p) + \phi] \tag{1.11}$$

where ϕ is a constant-phase angle. The T_p is the period of the oscillation; for every time interval $\Delta t = 2nT_p$ (where n is an integer) the cosine function goes through one complete cycle; $f = 1/T_p$ is the frequency of oscillation, and $\omega = 2\pi/T_p = 2\pi f$ is the angular frequency of the signal. The average of $v(t)$ is identically zero when the average is taken over an integral number of cycles; i.e.,

$$\begin{aligned}\overline{v(t)} &= \lim_{T\to\infty} \frac{1}{T}\int_{-T/2}^{T/2} v_0 \cos\left(\frac{2\pi t}{T_p} + \phi\right) dt \\ &= \frac{1}{T_p}\int_{-T_p/2}^{T_p/2} v_0 \cos\left(\frac{2\pi t}{T_p} + \phi\right) dt \\ &= 0\end{aligned} \tag{1.12}$$

The transition from an indefinite time of averaging T to the specific time T_p holds because when $T > T_p$ then T can be written as $nT_p + t$, where n is an integer and t is continuous in the interval of the period $-T_p/2 < t < T_p/2$. Since the function repeats itself for each n, nothing is gained by extending the average over more than one cycle. As before, the mean square does not vanish; i.e.,

$$\begin{aligned}\overline{v^2(t)} &= \frac{1}{T_p}\int_{-T_p/2}^{T_p/2} v_0^2 \cos^2\left(\frac{2\pi t}{T_p} + \phi\right) dt \\ \overline{v^2(t)} &= \frac{1}{2} v_0^2\end{aligned} \tag{1.13}$$

and it is useful to speak of the root mean square

$$\sqrt{\overline{v^2}} = v_{\text{rms}} = (1/\sqrt{2})v_0 \tag{1.14}$$

The autocorrelation function is defined as

$$\begin{aligned}\hat{R}_{vv}(\tau) &= \overline{v(t)v(t+\tau)} \\ &= \frac{1}{T_p}\int_{-T_p/2}^{T_p/2} v_0^2 \cos[\omega t + \phi]\cos[\omega(t+\tau) + \phi] dt\end{aligned} \tag{1.15}$$

and it evaluates to

$$\hat{R}_{vv}(\tau) = \frac{1}{2} v_0^2 \cos \omega\tau \tag{1.16}$$

The mean-square value is given by $\hat{R}_{vv}(\tau = 0)$. It is to be noted that phase information expressed here by ϕ does not appear in the autocorrelation function; only the amplitude and frequency information appears.

In the context of our example of the vibrating source, if the acceleration has a steady-state cosine dependence given by

$$A(t) = A_{\mathrm{m}} \cos \omega t$$

then the resulting pressure is

$$p(r,t) = P_{\mathrm{m}} \cos[\omega t - (r-a)/c_0]$$

where A_{m} and P_{m} are related through Eq. (1.1). The cross-correlation function of these signals is

$$\hat{R}_{Ap}(\tau) = \frac{1}{2} A_{\mathrm{m}} P_{\mathrm{m}} \cos\left[\omega(\tau - (r-a)/c_0)\right] \tag{1.17}$$

and as before it has a phase shift that is proportional to the distance $r - a$. In another perspective the phase represents the range expressed as multiples of the number of acoustic wavelengths; i.e.,

$$\phi = \frac{\omega}{c_0}(r-a) = k_0(r-a) = 2\pi(r-a)/\lambda_0$$

where $\lambda_0 = c_0/f$ is the acoustic wavelength and $k_0 = 2\pi/\lambda_0$ is the acoustic wave number.

1.4.2.2.2 Random Signals

Random signals are those that do not repeat with any definite sequence, but rather must be described in terms of some probability. Randomization of signals may be the result of a lack of organization of durations and repetition rates of pulses of a given amplitude, a randomization of amplitude of pulses, or an infinite variety of steady-state waveforms. Some examples of random signals and pulses are shown in Fig. 1.6.

The average of the random function is defined as a limit

$$\overline{v(t)} = \lim_{T\to\infty} \frac{1}{T} \int_{-T/2}^{T/2} v(t)dt \tag{1.18}$$

and similarly the mean square

$$\overline{v^2(t)} = \lim_{T\to\infty} \frac{1}{T} \int_{-T/2}^{T/2} v^2(t)dt \tag{1.19}$$

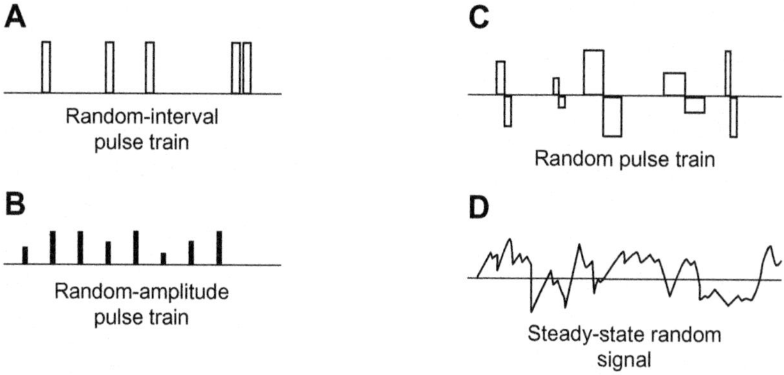

FIGURE 1.6 Examples of random signal functions.

and the autocorrelation function

$$\hat{R}_{vv}(\tau) = \overline{v(t)v(t+\tau)} = \lim_{T\to\infty} \frac{1}{T}\int_{-T/2}^{T/2} v(t)v(t+\tau)dt \tag{1.20}$$

One may also define cross-correlation functions between two random signals $v_1(t)$ and $v_2(t)$; i.e.,

$$\hat{R}_{v_1 v_2}(\tau) = \overline{v_1(t)v_2(t+\tau)} = \lim_{T\to\infty} \frac{1}{T}\int_{-T/2}^{T/2} v_1(t)v_2(t+\tau)dt \tag{1.21}$$

For signals C and D of Fig. 1.6 the mean $\overline{v}$ is zero and the mean square is the lowest-order statistical measure of the magnitude of the signal. In the theoretical limiting process $T\to\infty$ the steady-state signal is presumed to have all possible values so that the averages expressed in Eqs. (1.18)–(1.21) are complete. Even though the individual time series $v(t)$ cannot be defined, the averaged values $\overline{v}$, $\overline{v}^2$, and $\hat{R}_{vv}(\tau)$ all have definite values and therefore represent important measures of the behavior of $v(t)$. Just as the autocorrelation of the periodic and transient functions divulged important characteristics of the original signals, i.e., duration, periodicity, mean-square amplitude, and likelihood of positive and negative values, (Note that for the periodic wave and for the pulse correlation in Fig. 1.5 the integrals over τ were identically zero, showing equal likelihood for positive and negative values of the time series.) so too the correlation functions of random phenomena will give these measures.

In practice, the averaging time T is not infinite, but extends over a suitably long duration, say T_1, and, for example, we might replace the definition in Eq. (1.20) by

$$\hat{R}_{vv}(\tau; T_1, t_0) = \frac{1}{T_1}\int_{t_0}^{t_0+T_1} v(t)v(t+\tau)dt$$

and, similarly, for the other statistical measures. In this representation we have indicated a possible dependence of the correlation function on T_1 and the starting time t_0 of averaging. The object, of course, is to conduct the measurement in such a manner as to minimize dependence on T_1 because it is an artifact of how long the function was measured. The details of how T_1 is selected form part of the subject area of sampling theory [2–5]. A criterion for selecting T_1 will be described in Section 1.4.3.2 as it is dependent on the frequency limitations of the time function. In modern digital processing, the sample length T_1 is discretized into N increments each with time interval $\Delta t = T_1/N$. The time interval is also discretized giving (e.g., Refs. [2,5])

$$\hat{R}_{vv}(\tau_j; N, t_0) = \frac{1}{N}\sum_{i=1}^{N} v(i\Delta t)v(i\Delta t_i + \tau_j)$$

In general, $\hat{R}_{vv}(\tau; T_1, t_0)$ could be a function of the instant time, say, t_0, at which the averaging, or sampling, of data occurs. This dependence, too, is ideally minimized in data sampling. When $\hat{R}_{vv}(\tau; T_1, t_0)$ is independent of t_0, $v(t)$ is considered *statistically stationary*. If the process being measured has been going on so long and is of such duration that $\hat{R}_{vv}(\tau, T_1)$ is independent of how long the sample is accumulated, then in addition the duration T_1 is large enough that

$$\hat{R}_{vv}(\tau; T_1, t_0) = \hat{R}_{vv}(\tau) \tag{1.22}$$

thus the measured correlation function corresponds to the time-correlation function for the signal generated in the experiment. If the experiment is conducted over and over again and averaging is carried out for a variety of T_1 all of which are long enough to satisfy Eq. (1.22), and if all the samples of $\hat{R}_{vv}(\tau)$ and the other statistical measures are identical, then the process is said to be *ergodic*. Such averages are called ensemble averages, and under the ergodicity hypothesis we may replace Eqs. (1.18) through (1.21) by, e.g.,

$$\begin{aligned} \langle v(t)\rangle &= \overline{v(t)} = \overline{v} \\ \langle v^2(t)\rangle &= \overline{v^2(t)} = \overline{v^2} \\ \langle v_1(t)v_2(t+\tau)\rangle &= \overline{v_1(t)v_2(t+\tau)} = \hat{R}_{v_1v_2}(\tau) \end{aligned} \tag{1.23}$$

where the angle brackets denote an *ensemble* average of the quantity. Although this discussion is more of an operational or heuristic definition of ergodicity than a rigorous one, the reader will find more rigorous and complete discussions in the study of Bendat and Piersol [2], Lee [4], Cremer and Ledbedder [8], and Wiener [9].

To summarize, the autocorrelation function of a random variable has the following general properties:

a. $\overline{v^2} = \hat{R}_{vv}(0) \geq \hat{R}_{vv}(\tau)$

b. $\lim\limits_{t\to\infty} \hat{R}_{vv}(\tau) = (\overline{v})^2$

c. $\hat{R}_{vv}(-\tau) = \hat{R}_{vv}(\tau)$
d. $\hat{R}_{vv}(\tau)$ is continuous for all τ

The cross-correlation, or covariance, function of a *random* variable is continuous for all τ, but does not necessarily satisfy the limiting conditions a or c above. In fact the maximum value of the cross-spectral density may occur at a delay time $\tau_{\mathrm{m}} \neq 0$ such that

e. $\hat{R}_{v_1 v_2}(\tau_{\mathrm{m}}) \geq \hat{R}_{v_1 v_2}(\tau)$

The meaning of this inequality will be made more apparent in Section 1.4.3.2.

1.4.3 Review of Fourier Series and the Fourier Integral

The most extensively used powerful tool of random-data analysis and of theoretical formulations of steady-state sound and vibration problems is Fourier analysis. In this section we shall review many of the fundamentals of Fourier analysis as it is used in this book to relate statistical properties of signals to the frequency content. Readers desiring a more refined and complete discussion of specific points are referred to the monographs by Lighthill [10] and Titchmarsh [11]; those wishing to relate to the digital techniques and the Fast Fourier transform (FFT) are referred to texts on random vibration and signal processing, e.g., by Bendat and Piersol [2], Newland [5], and Pierce [7].

1.4.3.1 Periodic Signals

Periodic signals are known to be expressible as a summation of sine and cosine functions; i.e., if $v(t)$ is a periodic signal of period T_{p}, then we may write it as a sum of harmonics of the fundamental frequency $2\pi/T_{\mathrm{p}}$

$$v(t) = \frac{a_0}{2} + \sum_{n=1}^{\infty} a_n \cos\left(\frac{2\pi n t}{T_{\mathrm{p}}}\right) + b_n \sin\left(\frac{2\pi n t}{T_{\mathrm{p}}}\right) \tag{1.24}$$

where the coefficients

$$a_n = \frac{2}{T_{\mathrm{p}}} \int_{-T_{\mathrm{p}}/2}^{T_{\mathrm{p}}/2} v(t) \cos\left(\frac{2\pi n t}{T_{\mathrm{p}}}\right) dt \tag{1.25}$$

and

$$b_n = \frac{2}{T_{\mathrm{p}}} \int_{-T_{\mathrm{p}}/2}^{T_{\mathrm{p}}/2} v(t) \sin\left(\frac{2\pi n t}{T_{\mathrm{p}}}\right) dt \tag{1.26}$$

Among other conditions, these equations are convergent as long as

$$\int_{-T_p/2}^{T_p/2} |v(t)|dt$$

is finite, see, e.g., Refs. [11,12]. As an example, this harmonic analysis is particularly useful in describing the response of fan rotors to inflow distortions (see Chapter 12: Noise from Rotating Machinery). A rotating fan blade will respond to some nonuniformity of its inflow periodically each time it rotates one revolution. The periodicity T_p will be n_s^{-1}, where n_s is the rotation speed (revolutions/time). Therefore $v(t)$ will have a period $T_p = n_s^{-1}$ and will be expressible as a summation of harmonics of frequency mn_s.

The above representations may be given a little more general appearance by replacing the sine and cosine functions by their exponential equivalences:

$$\cos x = (e^{ix} + e^{-ix})/2$$

$$\sin x = (e^{ix} - e^{-ix})/2i$$

Then Eq. (1.24) is replaced by

$$v(t) = \frac{1}{2}\sum_{-\infty}^{\infty}(a_n + ib_n)e^{-in\omega_1 t}$$

where $\omega_1 = 2\pi/T_p$. The coefficients $a_n + ib_n$ may be written

$$a_n + ib_n = \frac{2}{T_p}\int_{-T_p/2}^{T_p/2} v(t)e^{in\omega_1 t}\,dt$$

and it is noted that $a_n = a_{-n}b_n = -b_{-n}$. Accordingly, letting

$$V_n = \frac{1}{2}(a_n + ib_n)$$

we can write the Fourier transform pair over the entire range of positive and negative values of n as

$$v(t) = \sum_{-\infty}^{\infty} V_n e^{-in\omega_1 t} \tag{1.27}$$

and

$$V_n = \frac{1}{T_p}\int_{-T_p/2}^{T_p/2} v(t)e^{in\omega_1 t}\,dt \tag{1.28}$$

where V_n is a complex number, which may be written

$$V_n = |V_n|e^{i\phi_n} \tag{1.29}$$

The correlation function of the periodic signal defined by Eq. (1.15) and expanded as in Eq. (1.27) is

$$\hat{R}_{vv}(\tau) = \frac{1}{T_{\mathrm{p}}}\int_{-T_{\mathrm{p}}/2}^{T_{\mathrm{p}}/2}\left[\left(\sum_{n} V_{n}^{*} e^{in\omega_1 t}\right)\left(\sum_{m} V_{m} e^{-in\omega_1 (t+\tau)}\right)\right] dt \tag{1.30}$$

where the asterisk represents the complex conjugate (i.e., $z = x + iy$ and $z^* = x - iy$). The correlation function then reduces to

$$\hat{R}_{vv}(\tau) = \sum_{-\infty}^{\infty} |V_n|^2 \, e^{-in\omega_1 \tau} \tag{1.31}$$

since

$$\frac{1}{T_{\mathrm{p}}}\int_{-T_{\mathrm{p}}/2}^{T_{\mathrm{p}}/2} \cos[(m \pm n)\omega_1 t] \, dt = \begin{cases} 1, & m \pm n = 0 \\ 0, & m \pm n \neq 0 \end{cases}$$

and likewise for the average value of $\sin[(m \pm n)\omega_1 t]$.

In Eq. (1.31) only the diagonal term of all the possible combinations of $V_m^* V_n$ values contribute to the autocorrelation, the off-diagonal terms, $m \neq n$, do not contribute. The inverse of Eq. (1.31) is, from Eq. (1.28),

$$|V_n|^2 = \frac{1}{T_{\mathrm{p}}}\int_{-T_{\mathrm{p}}/2}^{T_{\mathrm{p}}/2} \hat{R}_{vv}(\tau) e^{-in\omega_1 t} \, d\tau, \qquad -\infty < n < \infty \tag{1.32}$$

A similar Fourier transform pair may be defined for cross-correlation functions.

These relationships show that for a correlation function $\hat{R}_{vv}(\tau)$ there is a spectrum function $|V_n|^2$ that gives the contribution of each harmonic n of the period $\omega_1 = 2\pi/T_{\mathrm{p}}$. Certain of these functions will be discussed in Chapter 12, Noise from Rotating Machinery, when we discuss sound fields of rotating sources.

1.4.3.2 Random Signals

A random signal cannot be described by an explicit mathematical relation. Rather, the value of the signal at a particular time may be expressed only in terms of some describable probability of occurrence. Accordingly, there is a continuum of "frequencies" that may be used to characterize the function. The Fourier transform pair analogous to Eqs. (1.31) and (1.32) is

$$\hat{R}_{vv}(\tau) = \int_{-\infty}^{\infty} \Phi_{vv}(\omega) e^{-i\omega\tau} d\omega \tag{1.33}$$

$$\Phi_{vv}(\omega) = \frac{1}{2\pi}\int_{-\infty}^{\infty} \hat{R}_{vv}(\tau) e^{i\omega\tau} \, d\tau \tag{1.34a}$$

The mean square of v may be obtained by comparing Eqs. (1.19), (1.20), and (1.33), i.e.,

$$\overline{v^2} = \int_{-\infty}^{\infty} \Phi_{vv}(\omega)d\omega \tag{1.34b}$$

$\Phi_{vv}(\omega)d\omega$ represents the (now-continuous) distribution of $|V_n|^2$, which is now not restricted to integral values of n; it is often referred to as a "2-sided" spectrum in the sense that it is defined over the full frequency vector, $-\infty<\omega<\infty$. The cross-correlation function, gives a corresponding cross-spectral density

$$\Phi_{v_1v_2}(\omega) = \frac{1}{2\pi}\int_{-\infty}^{\infty} \hat{R}_{v_1v_2}(\tau)e^{i\omega\tau}\, d\tau \tag{1.35}$$

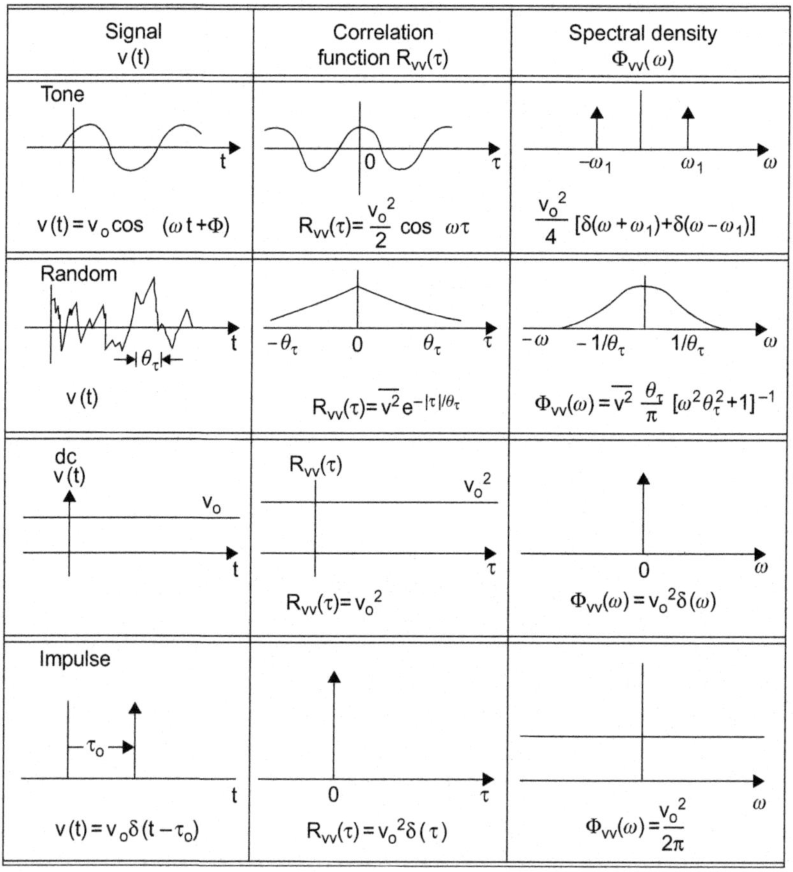

FIGURE 1.7 Illustration of common signal types, their autocorrelation functions, and frequency spectral density functions.

If the correlation function is not symmetrical about $\tau=0$, i.e., if $\hat{R}_{v_1v_2}(\tau) \neq \hat{R}_{v_1v_2}(-\tau)$ then $\Phi_{v_1v_2}(\omega)$ will be a complex number; i.e., it will be expressible with a magnitude and phase

$$\Phi_{v_1v_2}(\omega) = |\Phi_{v_1v_2}(\omega)|e^{i\phi(\omega)}$$

Fig. 1.7 shows examples of correlation functions and frequency spectra of various types of signals: random and periodic. The periodic signal is a special case of the generalized transform pair, Eqs. (1.33) and (1.34a); the correlation of such functions is also periodic, e.g., Eq. (1.16). We can see this more clearly by the following derivation. For periodic functions of period $T_\mathrm{p} = 2\pi/\omega_1$, the frequency spectrum is found by a substitution of Eq. (1.16) into Eq. (1.34a)

$$\Phi_{vv}(\omega) = \frac{1}{2\pi}\frac{v_0^2}{2}\int_{-\infty}^{\infty} e^{i\omega\tau} \cos\omega_1\tau \, d\tau$$

Since the period is T_p, the correlation function may be expanded as a sequence of functions in a finite interval equal to the period; i.e., as we did for Eq. (1.13),

$$\cos\omega_1\tau = \sum_{n=-\infty}^{\infty} \cos\omega_1(\tau - nT_\mathrm{p}), \quad -\frac{T_\mathrm{p}}{2} < \tau\frac{T_\mathrm{p}}{2}$$

The spectrum then reduces to a summation

$$\Phi_{vv}(\omega) = \frac{v_0^2}{2}\frac{1}{2\pi}\sum_{n=-\infty}^{\infty} e^{inT_\mathrm{p}\omega}\int_{-T_\mathrm{p}/2}^{T_\mathrm{p}/2} e^{i\omega\xi}\cos\omega\xi \, d\xi$$

Poisson's summation formula [10] gives the identity

$$\frac{T_\mathrm{p}}{2\pi}\sum_{n=-\infty}^{\infty} e^{i(nT_\mathrm{p})\omega} = \sum_{m=-\infty}^{\infty} \delta\left[\omega - m\left(\frac{2\pi}{T_\mathrm{p}}\right)\right] \tag{1.36}$$

where the function $\delta[\omega]$ is the Dirac delta function defined in Section 1.6.4. Accordingly, the spectrum has values only at integral multiples of ω_1; i.e., it is given by

$$\Phi_{vv}(\omega = n\omega_1) = \frac{v_0^2}{2}\left\{\frac{1}{T_\mathrm{p}}\int_{-T_\mathrm{p}/2}^{T_\mathrm{p}/2} e^{in\omega_1\xi}\cos\omega_1\xi \, d\xi\right\}$$

or for this special case of a pure tone continuous signal,

$$\Phi_{vv}(\omega) = \frac{v_0^2}{4}\delta(\omega \pm \omega_1) \tag{1.37}$$

This is precisely the result that would have been obtained if Eqs. (1.25) and (1.26) had been used, and it accounts for the two equal spikes at $\omega = \omega_1$ and $-\omega_1$ shown at the top of Fig. 1.7. Other examples shown in the Fig. 1.7 can be worked out by the reader.

In the cases of space–time correlations, the cross-correlation function will give a complex cross-spectral density when the cross-correlation is not symmetrical about $\tau = 0$. As example, in the case of the pulsating source the cross-spectrum of the traveling wave pulse shown in Fig. 1.3 the cross-spectrum function between $A_r(t)$ and $p(t,r)$ is found using Eq. (1.10) in Eq. (1.34a) producing

$$\Phi_{Ap}(r,\omega) = \rho_0\left(\frac{a}{r}\right)a\frac{1}{2\pi}\int_{-\infty}^{\infty} e^{i\omega\tau}\hat{R}_{AA}\left(\tau - \frac{r-a}{c_0}\right)d\tau$$

or changing variables with $\xi = \tau - (r-a)/c_0$

$$\Phi_{Ap}(r,\omega) = \rho_0\left(\frac{a}{r}\right)ae^{ik_0(r-a)}\frac{1}{2\pi}\int_{-\infty}^{\infty} e^{i\omega\xi}\hat{R}_{AA}(\xi)\,d\xi$$

and therefore

$$\Phi_{Ap}(\omega) = \rho_0\left(\frac{a}{r}\right)a\Phi_{AA}(\omega)e^{ik_0(r-a)} \tag{1.38}$$

in which the autospectrum of the surface acceleration $\Phi_{AA}(\omega)$ has been utilized. By Eq. (1.33) the integral of the autospectrum $\Phi_{AA}(\omega)$ overall frequencies gives $\hat{R}_{AA}(0)$, which is the mean square acceleration $\overline{A^2}$ in this case. The phase is in Eq. (1.38) a continuously increasing function of $\omega = k_0c_0$ as illustrated in the center of Fig. 1.8 which illustrates functional forms that typically found in statistics of turbulence. The other illustrations in Fig. 1.8 show a zero phase for correlation functions that are perfectly symmetrical about $\tau = 0$ and a varying slope of the phase–frequency line when the correlation function is not symmetrical about its maximum value at $\tau = \tau_m$. The function

$$\frac{\Phi_{Ap}(r,\omega)}{\Phi_{AA}(\omega)} = \rho_0\left(\frac{a}{r}\right)ae^{ik_0(r-a)} \tag{1.39}$$

is often called the "transfer function" between sound pressure and vibration of the surface. By comparing Eq. (1.39) with Eq. (1.1) we see that the transfer function includes all the important characteristics that "transfer" the acceleration of the surface into a pressure at r. The propagation velocity is given by the phase, i.e. $c_0 = \omega a[\mathrm{Real}(k_0a)]^{-1}$. The real part of this transfer function is also recognizably the magnitude of the free space Green function for the propagating medium. Eqs. (1.38) (or (1.39)) and (1.10) contain identical amounts of information. Ideally one could discern all the information about the physics of the source field with either measurement. However, when one is interested in the behavior of the source as a function of frequency (as this information may be subsequently used to assess acoustical properties in a variety of applications) we encounter the frequency autospectrum and cross-spectrum analysis more frequently than correlation analysis. Correlation analysis of pulsed signals is often used to identify energy transmission paths.

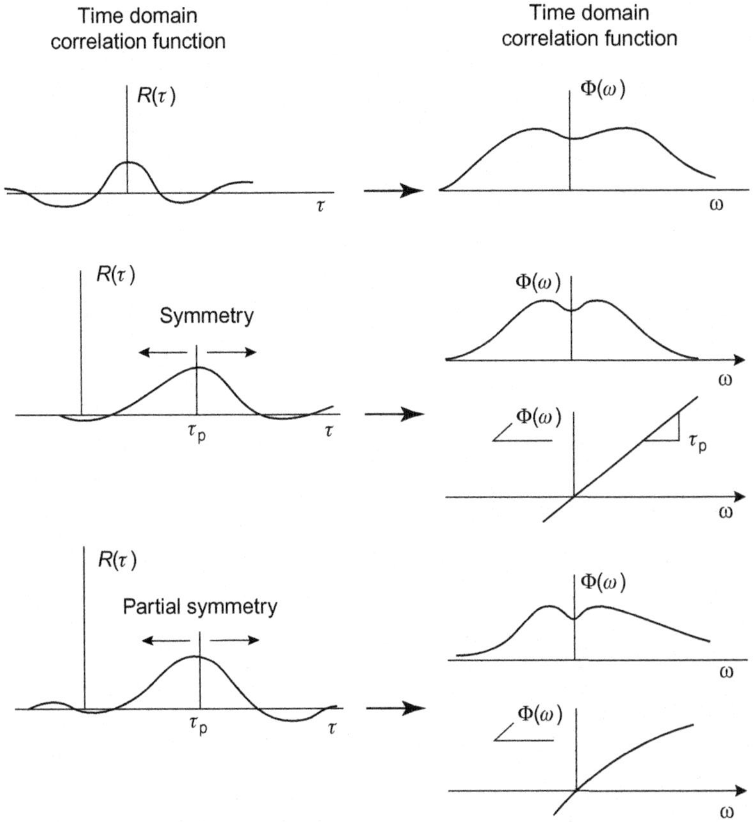

FIGURE 1.8 Illustrations of (A) symmetric and (B) nonsymmetric correlations and their spectral transformations.

1.4.4 Conclusions of Simple Filtering Theory

1.4.4.1 Descriptions of Linear Bandpass Filters

We now examine in a classical way the means of measuring auto- and cross-spectral densities. Fig. 1.9 is a diagram of the electronic operations done. Our discussion here will again be operational and explanatory rather than rigorous and will apply to nearly all signal analysis implied by the treatment of this book. References [2–7] give the classical treatments of filtering theory; Refs. [2] and [5] give discussions of high-speed digital sampling methods. Whatever processing is used (analog or FFT), the discussion here is applicable. The practical physical arrangement is illustrated in the top of the figure showing two sensors providing voltages $v_1(t)$ and $v_2(t)$. All the correlation and integral properties that were discussed in Sections 1.4.2 and 1.4.3 apply to the signals $v_1(t)$ and $v_2(t)$.

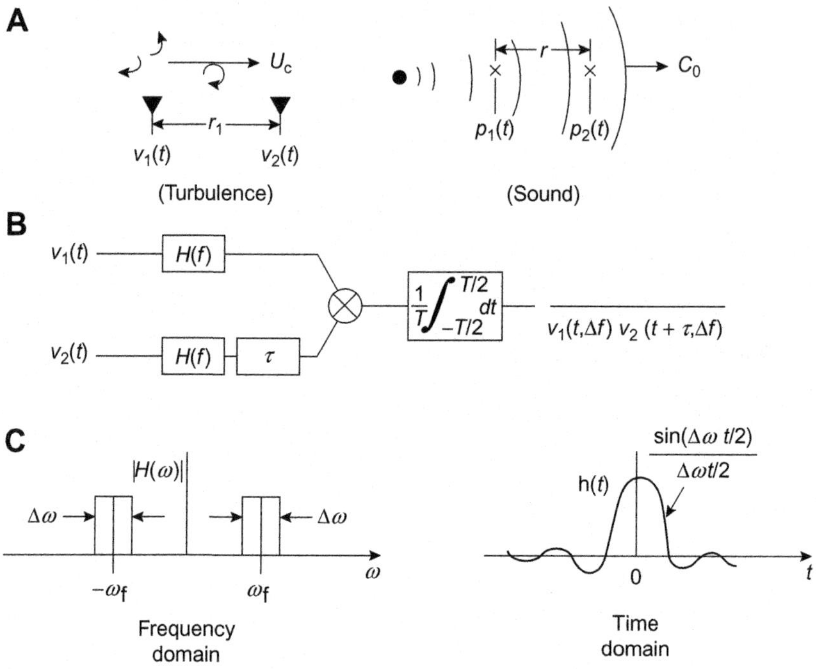

FIGURE 1.9 Diagram of (A) physical arrangements, (B) electronic components, and (C) functions which are used in measurements of cross-spectral densities.

In the following we shall relate the analytically-relevant 2-sided spectrum, $\Phi_{vv}(\omega)$, to the 1-sided autospectral density, $G_{vv}(f)$, that might be obtained with a linear band pass filter. The signals are passed through a bandpass filter tuned to frequency $f_f = \omega_f/2\pi$ with a bandwidth $\Delta f = \Delta\omega/2\pi$. The signals $v_1(t)$ and $v_2(t)$ are steady-state periodic or random with Fourier transforms (Conditions for validity of this transform for random variables are outside the scope of this chapter. However, as by Batchelor [13] and Cremer and Ledbedder [8], the transform exists in a generalized sense.)

$$v_1(\omega) = \frac{1}{2\pi}\int_{-\infty}^{\infty} v_1(t)e^{i\omega t}\,dt \tag{1.40}$$

and similarity for $v_2(t)$. It can be shown [9] (see also Chapter 2: Theory of Sound and its Generation by Flow) that the 2-sided cross-spectral density of $v_1(t)$ and $v_2(t)$ is related to the ensemble of the product $v_1(\omega)$ and $v_2(\omega)$ as

$$\Phi_{12}(\omega)\delta(\omega-\omega') = \langle v_1(\omega)v_2^*(\omega')\rangle, \quad -\infty < \omega, \quad \omega' < \infty \tag{1.41}$$

and each autospectrum as

$$\Phi_{vv}(\omega)\delta(\omega-\omega') = \langle v(\omega)v^*(\omega')\rangle \tag{1.42}$$

where $\delta(\omega - \omega')$ is unity for $\omega = \omega'$ and zero otherwise. The ideal rectangular filter has a response characteristic, shown in Fig. 1.9, that passes all signals with frequency content in the ranges—$\omega_f - \Delta\omega/2 < \omega < -\omega_f + \Delta\omega/2$ and $\omega_f - \Delta\omega/2 < \omega < \omega_f + \Delta\omega/2$ so that if the input is $v_1(t)$, the filtered output is written

$$v_1(t, \Delta f) = \int_{-\infty}^{\infty} e^{-i\omega t}\, v_1(\omega) H(\omega) d\omega \tag{1.43a}$$

where $\Delta f = \Delta\omega/2\pi$, and $H(\omega)$ is the response characteristic of the filter with an inverse Fourier transform $h(t)$ shown in the bottom of Fig. 1.9. The autocorrelation of the filtered output is

$$\langle v^*(t, \Delta f) v(t+\tau, \Delta f)\rangle = \left\langle \int_{-\infty}^{\infty} e^{i\omega' t} v_1^*(\omega') H^*(\omega') d\omega' \cdot \int_{-\infty}^{\infty} e^{-i\omega(t+\tau)} v_1(\omega) H(\omega) d\omega \right\rangle$$

$$\langle v^*(t, \Delta f) v(t+\tau, \Delta f)\rangle = \int_{-\infty}^{\infty} e^{-i\omega\tau} \Phi_{vv}(\omega) |H(\omega)|^2 d\omega \tag{1.43b}$$

where we have made use of Eq. (1.40) to eliminate ω'. The time delay is introduced in one of the filtered signals as shown in Fig. 1.9.

The mean square of the filtered signal, obtained without the time delay is given analogously to Eq. (1.34b) as

$$\overline{v^2}(f, \Delta f) = \int_{-\infty}^{\infty} \Phi_{vv}(\omega) |H(\omega)|^2 d\omega \tag{1.44}$$

To the extent that the 2-sided spectrum of $\Phi_{vv}(\omega)$ is invariant over a small interval in frequency $\Delta\omega$, the expression may be approximated

$$\overline{v^2}(f, \Delta f) \simeq 2\Phi_{vv}(\omega)\Delta\omega \tag{1.45}$$

in which $|H(\omega)|^2$ has been taken as unity over $\Delta\omega$ and since $\Phi_{vv}(-\omega) = \Phi_{vv}(\omega)$.

Measurement systems operate only in positive frequencies, i.e., from $0 < f < \infty$, so measured spectral densities are one-sided. Analogous to Eqs. (1.42)–(1.45) we define the 1-sided spectral density

$$G_{vv}(f) = \lim_{\Delta f \to 0} \frac{\overline{v^2}(f, \Delta f)}{\Delta f} \tag{1.46}$$

where, since $\Delta f = \Delta\omega/2\pi$, the measured spectral density is related to the "theoretical" two-sided spectrum by

$$G_{vv}(f) = 4\pi\Phi_{vv}(\omega) \quad f \geq 0 \tag{1.47}$$

and the mean square is an extension of Eq. (1.34b)

$$\overline{v^2} = \int_0^{\infty} G_{vv}(f) df \tag{1.48}$$

The cross-correlation function of the filtered signals $v_1(t)$ and $v_2(t)$ may be used to determine the cross-spectral density and is found using Eqs. (1.43)–(1.45) to find the equivalence for narrow bands

$$\hat{R}_{v_1 v_2}(r,\tau;f,\Delta f) = \lim_{\Delta\omega \to 0} 2|\Phi_{12}(\omega)|\cos(\phi(\omega) - \omega\tau)\Delta\omega \tag{1.49}$$

under the presumption that $|\Phi_{12}(\omega)| = |\Phi_{12}(-\omega)|$ and $\phi(\omega) = -\phi(-\omega)$. In the context of our example of the pulsating sphere, if v_1 is the surface acceleration and v_2 the field pressure, then the phase has the form introduced before, $\phi = \omega(r-a)/c_0$. A form of Eq. (1.49) that is consistent with Eq. (1.47) and gives a method of extracting the cross-spectrum from measurements is

$$\begin{aligned} |G_{12}(f)|\cos(\phi - 2\pi f\tau) &= \lim_{\Delta f \to \infty} \frac{\hat{R}_{v_1 v_2}(r,\tau;f,\Delta f)}{\Delta f} \\ &= 4\pi|\Phi_{12}(\omega)|\cos(\phi - \omega\tau) \quad f \geq 0 \end{aligned} \tag{1.50}$$

where the 4π includes G(*f*) being defined only over $f \geq 0$ as in Eq. (1.47). The cross-spectral density may also be written

$$\Phi_{12}(\omega) = |\Phi_{12}(\omega)|[\cos\phi + i\sin\phi] \tag{1.51a}$$

with a cospectrum

$$\mathrm{Co}[G_{12}(f)] = 4\pi|\Phi_{12}(\omega)|\cos\phi, \quad f \geq 0 \tag{1.51b}$$

and a quadrature spectrum

$$\mathrm{Quad}[G_{12}(f)] = 4\pi|\Phi_{12}(\omega)|\sin\phi, \quad f \geq 0 \tag{1.51c}$$

The cospectrum, can be obtained from a correlation function by setting $\tau = 0$, and the quadrature spectrum is achievable from Eq. (1.49) by setting a frequency-dependent time delay

$$\tau = \pi \times \omega/2 = \frac{1}{4}f$$

i.e., by setting a 90° phase shift between channels. The phase angle is then determined by

$$\phi = \tan^{-1}\left[\frac{\mathrm{Quad}[G_{12}(f)]}{\mathrm{Co}[G_{12}(f)]}\right] \tag{1.51d}$$

Although this method was used through the 1960s, current FFT technology now computes the complex spectrum directly. An appreciation of its physical significance, however, aids in the interpretation.

1.4.4.2 Spatial Filtering and Wave Number Transformations

The use of spatial transformations in experimental acoustics and vibration has become more common with the development of digital data acquisition technology. Multi-channel data acquisition using hydrophones and accelerometers, for example, have been used to acquire much of the experimental data that will be discussed in this book. The uses of these data get to the heart of the spatial interfacing between the fluid and the structure that governs both

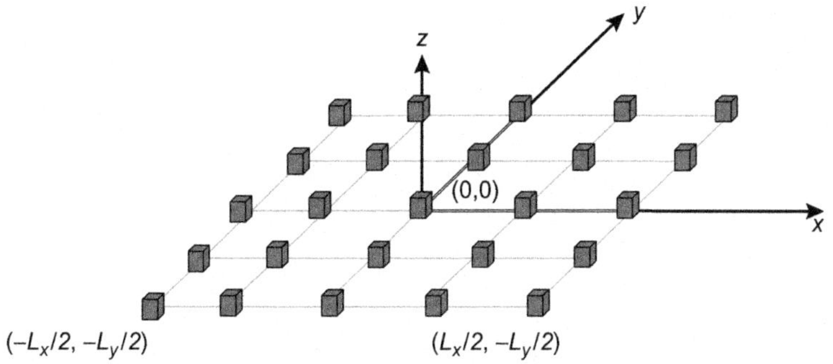

FIGURE 1.10 Diagram of a notional two-dimensional array of equally-spaced sensors on the (x,y) axis sensing a field due to disturbances in the $z\neq 0$ volumes or on the $z=0$ surface.

the structure–fluid acoustic coupling, the fluid to structure flow-excitation, or the fluid–fluid source efficiency. An example array of this type is often placed on a surface which can be described in terms of harmonic functions, i.e., plane, cylinder, sphere, prolate spheroidal, etc. In the example shown in Fig. 1.10 we see 25 equally-spaced sensors in the x, y plane. Consider that these are pressure sensors or some other form of motion, strain, or stress sensor sensing a signal, say $v(x_{i,j},\omega)$. We assume that $v(x_{i,j},\omega)$ has been obtained at the coordinate $x_{i,j}$ and filtered in frequency in the manner of Section 1.4.4.1. Due to the variation in the physical property overall $x_{i,j}$, each signal has some mixture of deterministic or random phase behavior.

Assume that $v(\overline{x},\omega)$ can be described as a two-dimensional Fourier transformation analogous to the inverse of Eq. (1.40) and its inverse

$$v(\mathbf{x},\omega)=\int_{-\infty}^{\infty}\int_{-\infty}^{\infty}V(\mathbf{k},\omega)e^{i\mathbf{k}\bullet\mathbf{x}}d\mathbf{k}$$

and its inverse

$$V(\mathbf{k},\omega)=\frac{1}{(2\pi)^2}\int_{-\infty}^{\infty}\int_{-\infty}^{\infty}v(\mathbf{k},\omega)e^{-i\mathbf{k}\bullet\mathbf{x}}d^2\mathbf{x}$$

The vectors $\mathrm{k_i}$ are defined over the entire axis $-\infty<\mathrm{k_i}<\infty$. The variables $v(x_{i,j},\omega)$ and $V(k_{i,j},\omega)$ are subject to the same restrictions of stationarity as are those in the time-frequency domain. Since the $x_{i,j}$ are evenly-spaced in the plane illustrated in the figure, the ensemble average of a covariance similar to that of Eqs. (1.43a,b) is

$$\langle v(x_{i,j},\omega)v^*(x'_{k,l},\omega)\rangle=\left\langle\int_{-\infty}^{\infty}\int_{-\infty}^{\infty}V(\mathbf{k},\omega)e^{i\mathbf{k}\bullet\mathbf{x}}d^2\mathbf{k}\bullet\int_{-\infty}^{\infty}\int_{-\infty}^{\infty}V^*(\mathbf{k},\omega)e^{-i\mathbf{k}'\bullet\mathbf{x}'}d^2\mathbf{k}'\right\rangle$$

The coordinate locations of the sensors are
$x = rd_x - L_x/2$ and $x' = pd_x - L_x/2$
$y = sd_y - L_y/2$ and $y' = qd_y - L_y/2$
Substituting these coordinates:

$$\langle v(x_{r,s},\omega)v^*(x'_{p,q},\omega)\rangle = \int_{-\infty}^{\infty}\int_{-\infty}^{\infty}\int_{-\infty}^{\infty}\int_{-\infty}^{\infty}\left\langle V(\mathbf{k},\omega)V^*(\mathbf{k}',\omega)\right\rangle$$
$$(e^{ird_xk_x}e^{-ipd_xk'_x})(e^{isd_yk_y}e^{-iqd_yk'_y})dk_xdk_ydk'_xdk'_y \tag{1.52}$$

We shall see in Sections 2.6.2 and 3.5.2 that if the variable $v(\overline{x},\omega)$ is continuous and homogeneous over $\overline{x}$ then the ensemble in the wave number domain is

$$\langle V(\mathbf{k},\omega)V^*(\rightharpoonup k',\omega)\rangle = \Phi_{vv}(\mathbf{k},\omega)\delta(\mathbf{k}-\mathbf{k}') \tag{1.53}$$

Where $\delta(\vec{k}-\vec{k}')$ is the Dirac delta function, see Section 1.6. Introducing this spectrum function and carrying out the integration over one of the wave numbers

$$\langle v(x_{r,s},\omega)v^*(x'_{p,q},\omega)\rangle = \int_{-\infty}^{\infty}\int_{-\infty}^{\infty}\Phi_{vv}(\boldsymbol{\kappa},\omega)(e^{-i(s-p)d_x\kappa_x})(e^{-i(r-q)d_y\kappa_y})d\kappa_xd\kappa_y \tag{1.54}$$

If the covariance matrix $\langle v(x_{i,j},\omega)v^*(x'_{k,l},\omega)\rangle$ is an N_x by N_x submatrix in the x-coordinate and with indices s and p and in the y-coordinate it is N_y by N_y with indices r and q which are analogs to Eq. 1.41 in the frequency domain. If we transform this matrix $\langle v(x_{r,s},\omega)v^*(x'_{p,q},\omega)\rangle$ we create a two-dimensional spatial filtering. To do this, we first note that the digital equivalent to the Fourier transform of Eq. 1.28 for spatial variables is for 1 spatial dimension

$$V(k_x,\omega) = \frac{1}{L_x}\int_{-L_x/2}^{L_x/2} v(x,\omega)e^{ik_xx}dx$$

The digital equivalent of this is derived invoking Simpson's rule for the integral and approximate dx by d_x and L_x by N_xd_x;

$$[V(k_x,\omega)]_{est} = \frac{1}{N_x}\sum_{r=0}^{N_x-1} v(rd_x - L_x/2,\omega)e^{ik_x(rd_x-L_x/2)}$$

Note that this integral approximation is made for convenience here; other approaches, such as fast Fourier transformations, may be used in practice.

Applying this transformation to each variable we form a spectral estimate

$$[\Phi_{vv}(\mathbf{k}\omega)]_{est} = \frac{1}{N_x{}^2N_y{}^2}\sum_{r=1}^{N_x}\sum_{p=1}^{N_x}\sum_{s=1}^{N_y}\sum_{q=1}^{N_y}\langle v(x_{r,s},\omega)v^*(x'_{p,q},\omega)\rangle(e^{i(r-p)d_xk_x})(e^{i(s-q)d_yk_y}) =$$
$$= \frac{1}{N_x{}^2N_y{}^2}\sum_{r=1}^{N_x}\sum_{p=1}^{N_x}\sum_{s=1}^{N_y}\sum_{q=1}^{N_y}\left[\int_{-\infty}^{\infty}\int_{-\infty}^{\infty}\Phi_{vv}(\boldsymbol{\kappa},\omega)(e^{-i(r-p)d_x\kappa_x})(e^{-i(s-q)d_y\kappa_y})dk_xdk_y\right]$$
$$(e^{i(r-p)d_xk_x})(e^{i(s-q)d_yk_y})$$

The summations rearrange to form separate functions that we represent by the symbol $A_x(k_xd_x)$, say, where

$$A_x(k_xd_x) = \frac{1}{N_x}\sum_{r=1}^{N_x}\exp(ir(k_x - \kappa_x)d_x) = \frac{\sin\left[\frac{1}{2}(k_x - \kappa_x)d_x\right]}{[N_x(k_x - \kappa_x)d_x]}e^{i(N_x-1)\frac{1}{2}(k_x-\kappa_x)d_x} \tag{1.55}$$

Substitution gives

$$[\Phi_{vv}(\mathbf{k},\omega)]_{est} = \int_{-\infty}^{\infty}\int_{-\infty}^{\infty}\Phi_{vv}(\boldsymbol{\kappa},\omega)|A_x((k_x-\kappa_x)d_x/2)|^2|A_y((k_y-\kappa_y)d_y/2)|^2 d\kappa_x d\kappa_y \tag{1.56}$$

This integral states that an estimation of the true wave number spectrum is given by a discrete Fourier transform of an array of interelement covariances. An example filter function,

$$|A_{xy}(\boldsymbol{\kappa}\bullet\mathbf{d})| = |A_x((\kappa_x)d_x/2)|^2|A_y((\kappa_y)d_y/2)|^2$$

is graphed in Fig. 1.11 for a specific case of 9×9 array of elements. On the left-hand side is a three-dimensional view of the lobe pattern near the origin $\kappa = 0$, showing the magnitudes of a major lobe at the origin and a series of lower-amplitude minor lobes surrounding the major lobe. The magnitudes of

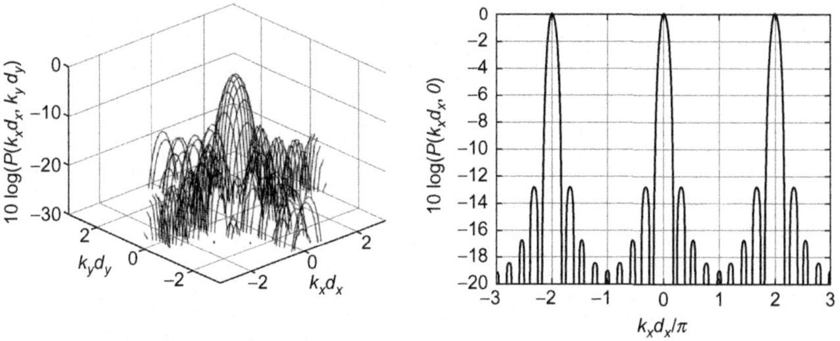

FIGURE 1.11 Diagrams of the wave number resolution function for an example 9×9 element of equally-spaced sensors in a rectangular grid pattern. On the left is a three-dimensional view near the origin showing the pattern of grating lobes; on the right is a view along k_x for a broader range of wave numbers showing aliasing lobes.

these minor lobes decrease as the number of elements in the array increase. They show an orthogonal pattern here because of the rectangular pattern of the array. A circular array would have a sequence of circular minor lobes encircling the central major lobe whose magnitude is always unity in the normalization on sensor number used here. Other distribution patterns are used for the sensors in order to achieve specific lobe structures for measurements, see Ref. [14].

The estimated wave number spectrum is obtained through integration over the covariance space separation increments; this wave number resolution function then "scans" the true spectrum. This behavior is completely analogous to that of a frequency-domain filter function. The wave number resolution, which has many applications, is called "point spread function" in optics and "array beam patterns" in underwater acoustics.

The spectral estimation integral of Eq. (1.56) is the key result of this subsection. Fig 1.11 shows the existence of both grating and aliasing lobes that will contaminate the spectral estimate. Accordingly, design of the spatial array will be contingent on the expected character of the actual wave number spectrum. For example, convection of signals by flow will introduce a wave number feature at wave number $\mathbf{k} = \omega/\mathbf{U}$; vibration will have a wave number feature at the inverse of some spatial scale of the structure; sound has a propagation wave number. Consideration of characteristics such as these will determine both the spacing of sensors and the length (aperture) of the array. Large arrays are appropriate to resolve low wave numbers; small spacings are driven by desire to resolve higher wave numbers.

1.4.4.3 Notes on Error Analysis

Although the general analysis of statistical errors in signal processing is a complex one [2], there is a simple rule for a "quick-look" assessment of measurement precision as far as averaging time T_1 is concerned. A sufficiently long value of T_1 permits a recognition of measured statistical properties in terms of analytically derived ones that generally presuppose ergodicity of the quantities involved. For continuous, steady-state random signals, the minimum requirements for averaging time will be summarized below.

Specifically, we are interested in setting a limit on the necessary averaging time T_1 in order that Eqs. (1.22) and (1.23) are achieved by an estimate of square averaging or correlation. Such estimates all replace an ensemble average with a time average

$$\langle b \rangle = \frac{1}{T_1}\int_{-T_1/2}^{T_1/2} b(t)dt$$

and we define a condition on T_1 to make this replacement valid.

Random signals are often assumed, and in fact, have been observed, to have a Gaussian probability distribution. Presuming a Gaussian (or normal) probability distribution, both simplifies the mathematical modeling of the statistical theories and has certain important statistical implications. With a Gaussian probability distribution, a broadband signal of frequency f in a frequency bandwidth Δf will have a standard error in the estimate of the spectrum $G(f)$ of

$$\epsilon = 1/\sqrt{T_1 \Delta f} \tag{1.57}$$

as long as the average $\overline{v}$ is zero. ε^2 is really the normalized variance of the estimate; i.e.,

$$\varepsilon^2 = \langle(\overline{G}(f) - G(f))^2\rangle/(G(f))^2$$

where $\overline{G}(f)$ is the measured or estimated value of the spectrum and $G(f)$ is the "true value." Loosely interpreted, for a Gaussian-distributed signal, errors less than ϵ, i.e.,

$$|\overline{G}(f) - G(f)|/G(f) < \varepsilon$$

can be expected with roughly 85% certainty, but errors less than 2ϵ may be expected with 98% certainty. A similar relation may be derived for the cross-spectral density, but

$$\varepsilon = 1/\sqrt{2T_1 \Delta f}$$

In the case of correlation functions, we see that there is a relationship between the time variable τ and the frequency variable ω. In fact, the lowest frequency f' at which the spectrum $G_{vv}(f)$ has a nonzero value is of order $1/\tau'$ where τ' is the largest time at which $\hat{R}_{vv}(\tau)$ is acceptably close to zero. The averaging time T_1 must certainly exceed τ', and it therefore also determines the frequency limit f'.

A practical criterion for the standard error of a correlation estimate at time delay τ is given by Bendat and Piersol [2] as

$$\varepsilon \simeq \frac{1}{\sqrt{2T_1 \Delta f}} \left[1 + \frac{\hat{R}(0)}{\hat{R}(\tau)}\right]^{1/2}$$

where Δf is the bandwidth of the frequency spectrum of the correlated signal, $\hat{R}(\tau)$ is the true correlation function, and $\hat{R}(0)$ is the mean square of the signal. From this expression we can see that when $\hat{R}(\tau) \rightarrow 0$ the error can be quite large. In a crude sense $\hat{R}(\tau)$ at large τ is controlled by the lowest-frequency components of the spectral density $G(f)$ or $\mathbf{\Phi}(\omega)$.

1.5 REPRESENTATIONS OF MEASURED SOUND

Although the physical nature of sound and its production will be examined in Chapter 2, Theory of Sound and its Generation by Flow, we shall here summarize the established means of quantifying acoustic variables.

1.5.1 Sound Levels

1.5.1.1 Sound Pressure Level

The principal measured property of sound is the pressure P at a point. Since sound is a dynamic phenomenon, the acoustically induced pressure is also a time-varying quantity. The measure of the acoustic pressure that is conventionally reported is the time average of a pressure squared; i.e.,

$$\overline{p^2} = \frac{1}{T}\int_{-T/2}^{T/2} p^2(t)\,dt$$

with zero time average, $\overline{p} = 0$. This is then simply related to sound intensity and power levels as we shall see in Chapter 2, Theory of Sound and its Generation by Flow. The sound pressure level is determined from the above as

$$L_s = 10\log(\overline{p^2}/p_{\text{ref}}^2)$$

where p_{ref} is 2×10^{-5} n/m^2 = 20 μPa for sounds in gases and 20^{-6} n/m^2 = 1 μPa for sounds in liquids.

Transmission of sound is generally considered on a power basis; the sound power level is defined as above as

$$L_{\text{n}} = 10\log(\mathbb{P}/\mathbb{P}_{\text{ref}})$$

where $\mathbb{P}$ is the sound power transmitted across a specified surface and $\mathbb{P}_{\text{ref}}$ is a reference quantity conventionally taken as 10^{-12} W. As we shall describe in Chapter 2, Theory of Sound and its Generation by Flow, the sound power radiated across a spherical surface of area A_s, from an omnidirectional source is related to the sound pressure as

$$L_{\text{n}} = L_s + 10\log(p_{\text{ref}}^2 A_s/\rho_0 c_0 \mathbb{P}_{\text{ref}})$$

The sound intensity level may be found from

$$L_I = 10\log(I/I_{\text{ref}})$$

where the acoustic intensity is related to the mean-square pressure by

$$I = \overline{p^2}/\rho_0 c_0$$

and $I_{ref} = 10^{-12}$ W/m^2. Actually the acoustic intensity is a vector property. However, as we shall see in the next chapter, far enough from the source the acoustic energy intensity across a spherical surface surrounding the source will be directed normal to the surface. Therefore in the far field the direction of I is radial from the acoustic center of the source.

The sound pressure can be extracted from the sound pressure level by

$$\overline{p^2} = p_{ref}^2 10^{L_s/10}$$

and similarly for the sound power level.

1.5.1.2 The Use of Dynamic Fluid Pressure in Nondimensionalizing Sound Pressure

Typically when dealing with flow-induced noise (and analogously, flow-induced vibration), it is appropriate to present radiated sound pressures in a form normalized on the fluid-dynamic pressure

$$q = \tfrac{1}{2}\rho_0 U^2$$

where U is a reference velocity of the flow. This normalization makes sense because the fluid forces that produce the sound increase quadratically with fluid velocity; therefore a fixed reference pressure will not collapse measurements made at different fluid velocities. Dimensionless sound pressures will be of the form

$$10\,\log(\overline{p^2}/q^2)$$

so that

$$L_s = 10\,\log(\overline{p^2}/q^2) + L_q \tag{1.58}$$

where

$$L_q = 20\,\log(q/p_{ref}).$$

Fig. 1.12 shows a nomograph for computing $20\,\log(q/p_{ref})$ for a given velocity of a fluid of specified density.

1.5.1.3 The Use of Transfer Functions

Since

$$10\,\log(AB) = 10\,\log A + 10\,\log B$$

the sound pressure level calculated from a product of quantities can often be easily calculated as a summation of logarithms. For example, we may be given the quantity

$$10\,\log\frac{\overline{p^2}}{q^2M^2A/r^2}$$

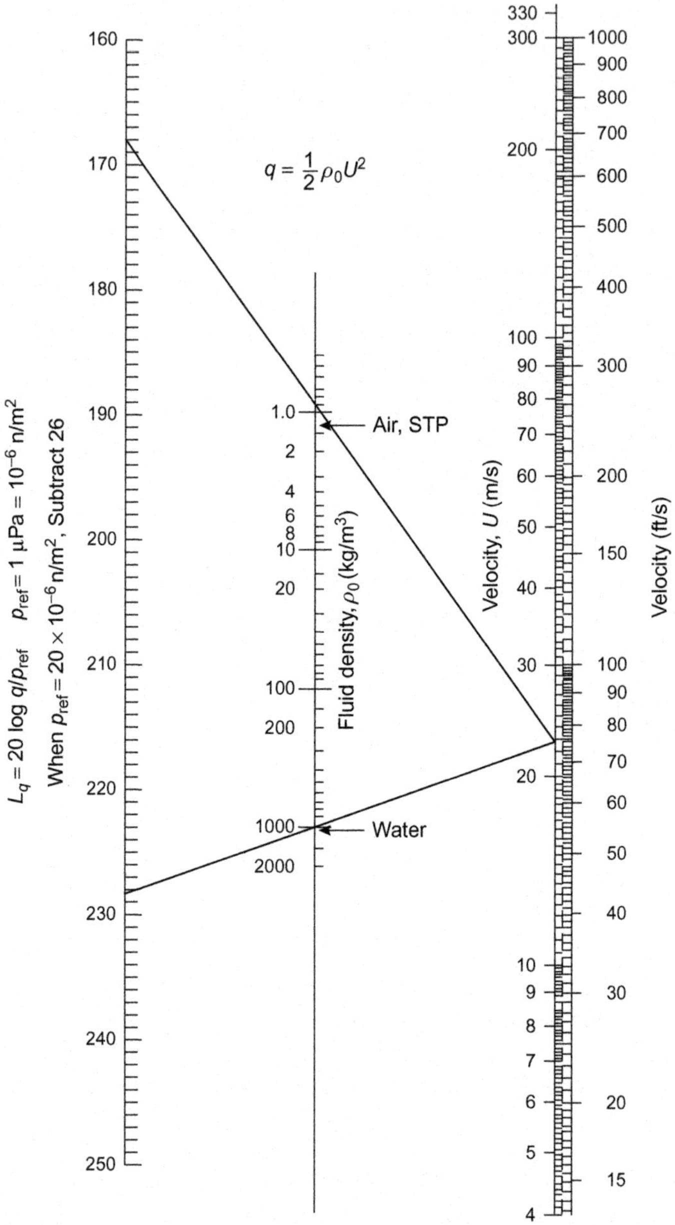

FIGURE 1.12 Nomograph for calculating $20 \log(q/p_{ref})$ for fluids of varying density and speed, for example, $L_q = 168$ at 23 m/s in air.

where A is an area factor, and $\overline{p^2}$ is the mean square sound pressure at a distance r from a flow-dependent source. For a new Mach number $M = U/c_0$ we can find

$$L_s = 10\log\frac{\overline{p^2}}{q^2M^2A/r^2} + 20\log\frac{q}{p_{\text{ref}}} + 20\log M + 10\log\frac{A}{r^2} \tag{1.59}$$

where the first factor is given, the second factor is found in Fig. 1.12, and the third and fourth are found by computation.

Alternatively if we are interested in a transmission loss, this may be expressed

$$\overline{p^2} = \overline{p^2_{\text{in}}}/\tau$$

where $\overline{p^2_{\text{in}}}$ is the mean-square pressure in the structure. The sound pressure level outside the structure is given by

$$(L_s) = (L_s)_{\text{in}} + TL.$$

where $TL = 10\log(1/\tau)$. Similar relationships follow from the example given as Eq. (1.39) and one-sided spectrum functions may be constructed as in Section 1.4.4.1 as long as their 2-sided counterparts are symmetric over $\pm\omega$.

1.5.2 An Example of the Use of Nondimensional Spectrum Levels for Scaling

The use nondimensionalization is at the heart of many analysis procedures used in this book. To illustrate and to fix ideas we shall examine various dimensionless forms of a 2-sided spectral density function for random signals of the type illustrated in Fig. 1.7. In our example, let

$$\Phi_p(\omega) = \frac{\overline{p^2}\theta_t}{\pi}\frac{1}{(\omega\theta_t)^2+1} \quad \text{for} \quad -\infty < \omega < \infty$$

which is a spectrum form that is qualitatively typical of those generated by many common random phenomena. Suppose that the overall sound pressure $\overline{p^2}$ and the time scale θ_t depend on velocity and length scale as, e.g.,

$$\frac{\overline{p^2}}{\left[\frac{1}{2}\rho_0U^2\right]^2M^2(L/r)^2} = \alpha^2$$

and

$$\theta_t\, U/L = \beta^{-1}$$

where α and β are constants, M is a Mach number, and L is a dimension of the sound source. These forms are typical of those that occur in flow-generated phenomena; in fact, this example relates directly to the material in Chapter 4, Dipole Sound from Cylinders. The above ratios account for both the dimensionality and the physical variability of p and ω (or f) with flow parameters. Therefore α and β being both constant and dimensionless, will be considered as constants that are universally applicable for the dynamically and geometrically similar flows of the type under consideration. They may be functions of dimensionless parameters of the flow, e.g., the Mach numbers and the Reynolds number, however.

Incorporating these parameters in the spectrum we can write an alternative representation of the spectrum given above

$$\Phi_p(\omega) = \frac{\alpha^2\beta}{\pi} q^2 \frac{L}{U}\left(\frac{U}{c_0}\right)^2\left(\frac{L}{r}\right)^2 \frac{1}{(\omega L/U)^2 + \beta^2} \tag{1.60a}$$

Since the dimensions of $\Phi_p(\omega)$ are (pressure)^2(time), a dimensionless form of it is

$$\frac{\Phi_p(\omega\, L/U)}{q^2 M^2 (L/r)^2} = \frac{\alpha^2\beta}{\pi}\left[\left(\frac{\omega L}{U}\right)^2 + \beta^2\right]^{-1} \tag{1.60b}$$

where

$$\Phi_p(\omega) \equiv L/U\; \Phi_p(\omega L/U) \tag{1.60c}$$

The spectrum can also be a shorthand representation that is fully dimensionless:

$$\frac{\Phi_p(\Omega)}{q^2 M^2 (L/r)^2} = \frac{\alpha^2\beta/\pi}{\Omega^2 + \beta^2} \tag{1.60d}$$

where Ω is a dimensionless frequency

$$\Omega = \omega L/U$$

Such nondimensionalizations have direct implications regarding speed dependence. In this example Eqs. (1.60a–c) show that when $\Omega \ll \beta$ (corresponding to $U \gg \omega L/\beta$) the dependence of $\Phi_p(\omega)$ on U at fixed ω is

$$\Phi_p(\omega) \propto U^5$$

when $\Omega \gg \beta$ (corresponding to $U \ll \omega L/\beta$) the spectrum level at a fixed frequency increases with speed as

$$\Phi_p(\omega) \propto U^7$$

the overall pressure level is found by integrating overall frequency and it has the speed dependence

$$\overline{p^2} \propto U^6$$

As an example of how this nondimensionalization is done, Fig. 1.13 shows various forms of the spectrum functions in both dimensional and dimensionless forms. From a practical viewpoint the construction of dimensionless spectra like $\Phi(\Omega)/q^2\, M^2(L/r)^2$ from band levels $G(f)\Delta f = \overline{p^2}(f, \Delta f)$ (or vice versa) is often an important step to be made in physical interpretations of measurements. These functions are related through Eq. (1.60c) and the other relations in this section.

By definition

$$L_s(f, \Delta f) = 10 \log[G(f)\Delta f / p_{\text{ref}}^2]$$

and substituting Eqs. (1.45) and (1.60a) we find

$$L_s(f, \Delta f) = L_q + 10 \log \frac{\Phi(\Omega)}{q^2} + 10 \log \frac{2\Delta\Omega\, L}{U}$$

or in expanded form

$$L_s(f, \Delta f) = L_q + L_M + 20 \log \frac{L}{r} + 10 \log \left\{ \frac{\Phi(\Omega)}{q^2 M^2 (L/r)^2} \right\} + \cdots$$
$$+ 10 \log \frac{4\pi L}{U} + 10 \log \Delta f$$

where

$$L_M = 20 \log M$$

and where the frequency and dimensionless frequency are related by

$$f = (\Omega/2\pi) U/L$$

The values of L_q may be found for a given velocity from the nomograph of Fig. 1.12. The overall sound pressure level is found from the formula that is analogous to Eq. (1.59)

$$L_s = L_q + L_M + 20 \log(L/r) + 20 \log \alpha$$

since the overall pressure is given by

$$\overline{p^2} = \alpha^2 q^2 (L/r)^2 M^2$$

1.6 MATHEMATICAL REFRESHER

In this section we shall set down a number of mathematical operations and definitions that occur repeatedly in this book. This is to refresh the reader

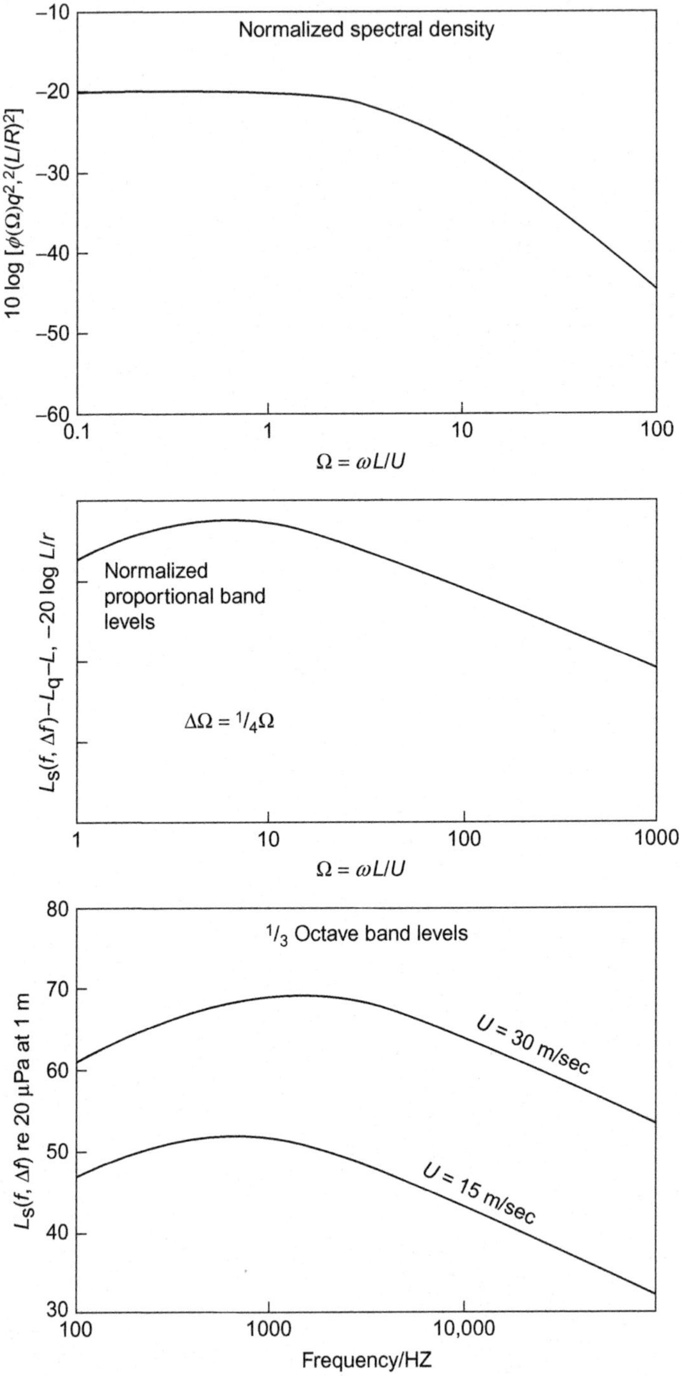

FIGURE 1.13 Illustrations of various spectrum functions for speed-dependent sounds in air. Parameters: $\alpha^2 = 1/10$, $\beta = 5$, $L = 2.5$ cm, $R = 1$ m, $U = 15$ m/s, 30 m/s.

with a number of notions from the subjects of advanced calculus and partial differential equations covered in texts such as Refs. [11,12, 15–17].

1.6.1 Coordinate Systems

Orthogonal coordinate systems will be used exclusively. Reference to a rectangular coordinate system specifically shall be done using (x, y, z) for the three coordinate axes or the subscripts (1,2,3), such as (x_1, x_2, x_3), to denote these three axes. The subscript notation will occur more predominantly because often one needs to refer to three vector quantities, say $\mathbf{x}$, $\mathbf{y}$, and $\mathbf{r} = \mathbf{x} - \mathbf{y}$. In situations involving flow, the 1 direction will nearly always correspond to the main flow direction. When referring to an unspecified orthogonal coordinate system an index, or tensor notation will be used, thus $\mathbf{x} = x_i, x_j, x_k$. The notation $\mathbf{x}_{1,2}$, for example, represents a two-dimensional vector in the 1, 2 plane. Bold face type will be used for vectors.

1.6.2 Differential Operators

Depending on which notation provides the more compact expressions either the del operator or a tensor notation will be used. Thus in Cartesian coordinates the gradient is

$$\nabla\phi = \frac{\partial\phi}{\partial x_1}\mathbf{i} + \frac{\partial\phi}{\partial x_2}\mathbf{j} + \frac{\partial\phi}{\partial x_3}\mathbf{k} \quad \text{or} \quad (\nabla\phi)_i = \frac{\partial\phi}{\partial x_i}$$

the divergence is

$$\nabla\cdot\mathbf{u} = \frac{\partial u_1}{\partial x_1} + \frac{\partial u_2}{\partial x_2} + \frac{\partial u_3}{\partial x_3} = \frac{\partial u_i}{\partial x_i}$$

and the curl is

$$\nabla\times\mathbf{u} = \left(\frac{\partial u_3}{\partial x_2} - \frac{\partial u_2}{\partial x_3}\right)\mathbf{i} + \left(\frac{\partial u_1}{\partial x_3} - \frac{\partial u_3}{\partial x_1}\right)\mathbf{j} + \left(\frac{\partial u_2}{\partial x_1} - \frac{\partial u_1}{\partial x_2}\right)\mathbf{k}$$

or

$$(\nabla\times\mathbf{u})_i = \frac{\partial u_k}{\partial x_j} - \frac{\partial u_j}{\partial x_k}$$

where $\mathbf{i}$, $\mathbf{j}$, $\mathbf{k}$ are unit vectors in the 1, 2, 3 directions. In the tensor system a three-dimensional vector is represented as

$$\mathbf{a} \equiv a_i \quad \text{and} \quad \mathbf{a}\times\mathbf{b} = a_j b_k - a_k b_j$$

where either the index appears singularly or appears in a combination that does not repeat. Repeating indices denote dot or scalar products

$$\mathbf{a}\cdot\mathbf{b} = a_i b_i$$

The Kronecker delta function

$$\delta_{ij} = \begin{cases} 1, & i = j \\ 0, & i \neq j \end{cases} \tag{1.61}$$

is often used to combine a scalar and a tensor quantity, e.g.,

$$\tau_{ij} = \tau'_{ij} + p\,\delta_{ij}$$

The Laplacian operator has the alternative representations

$$\nabla^2 a = \frac{\partial^2 a}{\partial x_i^2}$$

and the biharmonic operator, eg. in Cartesian coordinates

$$\nabla^4 = \frac{\partial^4}{\partial x^4} + 2\frac{\partial^4}{\partial x^2\,\partial z^2} + \frac{\partial^4}{\partial z^4}$$

appears in the Bernoulli–Euler plate equation (see Chapter 5: Fundamentals of Flow-Induced Vibration and Noise).

1.6.3 Integral Theorems

Frequently there will appear line integrals, say

$$I_L(x) = \int f(x, y)dy$$

and surface integrals with the alternative definitions, say

$$I_S(\mathbf{x}) = \iint_S f(\mathbf{x}, \mathbf{y})dS(\mathbf{y}) = \iint_S f(\mathbf{x}, \mathbf{y})d^2\mathbf{y}$$

and volume integrals with the forms, say

$$I_V(\mathbf{x}) = \iiint_V f(\mathbf{x}, \mathbf{y})dV(\mathbf{y}) = \iiint_V f(\mathbf{x}, \mathbf{y})d^3\mathbf{y}$$

In each case the number of integral signs denotes the number of dimensions over which integration exists. This notation permits a discrimination between two possible space variables as occur in certain influence functions. See references on the mathematics of fields and potentials, such as Refs. [7,15–17] for further details on these theorems.

Two common vector integral identities will be frequently used [7,15–17].

1. Gauss's theorem, or the divergence theorem:

$$\iiint_V (\nabla \cdot \mathbf{u}) dV(\mathbf{y}) = \iint_S (\mathbf{u} \cdot \mathbf{n}) dS(\mathbf{y}) = \iint_S = u_n \, dS(\mathbf{y})$$

or

$$\iiint_V \nabla\phi \, dV(\mathbf{y}) = \iint_S \phi \, d\mathbf{S} = \iint_S \phi \, \mathbf{n} \, dS(\mathbf{y})$$

where V is a closed region surrounded by the closed surface S and $\mathbf{n}$ is the outwardly directed unit normal vector to the surface.

2. Stokes's theorem:

$$\iint_S (\nabla \times \mathbf{u})_n \, dS(\mathbf{y}) = \oint \mathbf{u} \cdot d\mathbf{l}(\mathbf{y})$$

where S is a surface bounded by the closed curve C and $d\mathbf{l}$ is an element of the tangent to C. The component $(\nabla \times \mathbf{u})_n$ is normal to the surface. This theorem in fluid dynamics relates vorticity to velocity.

1.6.4 Dirac Delta Function

Dirac delta function, introduced previously in Eq. (1.36), is the most widely used generalized function in this book. It is defined formally as an integral

$$\int_{-\infty}^{\infty} G(t)\delta(t - t_0)dx = G(t_0) \tag{1.62}$$

The integral's limits may be finite and $-T \leq t_0 \leq T$. The delta function is also commonly regarded as a spike of indeterminate magnitude at $t = t_0$ but having an integral equal to unity. Thus we commonly see, e.g., Ref. [15–17]

$$\lim_{t \to t_0} \delta(t - t_0) = \infty$$

and

$$\delta(t) = 0 \quad t \neq 0$$

The above conditions imply the integral:

$$\int_0^{\infty} \delta(t - t_0)dt = 1 \tag{1.63}$$

The Fourier transform of the delta function is

$$F[\delta] = \frac{1}{2\pi} \int_{-\infty}^{\infty} e^{i\omega t} \, \delta(t - t_0)dt = \frac{1}{2\pi} e^{i\omega t_0} \tag{1.64}$$

The inverse transform then serves as an alternative definition of the delta function, which will be useful in future chapters, i.e.

$$\delta(t - t_0) = \frac{1}{2\pi}\int_{-\infty}^{\infty} e^{i\omega(t_0 - t)}\, d\omega \tag{1.65}$$

The multi-dimensional delta function has uses in formulating Green functions and depicting localized sources and force distributions. To this end, using Cartesian coordinates for example,

$$\delta(\boldsymbol{x} - \mathbf{x}_0) = \delta(x - x_0)\delta(y - y_0)\delta(z - z_0) \tag{1.66}$$

If $\mathbf{x}$ is contained in volume ΔV which also contains $\mathbf{x}_0$, then

$$\iiint_{\Delta V} \delta(\boldsymbol{x} - \boldsymbol{x}_0) d^3\mathbf{x} = 1 \tag{1.67}$$

and if $\mathbf{x}$ is not within ΔV, then

$$\iiint_{\Delta V} \delta(\boldsymbol{x} - \boldsymbol{x}_0) d^3\mathbf{x} = 0 \tag{1.68}$$

The Laplacian of $1/(\mathbf{x} - \mathbf{x}_0)$ can be expressed as a Dirac delta function

$$\nabla^2(1/(\boldsymbol{x} - \boldsymbol{x}_0)) = -4\pi\delta(\mathbf{x} - \mathbf{x}_0) \tag{1.69}$$

for which $\nabla^2(1/(\mathbf{x} - \mathbf{x}_0)) = 0$ for $\mathbf{x} \neq \mathbf{x}_0$ and the volume integral is

$$\iiint_{\Delta V} \nabla^2\left(1/(\boldsymbol{x} - \boldsymbol{x}_0)\right) d^3\mathbf{x} = -4\pi \tag{1.70}$$

REFERENCES

[1] Beranek L. Noise and vibration control. New York: McGraw-Hill; 1971.

[2] Bendat JS, Piersol AG. Random data analysis and measurement procedures. 4th ed. New York: Wiley; 2010.

[3] Davinport WB, Root WL. An introduction to the theory of random signals and noise. New York: McGraw-Hill; 1958.

[4] Lee YW. Statistical theory of communication. New York: Wiley; 1964.

[5] Newland DE. Introduction to random vibration, spectral and wavelet analysis. 4th ed. New York: Dover Publications; 2005.

[6] Papoulis A, Pillai SU. Probability, random variables, and stochastic processes. 4th ed. New York: McGraw-Hill; 2002.

[7] Pierce AD. Acoustics: an introduction to its physical principles and applications. New York: American Institute of Physics; 1989.

[8] Cremer H, Ledbedder MR. Stationary and related stochastic processes. New York: Wiley; 1967.

[9] Wiener N. The Fourier integral and certain of its applications. New York: Dover Publications; 1933.

[10] Lighthill MJ. Fourier analysis and generalized functions. London and New York: Cambridge University Press; 1964.

[11] Titchmarsh EC. Introduction to the theory of Fourier integrals. London: Oxford University Press; 1948.

[12] Jones DS. Generalized functions. New York: McGraw-Hill; 1966.

[13] Batchelor GK. The theory of homogeneous turbulence. London and New York: Cambridge University Press; 1960.
[14] Mueller TJ, editor. Aeroacoustic measurements. Springer; 2002.
[15] Kreyszig E. Advanced engineering mathematics. 10th ed. New York: Wiley Interscience; 2010.
[16] Jackson JD. Classical electrodynamics. 3rd ed. New York: Wiley; 1999.
[17] Kinsler LE, Frey AR, Coppens AB, Sanders JV. Fundamentals of acoustics. 4th Ed., New York: Wiley; 2000.

Chapter 2

Theory of Sound and its Generation by Flow

In this chapter we shall discuss the theories and derive the equations that are the foundations of theoretical hydroacoustics principally relating to subsonic flow and its excitation of elastic surfaces. The general relationships will be specialized in later chapters for application to experimental acoustics. First, the common relationships of linear acoustic theory will be explored to emphasize the fundamental qualities of multipole source types. The general theory of fluid-induced noise generation will then be derived with attention given to the classification of source types and noise mechanisms as well as to the influences of solid boundaries of various types on radiated intensity.

2.1 FUNDAMENTALS OF LINEAR ACOUSTICS THEORY

2.1.1 The Wave Equation

We begin with the equations of continuity and momentum for inviscid, fluid motion. The equations are derived in this form in a number of basic fluid mechanics texts, among them those of Milne-Thompson [1], Batchelor [2], Sabersky, Acosta, and Hauptmann [3], White [4], and Kundu, Cohen, and Dowling [5]. In tensor notation, these are

$$\frac{\partial \rho}{\partial t} + \frac{\partial}{\partial x_i}(\rho u_i) = 0 \tag{2.1}$$

and

$$\rho \frac{\partial u_i}{\partial t} + \rho u_j \frac{\partial u_i}{\partial x_j} = -\frac{\partial p}{\partial x_i} \tag{2.2}$$

respectively. The barotropic fluid is so defined because the density is a thermodynamic property that is a function of pressure alone. The instantaneous fluid density is ρ, the pressure is p, the three-dimensional local fluid velocity is u_i, and the space and time variables are x_i and t, respectively. We shall

Mechanics of Flow-Induced Sound and Vibration, Volume 1.
DOI: http://dx.doi.org/10.1016/B978-0-12-809273-6.00002-6

have occasion later in this chapter and elsewhere in the book to manipulate these equations in vector form for which we make use of the del operator, which is

$$\nabla = \frac{\partial}{\partial x}\mathbf{i} + \frac{\partial}{\partial y}\mathbf{j} + \frac{\partial}{\partial z}\mathbf{k}$$

in three dimensions, where **i**, **j**, **k** are the unit vectors in the (x, y, z) directions, respectively. In this notation the equations of continuity and momentum then are

$$\frac{\partial \rho}{\partial t} + \nabla(\rho \mathbf{u}) = 0 \tag{2.3}$$

and

$$\rho\frac{\partial \mathbf{u}}{\partial t} + \rho(\mathbf{u}\cdot\nabla)\mathbf{u} = -\nabla p \tag{2.4}$$

respectively, where $\mathbf{u} = (u_x\mathbf{i} + u_y\mathbf{j} + u_z\mathbf{k}) = (u_i)$. These equations apply to fluid regions which are free of the production of local mass or momentum and which are free of gravitational (or body) forces. For barotropic fluids we can write the pressure in terms of the density as

$$p - p_0 = \text{const}\left(\rho - \rho_0\right)^{\alpha}$$

in which α has special values depending on the thermodynamic equation of state of the fluid. The wave equation in acoustics has been derived in many acoustics texts, e.g., of Morse and Ingard [6], Pierce [7], Fahey et al. [9], and Kinsler et al. [8]. For ideal gases undergoing isothermal expansion $\alpha = 1$; conversely for adiabatic expansions (vanishing heat transfer among adjacent fluid elements) $\alpha = \gamma = c_p/c_v$, where c_p and c_v are the specific heats at constant pressure and volume, respectively. For liquids, the state equation takes on a more complex form; however, the variations in pressure and density are related through the fluid compressibility. For perfect diatomic gases such as air, $\gamma = 1.4$.

The adiabatic speed of sound in the fluid is determined by the rate of change of pressure with density at constant entropy using the relation

$$c_0^2 = \left(\frac{\partial p}{\partial \rho}\right)_s = \gamma p_0\left(\frac{\rho}{\rho_0}\right)^{\gamma-1} \tag{2.5}$$

For the present, we shall restrict considerations to lossless (or in viscid) fluids for which the acoustic compression–expansion process is adiabatic. The so-called linear acoustic approximation results from the assumption that local velocities u are much less than the speed of sound in the fluid. Even for real fluids if the disturbances are of long enough wavelength, then the fluid gradients are small so that nearly adiabatic expansions occur. Under these conditions small deviations of pressure and density from an equilibrium value can be expressed

$$p - p_0 = c_0^2(\rho - \rho_0) \tag{2.6}$$

where p_0 and ρ_0 are the equilibrium pressure and density, respectively.

The linear wave equation for low Mach number is obtained by taking the substantial time derivative of Eq. (2.1) and the gradient $\partial/\partial x_i$ of Eq. (2.2). Combining these equations, and neglecting second-order terms (see Section 2.3.1), we obtain the wave equation and define a differential wave operator for density fluctuations in a homogeneous fluid as

$$\left(\frac{\partial}{\partial t} + U \cdot \nabla\right)^2 \rho - c_0^2 \frac{\partial^2 \rho}{\partial x_i^2} \equiv \Box^2 \rho = 0 \tag{2.7}$$

or

$$\left(\frac{\partial}{\partial t} + \nabla \cdot U\right)^2 p - c_0^2 \frac{\partial^2 p}{\partial x_i^2} \equiv \Box^2 p = 0 \tag{2.8}$$

for the pressure fluctuations, where $\partial^2/\partial x_i^2 = \nabla^2$ is the Laplacian operator.

Specific functional forms of solutions of the wave equation depend on the geometric order (one, two, or three dimensions) of the fluid region. The solutions are also obviously dependent on whatever temporal and spatial character the boundary of the fluid has. In consideration of the sound field that is realized at some distance r from a vibrating body, the sound pressure at a given time is the linear superposition of the acoustic contributions from each of the spatial wave harmonics that are invoked to describe the surface motion and all the frequencies that describe the time variation of each of the spatial harmonics. In a few instances, however, consideration of acoustic energetics can be simplified. These are classified into two options. Either the motion is spatially uniform over the surface, i.e., of the zero-order spatial harmonics, or the motion has a prescribed spatial variation of a given harmonic and the time behavior has a single frequency. In all other more general cases the temporal wave forms at varying range will depend on range r.

2.1.2 Acoustic Plane Waves and Intensity

Far from a curved radiating surface, such that the range is much greater than both the size of the body and the acoustic wavelength, the sound pressure is locally one dimensional; i.e., the propagation is along a radius from the source zone. Accordingly, we shall examine below some fundamental characteristics of one-dimensional sound fields. The acoustic field variables in a fluid with no mean motion are the particle velocity $\mathbf{u}$ and the sound pressure p. The linearized form of Eq. (2.4) in which the products $|\mathbf{u}|^2$ are ignored in comparison p and $\mathbf{u}$ is, in three dimensions and for an isentropic acoustic disturbance in a fluid with no mean flow and uniform ambient pressure and density,

$$\rho_0 \frac{\partial \mathbf{u}}{\partial t} = -\nabla p \tag{2.9}$$

In this linearization of Eq. (2.4), the ignored nonlinear terms account for the convective acceleration of the acoustic disturbance by the acoustic field. For a simple-harmonic time dependence, the pressure may be written in the complex exponential form

$$p(\mathbf{x}, t) = p_0(\mathbf{x}) e^{-i2\pi ft}$$

or in the real part

$$p(\mathbf{x}, t) = p_0(\mathbf{x}) \cos 2\pi ft$$

where f is the frequency of the wave. For the exponential time dependence the linearized equation takes on the form

$$\nabla p = i(2\pi f)\rho_0 \mathbf{u} \tag{2.10}$$

and the wave equation takes the form

$$\frac{\partial^2 p}{\partial x_i^2} + k_0^2 p = 0 \tag{2.11}$$

where $k_0 = 2\pi f/c_0 = 2\pi/\lambda_a$ is the *acoustic wave number* and λ_a is the acoustic wavelength.

The utility of these relationships will now be examined further for the special case of a one-dimensional sound field, which is the most elementary of acoustic fields. One-dimensional field pressure or velocity disturbances are those that depend only on one space dimension, say x, and time t, and is independent of the other 2 coordinates. These one-dimensional disturbances are called *plane waves*. Such a field may be physically realized in a long duct at low frequency for which end reflections are prohibited by the use of a good absorber. Eq. (2.4) in its linearized form now becomes for one space dimension

$$\frac{\partial p}{\partial x} = -\rho_0 \frac{\partial u_x}{\partial t}$$

where u_x is the velocity now directed in the x direction. The wave equation for simple harmonic time dependence is

$$\frac{\partial^2 p}{\partial x^2} + k_0^2 p = 0$$

and we shall let the pressure be described as

$$p = p(x) \cos \omega t$$

in which we have adopted the use of the circular frequency $\omega = 2\pi f$.

One solution of the wave equation is

$$p(x) = A\cos kx$$

where the k is a wave number. The second derivative of the pressure with respect to x is

$$\frac{\partial^2 p}{\partial x^2} = -Ak^2 \cos kx = -k^2 p$$

so that the wave equation becomes

$$(k_0^2 - k^2)p = 0$$

which requires that the wave number k be identical to the acoustic number k_0.

The pressure is now given by the formula

$$p(x,t) = A\cos(k_0 x)\cos(\omega t)$$

which may be expanded into two terms

$$p(x,t) = \frac{1}{2}\left[\cos(k_0 x - \omega t) + \cos(k_0 x + \omega t)\right]$$

This, as we shall see below, represents a superposition of two waves. Either of the cosine functions is constant when $(k_0 x - \omega t)$ or $(k_0 x + \omega t)$ is held constant, thus the first term represents a wave traveling in the positive x direction while the second term represents a wave of equivalent amplitude traveling in the negative x direction. One term of the equation must thus be rejected when the problem pertains to a traveling wave that is generated by a single source in a medium without reflections. This is because sound that is radiated by a single source into a fluid region that does not reflect the sound is propagated only away from the source. For the one-dimensional problem under discussion we shall let the source be radiating to the right from the left; that is, the disturbance propagation is governed by phase fronts described by $(k_0 x - \omega t)$ as x and t increase from zero. Thus the magnitudes of the disturbance will be constant only when $k_0 x - \omega t = \phi$ remains constant. This requires that the disturbance be described with functions of $(k_0 x - \omega t)$ and the other combination must be dropped. In Section 2.2 this rejection of the incoming wave will be recognized as a special case of a more general radiation condition. Accordingly the solution is restricted to the function

$$p(x,t) = p_0 \cos(k_0 x - \omega t)$$

where p_0 now represents the amplitude of the pressure. The acoustic particle velocity $\mathbf{u}$ is given by the linearized form of Eq. (2.4) as

$$\frac{\partial p}{\partial x} = p_0 k_0 \sin(k_0 x - \omega t) = \rho_0 \frac{\partial u_x}{\partial t}$$

so that by integrating

$$\rho_0 u_x = p_0 \frac{k_0}{\omega}\cos(k_0 x - \omega t) = \frac{k_0}{\omega} p(x,t)$$

or

$$p(x,t) = \rho_0 c_0 u_x(x,t) \tag{2.12}$$

This formula gives the acoustic pressure in terms of the acoustic particle velocity that is directed in the direction of wave propagation. In this case the wave propagates along $x > 0$ and particle velocity is directed in the x direction. The $\rho_0 c_0$ is the *specific acoustic impedance* of the fluid.

In more general cases of three-dimensional fields the direction of propagation will be denoted by a unit vector $\mathbf{n}_r$ in the direction r of the radial vector from the source, say $\mathbf{r}$, i.e.,

$$\mathbf{r}/|\mathbf{r}| = \mathbf{n}_r$$

and the particle motion is $\mathbf{u}$ so that Eq. (2.12) may be rewritten as a solution of Eqs. (2.10) and (2.11),

$$p(\boldsymbol{x},t) = \rho_0 c_0 \mathbf{n}_r \cdot \mathbf{u}(\boldsymbol{x},t) \tag{2.13a}$$

or

$$p(\boldsymbol{x},t) = \rho_0 c_0 u_r(\boldsymbol{x},t) \tag{2.13b}$$

This relationship between the acoustic pressure and the acoustic particle velocity is fundamental to all far-field acoustics.

In a stationary, or nearly stationary ($U/c_0 \ll 1$) medium, the instantaneous acoustic intensity is a vector quantity that is the product of the acoustic pressure and particle velocity,

$$\mathbf{I}(\boldsymbol{x},t) = p(\boldsymbol{x},t)\mathbf{u}(\boldsymbol{x},t) \tag{2.14}$$

Intensity is an instantaneous power flux across a surface so that the acoustic energy transmitted, say, across a surface S_0 for a period of time T, is given by the integrals over both T and S_0, i.e.,

$$E_a = \int_0^T dt \iint_{S_0} \mathbf{I}(\boldsymbol{x},t) \cdot d\mathbf{S}(\boldsymbol{x})$$

where $d\mathbf{S}(\mathbf{x})$ is the elemental surface vector and is directed normal to the surface. Substituting either Eq. (2.13a) or (2.13b) into Eq. (2.14) gives an expression for the intensity

$$\mathbf{I}(\boldsymbol{x}) = \frac{\overline{p^2}(\boldsymbol{x})}{\rho_0 c_0}\mathbf{n}_r(\boldsymbol{x}) \tag{2.15}$$

where the $\mathbf{n}_r(\mathbf{x})$ is still the unit vector propagation direction. The time average acoustic power is defined as the average energy passing across S so that

$$\mathbb{P}_{\text{rad}} = \frac{1}{T} E_{\text{a}}$$

or

$$\mathbb{P}_{\text{rad}} = \frac{1}{T} \int_0^T dt \iint_{S_0} \frac{p^2(\boldsymbol{x}, t)}{\rho_0 c_0} \mathbf{n}_r(\boldsymbol{x}) \cdot d\mathbf{S}(\boldsymbol{x})$$

or

$$\mathbb{P}_{\text{rad}} = \iint_{S_0} \frac{\overline{p^2}(\boldsymbol{x})}{\rho_0 c_0} d\mathbf{S}_r \tag{2.16}$$

where S_0 represents the surface surrounding the source over which the power flux is of interest.

Since the use of Eq. (2.16) implies that Eq. (2.14) also holds, the surface must be in the distant far field of the source. $\mathbf{S}_r$ represents an element of S_0 normal to the direction of wave propagation and normal to the acoustic ray. The acoustic power is generally of interest for closed surfaces surrounding the source zone in the far field. Examples of such surfaces are shown in Fig. 2.1A and B for cylindrical and spherical coordinates. Generally, as illustrated for specific examples in Section 2.1.3, the far-field acoustic radiation is directed along the radius vector $\mathbf{r}$ measured from a point in the body (the acoustic center) to the point in the far field. This is illustrated in Fig. 2.1.

For a spherical surface surrounding the radiating source, the average power is found by evaluating

$$\mathbb{P}_{\text{rad}} = \iint_S \frac{\overline{p^2}(\boldsymbol{x})}{\rho_0 c_0} r^2 \sin\phi \, d\phi \, d\theta \tag{2.17}$$

integration is over the spherical surface shown in Fig. 2.1B.

2.1.3 Fundamental Characteristics of Multipole Radiation

In the analytical treatments throughout this chapter, formulations will be interpreted in terms of combinations of simple sources. In the following analysis it will be shown that these source combinations can represent the driving of the fluid by localized time-varying volumetric (or dilatational) changes, or forces, or force couples. We will consider a nearly stationary ($U/c_0 \ll 1$) medium.

2.1.3.1 Monopole Sources

To begin, the relationship for acoustic radiation from a volumetric pulsation is derived. Physically, this source can represent the radiated sound from

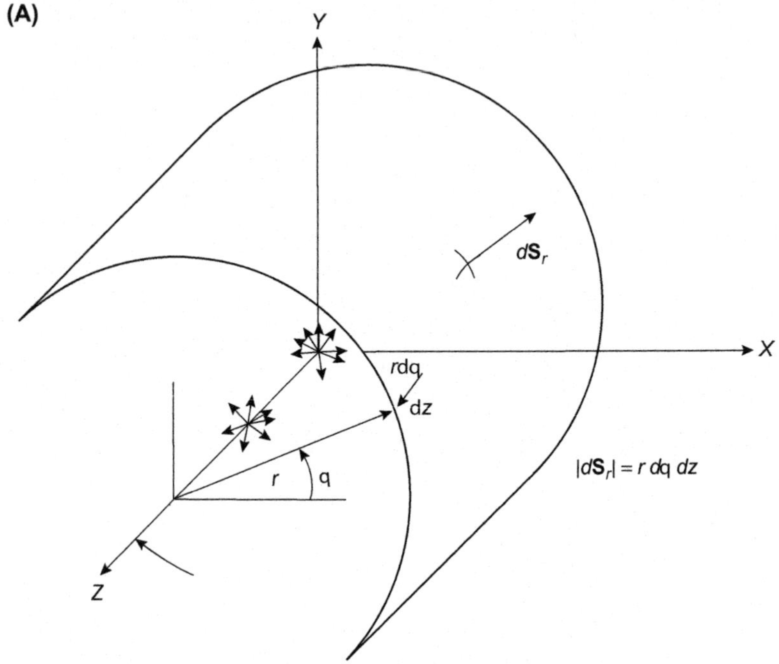

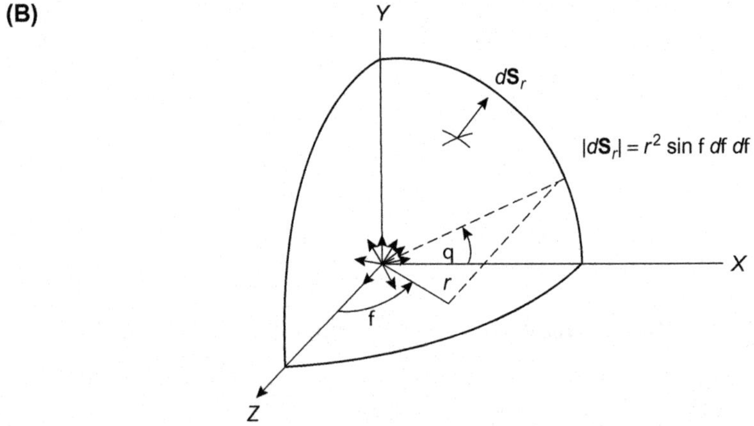

FIGURE 2.1 (A) Cylindrical and (B) spherical coordinate systems.

axisymmetric bubble vibration. The time-varying characteristic of the pressure at a single frequency, ω, is assumed to be given by

$$p(r,t) = \tilde{p}(r,\omega)e^{-i\omega t} \tag{2.18}$$

where $\tilde{p}(r,\omega)$ is a complex pressure amplitude. The source motion is completely radial so that the wave equation for the radiated acoustic pressure is found by introducing the Laplacian in axisymmetric spherical coordinates,

$$\frac{1}{r^2}\frac{\partial}{\partial r}\left[r^2\frac{\partial \tilde{p}(r,\omega)}{\partial r}\right] + \frac{\omega^2}{c_0^2}\tilde{p}(r,\omega) = 0 \tag{2.19a}$$

which is equivalent to

$$\frac{\partial^2[r\tilde{p}(r,\omega)]}{\partial r^2} + \left(\frac{\omega}{c_0}\right)^2\left[r\tilde{p}(r,\omega)\right] = 0 \tag{2.19b}$$

The acoustic wave number is

$$k_0 = \omega/c_0 \tag{2.20}$$

and a solution to Eq. (2.19b) that represents traveling waves is

$$\tilde{p}(r,\omega) = (A/r)e^{\pm ik_0 r} \tag{2.21}$$

As above in the case of plane waves, the positive sign on the root of $\sqrt{-1} = i$ is chosen for outward-traveling waves in consonance with the time dependence assumed in Eq. (2.18). The instantaneous volume of the sphere of radius a is

$$Q = \frac{4}{3}\pi a^3$$

and the (small, linear) amplitude of the time rate of volume change is

$$\dot{Q}(\omega) = 4\pi a^2 u_r(\omega) \tag{2.22}$$

for this linearized expression to be valid, the amplitude of radial motion of the surface must be small compared to the radius of the sphere. For simple harmonic motion of the source the linearized boundary condition on the surface is given by Eq. (2.9), i.e.,

$$i\omega\rho_0\,\mathbf{u} = \nabla p = \frac{\partial p}{\partial r} = i\omega\rho_0\,u_r(\omega)$$

Combining Eqs. (2.9), (2.21), and (2.22) yields

$$A\left[\frac{1}{a^2} - \frac{ik_0}{a}\right]e^{+ik_0 a} = i\omega\rho_0\,u_r(\omega)$$

so that

$$A = \frac{-i\omega\rho_0\,\dot{Q}(\omega)}{4\pi(1+ik_0 a)}e^{-ik_0 a} \tag{2.23}$$

and

$$p(r,t) = \frac{-i\omega\rho_0\,\dot{Q}(\omega)}{4\pi r} e^{+ik_0(r-a)-i\omega t} \tag{2.24a}$$

is the radiated pressure from the source when $k_0 a \ll 1$. This condition states that the diameter of the sphere is small compared to the acoustic wavelength, i.e., $\lambda_0 = 2\pi/k_0$, so that $2a/\lambda_0 \ll 1/\pi$.

From Eq. (2.24a) the linearized acoustic particle velocity anywhere in the fluid is given by Eq. (2.9) so that

$$\mathbf{u}(r,t) = u_r(r,t) = \frac{-i\omega\dot{Q}(\omega)}{4\pi c_0 r}\left[1 + \frac{i}{k_0 r}\right] e^{i[k_0(r-a)-\omega t]}$$

In the limit $k_0 r \gg 1$ this expression for the particle velocity reduces to that given by Eq. (2.13a,b). The far field of a point volume source is thus established at distances r such that $2\pi r/\lambda_0 \gg 1$.

For either periodic or aperiodic volume pulsations Eq. (2.24a) has an analog in the completely temporal domain that is found by inverse Fourier transformation (Eq. (1.40))

$$p(r,t) = \frac{\rho_0 \ddot{V}(t-(r-a)/c_0)}{4\pi r} \tag{2.24b}$$

where $\ddot{V}(t)$ is the instantaneous volumetric acceleration of the source and $t-(r-a)/c_0$ is a delayed, or retarded, time. The retardation accounts for the fact that the pressure pulse at time t is caused by a source motion $(r-a)/c_0$ earlier.

2.1.3.2 Dipole Sources

The next order of source complexity is the dipole, which can be represented by a pair of simple sources aligned with the z axis as shown in Fig. 2.2A. We shall consider this type source in detail because of its paramount importance in many practical instances. These sources may be supposed to oscillate in harmonic motion as above, but either in phase or out of phase. If they oscillate π out of phase, so at any instant there is no net influx of mass into the fluid space. This source–sink pair simply oscillates the fluid back and forth. The motion is symmetric about the z axis and unsymmetric about the angle ϕ.

The vector distance between the sources is $2d_z$. The field point from the centroid of the system is at the coordinates r, ϕ, with individual ranges r_1 and r_2. The resultant sound pressure is given by the sum of the individual contributions (the phase shift $k_0 a$ is assumed to be negligibly small)

$$\tilde{p}(r,\phi) = \frac{-i\omega\rho_0\,\dot{Q}(\omega)}{4\pi}\left[\frac{e^{+ik_0 r_1}}{r_1} \pm \frac{e^{+ik_0 r_2}}{r_2}\right] \tag{2.25}$$

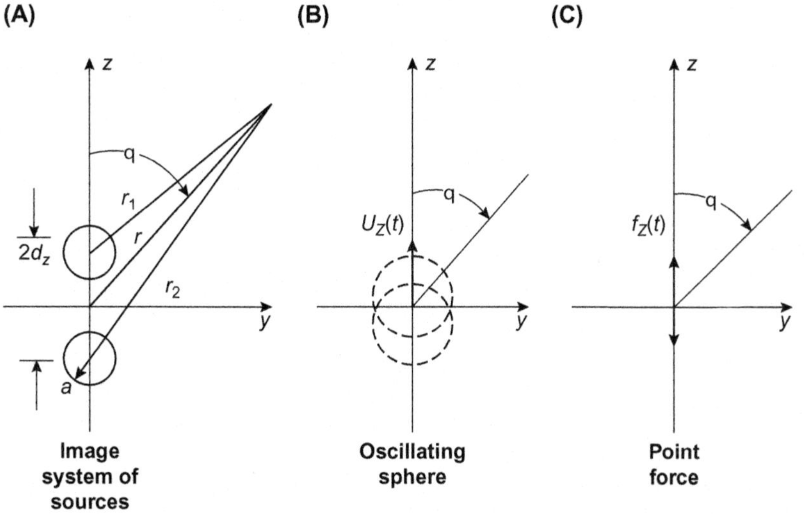

FIGURE 2.2 Equivalent dipole forms.

The sign applies to the phasing of the sources, either in phase (+) or out of phase (−). For separations $d_z \ll r$, we can invoke limiting expressions for the two ranges r_1 and r_2. Thus we can write

$$r_{\substack{1\\2}}^2 = r^2 + d_z^2 \mp 2rd_z\cos\phi = r^2\left(1 + \left(\frac{d_z}{r}\right)^2\right) \mp 2\left(\frac{d_z}{r}\right)\cos\phi$$

Since $r \gg d_z$ each range is of the form

$$r_i = r\sqrt{1 + \epsilon_i^2}$$

where ϵ_i is much smaller than unity. The radical has the approximate value

$$r_i \approx r(1 + \epsilon_i^2/2)$$

Accordingly, the two ranges are

$$r_{\substack{1\\2}} \approx r \mp d_z \cos\phi$$

where $(d_z/r)^2 \ll (d_z r)$ and has been neglected. Substituting for r_1 and r_2 the resultant far-field acoustic pressure is found to be

$$\tilde{p}(r,\phi) = \frac{-2\omega\rho_0\,\dot{Q}(\omega)}{4\pi r}\left\{\begin{matrix} +\cos(k_0 d_z \cos\phi) \\ -i\sin(k_0 d_z \cos\phi) \end{matrix}\right\} e^{+ik_0 r} \tag{2.26}$$

where alternate use of the cosine and sine apply to sources in phase and out of phase, respectively.

In the case that $k_0 d_z \ll 1$, the $\cos(k_0 d_z \cos\phi)$ is replaced by unity at all angles because the sources simply reinforce each other. The interesting function, from our point of view, is $\sin(k_0 d_z \cos\phi)$, which becomes simply $k_0 d_z \cos\phi$ for $k_0 d_z \ll 1$. In this case the resultant acoustic pressure is

$$\tilde{p}(r,\phi) \simeq \rho_0 c_0 k_0^3 \left[2d_z\, \dot{Q}(\omega)\right] \cos\phi \frac{e^{+i(k_0 r + \pi/2)}}{4\pi k_0 r} \tag{2.27}$$

and $2d_z\dot{Q}(\omega)$ is the dipole strength.

The pattern of far-field acoustic radiation from a dipole is determined by the $\cos\phi$ and therefore has the angular directivity pattern in the $z-y$ plane as shown in Fig. 2.3, i.e., in the plane containing the axis of the dipole. It is omnidirectional in the $x-y$ plane, i.e., in the plane perpendicular to the axis of the dipole. Thus the properties of the simple dipole field that differ from those of the monopole field, arising from the gradient along the axis of the dipole, are twofold: first, the two-lobed structure of the radiation field and second, and more important, the dependence of the sound pressure on the speed of sound as shown in Eq. (2.27).

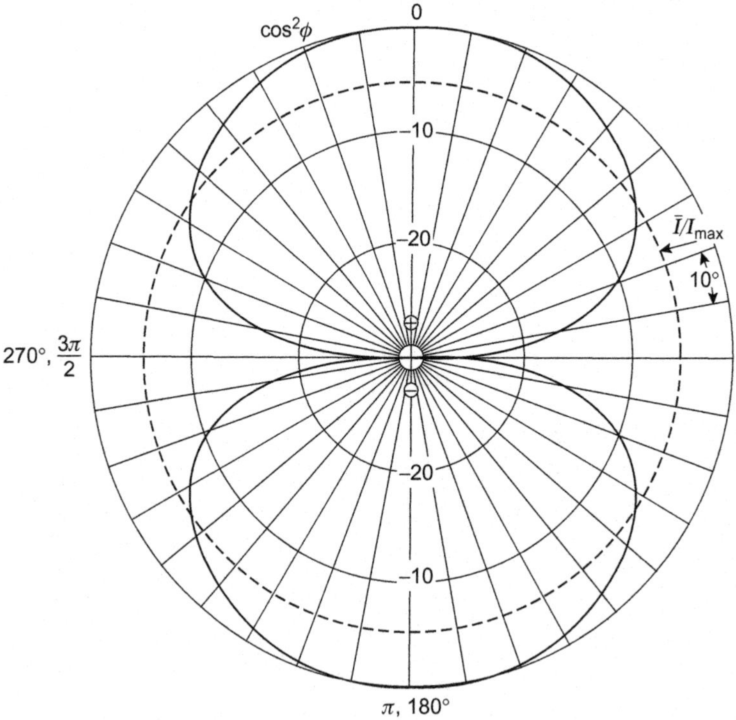

FIGURE 2.3 Directivity pattern, $10 \log \bar{I}/I_{\max}$, in far field of a compact dipole.

This result has important implications regarding sound pressures radiated by sources near boundaries [7,10]. In the case depicted in Fig. 2.2A the $x-y$ plane is a model of a rigid boundary when the sources are in phase. This can be seen by evaluating the tangential gradient $(1/r)[\partial\tilde{p}(r,\phi)/\partial\phi]$ as $\phi=\pi/2$ and noting that it vanishes there. Thus we see that the normal velocity vanishes everywhere in the $z=0$ plane as it would physically on a rigid surface. Alternatively sources in phase opposition give a vanishing pressure and a velocity maximum on the $z=0$ plane. This is as it would be on a free surface. Thus a compact simple source (i.e., one whose largest dimension is smaller than acoustic wavelength) in water near the surface would be expected to behave as a simple dipole as long as $4hk_0 \ll \pi$, where h is the depth of the source. This is a practical limit that allows the replacement of $\sin(k_0h\cos\phi)$ by $k_0h\cos\phi$ to within 10%. Stated in other terms, this limit is just $h/\lambda_0 < 1/8$, or the sources situated within 1/8 wavelength of the surface have their strengths altered by the reflection of the surface. Accordingly, as shall be derived below, their sound power outputs are also altered.

The time-averaged, far-field acoustic power spectral density (two-sided) for the simple source is obtained using Eqs. (1.13), (2.24a), and (2.17)

$$\begin{aligned}\mathbb{P}_{\mathrm{M}} &= \frac{1}{2}\int_0^{2\pi} d\theta \int_0^{\pi} d\phi \frac{\rho_0^2\omega^2|\dot{Q}(\omega)|^2}{\rho_0 c_0 16\pi^2}\sin\phi \\ &= \frac{\rho_0^2\omega^2|\dot{Q}(\omega)|^2}{8\pi\rho_0 c_0}\end{aligned}$$

where the factor of 1/2 accounts for the time averaging (see Eq. (1.13)). Similarly, for the dipole we use Eq. (2.27) to obtain

$$\begin{aligned}\mathbb{P}_{\mathrm{D}} &= \frac{1}{2}\int_0^{2\pi} d\theta \int_0^{\pi} d\phi \frac{\rho_0^2\omega^2|\dot{Q}(\omega)|^2[2k_0d_z]^2}{\rho_0 c_0 16\pi^2}\cos^2\phi\sin\phi \\ &= \frac{\rho_0^2\omega^2|\dot{Q}(\omega)|^2}{8\pi\rho_0 c_0}\frac{1}{3}[2k_0d_z]^2 \\ &= \mathbb{P}_{\mathrm{M}}\frac{1}{3}[2k_0d_z]^2 \qquad (2.28a) \\ &= \frac{1}{3}(4\pi r^2)\left[\frac{\rho_0 c_0 k_0^2|\dot{Q}(\omega)|^2(2k_0d_z)^2}{32\pi^2r^2}\right] \qquad (2.28b)\end{aligned}$$

The presence of the free (pressure-release) surface *reduces* the power output of the monopole by the factor $\frac{1}{3}(2k_0d_z)^2$. This reduction in sound power has been confirmed in a simple experiment in which a small spherical sound source was placed at various depths $H/2$ beneath a water surface (Fig. 2.4). When deeply immersed ($k_0H \gg 2$) the sound power from the source is $\mathbb{P}_{\mathrm{M}}$, which is the free field value. As k_0H was reduced to small

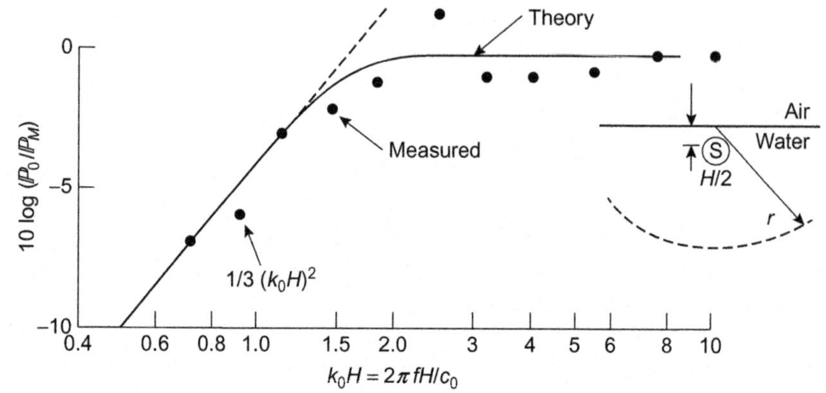

FIGURE 2.4 Reduction in sound power radiated by a simple source in close proximity to a free surface that may be interpreted as the dipole power to monopole power ratio.

values, the power radiated decreased in accordance with Eq. (2.28a). In contrast, Eq. (2.26) shows that the presence of a *rigid* surface *increases* the acoustic power output by 2^2 (or 6 dB) when $k_0 d_z < \pi/4$. Eq. (2.28b) expresses the total average acoustic power as a product of the far-field radiating area $4\pi r^2$, the maximum value of the time-averaged far-field intensity (i.e., at $\phi = 0$), and a numerical factor 1/3 which accounts for the spatial averaging of the directivity over a far-field control surface. This relationship also gives a simple formula for estimating far-field radiated power from an estimate of the maximum time-averaged intensity at a point; the 1/3 factor accounting for the area-averaging of the pressure squared.

A more general statement of the sound field of a dipole aligned with the y_i axis can be stated from Eq. (2.27); i.e., it is equivalent to

$$\tilde{p}(r,\phi) = \frac{-i\omega\rho_0\,\dot{Q}_i}{4\pi}\left[\frac{d}{dy_i}\left(\frac{e^{ik_0 r}}{r}\right)\right] \tag{2.29}$$

where Q_i represents the dipole strength, $Q_z = Qd_z$ in the above problem which is directed along the z axis. This alternative relationship brings out the origin of an important feature of the dipole sound: its dependence on the speed of sound. In contrast to the sound pressure from the simple source, the dipole sound pressure is proportional to ${c_0}^{-1}$ as may be gleaned from substituting ω/c_0 for k_0 in Eq. (2.27) and collecting like terms. This dependence is the direct result of the spatial gradient indicated in Eq. (2.29), which was also the essence of Eq. (2.25). The gradient also accounts for the cos ϕ dependence of the directivity.

We can use Eq. (2.29) to determine the sound radiated from a heaving sphere as illustrated in Fig. 2.2B. The surface is assumed to be impervious. The velocity of the fluid in the r direction (u_r) is related to the pressure by the linearized Eq. (2.9)

$$i\omega u_r = \frac{1}{\rho_0}\frac{\partial p}{\partial r} = i\omega u_z \cos\phi \tag{2.30}$$

where u_z is the z-directed particle velocity anywhere in the fluid and on the sphere. From above, since $dr/dz = z/r = \cos\phi$ the pressure is

$$\tilde{p}(r,\phi) = \frac{-i\omega\rho_0\,\dot{Q}_i}{4\pi}\frac{e^{ik_0 r}}{r^2}(ik_0 r - 1)\cos\phi$$

and the radially directed velocity at a distance r from the surface is

$$u_r(r,\phi) = \frac{-\dot{Q}_i}{4\pi}\frac{e^{ik_0 r}}{r}\left[\frac{2}{r^2} - \frac{2ik_0}{r} - k_0^2\right]\cos\phi$$

then by rearranging we find the velocity on the surface $U_z = u_z(a)$ is related to $\dot{Q}_i$ by

$$U_z = \frac{-\dot{Q}_i e^{ik_0 a}}{2\pi a^3}\left[1 - \frac{1}{2}(k_0 a)^2 - ik_0 a\right] \tag{2.31}$$

so that substituting for $\dot{Q}_i$ we have the pressure in the fluid surrounding the heaving sphere:

$$\tilde{p}(r,\phi) = \frac{i\omega\rho_0 a^3 U_z}{2}\cdot\frac{e^{ik_0(r-a)}}{r}\frac{[ik_0 r - 1]\cos\phi}{\left[1 - \frac{1}{2}(k_0 a)^2 - ik_0 a\right]} \tag{2.32}$$

This expression agrees with that often derived by other methods [7,10].

The resultant force on the fluid in the z direction can be found by evaluating (with $dS(a,\phi) = a^2 \sin\phi\, d\theta\, d\phi$)

$$F_z = \iint_S \tilde{p}(a,\phi) n_z\, dS(a,\phi)$$

$$= \int_{\theta=0}^{2\pi}\int_{\phi=0}^{\pi}\left\{\frac{ia\omega U_z\cos\phi}{2}\frac{(ik_0 a - 1)}{\left[1 - \frac{1}{2}(k_0 a)^2 - ik_0 a\right]}\right\}\cos\phi\left[a^2\sin\phi\, d\theta\, d\phi\right]$$

In the limit of $k_0 a \ll 1$ (i.e., for acoustic wavelengths much less than the radius of the sphere)

$$F_z = \frac{2\pi}{3}\rho_0\omega a^3 U_z\left[-i + (k_0 a)^3\right]$$

Now, we can define a fluid impedance $Z_z = F_z/U_z$

$$Z_z = \left(\rho_0\frac{4}{3}\pi a^3\right)\left(-\frac{1}{2}i\omega\right) + \frac{1}{6}(4\pi a^2)\rho_0 c_0 (k_0 a)^4$$

The first term is mass-like, giving the added mass of the sphere $2\pi\rho_0 a^3/3$. The second term is resistive and represents a radiation damping as it accounts for energy taken from the sphere as sound radiated. If we define an *acoustic impedance* per unit area as the real or resistive component of Eq. (2.32), i.e.,

$$Z_a = \mathrm{Re}\{p(a,\phi)/u_n(\phi)\} \tag{2.33}$$

where Re{ } stands for "real part of." then when $k_0 a \ll 1$

$$Z_a = \frac{1}{4}\rho_0 c_0 (k_0 a)^4$$

This is the radiation impedance per unit area of the sphere for the mode of oscillation such that normal surface velocity is

$$u_n(\phi) = u_r(a,\phi) = U_z n_z(\phi) = U_z \cos\phi$$

The force impedance defined above is related to this acoustic impedance by relationship

$$\mathrm{Re}\{Z_z\} = Z_a \iint_S n_z(\phi) d^2S(a,\phi)$$

which can be seen easily by making the necessary substitutions.

The acoustic impedance Z_a shows that the sphere becomes a more "efficient" radiator as the product $(k_0 a)$ increases. The acoustic power radiated in the above problem (Eq. (2.28a,b)) is also simply written as

$$\mathbb{P}_{\mathrm{D}} = \frac{1}{2} Z_a \overline{u_n^2} S \tag{2.34}$$

where $S = 4\pi a^2$ is the surface area and $\overline{u_n^2}$ is the surface average velocity squared, or

$$\overline{u_n^2} = \frac{1}{S}\iint u_n^2(\phi) d^2S(a,\phi)$$

and

$$\overline{u_n^2} = \frac{1}{3} U_z^2$$

in this example. Both Eqs. (2.33) and (2.34) are general statements of the specific results of this section and will be used and discussed more fully in Chapter 5, Fundamentals of Flow-Induced Vibration and Noise.

2.1.3.3 Quadrupole Sources

Compositions of quadrupoles with dipole pairs are shown in Fig. 2.5. In sketch A, the quadrupole is represented as an array of four simple sources, or two dipoles in the $z-y$ plane separated by a distance $2d_y$. In sketch B the quadrupole is shown as a pair of force dipoles separated by a distance $2d_x$ and $2d_y$. This pair of force dipoles imposes no net moment. These two

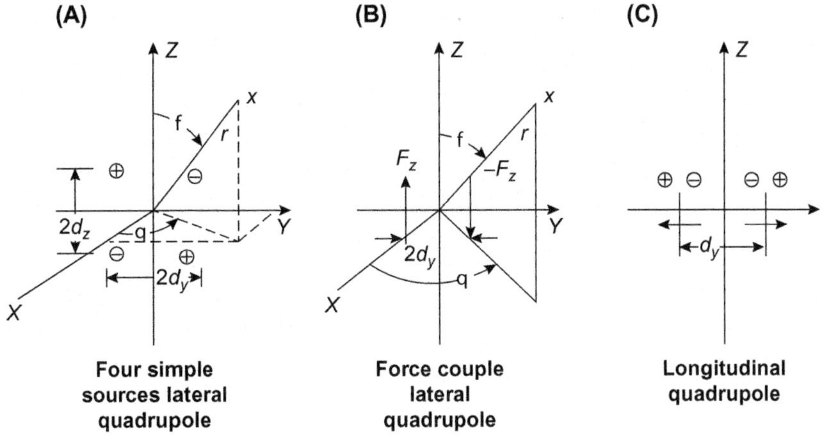

FIGURE 2.5 Equivalent quadrupole forms.

orientations of dipoles, which impose a fluid moment-pair, are called lateral quadrupoles. Another orientation of dipoles in which the forces are in-line is called a longitudinal quadrupole.

Of importance in many fluid applications is the lateral quadrupole for which the far-field directivity will now be derived. The far-field pressure amplitude from the dipole system of Fig. 2.5A can be written in the terms of Eq. (2.34)

$$\tilde{p}(r,\phi,\theta) \simeq \frac{\rho_0 c_0 k_0^3}{4\pi k_0}\left[2d_z\dot{Q}(\omega)\right]\cos\phi\left[\frac{e^{+ik_0 r_1}}{r_1} - \frac{e^{+ik_0 r_2}}{r_2}\right] \tag{2.35}$$

where, in the frequency domain, Eq. (2.18) applies. As in the analysis of the dipole, we write

$$r_{\substack{1\\2}} \simeq r \pm d_y \sin\phi \sin\theta$$

for $r \gg d_y$. Substitution into Eq. (2.35) yields

$$\tilde{p}(r,\phi,\theta) \simeq \frac{i}{2}\rho_0 c_0 k_0^4\left[2d_z 2d_y\dot{Q}(\omega)\right]\sin 2\phi \sin\theta \frac{e^{+ik_0 r}}{4\pi k_0 r} \tag{2.36}$$

where $2d_z 2d_y\dot{Q}(\omega)$ is called the quadrupole strength. This can be written in terms of the dipole force, Eq. (2.31),

$$2d_z 2d_y\dot{Q}(\omega) = \frac{-3i}{\rho_0 k_0 c_0} 2d_y F_z$$

so that

$$\tilde{p}(r,\phi,\theta) \simeq -\frac{3}{2}k_0^2\left[2d_y F_z\right]\sin 2\phi \sin\theta \frac{e^{+ik_0 r}}{4\pi r} \tag{2.37a}$$

This expression shows that the quadrupole pressure is of order $2k_0 d_y$ less than the equivalent dipole pressure. Thus the spatial gradient represented by two closely spaced dipoles gives rise to an additional k_0 dependence compared to dipole radiation. In other terms, the two spatial gradients represented by four monopoles give a k_0^2 dependence compared to simple monopole radiation. The net force on the fluid is also instantaneously zero, yet since the fluid disturbances emitted from each dipole do not cancel identically: sound is still radiated. Statements made above for the monopole imaging apply equally well to the imaging of the dipole. The directivity of the sound from the quadrupole is concentrated on four lobes, which are oriented at $\phi = (2n+1)\pi/4$, $n = 1, 2, 3, 4$. The sound pressure level is zero on the $x-y$ and $x-z$ planes.

Eq. (2.35) can be rewritten for the cases $d_z/r\ d/r, \ll 1$

$$\tilde{p}(r,\phi,\theta) = \frac{-i\rho_0\omega\dot{Q}(\omega)}{4\pi}\frac{d^2}{dy_i\,dy_j}\left\{\frac{e^{ik_0 r}}{r}\right\}(2d_i 2d_j) \tag{2.37b}$$

for source poles displaced along the i and j axes. This becomes equal to Eqs. (2.36) and (2.37a) once the spatial derivatives are evaluated. Note again that the range r is equal to $|\mathbf{x}\text{-}\mathbf{y}|$, where $\mathbf{x}$ is in the far field. The reader should compare Eqs. (2.24a), (2.29), and (2.37b) as well as Eqs. (2.24a), (2.27), and (2.36), to see clearly the progressive dependence on speed of sound that parallels the spatial gradients of the multipoles. This behavior will be shown in later chapters to determine the dependence of radiated sound on speed for flow-induced dipole and quadrupole sources.

Two radiation patterns may be formed from simple quadrupoles, which depend on whether the constituent dipoles are oriented either parallel (Fig. 2.5A) or in-line (Fig. 2.5C). These alternative patterns are shown in Fig. 2.6. The parallel form (lateral) quadrupole has the four lobes in the $y-z$ plane ($\theta = \pm\pi/2$); on the surface of a cone ($\phi = \pi/4$), as Eq. (2.37a) shows, the directivity pattern will depend on θ as shown in Fig. 2.3. The in-line (longitudinal) form has but two lobes, and it has uniform directivity in planes parallel to the z plane.

2.1.3.4 Average Acoustic Intensity

In all of the above cases the acoustic far-field intensity can be written in the general form

$$I = I_{\max} g(\theta, \phi)$$

where the directivity factor $g(\theta, \phi)$ is independent of distance from the source; $g(\theta, \phi) = 1$ for monopole radiation; and for multipole radiation $g(\theta_m, \phi_m) = 1$ at the angles θ_m, ϕ_m of maximum intensity, $I_{\max}$. Often, as previously discussed in connection with the compact dipole, we shall be interested in a spatial average of the intensity or of the mean-square sound

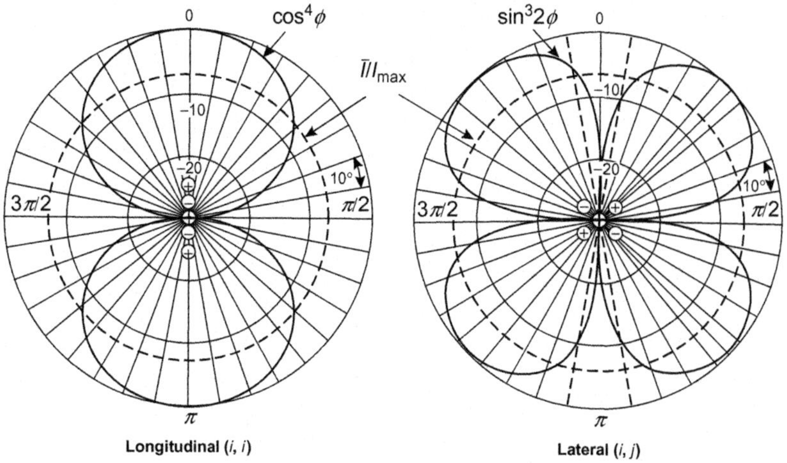

FIGURE 2.6 Directivity patterns, $10 \log \boldsymbol{I}/I_{max}$, for lateral and longitudinal quadrupole sources.

TABLE 2.1 Directivity Factors[a] for Multipole Sources

Source	$\bar{I}/I_{max}$	$I(\theta, \phi)/I_{max} = g(\theta, \phi)$
Monopole	1.0	1.0
Dipole, compact	1/3	$\cos^2 \phi$
Quadrupole, lateral	4/15	$\sin^2 2\phi \sin^2 \theta$
Quadrupole, longitudinal	1/5	$\cos^4 \phi$
Dipole, half plane	$\pi/8$	$\sin^2 \frac{\theta}{2} \lvert\sin \phi\rvert$

[a] *See Fig. 2.1B for coordinate system.*

pressure from a multipole source. The maximum intensity at a field point, I_{max}, will be given in terms of a source strength and the spatial average I of $\boldsymbol{I}$ will be some fraction of that maximum. Using Eqs. (2.15) and (2.17) the spatial averages can be obtained for these quantities. The spatial average mean-square sound pressure is related to the spatial average intensity $\bar{I}$ by Eq. (2.15)

$$\overline{p^2} = \bar{I}\rho_0 c_0$$

and the sound power is related to $\bar{I}$ by Eq. (2.16)

$$\mathbb{P}_{rad} = 4\pi r^2 \bar{I}$$

Table 2.1 summarizes the directivity factors for far-field acoustic intensities of the basic source types. These calculations show that the space average acoustic intensity is generally within a factor 1/3 to 1/5 (5–7 dB) of the local maximum value.

2.2 SOMMERFELD'S RADIATION CONDITION

In Section 2.1.2 the direction of sound propagation was taken into consideration in order to select a solution to the wave equation. This solution was selected according to a condition of causality that required sound to propagate from the source to the receiver such that as time t increased the wave progressed in the positive x direction. Thus of the two possible solutions of the wave equations $\cos(k_0x - \omega t)$ and $\cos(k_0x + \omega t)$ only the former applied. This was just a sample case of the use of a radiation condition that will be generalized below.

The far-field radiation from multipole sources in the absence of reflecting surfaces in the acoustic field that lies outside the source zone has been shown to be dependent on the distance from the source as $e^{ik_0r}/4\pi r$. This can be seen by reference to Eqs. (2.24a), (2.36), and (2.37a,b), which show for outward-traveling waves the general spherical wave propagation is given by

$$g(r - c_0t) = \frac{1}{4\pi r}e^{ik_0(r-c_0t)}$$

Therefore for these outward-traveling waves we have the far-field condition that

$$\lim_{r\to\infty} r\left(\frac{\partial p}{\partial r} - ik_0p\right) = 0 \tag{2.38}$$

Alternatively, for inward-traveling waves we have

$$g(r + c_0t) = \frac{1}{4\pi r}e^{ik_0(r+c_0t)}$$

so that

$$\lim_{r\to\infty} r\left(\frac{\partial p}{\partial r} + ik_0r\right) = 0 \tag{2.39}$$

This latter condition has been termed the "absorption condition" by Sommerfeld [11].

These radiation conditions have been shown by Sommerfeld [11], see also the work of Pierce [7] and Howe [12], to be necessary for the uniqueness of a solution of the wave equation. In two dimensions, the radiation condition is

$$\lim_{r\to\infty} \sqrt{r}\left(\frac{\partial p}{\partial r} - ik_0r\right) = 0$$

and similarly for the absorption condition. Quite simply, the radiation condition ensures that for a single-source distribution that radiates to the far field the solution of the wave equation excludes inward radiation. It amounts to a far-field boundary condition. Similarly, in the case of a surface-source distribution that radiates to an interior, the absorption condition rules out any internal sources that radiate outward.

The use of the retarded time, as introduced in Eq. (2.24b) also implies a statement of the radiation condition. The retarded time just quantifies the time delay required for the propagation of the disturbance over the distance r at the speed c_0.

2.3 LIGHTHILL'S THEORY OF AERODYNAMIC NOISE

2.3.1 The Wave Equation

We shall now determine the wave equation for the acoustic pressure that results from turbulent motion. For a spatially concentrated region of turbulent fluid motion, Lighthill's [13–15] formulation (see also Refs. [6,7]) is unique in that it considers this region as an acoustic source that drives the surrounding fluid. The starting point of the analysis will again be the equations of continuity and momentum applied to a fluid region such as depicted in Fig. 2.7. However, the velocity disturbance u_i that we are now considering includes both acoustic and, in a restricted region, hydrodynamic contributions. We shall not assume inviscid motion in the region of the turbulence.

In this case, the equation of continuity is, again,

$$\frac{\partial \rho}{\partial t} + \frac{\partial}{\partial y_i}(\rho u_i) = q \tag{2.40}$$

where q $(=\partial \rho/\partial t)$ has been added as the rate of mass injection per unit volume, and the momentum equation is written in the following form as long as it can be said that mass is injected into the turbulent medium at zero velocity

$$\frac{\partial \rho u_i}{\partial t} = + \frac{\partial \tau_{ij}}{\partial y_j} - \frac{\partial(\rho u_i u_j)}{\partial y_j} \tag{2.41}$$

where τ_{ij} is the Stokes stress tensor, e.g., [2–5, 12,16]. This stress tensor is written

$$\tau_{ij} = p\delta_{ij} - \tau'_{ij} \tag{2.42}$$

$$\tau'_{ij} = 2\mu\left(\varepsilon_{ij} - \frac{1}{3}\varepsilon_{kk}\delta_{ij}\right)$$

where

$$\epsilon_{ij} = \frac{1}{2}\left(\frac{\partial u_i}{\partial y_j} + \frac{\partial u_j}{\partial y_i}\right) \tag{2.43}$$

As before, the fluid pressure is p and ϵ_{ij} are the fluid rates of strain in a fluid of zero bulk viscosity . The momentum equation can be rewritten in the form of Eq. (2.2)

$$u_i q + \rho \frac{\partial u_i}{\partial t} = -\frac{\partial \tau_{ij}}{\partial y_j} - \rho u_j \frac{\partial u_i}{\partial y_j} \tag{2.44}$$

where the term $u_i q$ represents the convective acceleration of the injected mass per unit volume. Eq. (2.44) reverts to Eq. (2.2) when the fluid is inviscid and $q = 0$. Thus, taking the divergence of Eq. (2.41) and the time derivative of Eq. (2.1), we have

$$\frac{\partial^2 \rho}{\partial t^2} = \frac{\partial^2 \tau_{ij}}{\partial y_i \partial y_j} + \frac{\partial^2}{\partial y_j \partial y_i}\left(\rho u_i u_j\right) + \frac{\partial q}{\partial t}$$

Now by the identity

$$c_0^2 \nabla^2 \rho = \frac{\partial^2 [\rho c_0^2 \delta_{ij}]}{\partial y_i y_j}$$

we have

$$\frac{\partial^2 \rho}{\partial t^2} - c_0^2 \nabla^2 \rho = \frac{\partial^2}{\partial y_i \partial y_j}\left[\tau_{ij} + \rho u_i u - c_0^2 \rho \delta_{ij}\right] + \frac{\partial q}{\partial t} \tag{2.45}$$

$$\tau_{ij} = -p\delta_{ij} + \tau'_{ij}$$

Using Eq. (2.42) we can separate the contributions of the viscous stresses and the pressures, or normal stresses, to obtain the wave equation in the final form

$$\frac{\partial^2 \rho}{\partial t^2} - c_0^2 \nabla^2 \rho = \frac{\partial^2 T_{ij}}{\partial y_i \partial y_j} + \frac{\partial q}{\partial t} \tag{2.46}$$

where

$$T_{ij} = \rho u_i u_j + (p - c_0^2 \rho)\delta_{ij} - \tau'_{ij} \tag{2.47a}$$

$$T_{ij} = \rho u_i u_j + \left[(p - p_0) - (\rho - \rho_0)c_0^2\right]\delta_{ij} - \tau'_{ij} \tag{2.47b}$$

is Lighthill's stress tensor. The tensor $\rho u_i u_j$ is called the Reynolds stress, and it expresses the intensity of the turbulence in the source region. The term τ'_{ij} is just the viscous part of the Stokes stress tensor, Eq. (2.42), and the term $p - {c_0}^2\rho$ expresses the differential between the actual pressure fluctuations in the ambient fluid medium which has been characterized by c_0 and it accounts for heat conduction sources. The pressure p and density ρ are the local instantaneous pressure and density of the fluid.

The noise-producing character of the fluid field is such that outside a specified region of the disturbances we have

$$\frac{\partial^2 T_{ij}}{\partial y_i \partial y_j} \equiv 0$$

Now, the pressure and density in the far-field ambient undisturbed fluid are p_0 and ρ_0. These quantities are constant so that

$$\frac{\partial \rho_0}{\partial t} \quad \text{and} \quad \frac{\partial^2 \rho_0}{\partial y_i \partial y_j}$$

are both zero and similarly for p_0. Thus we can write the wave equation for the instantaneous *density fluctuation* to obtain the Lighthill equation in its final form

$$\frac{\partial^2}{\partial t^2}(\rho-\rho_0)-c_0^2\nabla^2(\rho-\rho_0)=\frac{\partial^2}{\partial y_i\partial y_j}\{T_{ij}\}+\frac{\partial q}{\partial t} \tag{2.48}$$

where we now introduced the fluctuation of pressure $p-p_0$ into the stress tensor. If everywhere in the fluid region the fluctuation in pressure is a thermodynamic variable with isentropic fluctuations with speed of sound c_0 then the pressure and density fluctuations are balanced by Eq. (2.6). Under this circumstance the pressure and density terms cancel identically in Eq. (2.48). Often the magnitudes of the turbulent Reynolds stresses dominate the viscous stresses in turbulent motion so the latter may be neglected. The wave equation in the absence of mass injection is now finally reduced to the more simplified form

$$\frac{\partial^2(\rho-\rho_0)}{\partial t^2}-c_0^2\nabla^2(\rho-\rho_0)=\frac{\partial^2(\rho u_i u_j)}{\partial y_i\partial y_j} \tag{2.49}$$

which shows that the acoustic field is driven by the region of fluctuating Reynolds stresses. Outside the region of the Reynolds stress fluctuations, the velocity fluctuations are acoustic.

Thus, outside the region of turbulent fluid motion, Lighthill's equation reduces to the wave equation of linear acoustics theory. As we shall see later in this chapter, the knowledge of the behavior of the stress tensor T_{ij} in the source region is crucial to the analytical modeling of acoustic radiation. In this regard Eq. (2.46) is simply a restatement of the equations of motion recast in a form which looks like an inhomogeneous wave equation. The source terms, so cast, are really the nonlinear fluid motions. This is clearly the case in the simplified Eq. (2.49). In this casting, the nonlinear turbulence fluctuations are presumed to act as sources to the linearly responding ambient acoustic medium. If we compare the derivations of Eqs. (2.7) and (2.9) with that of Eq. (2.46), this further meaning of the "source" T_{ij} may be deduced. For inviscid, isentropic fluid motions, then, the T_{ij} reduces to the $\rho u_i u_j$, which is just the nonlinear residue of the combination of the momentum and continuity equations which had been neglected in Sections 2.1 and 2.2. Applying Eq. (2.49) to describe the sound from a region of turbulent motion may thus involve a considerable assumption that the nonlinear terms in fact represent acoustic sources. This point will be touched on further in Chapter 3, Shear Layer Instabilities, Flow Tones, and Jet Noise when we discuss unsteady mass injection.

2.3.2 Kirchhoff's Integral Equation and the Retarded Potential

The acoustic field radiated outward to a point in free space from a distributed region of sources is the summation of the individual contributions that result from each of the sources composing the region. In the physical summation process that occurs, contributions reinforce and interfere depending on the instantaneous phase relationships among the various sources. In a rudimentary sense, this summation process has already been demonstrated in Section 2.1.2 to determine the acoustic fields of dipole and quadrupole source distributions. In more complicated physical situations the acoustic radiation is determined as a weighted integral of the source distribution. The integral must account for the acoustic phase interactions among elements of the source distribution as well as the propagation to the far field. In the mathematical integral formulation described below, this is accomplished by the use of the retarded potential. Also in the analysis below, we will assume that the sources are stationary with respect to the field medium, or at least moving at a Mach number that is so small as to be neglected. We will discuss the effects of source convection relative to the medium in Section 2.5 *Effects of Source Motion on Flow Noise*.

We begin by deriving an integral equation for the density fluctuation $\rho_a = \rho - \rho_0$, in the manner of Bateman [17], although Stratton [18], Jackson [19], and Jones [20] have alternative derivations. (The author is indebted to the Late Professor Patrick Leehey of MIT for his contribution to this derivation.) The wave equation for an adiabatic acoustic field in an infinite homogeneous fluid is written as

$$\nabla^2 \rho_a - \frac{1}{c_0^2}\frac{\partial^2 \rho_a}{\partial t^2} = \frac{\sigma(\mathbf{y}, t)}{c_0^2} \tag{2.50}$$

where $\sigma(\mathbf{y}, t)$ is the source term of Eq. (2.48) or (2.49) and is specified over a volume V_0 contained in the control volume V as illustrated in Fig. 2.7. The field point $\mathbf{x}$ is considered to be surrounded by a small surface S_x, while the point $\mathbf{y}$ is located somewhere within the region V, which is surrounded by surface Σ. A function that is defined as $v(x, y, z, t) = \rho_a(x,y,z,t - r/c_0)$, where $r = |\mathbf{x} - \mathbf{y}|$, can be shown [11] to satisfy the equation

$$\nabla^2 v + \frac{2r}{c_0}\left\{\frac{\partial}{\partial y_i}\left(\frac{r_i}{r^2}\frac{\partial v}{\partial t}\right)\right\} + \frac{\sigma(\mathbf{y}, t - r/c_0)}{c_0^2} = 0 \tag{2.51}$$

by substitution into Eq. (2.50). Multiplication of Eq. (2.51) by $1/r$ and integration of that relationship throughout the entire control volume contained within the surface Σ yields

$$\iiint_V \frac{1}{r}\nabla^2 v \, dV(\mathbf{y}) + \frac{2}{c_0}\iiint_V \frac{\partial}{\partial y_i}\left[\frac{r_i}{r^2}\frac{\partial v}{\partial t}\right] dV(\mathbf{y}) + \frac{1}{c_0^2}\iiint_V \frac{\sigma(\mathbf{y}, t - r/c_0)}{r} dV(\mathbf{y}) = 0$$

Green's theorem yields the relation

$$\iiint_V \left\{\frac{1}{r}\nabla^2 v - v\nabla^2\left(\frac{1}{r}\right)\right\} dV(\mathbf{y}) = \iint_{\Sigma + S_x}\left\{\frac{1}{r}\frac{\partial v}{\partial n} - v\frac{\partial}{\partial n}\left(\frac{1}{r}\right)\right\} dS(\mathbf{y})$$

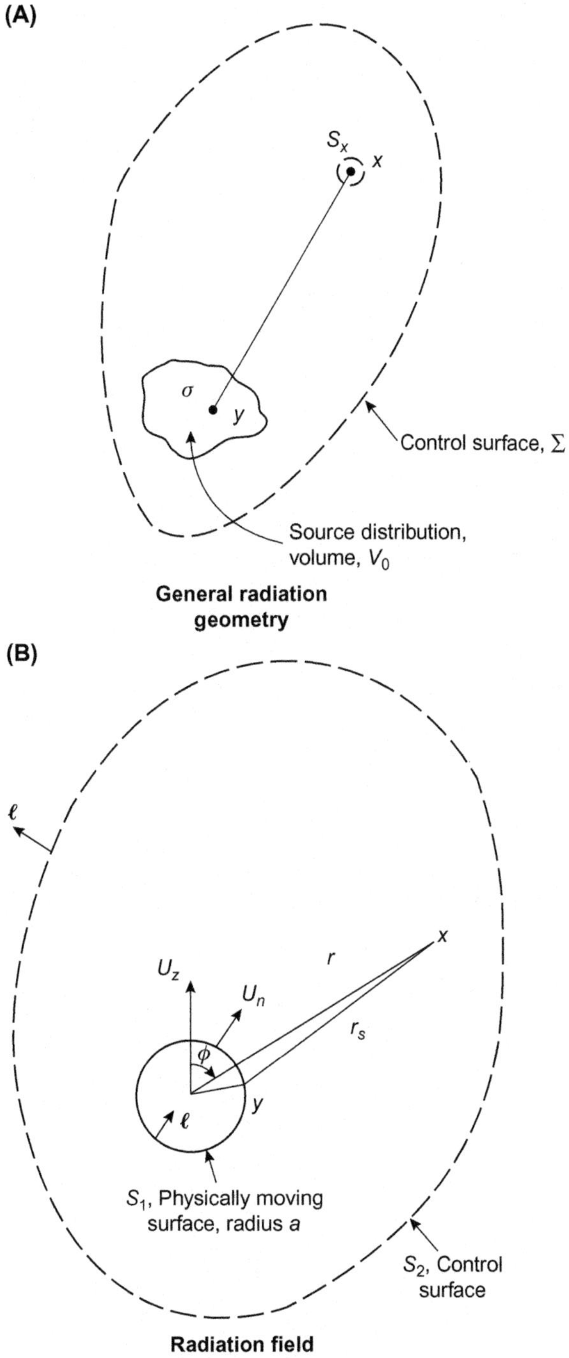

FIGURE 2.7 Illustrations of control volumes surrounding source regions and fields of interest.

Therefore, in conjunction with Eq. (2.51), we obtain

$$\iiint_V \left\{ \frac{1}{r}\nabla^2 \upsilon - \upsilon \nabla^2 \left(\frac{1}{r}\right) \right\} dV(\mathbf{y}) = \iint_{\Sigma+S_x} \left\{ \frac{1}{r}\frac{\partial \upsilon}{\partial n} - \upsilon \frac{\partial}{\partial n}\left(\frac{1}{r}\right) \right\} dS(\mathbf{y})$$

$$= \frac{-1}{c_0^2} \iiint_V \frac{\sigma(x,y,z,t-r/c_0)}{r} dV(\mathbf{y}) - \frac{2}{c_0} \iint_{\Sigma+S_x} n_i \frac{r_i}{r^2} \left(\frac{\partial \upsilon}{\partial t}\right) dS(\mathbf{y})$$

The surface integrals include all surfaces. If **x** lies within the circumscribing surface Σ and if this point is surrounded by a surface S_x of vanishing radius, then

$$\iint_{S_x} \upsilon \frac{\partial}{\partial n}\left(\frac{1}{r}\right) dS(\mathbf{y}) = \upsilon(\mathbf{x}) \lim_{r\to 0} \iint_{\Omega} \frac{\partial}{\partial n}\left(\frac{1}{r}\right) r^2\, d\Omega$$

$$= 4\pi \upsilon(\mathbf{x})$$

However, if x does not lie inside $S_{\mathbf{x}}$, then

$$\iint_{S_x} \upsilon \frac{\partial}{\partial n}\left(\frac{1}{r}\right) dS(\mathbf{y}) \equiv 0$$

Thus we have

$$4\pi \upsilon(\mathbf{x}) = \frac{1}{c_0^2} \iiint_V \frac{\sigma(\mathbf{y}, t-r/c_0)}{r} dV(\mathbf{y})$$

$$+ \iint_{\Sigma} \left\{ \frac{2}{c_0}\frac{1}{r}\frac{\partial r}{\partial n}\left(\frac{\partial \upsilon}{\partial t}\right) + \frac{1}{r}\frac{\partial \upsilon}{\partial n} - \upsilon \frac{\partial}{\partial n}\left(\frac{1}{r}\right) \right\} dS(\mathbf{y})$$

for **x** inside Σ and

$$0 = \frac{1}{c_0^2} \iiint_V \frac{\sigma(\mathbf{y}, t-r/c_0)}{r} dV(\mathbf{y})$$

$$+ \iint_{\Sigma} \left\{ \frac{2}{c_0}\frac{1}{r}\frac{\partial r}{\partial n}\left(\frac{\partial \upsilon}{\partial t}\right) + \frac{1}{r}\frac{\partial \upsilon}{\partial n} - \upsilon \frac{\partial}{\partial n}\left(\frac{1}{r}\right) \right\} dS(\mathbf{y})$$

for **x** outside Σ. Now, considering only that **x** lies with Σ, we have that $\upsilon(\mathbf{x}, t) = \rho_a(\mathbf{x}, t)$, and that

$$\frac{\partial \upsilon}{\partial n} = \frac{\partial}{\partial n}\left[\rho_a\left(\mathbf{y}, t - \frac{r}{c_0}\right)\right] = \left[\frac{\partial \rho_a}{\partial n}\right] - \frac{1}{c_0}\frac{\partial r}{\partial n}\left[\frac{\partial \rho_a}{\partial t}\right]$$

where the brackets denote that the function is evaluated at the retarded time $t - r/c_0$, i.e., $[f] = f(t - r/c_0)$. Use of this function now yields the instantaneous density fluctuation

$$4\pi\rho_a(\boldsymbol{x}, t) = \frac{1}{c_0^2}\iiint \frac{[\sigma(\mathbf{y})]}{r} dV(\mathbf{y}) + \iint_{\Sigma}\left\{\frac{1}{c_0 r}\frac{\partial r}{\partial n}\left[\frac{\partial \rho_a}{\partial t}\right] - [\rho_a]\frac{\partial(1/r)}{\partial n} + \frac{1}{r}\left[\frac{\partial \rho_a}{\partial n}\right]\right\} dS(\mathbf{y}) \tag{2.52}$$

which is Kirchhoff's equation for the fluctuating fluid density. The surface integral is taken over all surfaces that are contiguous to the subject volume, V; the volume V_0 is occasionally introduced below since the source density is presumed to vanish outside V_0. Unless other surfaces are present, the bounding Σ may be expanded indefinitely far from both V_0 and $\mathbf{x}$ so that the surface integral vanishes. This can be seen by noting that if the disturbance is initiated at $t = t_0$, a surface Σ can be selected suitably far from the sources that $\rho_a(t_0 - r/c_0)$ and its derivatives vanish identically. The condition is equivalent to the radiation condition. Thus the instantaneous density fluctuation is given by the volume integral

$$4\pi\rho_a(\boldsymbol{x}, t) = \frac{1}{c_0^2}\iiint_{V_0} \frac{[\sigma(\mathbf{y})]}{r} dV(\mathbf{y}) \tag{2.53}$$

If the source volume V_0 is finite, then this equation shows that $\rho_a(\mathbf{x}, t) \sim 1/r$ sufficiently far from V_0 in accordance with the classical law of spherical spreading.

Now, using a comparison of Eqs. (2.46), (2.50), and (2.53), we find that the Kirchhoff formulation from Lighthill's wave equation is

$$4\pi\left[\rho(\boldsymbol{x}, t) - \rho_0\right] = \frac{1}{c_0^2}\iiint_{V_0} \frac{1}{r}\left[\frac{\partial^2 T_{ij}}{\partial y_i \partial y_j}\right] dV(\mathbf{y}), \tag{2.54}$$

where $c_0^2(\rho(\mathbf{x}, t) - \rho_0) = p_a(\mathbf{x}, t)$ by Eq. (2.6) under the assumed condition of no unsteady mass injection to the volume. As before, the brackets in the integrand denote that the retarded time is used. This equation is the central result of this section because the integrand includes the required retardation or phase effects that give rise to the multipole nature of complex source regions.

The source term in Eqs. (2.46), (2.49), and (2.54) involves two spatial gradients. Now, in Section 2.1.2 it was shown that the quadrupole radiation results from two spatial fluid gradients; see Eq. (2.37b). Lighthill's source term is therefore interpreted as having a quadrupole nature. Furthermore, the source term is determined by correlated fluid velocities that give rise to the stress tensor T_{ij}. Physically, those local stresses are either in-line or lateral, as illustrated in Fig. 2.5. Thus the compressive stresses T_{ii}, T_{jj}, T_{kk} represent longitudinal quadrupoles, while the shearing stresses T_{ij} $(i \neq k)$ represent lateral quadrupoles.

Looking forward to the examination of dipole fields in Section 2.4.2, we note that Eq. (2.54), in its current form, is difficult to evaluate because the source function includes both the acoustic retardation effects and the spatial gradients. The equation can be modified to separate the spatial gradients from the retardation effects when the sources are in a free space, i.e., not enclosed by a reflecting surface. To do this we recast Eq. (2.54) into a form that permits an estimate of the radiated sound intensity from a restricted region of sources *when solid surfaces are not present*. This estimate will still depend on an appropriate statistical representation of the turbulent sources as discussed in subsequent sections. Consider the following retarded function $[F] = F(\mathbf{y}, t - r/c_0)$, and its derivatives, so

$$\frac{\partial}{\partial y_i}\left(\frac{1}{|\mathbf{x}-\mathbf{y}|}\right) = -\frac{(x-y)_i}{|\mathbf{x}-\mathbf{y}|^3} \tag{2.55a}$$

Then

$$\frac{\partial}{\partial x_i}\iiint \frac{F_i(\mathbf{y}, t - r/c_0)}{r} dV(\mathbf{y}) = \iiint\left\{\left[\frac{\partial F_i}{\partial t}\right]\left(\frac{-1}{c_0}\right)\frac{r_i}{r^2} - \frac{[F_i]}{r^2}\frac{r_i}{r}\right\} dV(\mathbf{y}) \tag{2.55b}$$

and

$$\iiint \frac{\partial}{\partial y_i}\left[\frac{F_i}{r}\right] dV(\mathbf{y})$$

$$= \iiint \left[\frac{\partial F_i}{\partial y_i}\right]\frac{dV(\mathbf{y})}{r} dV(\mathbf{y}) + \iiint\left\{\left[\frac{\partial F_i}{\partial t}\right]\frac{1}{c_0}\frac{r_i}{r^2} + \frac{[F_i]}{r^2}\frac{r_i}{r}\right\} dV(\mathbf{y})$$

so that combining we have

$$\iiint \frac{\partial}{\partial y_i}\left[\frac{F_i}{r}\right] dV(\mathbf{y}) = \iiint \left[\frac{\partial F_i}{\partial y_i}\right]\frac{dV(\mathbf{y})}{r} - \frac{\partial}{\partial x_i}\iiint \frac{[F_i]}{r} dV(\mathbf{y}) \tag{2.55c}$$

Now, Gauss's theorem provides that

$$\iiint \frac{\partial}{\partial y_i}\left[\frac{F_i}{r}\right] dV(\mathbf{y}) = \iint_{\Sigma} n_i \frac{[F]}{r} dS(\mathbf{y})$$

where n_i is the unit normal to the surface Σ pointing outward from the control volume. Since this surface can be arbitrarily selected to be a distance r_Σ far enough from the source region that outward-traveling waves have not reached a range $r_\Sigma > c_0 t$, we can set the surface integral equal to zero. Thus we have the identity (in free space, no reflecting surfaces)

$$\iiint \left[\frac{\partial F_i}{\partial y_i}\right] \frac{dV(\mathbf{y})}{r} = \frac{\partial}{\partial x_i} \iiint \frac{[F_i]}{r} dV(\mathbf{y}) \tag{2.56}$$

This relationship is a formal mathematical statement of the operation leading to Eq. (2.27), and it will be frequently used in other derivations in this book. Recall that expression for the far-held radiation from two closely spaced sources in phase opposition involved a gradient of the expression for radiation from the single simple source. Similarly, by the repeated application of these operations, we have

$$4\pi(\rho(\mathbf{x},t) - \rho_0) = \frac{1}{c_0^2}\frac{\partial^2}{\partial x_i \partial x_j} \iiint_{V_0} \frac{[T_{ij}]}{r} dV(\mathbf{y}) \tag{2.57}$$

as an alternative form of Lighthill's equation that applies only when there are no solid surfaces enclosed by the surface Σ. The reader should now compare Eqs. (2.37b) and (2.57) with regard to the presence of the two spatial gradients. Under the further assumption that the largest linear dimension of the source volume is small relative to the range r and that the velocity of the source relative to the receiver is small compared to c_0, the derivative in Eq. (2.56) yields

$$\begin{aligned} 4\pi(\rho(\mathbf{x},t) - \rho_0) &= \frac{1}{c_0^4} \iiint_{V_0} \frac{(x_i - y_i)(x_j - y_j)}{r^3} \left[\frac{\partial^2 T_{ij}}{\partial t^2}\right] dV(\mathbf{y}) \\ &= \frac{1}{c_0^4}\frac{x_i x_j}{r^3} \iiint_{V_0} \left[\frac{\partial^2 T_{ij}}{\partial t^2}\right] dV(\mathbf{y}) \end{aligned} \tag{2.58}$$

since $\partial^2 \overline{T}_{ij}/\partial t^2 = 0$. Near-field terms of the order r^{-3} have been neglected compared to terms of the order r^{-1}.

2.3.3 Acoustic Radiation From a Compact Region of Free Turbulence

This topic will be considered in detail in Chapter 3, Shear Layer Instabilities, Flow Tones, and Jet Noise but here we use Eq. (2.58) to heuristically develop certain general rules of similitude that govern sound power radiated from a small region of subsonic turbulence. The first treatment of this nature was that of Proudman [21]. Although the following derivation lacks rigor, it illustrates the marriage of acoustic theory with the field of statistical turbulence theory. This development further illustrates a classical dimensional analysis of flow quadrupoles.

The time average intensity is (see Eqs. (2.6) and (2.15))

$$I(\mathbf{x}) = \frac{c_0}{\rho_0}\overline{(\rho - \rho_0)^2}$$

or, using Eq. (2.57), we have

$$I(\mathbf{x}) = \frac{1}{(4\pi)^2 \rho_0 c_0^5}\left(\frac{x_i x_j}{r^3}\right)\left(\frac{x_k x_j}{r^3}\right) \times \iiint_{V_0} \cdot \iiint_{V_0} \left\{ \overline{\frac{\partial^2 T'_{ij}(\mathbf{y}_1, t - r_1/c_0)}{\partial t^2} \frac{\partial^2 T'_{kl}(\mathbf{y}_2, t - r_2/c_0)}{\partial t^2}} \right\} dV(\mathbf{y}_1) dV(\mathbf{y}_2) \tag{2.59}$$

where

$$T'_{ij} = T_{ij} - \overline{T}_{ij}$$

and T'_{ij} must approach zero at least as fast as $|\mathbf{y}|^{-3}$ as $|\mathbf{y}| \to \infty$ in order for the volume integral to converge (see also Crow [22]). The term in brackets is the *spatial covariance* of the retarded stress tensor, which is a function of both position vectors $\mathbf{y}_1$ and $\mathbf{y}_2$. The integration with respect to both of these vectors extends over the source volume V_0. The covariance of the stress tensor involves products of velocity fluctuations of the form

$$\overline{\left[\frac{\partial^2 T'_{ij}}{\partial t^2}\right]\left[\frac{\partial^2 T'_{kl}}{\partial t^2}\right]} = \overline{\left[\frac{\partial^2}{\partial t^2}(\rho u_i u_j)_{y=y_1}\right]\left[\frac{\partial^2}{\partial t^2}(\rho u_k u_l)_{y=y_2}\right]}$$

where $(\rho u_i u_j) = \rho u_i u_j - \rho_0 \overline{u_i u_j}$ and where we have used the source terms introduced in Eq. (2.49)

We shall consider the mathematical consequences of specific forms of covariance functions in future chapters. However, for now we shall develop certain general notions as they apply to turbulent fluid flow. Let us consider that the disturbances are the result of the irregular motion of a collection of eddies of typical correlation length Λ. This length is interpreted as a limiting separation of two velocity sensors in the flow so that the temporal average of the product of the signals from the sensors is considered negligible compared to the temporal mean square of each signal separately. For example, letting the velocities in the i and k directions at two points in the volume V_0 be $u_i(y_1, t)$ and $u_k(y_2, t)$ we can write, with $\mathbf{y}_2 = \mathbf{y}_1 + \boldsymbol{\xi}$

$$\overline{u_i^2(y_1, t)} \simeq \overline{u_k^2(y_1 + \Lambda, t)} \gg \overline{u_i(y_1, t) u_k(y_1 + \Lambda, t + \tau_r)} \tag{2.60}$$

i.e., the mean squares of the separate signals exceed the covariance of the signals for $\xi_i \geq \Lambda_i$. In this equation τ_r represents the difference in retarded times $(r_2 - r_1)/c_0$. For small separations $\xi \ll \Lambda$ the covariance approaches the product of the root mean squares of the separate signals; i.e.,

$$\lim_{|\xi| \to 0} [\overline{u_i(y_1, t) u_k(y_1 + \boldsymbol{\xi}, t + \tau_r)}] \simeq [\overline{u_i^2(y_1, t) u_k^2(y_1 + \boldsymbol{\xi}, t)}]^{1/2} \tag{2.61}$$

The covariance of u_i and u_k is a continuous function of ξ. Now, we shall assume that the turbulent patch is translated at a constant velocity U_c that is uniform throughout V_0 (yet still at vanishingly small Mach number). Under this assumption the individual eddies are locally translated in a wavelike manner according to the formula $x - U_c t = \text{constant}$. Thus a measure of the time variation is

$$\frac{\partial}{\partial t} \sim \frac{U_c}{\Lambda}$$

Now, we further assume that $\overline{u_i^2} \sim U_c^2$, that $U_c \ll c_0$, and that all the space–time statistical characteristics scale on Λ and U_c. Under these simplifications Eq. (2.59) can be written to depend parametrically on flow properties as

$$\mathrm{I}(\boldsymbol{x}) \sim \frac{1}{(4\pi)^2 \rho_0 c_0^5} \frac{1}{r^2} \left(\frac{U_c}{\Lambda}\right)^4 (\rho_0 U_c^2)^2 \Lambda^3 V_0 \tag{2.62}$$

This result follows directly from Eq. (2.60), which implies the representation

$$\overline{T'_{ij}\left(y_1, t - \frac{|x - y_1|}{c_0}\right) T'_{kl}\left(y_1 + \xi, t - \frac{|\xi|}{c_0}\right)} \sim \rho_0^2 \overline{u_k^2}\,\overline{u_k^2} R(\mathbf{y}, \boldsymbol{\xi}) \tag{2.63}$$

where $R(\mathbf{y}, \xi)$ is a correlation function of the stress tensor fluctuation and the separation variable ξ also determines the magnitude of the retarded time delay. By definition $R(\mathbf{y}, 0) \equiv 1$ and $\lim_{L \to \infty} R(\mathbf{y}, \mathbf{L}) \simeq 0$. Finally, the integration over ξ includes the integrated retardation effects.

$$\iiint \overline{\frac{\partial^2}{\partial t^2}(T_{ij})_1 \frac{\partial^2}{\partial t^2}(T_{kl})_2} dV(\boldsymbol{\xi}) \sim \rho_0^2 \left(\frac{U_c}{\Lambda}\right)^4 \overline{u_i^2 u_k^2} \iiint R(\mathbf{y}, \boldsymbol{\xi}) dV(\boldsymbol{\xi})$$
$$\sim \rho_0^2 \left(\frac{U_c}{\Lambda}\right)^4 \overline{u_i^2 u_k^2} \Lambda^3$$

This parametric expression does not include the sound speed; it is assumed that as long as the eddy correlation length Λ and the characteristic length scale of V_0 are much less than an acoustic wavelength and U_c is much less than the acoustic propagation velocity, the integrated value of the correlation depends only on Λ. Eq. (2.62) is rearranged as

$$\mathrm{I}(\boldsymbol{x}) \sim \frac{1}{(4\pi)^2} \rho_0 \frac{U_c^8}{c_0^5} \frac{V_0}{\Lambda r^2} \tag{2.64}$$

in order to emphasize that the acoustic intensity from free convecting turbulence as a distribution of free quadrupoles increases as the eighth power of the convection velocity and is linearly proportional to the size of the acoustically compact volume of the turbulence. A review of the derivation leading to Eq. (2.58) will show that the existence of the two spatial gradients in the compact source give rise to the c_0^{-4} dependence in the acoustic intensity. The high exponent on the Mach number is therefore set by the double gradient in the source term.

This fundamental result gives the often-quoted eighth-power velocity dependence of radiated power from free turbulence. Eq. (2.64) is a much simplified relationship that gives some variable dependencies that apply to noise from free jets and wakes. Chapter 3, Shear Layer Instabilities, Flow Tones, and Jet Noise will consider some more exact theories that account for explicit forms of the correlation function $R(\mathbf{y}, \xi)$, effects of turbulence convection on radiation efficiency and kinematic scaling.

2.4 EFFECTS OF SURFACES ON FLOW-INDUCED NOISE

It must be emphasized that the expression for the acoustic radiation from a restricted zone of fluid stress fluctuations given in the last section applies only when there are no reflecting boundaries in the field of consideration. When the boundary exists and its surface impedance is not identically equal to that of the fluid, its effect is to alter the sound field physically by causing acoustic reflections. It may also disturb the flow locally, causing surface pressures that act as radiating dipoles. The mathematical fundamentals of this class of situations have been developed by Curle [23] and Powell [24].

2.4.1 Curle's Development of Lighthill's Wave Equation

Provision has already been made in our discussions for considering these effects. Eq. (2.52) is a general formulation that applies as long as the observation point $\mathbf{x}$ remains somewhere within the control surface fixed with respect to the acoustic medium, which we have designated as Σ, as shown in Fig. 2.7A. We continue with our assumption of low Mach number for all fluid and surface velocities.

Let us now let Σ *not* necessarily be so far from the source region that disturbances have not reached $r_\Sigma = c_0 t$. This is a relaxation of our former condition on Σ, and it allows for reflections from some surfaces in the control volume. Eq. (2.52) expresses the acoustic density fluctuation as a volume integral of the source region plus surface integrals over Σ of the density fluctuations. The surface integrals can represent the effects of reflections if the surfaces of integration coincide with physical boundaries. Eq. (2.52) for density fluctuations in the fluid is now rewritten

$$4\pi(\rho(\mathbf{x},t)-\rho_0) = \frac{1}{c_0^2}\iiint_V \left[\frac{\partial^2 T_{ij}}{\partial y_i\,\partial y_j}\right]\frac{dV(\mathbf{y})}{r} + \iint_\Sigma \left\{\frac{1}{c_0 r}\frac{\partial r}{\partial n}\left[\frac{\partial \rho}{\partial t}\right] - [\rho]\frac{\partial(1/r)}{\partial n} + \frac{1}{r}\left[\frac{\partial \rho}{\partial n}\right]\right\} dS(\mathbf{y}) \tag{2.65}$$

to reintroduce the control volume V and to denote the source term explicitly. Since all derivatives of the ambient density ρ_0 are necessarily 0, we can use ρ_a and ρ interchangeably. We have let Σ be any closed region that includes both V_0 and the observation point $\mathbf{x}$. Applying the divergence theorem to the volume integral as given by Eq. (2.55a–c) we obtain

$$\iiint_V \left[\frac{\partial^2 T_{ij}}{\partial y_i \partial y_j}\right]\frac{dV(\mathbf{y})}{r} = \frac{\partial^2}{\partial x_i \partial x_j}\iiint_{V_0}\frac{[T_{ij}]}{r}dV(\mathbf{y}) + \iint_\Sigma l_i\left[\frac{\partial T_{ij}}{\partial y_j}\right]\frac{dS(\mathbf{y})}{r} + \frac{\partial}{\partial x_i}\iint_\Sigma l_j[T_{ij}]\frac{dS(\mathbf{y})}{r} \quad (2.66)$$

where the surfaces Σ include both physical surfaces and control volume bounding surfaces and where the only part of the region within V for which $T_{ij} \neq 0$ if V_0. The direction cosines l_i are for the unit normal to the surface pointing out of the control volume.

Substitution of Eq. (2.66) into Eq. (2.65) yields

$$\begin{aligned} 4\pi(\rho(\boldsymbol{x},t)-\rho_0) = {} & \frac{1}{c_0^2}\frac{\partial^2}{\partial x_i \partial x_j}\iiint_{V_0}\frac{[T_{ij}]}{r}dV(\mathbf{y}) \\ & + \frac{1}{c_0^2}\iint_\Sigma \frac{l_i}{r}\left[\frac{\partial}{\partial y_i}(T_{ij}+\rho c_0^2\delta_{ij})\right]dS(\mathbf{y}) \\ & + \frac{1}{c_0^2}\frac{\partial}{\partial x_i}\iint_\Sigma \frac{l_j}{r}\left[T_{ij}+\rho c_0^2\delta_{ij}\right]dS(\mathbf{y}) \end{aligned} \quad (2.67)$$

since the surface integral in Eq. (2.65) can be rewritten (with $\partial r/\partial y_i = -\partial r/x_i$)

$$\begin{aligned} & \iint_\Sigma\left\{\frac{l_i}{c_0 r}\frac{\partial r}{\partial y_i}\left[\frac{\partial \rho}{\partial t}\right] - l_i[\rho]\frac{\partial(1/r)}{\partial y_i} + \frac{l_i}{r}\left[\frac{\partial \rho}{\partial y_i}\right]\right\}dS(\mathbf{y}) \\ & = \iint_\Sigma\left\{\frac{-l_i}{c_0 r}\frac{\partial r}{\partial x_i}\left[\frac{\partial \rho}{\partial t}\right] + l_i[\rho]\frac{\partial(1/r)}{\partial x_i} + \frac{l_i}{r}\left[\frac{\partial \rho}{\partial y_i}\right]\right\}dS(\mathbf{y}) \\ & = \iint_\Sigma l_i\left\{\frac{\partial}{\partial x_i}\left(\frac{1}{r}[\rho]\right) + \frac{1}{r}\left[\frac{\partial \rho}{\partial y_i}\right]\right\}dS(\mathbf{y}) \\ & = \iint_\Sigma l_j\left\{\frac{\partial}{\partial x_i}\left(\frac{1}{r}[\rho\delta_{ij}]\right) + \frac{1}{r}\left[\frac{\partial \rho\delta_{ij}}{\partial y_i}\right]\right\}dS(\mathbf{y}) \end{aligned} \quad (2.68)$$

Now, since Lighthill's stress tensor is given by Eq. (2.47), Eq. (2.67) becomes by substitution

$$4\pi c_0^2(\rho(\mathbf{x},t)-\rho_0)=\frac{\partial^2}{\partial x_i\partial x_j}\iiint_V\frac{[T_{ij}]}{r}dV(\mathbf{y})$$
$$+\iint_\Sigma\frac{l_i}{r}\left[\frac{\partial}{\partial y_i}\left(\rho u_iu_j-\tau'_{ij}+p\delta_{ij}\right)\right]dS(\mathbf{y}) \qquad (2.69)$$
$$+\frac{\partial}{\partial x_i}\iint_\Sigma\frac{l_i}{r}\left[\rho u_iu_j-\tau'_{ij}+p\delta_{ij}\right]dS(\mathbf{y})$$

which is Curle's [23] result. Eq. (2.69) states that the acoustic pressure is directly radiated from a volume distribution of quadrupoles plus a contribution from motions and stresses existing on any surfaces present. The surface effect can be interpreted as a distribution of dipoles as can be deduced by comparing the surface integrals to the model of the dipole in Eq. (2.29). Again we see that the operation leading to Eq. (2.37b), i.e., the determination of the gradient of the free space Green's function in the direction of the vector between the source centers, is a limiting form of the operations on the surface integrals in Eq. (2.69). These integrals provide contributions that are proportional to the resultant fluid forces on the surfaces.

The momentum theorem, Eq. (2.41), rewritten as

$$l_i\frac{\partial}{\partial y_i}\left[\rho u_iu_j-\tau'_{ij}+p\delta_{ij}\right]=-l_i\frac{\partial(\rho u_i)}{\partial t} \qquad (2.70)$$

is used to change the integrand in the first surface integral of Eq. (2.69). Thus

$$4\pi c_0^2(\rho(\mathbf{x},t)-\rho_0)=\frac{\partial^2}{\partial x_i\partial x_j}\iiint_V\frac{[T_{ij}]}{r}dV(y)-\iint_\Sigma\frac{l_i}{r}\left[\frac{\partial(\rho u_i)}{\partial t}\right]dS(\mathbf{y})$$
$$+\frac{\partial}{\partial x_i}\iint_\Sigma\frac{l_j}{r}\left[\rho u_iu_j-\tau'_{ij}+p\delta_{ij}\right]dS(\mathbf{y}) \qquad (2.71)$$

The second dipole term is thus a contribution from the acceleration of the body surface in a direction normal to its surface.

Therefore the sound pressure is the resultant of three contributions: the radiation from the turbulent domain, radiation due to the instantaneous contiguous surface motion with phase cancellations included, and radiation from a distribution of forces acting on the region. Eq. (2.71) could as well have been derived from direct use of Eq. (2.53), but with the source term representing a superposition of monopole, dipole, and quadrupole sources. An equivalent inhomogeneous wave equation may accordingly be written with Eq. (2.6)

$$\frac{1}{c_0^2}\frac{\partial^2 p_a}{\partial t^2}-\nabla^2p_a=+\frac{\partial q}{\partial t}-\frac{\partial F_i}{\partial y_i}+\frac{\partial^2T_{ij}}{\partial y_i\partial y_j} \qquad (2.72)$$

where q is the rate of mass injection per unit volume, F_i is the ith component of the force vector per unit volume normal to i, and T_{ij} is the stress tensor. In using Eq. (2.72) q denotes all the mass flux and F_i denotes all the forces acting on the region. In the context of the integral form above, F_i includes all of the $l_j(\rho u_i u_j + \tau'_{ij} + p\delta_{ij})$. As we shall see below *the efficacy of Curle's result and of* Eq. (2.72) *in describing the nature of flow-induced noise is particularly realized when the surface in question is much smaller than an acoustic wavelength; then $\partial F_i/\partial y_i$ represents a divergence of the local interaction force per unit volume exerted on the fluid at the body surface.*

2.4.2 Illustration I of Curle's Equation: Radiation from a Concentrated Hydrodynamic Force

We consider here the far field of an acoustically compact rigid body on which is imposed a known time-dependent surface pressure field. This field results in a time-dependent concentrated force being applied to the fluid. The geometry of Fig. 2.7B applies to this problem, except that now S_1 is a rigid surface on which $u_i l_i = u_n = 0$. We may envision the surface pressure field as generated by a flow around the surface that is presumed to be steady, generating a steady-state dynamic pressure on S_1. For a closed rigid surface Eq. (2.71) reduces to

$$4\pi c_0^2(\rho(\boldsymbol{x},t)-\rho_0)=\frac{\partial}{\partial x_i}\iint_{S_1} l_i p\left(\boldsymbol{y},t-\frac{r}{c_0}\right)\frac{dS(\boldsymbol{y})}{r} \tag{2.73a}$$

if we ignore the viscous surface stresses τ'_{ij} and the contribution of Reynold's stresses in the wake. (This latter simplification is considered in more detail in Chapter 4, Dipole Sound from Cylinders.)

For any surface whose dimension is substantially less than an acoustic wavelength, Eq. (2.73) reduces to give the acoustic pressure fluctuation as

$$4\pi c_0^2\left(\rho(\boldsymbol{x},t)-\rho_0\right)=-\frac{\partial}{\partial x_i}\left[\frac{F_i}{r}\right] \tag{2.73b}$$

a result due to Lamb in the 1916 edition of Ref. [45]. With Eqs. (2.6) and (2.55a) it expands to

$$4\pi p_{\mathrm{a}}(\boldsymbol{x},t)=\frac{1}{c_0}\frac{x_i}{r^2}\left[\frac{\partial F_i}{\partial t}\right] \tag{2.74}$$

as the radiated pressure resulting from the concentrated force F_i exerted on the fluid. Note that the pressure is created in the direction of the force since $x_i = r\cos\phi$, where ϕ is the angle measured from the force direction; see Fig. 2.2C.

In the special case that the force is simple harmonic, then

$$[F_i(t)] = F_i e^{-i\omega(t-r/c_0)}$$

and

$$\frac{\partial F_i(t)}{\partial t} = -i\omega F_i e^{-i\omega(t-r/c_0)}$$

so that the simple harmonic acoustic pressure in the far field has the canonical dipole form

$$p_a(\mathbf{x}, t) = -ik_0 F_i \frac{x_i}{r} \frac{e^{-i\omega(t-r/c_0)}}{4\pi r} \tag{2.75}$$

Eqs. (2.74) and (2.75) are equivalent forms of the classical aerodynamic dipole sound from a compact source and we shall have cause to use them many times in this book. Application of these results is limited to a sufficiently precise knowledge of the force field F_i as well as to low enough frequencies that the spatial extent of the radiator is less than one-quarter of the acoustic wavelength.

Note that we could have derived this result by replacing $\sigma(\mathbf{y}, t)$ in Eq. (2.50) by the concentrated force gradient; i.e.,

$$\nabla^2 p_a - \frac{1}{c_0^2}\frac{\partial^2 p_a}{\partial t^2} = \frac{\partial F_i(t)}{\partial y_i}\delta(\mathbf{y} - \mathbf{y}_0)$$

Incorporating Eqs. (2.53) and (2.56) yields Eq. (2.74) directly. The result could also have been derived from Eq. (2.72).

2.4.3 Illustration II of Curle's Equation: Radiation from a Heaving Sphere

The far-field radiation from a sphere of radius a heaving harmonically at very low frequencies such that $\omega a/c_0 \ll 1$ is dominated by two terms of Eq. (2.71). In the case of an inviscid fluid these are

$$4\pi c_0^2(\rho(\mathbf{x}, t - \rho_0)) = -\iint_{\Sigma} \frac{l_i}{r_s}\left[\frac{\partial(\rho u_i)}{\partial t}\right] dS(\mathbf{y}) + \frac{\partial}{\partial x_i}\iint_{\Sigma} \frac{l_i}{r_s}[p] dS(\mathbf{y})$$

The first integral is due to the direct motion illustrated in Fig. 2.7B, the second integral is due to the local pressure field induced around the sphere by the fluid reacting to the vibration. This term is evaluated in a manner identical to that described above, and all that is needed now is to relate the forces on the fluid to the surface motion. As shown in Fig. 2.2B the motion of the center of the sphere is along the z axis and given by

$$U_z(t) = U_z e^{-i\omega t}$$

For such low-frequency oscillations and at low amplitudes, $U_z/c_0 \ll 1$, the quadrupole term is neglected since only weak Reynolds stresses are generated by the motion. The contribution p_{a_u} from the first surface term is therefore

$$4\pi p_{a_u}(\boldsymbol{x},t) = -\iint_{S_1} \frac{l_i}{r_s}\left[\rho_0 \frac{\partial u_i(y,t)}{\partial t}\right] dS(y)$$

$$= +\iint_{S_1} \frac{1}{r_s}\left[\rho_0 \frac{\partial u_n(y,t)}{\partial t}\right] dS(y)$$

$$= -\iint_{S_1} \frac{i\omega\rho_0\, U_z \cos\phi(\mathbf{y})}{r_s} e^{-i\omega(t-r_s/c_0)} dS(\mathbf{y})$$

since $l_i U_i = u'_n$ is the fluctuating velocity normal to the surface and directed out of the fluid. The normal to the surface directed into the fluid is $u_n = -u'_n$. The retardation effect is approximated by writing for $r \ll a$

$$r_s \simeq r - a\cos(\phi - \phi(\mathbf{y}))$$

so that the exponential becomes

$$e^{-i\omega(t-r_s/c_0)} \simeq e^{-i\omega(t-r/c_0)}[1 - ik_0\, a\cos(\phi - \phi(\mathbf{y}))] \tag{2.76}$$

where ϕ is the angle made by the position vector $\boldsymbol{x}$ with the direction of motion and $\phi(\mathbf{y})$ is that angle made with $\mathbf{y}$. This allows the acoustic pressure to be written

$$p_{a_u}(\boldsymbol{x},t) = \frac{-\rho_0 c_0 k_0^2 a^3 U_z}{4\pi r}\cos\phi e^{-i(\omega t - k_0 r)} 2\pi \int_0^{\pi} \cos^2\phi(y)\sin\phi(y)d\phi(y)$$

$$p_{a_u}(\boldsymbol{x},t) = \frac{-1}{3}\rho_0 c_0 k_0^2 U_z\, a^3 \cos\phi \frac{e^{-i(\omega t - k_0 r)}}{r}$$

The contribution from the reaction forces p_{a_p} from the second surface integral is written from Eq. (2.73) directly. The simple harmonic motion provides a simple harmonic force f_2 *on the liquid* giving

$$4\pi p_{a_p}(\boldsymbol{x},t) = -ik_0 f_2 \cos\phi \frac{e^{-i(\omega t - k_0 r)}}{r}$$

The added mass of the entrained fluid is

$$m_a = \frac{2}{3}\rho_0 \pi a^3$$

Since the sphere is acoustically compact, the fluid impedance $Z_z = Z_2$ preceding Eq. (2.33) is principally mass-like.

The force (f'_2) acting *on the sphere* is accordingly

$$(f'_2) = -i\omega\left(\frac{2}{3}\rho_0\pi a^3\right)U_z$$

so that since $f_2 = -f'_2$

$$p_{a_p}(\mathbf{x},t) = -\frac{1}{6}\rho_0 c_0 k_0^2 U_z a^3 \cos\phi \frac{e^{-i(\omega t - k_0 r)}}{r}$$

The net radiated sound pressure from the heaving sphere is sum of both contributions, $p_a = p_{a_u} + p_{a_p}$,

$$p_a(\mathbf{x},t) = -\frac{1}{2}\rho_0 c_0 k_0^2 a^3 U_z \cos\phi \frac{e^{-i(\omega t - k_0 r)}}{r}$$

This result is identical to that obtained in Section 2.2; compare Eq. (2.32) and take the limits as $k_0 a \to 0$ and $k_0 r \to \infty$.

Note that if we had carelessly ignored the retardation effect, the resulting surface integral involving U_z would have been exactly zero. Thus the dipole radiation emerges here as a second-order effect of the motion of the sphere; it is a mathematical consequence of the series expansion of Eq. (2.73a) and a physical consequence of the fact that the excess pressure and suction at opposite poles of the sphere do not instantaneously cancel identically because of acoustic delay around the surface.

2.4.4 Powell's Reflection Theorem

In Fig. 2.8, we specify the bounding surface to come in contact with the source region; we have segmented Σ as shown in Fig. 2.8B. Now, we have dissected Σ to be

$$\sum = S_0 + S_1 + S_2$$

where S_2 is the control surface of the region of interest and S_1 is an impedance boundary that can reflect sound and that intersects S_2 far from V_0; S_0 is adjacent to the disturbance region, and it does not necessarily have the same impedance as S_1. As before we select S_2 to lie far enough from V_0 that disturbances have not yet reached S_2. Since $T_{ij} = 0$ outside V_0, it vanishes on S_1 but not on S_0. Therefore, using Eqs. (2.57) and (2.70), Eq. (2.69) can be written (since $l_i u_i = l_n u_n = u_n$)

$$\begin{aligned} 4\pi p_a(\mathbf{x},t) = {} & \frac{\partial^2}{\partial x_i \partial x_j}\iiint_{V_0} \frac{[T_{ij}]}{r} dV(\mathbf{y}) \\ & - \iint_{S_0} \frac{l_n}{r}\left[\rho\frac{\partial u_n}{\partial t}\right] dS(\mathbf{y}) + \frac{\partial}{\partial x_i}\iint_{S_0} \frac{1}{r}\left[\rho u_i u - \tau'_{in} + p\delta_{in}\right] dS(\mathbf{y}) \\ & - \iint_{S_1} \frac{l_n}{r}\left[\rho\frac{\partial u_n}{\partial t}\right] dS(\mathbf{y}) + \frac{\partial}{\partial x_i}\iint_{S_1} \frac{l_i}{r}[p_a(\mathbf{y},t)] dS(\mathbf{y}) \end{aligned} \qquad (2.77)$$

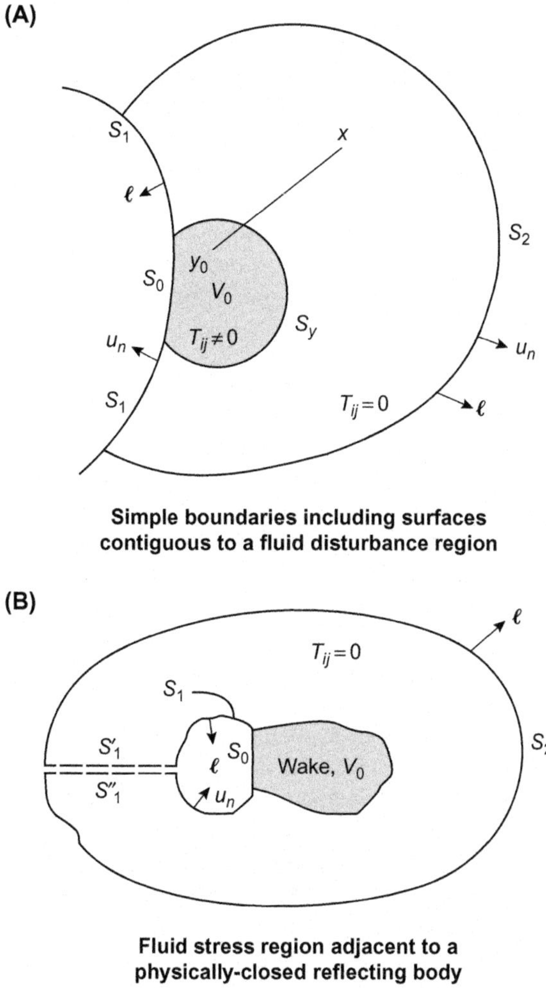
(A)
S_1
x
ℓ
S_2
y_0
S_0
V_0
S_y
u_n
$T_{ij} \neq 0$
u_n
S_1
$T_{ij}=0$
ℓ
Simple boundaries including surfaces contiguous to a fluid disturbance region
(B)
ℓ
$T_{ij}=0$
S_1
S'_1
S_0
S_2
ℓ
Wake, V_0
S''_1
u_n
Fluid stress region adjacent to a physically-closed reflecting body

FIGURE 2.8 Surface geometries used to illustrate Powell's analysis of the influences of surfaces on radiation.

This expression, which is really just a restatement of the Eq. (2.71), emphasizes the multiple effects of the adjacent boundary.

This integral relation of Eq. (2.77), originally derived by Powell [24], is an extension of Curle's [23] result that emphasizes the influences of boundaries adjacent to the region of the turbulent stresses. The result is general, and it includes all of the acoustic and hydrodynamic effects on the fluid region. Although the result applies for boundaries of any shape, it will be seen in later chapters that evaluations of Eq. (2.77) are not trivial. That is, in Chapter 9, Response of Arrays and Structures to Turbulent Wall Flow and

Random Sound this integral relationship will be further examined for application to turbulent boundary layer noise. When the turbulent region encloses a physical body, as shown in Fig. 2.8B, the surface of the body consists of S_0, and part of S_1 which is extended on both sides of a strip of vanishing thickness to connect to control surface S_2. It is clear the contributions from the connecting surfaces S_1' and S_1'' must cancel. This analytical situation could apply to noise from a wake behind a body in flow. On the boundary contiguous to the source volume, a contribution is emitted by the acceleration in the direction normal to the surface, $\partial u_n/\partial t$. Another contribution arises from the distributed stresses on the surface, which are $l_j\left[\rho u_i u_j - \tau_{ij}' + p\delta_{ij}\right] = \rho u_i u_n - \tau_{in}' + p\delta_{in}$. The fluid pressure fluctuations include both hydrodynamic and acoustic contributions, and they account for the normal stresses on the surface. Viscous (τ_{ij}') and Reynolds stresses ($\rho u_i u_j$) that involve normal motions and gradients on the surface will also radiate. The contributions from the adjacent surface S_1 involve the normal motion of that surface as well as the scattering of the acoustic pressure p_a

Another situation could arise in which surfaces S_1' and S_1'' coincide with the physical boundaries of a wedge. In this case the surface $S_1 + S_0$ in Fig. 2.8 reduces to a point at the apex of the wedge and the turbulent region could be situated adjacent to the wedge surface. This problem includes acoustic diffraction about the sharp apex and will be discussed further in Chapter 11 *Noncavitating Lifting Surfaces.*

A situation which consists of V_0 being adjacent to a *plane* boundary has application to boundary layer induced noise. This problem was considered by Powell [24], and its result has important general implications for any flow region adjacent to a boundary of large radius of curvature.

An illustration of Powell's problem is given in Fig. 2.9. The plane surface $S_0 + S_1$ separates a real fluid region from its virtual image, denoted by primes. This image system is provided to account for reflections at the boundary $S_0 + S_1$. For the image stress system, T_{ij}', enclosed by the surface $S_0' + S_1' + S_2'$ the acoustic field *outside* at position $\mathbf{x}$ vanishes identically. Velocity fluctuations in the plane of the surface are designated by u_s. Thus using Eq. (2.77) we have

$$0 = \frac{\partial^2}{\partial x_i' \partial x_j'} \iiint_{V_0'} \frac{\left[T_{ij}'\right]}{r'} dV(\mathbf{y}')$$

$$- \iint_{S_0'} \frac{l_n'}{r} \left[\rho' \frac{\partial u_n'}{\partial t}\right] dS(\mathbf{y}') + \frac{\partial}{\partial x_i} \iint_{S_0'} \frac{1}{r'} \left[\rho u_i' u_n' - \tau_{in}' + p' \delta_{in}\right] dS(\mathbf{y}') \quad (2.78)$$

$$- \iint_{S_1'} \frac{l_n'}{r} \left[\rho' \frac{\partial u_n'}{\partial t}\right] dS(\mathbf{y}') + \frac{\partial}{\partial x_i} \iint_{S_1'} \frac{l_i'}{r'} \left[p_a'(\mathbf{y}', t)\right] dS(\mathbf{y}')$$

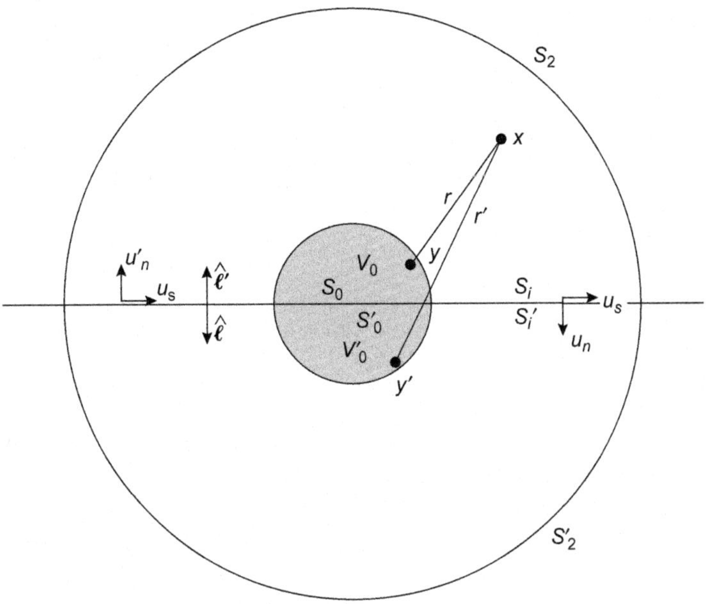

FIGURE 2.9 An illustration for Powell's reflection theorem.

Eq. (2.78) will be added to Eq. (2.77). On the boundary and in the volume $V_0 + V_0'$

$$l = -l', \quad u_n = -u_n', \quad u_s = u_s'$$
$$p = p', \quad \frac{\partial}{\partial y_n} = -\frac{\partial}{\partial y_n'}$$

but

$$l_n u_n = l_n' u_n'$$

Thus we have

$$\frac{\partial}{\partial x_i}\int_S \left[\rho u_i u_n - \tau_{in}'\right]\frac{dS(\mathbf{y})}{r} = \frac{\partial}{\partial x_i}\int_{S'} \left[\rho u_i' u_n' - (\tau_{in}')'\right]\frac{dS(\mathbf{y}')}{r'} i \neq j$$

and

$$\frac{\partial}{\partial x_n}\int_S \left[\rho u_n^2 - \tau_{nn}' + p\right]\frac{dS(\mathbf{y})}{r} = -\frac{\partial}{\partial x_n}\int_{S'} \left[\rho u_n^2 -, \tau_{nn}', +, p\right]'\frac{dS(\mathbf{y}')}{r'}$$

so that

$$4\pi p_a(\mathbf{x},t) = \frac{\partial^2}{\partial x_i \partial x_j}\iiint_{V_0+V_0'}\frac{[T_{ij}]}{r}dV_0(\mathbf{y})$$
$$-\iint_{S_0}\frac{2}{r}\left[\rho\frac{\partial u_n}{\partial t}\right]dS(\mathbf{y}) - \iint_{S_1}\frac{2}{r}\left[\rho\frac{\partial u_n}{\partial t}\right]dS(\mathbf{y}) \tag{2.79}$$
$$+\frac{\partial}{\partial x_s}\iint_{S_0}\frac{2}{r}\left[\rho u_s u_n - \tau'_{sn}\right]dS(\mathbf{y})$$

When the surface $S_1 + S_0$ is rigid and smooth and the fluid is in viscid ($\tau'_{sn} = 0$), we have the simple result that

$$4\pi p_a(\mathbf{x},t) = \frac{\partial^2}{\partial x_i \partial x_j}\iiint_{V_0+V_0'}\frac{[T_i]}{r}dV_0(\mathbf{y}) \tag{2.80}$$

which is the statement of Powell's [24] reflection theorem. This theorem states that:

> *The pressure dipole distribution on a plane, infinite, and rigid surface accounts for the reflection in that surface of the volume distribution of acoustic quadrupole generators of a contiguous inviscid fluid flow, and for nothing more, when these distributions are determined in accordance with Lighthill's concept of aerodynamic noise generation and its natural extension.*

We saw a trivial example of this in Fig. 2.4 in which the effect of a planar pressure release surface was to introduce a destructive interference between a monopole and its image when $k_0H < \pi/2$ and the surface did not introduce any other sources.

Powell goes on to state that the result, Eqs. (2.79) and (2.80), is independent of the Mach number of the mean flow as well as the wavelength of the sound. The stress tensor itself accounts for all effects of refraction and scattering by fluid inhomogeneities. The above result serves to emphasize the confusion that could arise from the interpretation of Curle's result Eq. (2.71) as indicating that the presence of a surface contiguous to a turbulent region results in the generation of three physically distinct acoustic sources. This is patently not the case when the contiguous surface is large compared with an acoustic wavelength whether or not the surface may react to the stresses induced on it by the turbulence.

In more general cases, Eq. (2.79) isolates the importance of dipoles oriented in the plane of the surface whose strengths become doubled by the plane. For example, Ffowcs Williams [26] considered the case of a flow source region over nonrigid, plane, homogeneous boundaries. To consider the simple case of a very soft boundary, the difference between Eqs. (2.77) and (2.78) is taken, which gives

$$4\pi p_a(\boldsymbol{x}, t) = \frac{\partial}{\partial x_i \partial x_j} \iiint_{V_0} \frac{[T_{ij}]}{r} dV_0(\boldsymbol{y})$$
$$+ \frac{\partial}{\partial x_n} \iint_{S_0} \frac{2}{r} \left[\rho u_n^2 - \tau'_{nn} + p\right] dS(\boldsymbol{y}) \tag{2.81}$$
$$+ \frac{\partial}{\partial x_n} \iint_{S_1} \frac{2}{r} [p] dS(\boldsymbol{y})$$

If the surface is limp enough so that no normal stresses may be maintained, then $p = 0$ (i.e., it is pressure release) on S_0. The surface S_1 Ffowcs Williams takes far enough from the source region that the integration over it vanishes. The sound field then consists of the interference of the primary source field with its negative image except for the addition of the term involving the induced surface motion (as noted above). This term Ffowcs Williams speculates is second order. For more complicated boundaries whose impedances are intermediate between hard and soft, Ffowcs Williams shows that the effect is still only to modify the sound field by adding to the primary wave field, given by the integral over V_0, a reflected wave, given above by the integrals over V'_0. A reflection coefficient appropriate to the surface impedance therefore causes a phase shift but no resonances. Thus the sound resulting from a turbulent flow over any *plane homogeneous* surface is essentially quadrupole or higher order with no alteration in the physical mechanism of the radiation, barring the possible contribution of shear stress dipoles. This has broad implications since what is required for enhancement of the sound field is inhomogeneities in the surface impedance (scatterers) or inhomogeneities in the surface stresses. Applications of Eq. (2.81) to any real flow–surface interaction must start with careful consideration of first principles in order to identify unambiguously the correct monopole–dipole–quadrupole analogies. This follows from the essential (and occasionally subtle) unbalances in magnitudes and phases of disturbances that constitute the sources. These implications will be discussed at greater length in Chapter 9, Response of Arrays and Structures to Turbulent Wall Flow and Random Sound.

Eqs. (2.80) and (2.81), bring out an interesting aspect of imaging multipole sources. Consider the thickness of a stress layer in a direction normal to the surface to be much smaller than an acoustic wavelength. Then, using the idealizations for the lateral and longitudinal quadrupoles shown in Fig. 2.5, it is easy to see that a rigid surface causes destructive interference of lateral quadrupoles (call the result an octupole!) but a doubling of the sound for longitudinal quadrupoles. The converse holds true for quadrupole sources near a soft boundary. Similarly, one can deduce alternative reinforcements for dipoles near either hard or soft surfaces. Therefore in Eq. (2.81) the volume integrals *do not* cancel, but multipoles of certain orientations may be enhanced over others of different orientations by the surface reflectivity.

2.5 EFFECTS OF SOURCE MOTION ON FLOW-INDUCED NOISE

The preceding discussions have examined the acoustic fields of sources and mean positions of surfaces which are stationary with respect to the observer. However a large number of applications involve translation or convection of sources. Prime examples include:

1. "Gutin" sound, see Ref. [27], from propellers in clean, inviscid, steady flow in which the motion of the aerodynamic loading pressure, which is steady and fixed with the rotating blade, is nonetheless rotating relative to a distant fixed observer, see Chapter 12, Noise from Rotating Machinery.
2. Jet noise in which the sound sources are convected relative to the observer by the local mean flow in the jet, see Chapter 3, Shear Layer Instabilities, Flow Tones, and Jet Noise.
3. Motion of turbulent sources fixed to a surface which is moving relative to a fixed-position observer.
4. Thickness noise occurring with high rotation-rate propellers due to the expansion of air by volumetric displacement of the moving blade.

This subject is complicated because motion of the source influences the propagation of sound through well-known Doppler amplification and it can affect the source mechanism, as well. In this section we shall discuss some of the basic features of motion and illustrate some examples so that the reader is able to roughly anticipate its effects. Greater detailed derivations of the theory are provided by Howe [12], Goldstein [28], and Glegg and Devenport [29]. Perhaps the most general treatment of the effect of source motion, and that forming the starting point for the work that follows 1968 is that of Ffowcs Williams and Hawkings [30] which provides extensions over the first systematic treatment of Lowson [31]. Dowling [32] and Crighton et al. [33] provide useful examples which bring clarity to distinctions between propagation effects and the possible alterations of source strength.

In the original and most general context [30] the theory accounts for sources which are not rigid, may not be impervious to flow, and are not in steady motion relative to observer. In contrast, the Lighthill–Curle theory (e.g., Eqs. (2.72) and (2.81)) holds for bodies which are stationary with respect to an observer and which may or may not be rigid. The Ffowcs Williams–Hawkings equation is

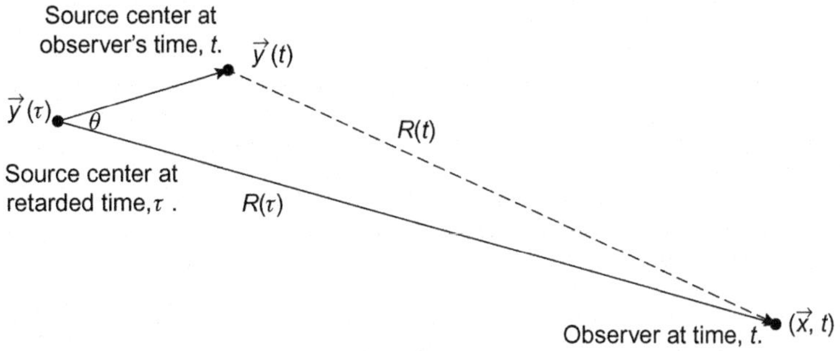

FIGURE 2.10 Coordinate system used to describe the moving source–observer geometry.

$$\begin{aligned} 4\pi c_0^2(\rho(\boldsymbol{x},t)-\rho_0) &= \frac{\partial}{\partial x_i \partial x_j}\iiint_{V_0}\frac{[T_{ij}]}{r}dV_0(\eta) \\ &+\frac{\partial}{\partial x_i}\iint_{S_0}\frac{1}{r|1-M_r|}\left[\rho u_i(u_j-u_j)-\tau'_{ij}+p\delta_{ij}\right]n_j dS(\eta) \\ &-\frac{\partial}{\partial t}\iint_{S_1}\frac{2}{r|1-M_r|}\left[\rho u_j\right]n_j dS(\eta) \end{aligned} \tag{2.82}$$

where it is assumed that the amplitude of motion is small with respect to the size of the moving body and that the ambient fluid density is constant. Again, the [] bracket denotes that the function within is evaluated at a retarded time, $\tau = t - r(\boldsymbol{x},t)/c_o$, the variable η represents a moving coordinate system that is fixed with the body, the center of the body is assumed to translate at velocity $c_0\mathbf{M}$. The coordinate $\boldsymbol{y}(\tau) = \boldsymbol{y}(t) - \boldsymbol{U}\left(r(\boldsymbol{x},t)/c_0\right)\boldsymbol{y}(\tau) = \boldsymbol{y}(t) - \boldsymbol{M}\,r(\boldsymbol{x},t)$ is the source's initial position; $\boldsymbol{y}(t)$ is the source's position at the observer's time, t; the third term is the source's position change that has taken place over the propagation time from the source to the observer, $r(\boldsymbol{x},t)/c_0$. Fig. 2.10 shows the geometry. The Mach number projected to the observer direction is subsonic with $M_r = M\cos\theta$.

The integration is therefore over the surface, S_0, and volume, V_0, attached to the moving body. When the surface is rigid, yet moving

$$\begin{aligned} 4\pi c_0^2(\rho(\boldsymbol{x},t)-\rho_0) &= \frac{\partial}{\partial x_i \partial x_j}\iiint_{V_0}\left[\frac{T_{ij}}{r(1-M_r)}\right]dV_0(\eta) \\ &+\frac{\partial}{\partial x_i}\iint_{S_0}\left[\frac{1}{r(1-M_r)}\left(p\delta_{ij}-\tau'_{ij}\right)\right]n_j dS(\eta) \\ &-\frac{\partial}{\partial t}\iint_{S_1}\left[\frac{2}{r(1-M_r)}\rho u_j\right]n_j dS(\eta) \end{aligned} \tag{2.83}$$

These relationships are the motion-dependent analogs of Eqs. (2.69) and (2.71) in the inclusion of the Doppler factor, $(1 - M_r)$ which we have regarded as an effect of "rigid-body" motion distinct from any the effects that motion might have on the strength of the source. These effects add complexity to the field at the observer as illustrated in examples presented in Refs. [12,30,32]. Each of the component terms in the above equations includes the derivatives out front of the integrals which, when carried out, cause additional powers of the Doppler factor. It is useful to provide some detail of one example, that being the case of the dipole term, for which we let $\left[p\delta_{ij} - \tau'_{ij}\right] l_j = F_i(t)\delta(y(t) - \eta(t))$ be a compact force dipole then

$$4\pi c_0^2(\rho(\mathbf{x},t) - \rho_0) = \frac{\partial}{\partial x_i}\iint_{S_0} \frac{F_i(\tau = t - |\mathbf{x} - \mathbf{y}(\tau)|/c_0)\delta(\eta - \mathbf{y}(\tau))}{|\mathbf{x} - \mathbf{y}(\tau = t - |\mathbf{x} - \mathbf{y}(\tau)|/c_0)||1 - M_r|} n_j dS(\eta)$$

or, letting $\tau = t - |\mathbf{x} - \mathbf{y}(\tau)|/c_0$ and $r(t) = |\mathbf{x} - \mathbf{y}(\tau)|$ we can write this equation in a more compressed form

$$4\pi c_0^2(\rho(\mathbf{x},t) - \rho_0) = \frac{\partial}{\partial x_i}\left(\frac{F_i(\tau)}{r(t)(1 - M_r)}\right)$$

and expand the derivative with respect to x_i

$$\frac{\partial}{\partial x_i}\left(\frac{F_i(\tau)}{r(t)(1 - M_r)}\right) = \frac{\partial F_i}{\partial \tau}\frac{\partial \tau}{\partial x_i}\frac{1}{r(t)(1 - M_r)} - F_i\left(\frac{1}{r^2}\frac{\partial(r(1 - M_r))}{\partial x_i}\right)$$

The second term on the right will vanish under an assumption that the observer is in the far field and that the motion of the source is steady with respect to the observer. That leaves the first term which expands to

$$\frac{\partial}{\partial x_i}\left(\frac{F_i(\tau)}{r(t)(1 - M_r)}\right) = F_i(\tau)\left(\frac{\partial \tau}{\partial x_i}\right)\frac{1}{r(t)(1 - M_r)}$$

but

$$\frac{\partial \tau}{\partial x_i} = \frac{-1}{c_0}\frac{r_i}{r}\left(1 - U_i\frac{\partial \tau}{\partial x_i}\right) = \frac{-1}{c_0}\cos\theta_i(1 - c_0 M_r)$$

On rearrangement we obtain $\frac{\partial \tau}{\partial x_i}$ and substitute it to obtain

$$\frac{\partial}{\partial x_i}\left(\frac{F_i(\tau)}{r(t)|1 - M_r|}\right) = F_i(\tau)\left(\frac{-\cos\theta_i}{c_0(1 - M_r)}\right)\frac{1}{r(t)(1 - M_r)}$$

giving the desired result

$$4\pi c_0^2(\rho(\mathbf{x},t) - \rho_0) = \frac{\mathbf{F}(t - r/c_0)\cdot\mathbf{r}}{c_0 r^2(1 - M_r)^2} \tag{2.84}$$

in the far field, $r \to \infty$.

For a compact convected quadrupole we apply the derivatives with respect to i and j to obtain

$$4\pi c_0^2\left(\rho(\boldsymbol{x},t)-\rho_0\right)=\frac{r_i r_j}{r^3}\frac{1}{c_0^2}\iint_V\int\frac{\partial^2 T_{ij}(\boldsymbol{y},\tau=t-r/c_0)}{\partial t^2}\frac{dV(\boldsymbol{y})}{(1-M_r)^3} \tag{2.85}$$

in the far field, $r\to\infty$. Both these expressions can be found in Ref. [12] with similar derivations. Eqs. (2.84) and (2.85) apply to compact sources in non-accelerated motion with respect to the observer in the far field assuming that the motion does not affect the source characteristics. These assumptions are that same as often made with turbulent sources in which the turbulence for both the source at rest and in motion is the same and that the correlation volume of the source is small.

The convection of other compact sources, whose source distribution and strengths are unaffected by source convection give rise to similar inverse powers of Doppler factor making the ratio of sound pressure with motion to that without motion of order

$$\frac{\{p_a(\boldsymbol{x},t)\}_M}{\{p_a(\boldsymbol{x},t)\}_{M=0}}\sim\frac{1}{(1-M_r)^n} \tag{2.86}$$

where, for this elementary theory, n is 2 for a convected pulsating dipole and 3 for a convected quadrupole [12]. These results ignore any potential field interaction between the source and the mean flow that add additional sources.

Benchmark examples of more complex, yet notionally-simple, sources are given exact solutions by Dowling [32,33] who has examined the sound produced by acoustically-compact spheres in volumetric breathing pulsation or in heaving rigid-body oscillation each with a subsonic radial surface velocity. In the case of the breathing-mode oscillation the propagation effect appears as the Doppler factor $|1-M_r|^{-3}$ plus a new dipole source whose strength is proportional to the product of Mach number and the entrained inertia of the pulsing sphere. These sources are illustrated in the result

$$\begin{aligned}4\pi c_0^2(\rho(\boldsymbol{x},t)-\rho_0)&\sim\frac{\rho_0[\ddot{V}]}{(|1-M\cos\theta|)^3}\left\{1+\frac{x_i}{r}\alpha M\right\}+O(M^2)\\&=\frac{\rho_0[\ddot{V}]}{r(1-M\cos\theta)^{7/2}}+O(M^2)\end{aligned} \tag{2.87}$$

where $[\ddot{V}]$ is the retarded-time volumetric acceleration of the sphere, as in Eq. (2.24a,b) and $\alpha=1/2$ is the entrained mass coefficient related to the displaced volume of the sphere's oscillation in the moving stream.

Dowling's [32] second example is that of the translating sphere in rigid-body heave oscillation, which we examined as a stationary source in Sections 2.1.3.2 and 2.4.2. The field from a steady moving source of this type is the sum of two dipole components:

$$4\pi c_0^2(\rho(\boldsymbol{x},t)-\rho_0)=\frac{\rho_0 V_0[\ddot{U}_r](1+\alpha)\cos\theta}{c_0 r}\left\{\frac{1}{|1-M\cos\theta|^4}-\frac{\alpha}{1+\alpha}M\right\} \tag{2.88}$$

The factor $\alpha = 1/2$ (again, see Section 2.4.2) is the added mass coefficient. The term outside the bracket is the field from the stationary source, Eq. (2.32) and the terms within the bracket are the effects of motion of the source. The first term in the bracket mimics the convection amplification given by Eq. (2.84), but the Doppler factor is raised to the fourth power. The second term has no directivity and increases in proportion to Mach number. Both of these results which specify the source motion, but do not specify the dipole strength in the moving frame, disclose a stronger Doppler effect than when the convected dipole strength is specified.

Examples which we will consider in later chapters, and enumerated at the opening of this section include jet noise in Chapter 3, Shear Layer Instabilities, Flow Tones, and Jet Noise for which the result will be similar to Eq. (2.85) and thickness noise and the "Gutin" sound [27] in Chapter 12, Noise from Rotating Machinery. As these few examples have shown, the amplification effects of convection can be dependent on the specifics of the source with no single result being applicable to all cases. Accordingly an overall a rough order of magnitude indicator of when effects of source convection might become relevant, without considering details, can be inferred from the correction given for steadily convecting quadrupoles which is also typical of subsonic jet noise $n = 3$ is shown in Fig. 2.11. Beyond the selected range of angles and Mach numbers, other effects such as refraction will further alter magnitudes. However, this chart gives an indication of when convection may or may not be important. It can be seen that effects in excess of 3 dB might be observable at within 45° of grazing at Mach numbers above about 0.125.

2.6 POWELL'S THEORY OF VORTEX SOUND

2.6.1 General Implications

The formulation of a region of eddy motion as an acoustic source was a major step toward a physical understanding of turbulence-induced noise.

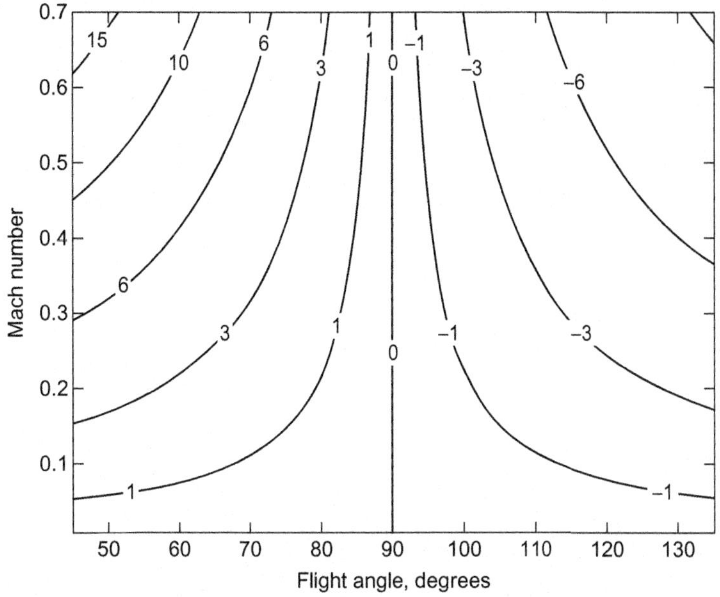

FIGURE 2.11 Doppler factor $-20 \log[1-M \cos \theta]^3$ for angles between $\pm 45°$ of the normal to the flight direction.

However, the consideration of the acoustic sound pressure as a volume integral of the Reynolds stress sheds little light on the details of vortex dynamics that are noise producing. It seems that Lighthill was motivated toward a description of sound intensity given the statistical characteristics of the turbulent source region. It was necessary to use representations that could be substituted by measured velocity correlations and length scales. Furthermore, the identification of the acoustic character of the source field as a distribution of quadrupoles and the subsequent establishment of the effects of eddy convection on the acoustic pressure were directed at an understanding of the observed acoustic aspects of jet noise.

Powell [34], on the other hand, was apparently interested in the aerodynamic (hydrodynamic) aspects of the flow structure that caused the noise and he desired an understanding of what characteristics of the eddy motion actually produce the noise. From this perspective Powell examined the connection between vortex motion and sound generation. The identification of the formation of vortices in the flow as the fundamental noise-producing mechanism is the result of this analysis.

2.6.2 Derivation of the Wave Equation With Vortical Sources

In our discussion we shall first derive the source term in Powell's [32] form and then discuss its physical implications. As before, we consider the fluid motions to be isentropic. Powell makes use of the well-known vector identities:

$$\begin{aligned}\nabla\left(\frac{1}{2}u^2\right) &= u_j\frac{\partial u_i}{\partial x_j} - \left[\left(\frac{\partial u_k}{\partial x_i} - \frac{\partial u_i}{\partial x_k}\right)u_k - \left(\frac{\partial u_j}{\partial x_i} - \frac{\partial u_i}{\partial x_j}\right)u_j\right] \\ &= (\mathbf{u}\cdot\nabla)\mathbf{u} - (\nabla\times\mathbf{u})\times\mathbf{u} \\ &= (\mathbf{u}\cdot\nabla)\mathbf{u} - \boldsymbol{\omega}\times\mathbf{u}\end{aligned} \tag{2.89}$$

and

$$\nabla^2 u_i - \frac{\partial}{\partial x_i}\left(\frac{\partial u_j}{\partial x_j}\right) = -\frac{\partial}{\partial x_j}\left(\frac{\partial u_j}{\partial x_i} - \frac{\partial u_i}{\partial x_j}\right) - \frac{\partial}{\partial x_k}\left(\frac{\partial u_k}{\partial x_i} - \frac{\partial u_j}{\partial x_k}\right) \tag{2.90}$$

or

$$-\nabla^2\mathbf{u} + \nabla(\nabla\cdot\mathbf{u}) = \nabla\times(\nabla\times\mathbf{u}).$$

The curl of the velocity, $\nabla\times\mathbf{u}$, is the vorticity vector $\boldsymbol{\omega}$. These relationships are used to transform the equations of continuity and momentum to

$$\frac{\partial\rho}{\partial t} + (\mathbf{u}\cdot\nabla)\rho + \rho\nabla\cdot\mathbf{u} = 0 \tag{2.91}$$

and

$$\rho\frac{\partial\mathbf{u}}{\partial t} + \rho(\boldsymbol{\omega}\times\mathbf{u}) + \nabla\left(\frac{\rho u^2}{2}\right) = -\nabla p. \tag{2.92}$$

(Both the vector and the tensor notations have been shown in order to effect for the reader an easy transition from one notation to the other. In the following we shall make liberal use of vector notation in order to compress expressions dealing with the curl operation.) The combination of Eqs. (2.91) and (2.92) can be performed in the same manner as in Section 2.2 to obtain the wave equation for the density as

$$\frac{\partial^2\rho}{\partial t^2} - c_0^2\nabla^2\rho = \nabla\cdot\left\{\rho(\boldsymbol{\omega}\times\mathbf{u}) + \nabla\left(\rho\frac{u^2}{2}\right) - \mathbf{u}\frac{\partial\rho}{\partial t} - \frac{u^2}{2}\nabla p + \nabla(p - \rho c_0^2)\right\} \tag{2.93}$$

This equation is an analog to Lighthill's equation (2.46), with $\dot{q} = 0$, where the divergence term is identical to Lighthill's source term. The equation is also exact except for the neglect of the viscous stress tensor τ'_{ij}, but the above representation exposes the influence of changes in vorticity on the radiated density fluctuations. The term $\boldsymbol{\omega}\times\mathbf{u}$ incorporates the sound due to the stretching of vortex filaments by an imposed velocity $\mathbf{u}$. The term $\mathbf{u}\,\partial\rho/\partial t$ is a contribution caused by local convection of density disturbances, and it is generally second order when $|\mathbf{u}| \ll c_0$. The term $\nabla(p + \rho u^2/2)$ includes both the hydrodynamic as well as the acoustic pressure. Inside the source region if the flow is perfectly irrotational, i.e., $\boldsymbol{\omega} = 0$ everywhere, the Bernoulli equation for the hydrodynamic pressure in the source zone is

$$\frac{p}{\rho} + \frac{u^2}{2} + \frac{\partial \phi_h}{\partial t} = \text{const} \tag{2.94}$$

where ϕ_h is the fluid potential. Thus

$$\nabla\left(\frac{p}{\rho} + \frac{u^2}{2}\right) = -\frac{\partial}{\partial t}\nabla\phi_h$$

and since $\nabla^2\phi_h = 0$ only for irrotational incompressible flow this term may be legitimately neglected only when $\boldsymbol{\omega} = 0$. Regarding the acoustic pressure, we note that when $c^2{}_0 \gg u^2/2$ (as is found in most situations involving flow of low Mach number) the acoustic pressure exactly balances ρc_0^2 Therefore Eq. (2.93) can be now written to include only the first-order terms:

$$\nabla^2 p_a - \frac{1}{c_0^2}\frac{\partial^2 p_a}{\partial t^2} = -\nabla\cdot\left\{\rho(\boldsymbol{\omega}\times\mathbf{u}) + \nabla\left(p - c_0^2\rho + \rho\frac{u^2}{2}\right)\right\} \tag{2.95}$$

to give the wave equation for the acoustic pressure. Eq. (2.95) should be compared to Eq. (2.46) (with $p_a = (\rho - \rho_a)c_0^2$ and $q = 0$); it is essentially that which was derived by Powell, and, in terms of exactness, it departs from Lighthill's neglect of the viscous stresses, τ'_{ij}, and of terms for which $u \ll c_0$.

A similar form of Eqs. (2.93) and (2.95) was subsequently derived by Howe [35] for the somewhat more general cases of convected sources in nonisentropic flow; it is derived by combining the equations of continuity, momentum, and the first law of thermodynamics and redefining the enthalpy as the acoustic variable,

$$B = \int \frac{dp}{\rho} + \frac{1}{2}u^2$$

to obtain

$$\left\{\frac{D}{Dt}\left(\frac{1}{c_0^2}\frac{D}{Dt}\right) + \frac{1}{c_0^2}\frac{D\mathbf{u}}{Dt}\cdot\nabla - \nabla^2\right\}B = \nabla\cdot\{\boldsymbol{\omega}\times\mathbf{u} - T\nabla S\}$$
$$-\frac{1}{c_0^2}\frac{D\mathbf{u}}{Dt}\cdot\{\boldsymbol{\omega}\times\mathbf{u} - T\nabla S\} + \frac{D}{Dt}\left(\frac{T}{c_0^2}\frac{DS}{Dt}\right) + \frac{\partial}{\partial t}\left(c_p\frac{DS}{Dt}\right)$$

In this equation $D/Dt = \partial/\partial t + \mathbf{u}\cdot\nabla$, S is entropy, c_p is specific heat at constant pressure and T is temperature. A derivation of Howe's equation is outside the scope of this chapter, requiring a number of vector substitutions and the introduction of the more general state equation of the ideal gas than that used in Section 2.1.1.

In an isentropic, slowly convected subsonic source field, such that $|\mathbf{u}|/c_0 \to 0$ Howe's equation linearizes to

$$\frac{1}{c_0^2}\frac{\partial^2 B}{\partial t^2} - \nabla^2 B = \nabla \cdot \boldsymbol{\omega} \times \mathbf{u} \tag{2.96}$$

where the enthalpy is

$$B = p/\rho + \frac{1}{2}u^2$$

which closely resembles Powell's equation, but the "source term" includes only the vorticity. Eq. (2.96) may be easily derived for isentropic flow by using Eqs. (2.91) and (2.95) and neglecting all terms of fluctuating quantities multiplied by $p/\rho c^2{}_0$ or u/c_0. To the extent that $p/\rho \gg u^2$ in further linearization of low-speed flows, Eqs. (2.95) and (2.96) may be regarded as approaching each other.

2.6.3 The Physical Significance of the Vorticity Source

In order to appreciate the importance of the vorticity as a source of sound, Powell considered the generation of fluid disturbances by the motion of an isolated ring. An equivalence is sought between changes in vorticity and changes in fluid momentum that can be interpreted as dipole and quadrupole acoustic sources.

The well-known relationship for the incompressible fluid velocity $\mathbf{u(x)}$ ([25], article 149) generated by a vortex filament is

$$\mathbf{u}(\mathbf{x}) = -\frac{1}{4\pi}\oint \Gamma \frac{\mathbf{r} \times d\mathbf{l}(\mathbf{y})}{r^3} = -\frac{1}{4\pi}\oint \nabla_y\left(\frac{\Gamma}{r}\right) \times d\mathbf{l}(\mathbf{y}) \tag{2.97}$$

As shown in Fig. 2.12A the vector from the source point to the field point $\mathbf{r} = \mathbf{x} - \mathbf{y}$, $d\mathbf{l}(\mathbf{y})$ is an increment of the vortex filament of vorticity (note that $\boldsymbol{\omega} \times d\mathbf{l} \equiv 0$). By definition the *circulation* is

$$\Gamma = \oint_{C'} \mathbf{u} \cdot d\mathbf{C}'$$

where C' is a *closed* circuit in the fluid. As long as C' encircles a vortex filament, as shown in Fig. 2.12A, the circulation is nonzero. Otherwise it is identically zero. Stokes' theorems state that for a vector quantity $\mathbf{A}$ the line integral $d\mathbf{l}$ and the area integral over $\mathbf{S}$ enclosed by C are related by the following integral identities:

$$\oint_C d\mathbf{l} \cdot \mathbf{A} = \iint_S (d\mathbf{S} \times \nabla) \cdot \mathbf{A} = \iint_S d\mathbf{S} \cdot (\nabla \times \mathbf{A})$$

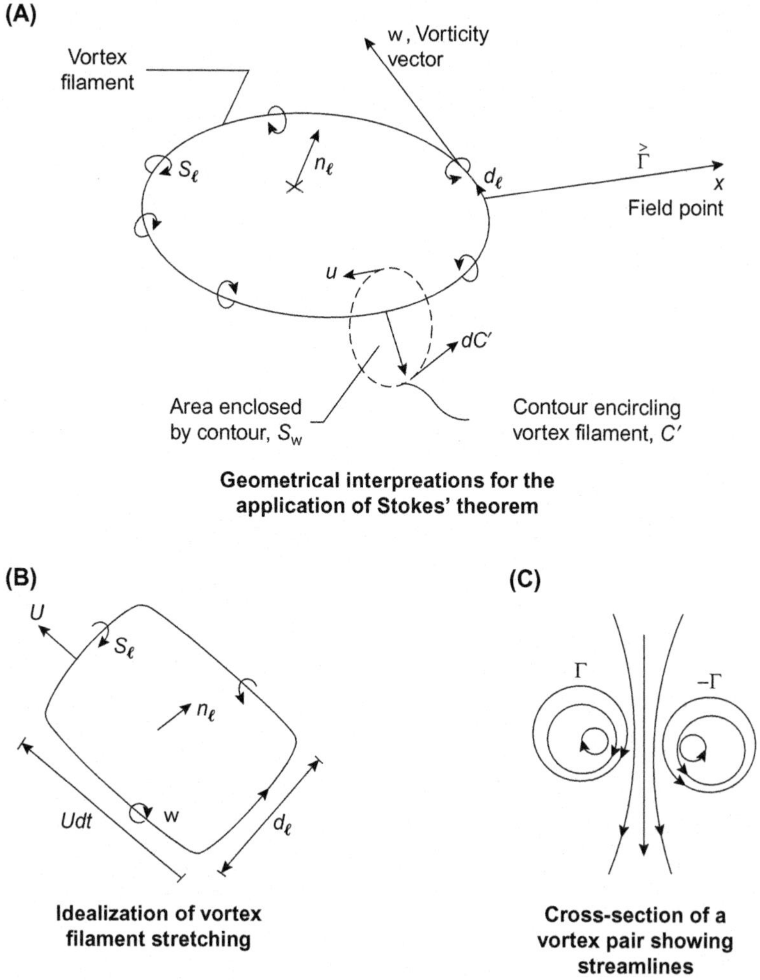

FIGURE 2.12 Geometry of a vortex filament.

and

$$\oint_C d\mathbf{l} \times \mathbf{A} = \iint_S (d\mathbf{S} \times \nabla) \times \mathbf{A}$$

Therefore

$$\Gamma = \oint_{C'} \mathbf{u} \cdot d\mathbf{C}' = \iint_{S_\omega} \mathbf{n}_\omega \cdot (\nabla \times \mathbf{u}) dS \tag{2.98}$$

$$= \iint_{S_\omega} \mathbf{n}_\omega \cdot \boldsymbol{\omega}\, dS = \int_{S_\omega} |\boldsymbol{\omega}| dS_\omega \tag{2.99}$$

where dS_ω is an element in the surface enclosed by the circuit C', $\mathbf{n}_\omega$ is the component of the normal vector to this surface in the direction of $d\mathbf{l}$ (or of $\boldsymbol{\omega}$), and S_ω is a component of S projected in the plane that is perpendicular to the vorticity vector $\boldsymbol{\omega}$. We shall assume that Γ is constant along the vortex filament $\mathbf{I}(\mathbf{y})$. Also, by Stokes's theorem we have

$$u(\boldsymbol{x}) = \frac{1}{4\pi}\oint d\mathbf{l}(\mathbf{y}) \times \nabla_y\left(\frac{\Gamma}{r}\right) = \frac{1}{4\pi}\iint_{S_\omega} (\mathbf{n}_l \times \nabla_y) \times \nabla_y\left(\frac{\Gamma}{r}\right) dS(\mathbf{y})$$
$$= -\frac{1}{4\pi}\iint_{S_l} (\nabla_y \times \mathbf{n}_l) \times \nabla_y\left(\frac{\Gamma}{r}\right) dS(\mathbf{y}) \tag{2.100}$$

where S_l is circumscribed by the vortex filament and $\mathbf{n}_l$ is the normal to the surface; see Fig. 2.12B. Now since by vector identity

$$(\mathbf{n}_l \times \nabla_y) \times \nabla_y\left(\frac{\Gamma}{r}\right) = \nabla_y\left(\mathbf{n}_l \cdot \nabla_y\left(\frac{\Gamma}{r}\right) - \mathbf{n}_l \nabla^2\left(\frac{\Gamma}{r}\right)\right),$$
$$= \nabla_y\left(\mathbf{n}_l \cdot \nabla_y\left(\frac{\Gamma}{r}\right)\right)$$

Eq. (2.100) becomes

$$u(\boldsymbol{x}) = \frac{1}{4\pi}\iint_{S_l} \nabla_y\left(\mathbf{n}_l \cdot \nabla_y\left(\frac{\Gamma}{r}\right)\right) dS(\overleftarrow{\mathbf{y}})$$
$$= -\frac{1}{4\pi}\nabla_x \iint_{S_l} -\mathbf{n}_l \cdot \nabla_x\left(\frac{\Gamma}{r}\right) dS(\mathbf{y}) \tag{2.101}$$

If the distance to the observation point is large compared to the dimension of the vortex ring r is essentially invariant throughout the integration over S_l. Eq. (2.101) becomes

$$u(\boldsymbol{x}) = \nabla_x\left\{\frac{1}{4\pi}\mathbf{n}_l \cdot \nabla_x\left(\frac{\Gamma S_l}{r}\right)\right\} \tag{2.102}$$

where ΓS_l is the *strength* of the vortex and n_l is the average normal to the surface area S_l. Now, Eq. (2.102) is of the form of a potential gradient $\mathbf{u} = \nabla_x(\phi)$, and so we recognize the term in brackets to be the far-field potential due to a concentrated vortex filament.

The important step in the analysis was Powell's recognition that the analog of Eq. (2.102) for a slightly compressible flow is

$$u(\boldsymbol{x},t) = \nabla_x \left\{ \frac{1}{4\pi} \nabla_x \cdot \frac{[\mathbf{n}_l \Gamma S_l]}{r} \right\} \tag{2.103}$$

where the term in brackets [ ] is now evaluated at the retarded time $t - r/c_0$. Eq. (2.103) holds as long as the wavelength of sound is much larger than the vortex ring, and it implies that the vortex streamlines in the slightly compressible flow are the same as if the fluid were incompressible. This is an important notion since it states that the flow field is established hydrodynamically and the sound is a by-product of the hydrodynamic motion.

The pressure disturbance in the far field is found by carrying out the indicated operations; and noting that $p_a = \rho_0 c_0 u_r$ where u_r is the velocity along $\mathbf{r}$

$$p_a(\boldsymbol{x},t) = \frac{\rho_0}{4\pi c_0} \frac{\mathbf{r}}{r^2} \cdot \left[\frac{\partial^2 \mathbf{n}_l \Gamma S_l}{\partial t^2} \right] \tag{2.104}$$

The far-field pressure is thus shown to be proportional to the time differential of the rate of change of the vortex strength. Also, since the fluid momentum $\mathbf{M}$ associated with the vortex ring is

$$\mathbf{M} = \rho_0 \Gamma \mathbf{n}_l S_l$$

the velocity perturbation is also seen to be proportional to the time derivative of the rate of change of fluid momentum in the direction of observation. The force applied to the fluid by the motion of the vortex ring is related to the rate of change of momentum by

$$\mathbf{F} = \frac{\partial \mathbf{M}}{\partial t}$$

Thus Eq. (2.104) can be rewritten

$$p_a(\boldsymbol{x},t) = \frac{\rho_0}{4\pi c_0} \left[\frac{\mathbf{r}}{r^2} \cdot \frac{\partial \mathbf{F}}{\partial t} \right] \tag{2.105}$$

which is the same as Eq. (2.74). The radiated pressure is determined by the rate of change of the force applied to the fluid by the vortex motion. This relationship will be considered extensively in Chapter 4, Dipole Sound from Cylinders; it is the fundamental relationship for dipole sound radiation.

If the area of the vortex ring S_l remains constant, the strength changes with the circulation so that the pressure perturbation is given by

$$p_a(\boldsymbol{x},t) = \frac{\rho_0}{4\pi c_0} \frac{\mathbf{n}_l \cdot \mathbf{r} S_l}{r^2} \left[\frac{\partial^2 \Gamma}{\partial t^2} \right]$$

In the alternative instance of a constant circulation Γ yet changing area S_l the velocity perturbation is proportional to $\partial^2 S_l / \partial t^2$. This change can arise from vortex-line stretching by flow as depicted in Fig. 2.10B. The vortex line stretches due to translation at velocity **u** so that the change in the enclosed vector area $\mathbf{n}_l\delta S_l$ in the time interval δt is $\mathbf{n}_l\delta S_l = (\mathbf{u}\delta t)\times d\mathbf{l}$. Thus the Eq. (2.104) becomes

$$p_a(\mathbf{x},t) = \frac{\rho_0}{4\pi c_0}\frac{\mathbf{r}}{r^2}\cdot\left[\frac{\partial}{\partial t}\oint \Gamma\mathbf{u}\times d\mathbf{l}\right]$$

If we consider the vortex lines to exist throughout the region of flow and introduce Eq. (2.99) we find

$$\begin{aligned} p_a(\mathbf{x},t) &= \frac{\rho_0}{4\pi c_0}\frac{\mathbf{r}}{r^2}\cdot\left[\frac{\partial}{\partial t}\oint\iint(\mathbf{n}_\omega\cdot\boldsymbol{\omega}\,dS_\omega)\mathbf{u}\times d\mathbf{l}\right] \\ &= -\frac{\rho_0}{4\pi c_0^2}\frac{\mathbf{r}}{r^2}\cdot\left[\frac{\partial}{\partial t}\iiint_V(\boldsymbol{\omega}\times\mathbf{u})dv\right] \end{aligned} \tag{2.106}$$

since $(\mathbf{n}_\omega\cdot\boldsymbol{\omega}\,dS_\omega)\mathbf{U}\times d\mathbf{l} = (\mathbf{U}\times\boldsymbol{\omega})\mathbf{n}_\omega\cdot d\mathbf{l}\,dS_\omega$ because the $d\mathbf{l}$ and to $\boldsymbol{\omega}$ vectors are coincident. This shows that the pressure perturbation is proportional to the change in the rate of vortex stretching by the fluid. The intensity is maximum in the plane normal to the vector $\boldsymbol{\omega}\times\mathbf{u}$, i.e., normal to the surface enclosed by the vortex ring of vorticity $\boldsymbol{\omega}$. Powell gives the term "vortex sound" to the far-field sound radiation from a finite region of vorticity since it emanates from changes in the net vortex strength of the region.

The integrand in Eq. (2.106) is also recognized as the first of the source terms of Eqs. (2.93) and (2.95). This equivalence complements the interpretation, given at the end of Section 2.4.1, of the source term so that we now have a complete physical explanation. The sound radiation from a localized region of turbulent unsteadiness is caused by the stretching of vortex lines and the rate of change of fluid potential associated with the dynamics of the region.

An important physical example of the generation of vortex sound is the Aeolian tone radiated from a circular cylinder in a cross-flow. In this situation diagramed in Fig. 2.13 (dealt with in detail in Chapter 4: Dipole Sound from Cylinders) fluid flows steadily past a cylinder so that the direction of fluid motion is perpendicular to the axis of the cylinder. Vortices are shed downstream of the cylinder with alternately varying changes in sign. Since the circulation of an element of the fluid incident on the cylinder is zero, the net circulation in the flow–cylinder system must remain zero downstream of the cylinder. This requires that for every vortex formed in the fluid an image vortex must be formed in the cylinder. The resulting vortex pair is composed of two legs of a closed ring as shown in Fig. 2.12C and in Fig. 2.13. In this manner we can see that the periodic formation of vortex pairs, one bound to the cylinder and the other formed in the wake and convected downstream, results in a similarly periodic change in vortex strength and therefore sound radiated in a direction normal to the plane of the vortex ring. This direction is also the perpendicular to the plane formed by the flow vector and the axis

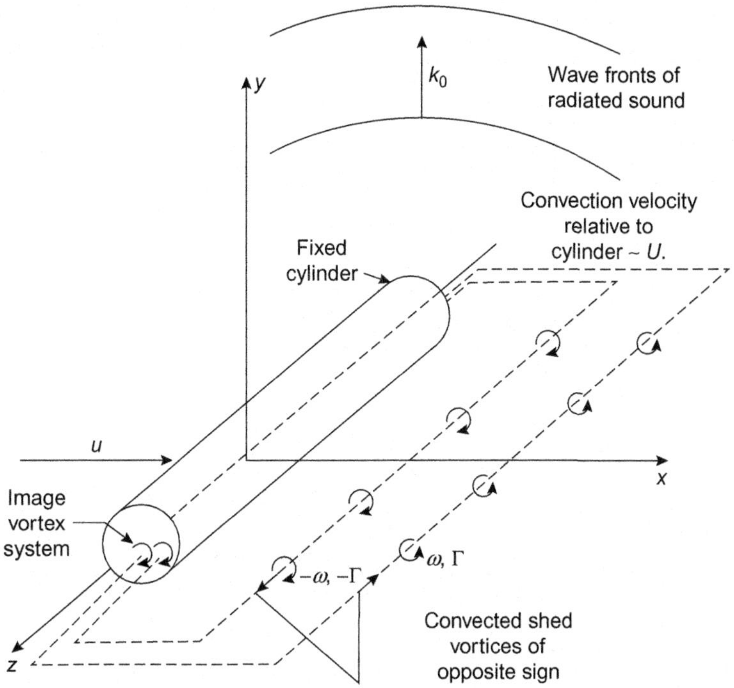

FIGURE 2.13 Diagram of a cylinder and its vortex system in a crosswind.

of the cylinder. The magnitude of the sound in this direction is proportional to the circulation ($\pm\Gamma$) of the vortices formed and to the square of the reciprocal of the temporal period of the vortex formation. Another derivation of Eq. (2.106), which pertains specifically to this problem, will be discussed in Chapter 4, Dipole Sound from Cylinders.

2.6.4 The Effect of Solid Boundaries on Vortex Sound

We now turn our attention to the integral form of the wave equation analogous to Eqs. (2.69) and (2.77), yet incorporating Powell's source term. Combining Eqs. (2.65) and (2.95) we obtain the acoustic pressure

$$\begin{aligned} 4\pi p_{\mathrm{a}}(\boldsymbol{x},t) &= \iiint_V \left[\frac{\partial}{\partial y_i}(\rho(\boldsymbol{\omega}\times\mathbf{u})_i)\right]\frac{dV(\mathbf{y})}{r} \\ &+ \iiint_V \frac{\partial^2}{\partial y_i^2}\left[p-\rho c_0^2+\frac{1}{2}\rho u^2\right]\frac{dV(\mathbf{y})}{r} \\ &+ \iint_\Sigma \left\{\frac{1}{c_0 r}\frac{\partial r}{\partial n}\left[\frac{\partial \rho c_0^2}{\partial t}\right]-\frac{\partial(1/r)}{\partial n}\left[\rho c_0^2\right]+\frac{1}{r}\left[\frac{\partial \rho c_0^2}{\partial n}\right]\right\}dS(\mathbf{y}) \end{aligned} \tag{2.107a}$$

where Σ is the total surface enclosing both the source volume and the observation point as in Fig. 2.7A. The first 2 terms are equivalent to the integrated effect of a volume distribution of quadrupoles and the surface integral mimics that considered above. Noting that $\partial r/\partial n = \partial r/\partial y_n = -\partial r/\partial x_n$, using the divergence theorem as in the beginning of Section 2.3.3 and in Eq. (2.65), and using Eq. (2.83) it is a simple matter to write down the far-field acoustic pressure

$$4\pi p_a(\mathbf{x},t) = \frac{\partial}{\partial x_i}\iiint_V \frac{[\rho(\boldsymbol{\omega}\times\mathbf{u})_i]}{r}dV(\mathbf{y}) + \frac{1}{c_0^2}\frac{\partial^2}{\partial t^2}\iiint_V \left[p - \rho c_0^2 + \frac{1}{2}\rho u^2\right]\frac{dV(\mathbf{y})}{r} + \frac{\partial}{\partial x_n}\iint_\Sigma \left[p + \frac{1}{2}\rho u^2\right]\frac{dS(\mathbf{y})}{r} - \iint_\Sigma \left[\rho\frac{\partial u_n}{\partial t}\right]\frac{dS(\mathbf{y})}{r} \quad (2.107b)$$

Eq. (2.107b) should be compared with Eq. (2.71) with the integrals being the same in both equations and the (small) viscous stresses ignored. The surface Σ may be interpreted to pertain to specific applications, e.g., as shown in Fig. 2.8. As Powell [34] explains, Eq. (2.107b) gives the acoustic pressure as the sum of four contributions:

1. a volume distribution of dipoles proportional to $\boldsymbol{\omega}\times\mathbf{u}$,
2. a volume distribution of nondirectional sources

$$\partial^2\left(p - \rho c_0^2 + \frac{1}{2}\rho u^2\right)/\partial t^2,$$

   $p = \rho c_0^2$ for isentropic sound,
3. a surface distribution of dipoles whose strength is proportional to the Bernoulli pressure on the surface, $p + \frac{1}{2}\rho u^2$,
4. a monopole distribution whose strength is proportional to the acceleration of the surface normal to itself, $\partial u_n/\partial t$.

Furthermore, the surface-integral terms are equivalent to those in Eqs. (2.69) and (2.77) (note being taken of the inviscid nature of the current problem), while the volume distribution has been reexpressed. The second volume integral involves quantities of order $(u/c_0)^2$ and $p(\rho c_0^2)^{-1}$ compared to the other terms, which can be safely ignored in favor of the other terms for low Mach-number flows. The pertinent nature of the Reynolds stress distribution as involving a change in vortex strength is thus exposed.

2.6.5 Relationships Between the Powell and the Lighthill–Curle Theories

To indicate formal relationships between the three theoretical formulations, alternative expansions of the retarded source will be used in order to expose either the Lighthill or the Curle form explicitly.

In the linear acoustic limit of an isentropic fluid and in the absence of any surfaces Eq. (2.107a) reduces in linearized form to

$$p_a(\mathbf{x},t) = \frac{1}{4\pi}\iiint_V \left[\frac{\partial}{\partial y_i}(\rho(\boldsymbol{\omega}\times\mathbf{u})_i)\right]\frac{dV(\mathbf{y})}{r} \tag{2.108}$$

and one questions how this form of the vortex-sound theory relates to Lighthill's theory in the form of Eq. (2.54). The above linearization is justified in isentropic flow in which the speed of sound in the source zone is also identical to that in the medium (where $p_a = (\rho - \rho_0)c_0^2$) and for which variations in ρu^2 are of order Mach number squared compared with variations in the acoustic pressure. The factor $\rho(\boldsymbol{\omega}\times\mathbf{u})$, without any retardation in an incompressible fluid ($c_0 \to \infty$), represents a rate of change of momentum in an element volume δv of fluid; i.e.,

$$\delta\left\{\frac{\partial \mathbf{M}}{\partial t}\right\} = \rho_0(\boldsymbol{\omega}\times\mathbf{u})\delta v$$

where $\mathbf{M}$ is the momentum of the fluid. In a free and unbounded fluid (such as representing a zone of free quadrupoles in a mixing shear layer) there is no net force on the fluid. Thus the net rate of momentum exchange to the fluid will be zero; i.e.,

$$\frac{\partial \mathbf{M}}{\partial t} = \iiint_V \rho_0(\boldsymbol{\omega}\times\mathbf{u})dV(\mathbf{y}) = 0$$

Thus neglecting retardation effects, the integral of the source term in the acoustic equation will yield a null pressure.

As shown by Powell [34] a mathematical connection with Lighthill's source therefore may be made by expanding $[\rho(\boldsymbol{\omega}\times\mathbf{u})]$ into a Taylor Series in $\mathbf{y}$ about $\mathbf{y}=0$ in the retardation factor; i.e., letting $L_i = \rho(\boldsymbol{\omega}\times\mathbf{u})_i$, we have

$$\begin{aligned}[L_i] &= L_i\left(\mathbf{y}, t-\frac{r}{c_0}\right)\\ &\simeq L_i\left(\mathbf{y}, t-\frac{x}{c_0}\right) + \frac{\partial L_i(\mathbf{y}, t-x/c_0)}{\partial t}\left(\frac{-1}{c_0}\frac{\partial r}{\partial y_j}\right)y_j\end{aligned}$$

recalling that

$$\frac{\partial r}{\partial y_j} = -\frac{x_j - y_j}{r} \simeq \frac{-x_j}{r}$$

$$[L_i] \simeq L_i\left(\mathbf{y}, t-\frac{x}{c_0}\right) + L_i'\left(\mathbf{y}, t-\frac{x}{c_0}\right)\frac{y_j}{c_0}\frac{x_j}{r}$$

where the prime denotes differentiation with respect to time. Invoking Eq. (2.56) and noting that derivatives with respect to x may be translated into derivatives with respect to time we obtain the parallel to Eq. (2.58)

$$p_{\mathrm{a}}(\boldsymbol{x},t)=\frac{1}{4\pi}\frac{x_i}{c_0 r}\iiint_V\left[\frac{\partial}{\partial t}\rho(\boldsymbol{\omega}\times\mathbf{u})_i\right]\frac{dV(\mathbf{y})}{r}$$

If the distance $|\boldsymbol{x}|$ is much larger than the size of the source region the substitution of the Taylor series yields

$$p_{\mathrm{a}}(\boldsymbol{x},t)\simeq\frac{1}{4\pi}\frac{x_i}{c_0 r}\left\{\frac{1}{r}\frac{\partial}{\partial t}\iiint_V L_i\left(\mathbf{y},t-\frac{x}{c_0}\right)dV(\mathbf{y})+\frac{1}{c_0}\iiint_V y_j\frac{x_j}{r}L'\left(\mathbf{y},t-\frac{x}{c_0}\right)\frac{dV(\mathbf{y})}{r}\right\}$$

Since there is presumed no instantaneous net force on the fluid, the first term is identically zero. The integrand in the second term includes terms of the type

$$y_jL_i'=\rho y_j\omega_j u_l=\rho y_j\frac{\delta u_k}{\partial y_j}u_i=\rho u_k u_l$$

The acoustic pressure then, in the limit of vanishing Mach number, reduces to

$$p_{\mathrm{a}}(\boldsymbol{x},t)\simeq\frac{1}{4\pi}\frac{x_i x_j}{r^3c_0^2}\iiint_V\left[\frac{\partial^2}{\partial t^2}(\rho u_i u_j)\right]dV(\mathbf{y})$$

which resembles the Eq. (2.58) that results from Lighthill's equation (with due regard to Eq. (2.6)). In the Taylor series the second term is of order Mach number M with respect to the first and additional order of terms increase as M^2, M^3, etc. This may be seen in the limit of $\Delta y_i\rightarrow 0$,

$$\frac{1}{c_0}\Delta y_i\frac{\partial L_i'}{\partial t}\sim M_i[\delta L_i]$$

Thus the above approximation amounts to the second term of a Mach number expansion in which the first term in the expansion vanishes under the condition of zero net force on the unbounded medium. There is a lack of formal identity, however, in the simple analysis shown above because proof has not been given that the neglected terms are fully consistent with one another. This point, discussed by Lauvstad [36], centers on the fact that at low Mach number the evaluation of integrals is controlled by careful considerations of the balances among high-order terms. Other approaches to the theory of aerodynamic sound as an expansion in orders of Mach number are those of Crow [22], Obermeier [37], and Möhring et al. [38,39].

The starting point for comparing the Powell and the Curle theories for low Mach number flow past a rigid body is a reduced form of Eq. (2.108), i.e.,

$$4\pi p_{\mathrm{a}}(\boldsymbol{x},t)=\frac{\partial}{\partial x_i}\iiint_V\frac{[\rho_0(\boldsymbol{\omega}\times\mathbf{u})_i]}{r}dV(\mathbf{y})+\frac{\partial}{\partial x_n}\iint_\Sigma\left[p+\frac{1}{2}\rho_0u^2\right]\frac{dS(\mathbf{y})}{r}$$

Substituting for $\rho_0(\boldsymbol{\omega} \times \mathbf{u})$ using Eq. (2.92) and using the special form of the divergence theorem (2.55c) we find

$$\iiint_V [\rho_0(\boldsymbol{\omega} \times \mathbf{u})_i] \frac{dV(y)}{r} = -\frac{\partial}{\partial x_i} \iiint_V \left[p + \frac{1}{2}\rho_0 u^2\right] \frac{dV(y)}{r}$$
$$- \iint_\Sigma n_i \left[p + \frac{1}{2}\rho_0 u^2\right] \frac{dS(y)}{r}$$
$$- \iiint_V \left[\rho_0 \frac{\partial u_i}{\partial t}\right] \frac{dV(y)}{r}$$

so that only two terms are retained because the two surface integrals cancel.

Thus

$$4\pi p_a(\boldsymbol{x}, t) = -\frac{\partial}{\partial x_i} \iiint_V [\rho_0 \dot{u}_i] \frac{dV(y)}{r} - \frac{1}{c_0^2} \iiint_V \frac{\partial^2}{\partial t^2} \left[p + \frac{1}{2}\rho_0 u^2\right] \frac{dV(y)}{r}$$

The second term is of order M with respect to the first and may be ignored because it pertains to the quadrupole source derived above. In the first term, the retarded variable is expanded in a Taylor series to obtain

$$4\pi p_a(\boldsymbol{x}, t) \simeq -\frac{\partial}{\partial x_i} \iiint \rho_0 \dot{u}_i \left(\boldsymbol{y}, t - \frac{|\boldsymbol{x}|}{c_0}\right) \frac{dV(y)}{r}$$
$$- \frac{\rho_0}{c_0} \iiint \boldsymbol{y} \cdot \frac{\partial^3 \mathbf{u}}{\partial t^3} \left(\boldsymbol{y}, t - \frac{|\boldsymbol{x}|}{c_0}\right) \frac{dV(y)}{r}$$

where the Taylor series is

$$u_i\left(\boldsymbol{y}, t - \frac{\boldsymbol{r}}{c_0}\right) = u_i\left(\boldsymbol{y}, t - \frac{\boldsymbol{x}}{c_0}\right) - \frac{\partial u_i}{\partial t}\left(\boldsymbol{y} - \frac{\boldsymbol{x}}{c_0}\right)\left(\frac{x_i}{r}\right)\frac{y_i}{c_0} + \cdots$$

Now, the second term is of order

$$\rho_0 \frac{u^2}{c_0^2} u^2 \sim \rho_0 u^2 M^2$$

since $y_i \partial/\partial t \sim u_i$ and it also contributes as a usual quadrupole source, while the first term is of order

$$\rho_0 \frac{u}{c_0} u^2 \sim \rho_0 u^2 M$$

and is the usual dipole source contribution. The first term is nonzero when a time-dependent force is applied to the fluid as with a flow–body interaction by the momentum equation for the vortex-free surrounding fluid

$$F_i\left(t - \frac{|\boldsymbol{x}|}{c_0}\right) = \iiint_V \rho_0 \frac{\partial u_i}{dt}\left(\boldsymbol{y}, t - \frac{|\boldsymbol{x}|}{c_0}\right) dV(\boldsymbol{y})$$

Thus for a compact source zone surrounding a rigid surface, the radiated sound is given by the first-order term

$$p_{\mathrm{a}}(\boldsymbol{x}, t) = \frac{-1}{4\pi} \frac{\partial}{\partial x_i} \left[F_i \left(t - \frac{|\boldsymbol{x}|}{c_0} \right) \right] (1 + O[M])$$

which is Eq. (2.74).

2.7 REPRESENTATIONS IN THE FREQUENCY AND WAVE NUMBER DOMAINS

When the integral relationships of this chapter are used for the solution of specific physical problems, it is often convenient to invoke various types of harmonic analyses. In this text the Fourier transform, applied for both time and space variables, will be used exclusively of other transforms when harmonic analysis is necessary. We have already used a simplified form of Fourier analysis when we specified time dependence to be of the form $e^{-i\omega t}$.

2.7.1 The Helmholtz Integral Equation

We begin by deriving analogous forms of Eqs. (2.50) and (2.65) using the time Fourier transform. The Fourier transform $V(\omega)$ (see Section 1.4.3: Review of Fourier Series and the Fourier Integral for a brief review, or Tichmarsh [40] for a complete treatise) of a variable $v(t)$ is

$$V(\omega) = \frac{1}{2\pi} \int_{-\infty}^{\infty} e^{+i\omega t} v(t)\, dt \tag{2.109}$$

and its inverse is

$$v(t) = \int_{-\infty}^{\infty} e^{-i\omega t} V(\omega)\, d\omega \tag{2.110}$$

Then letting $p_a(\mathbf{x}, t) = c_0^2 \rho_{\mathrm{a}}(\mathbf{x}, t)$, the transform of the inhomogeneous wave equation, Eq. (2.50) is

$$\nabla^2 P_{\mathrm{a}}(\boldsymbol{y}, \omega) + k_0^2 P_{\mathrm{a}}(\boldsymbol{y}, \omega) = -\tilde{\sigma}(\boldsymbol{y}, \omega) \tag{2.111}$$

where $k_0 = \omega/c_0$ is the magnitude of the acoustic wave number and $\tilde{\sigma}(\mathbf{y}, \omega)$ is the Fourier transform of $\sigma(\mathbf{y}, t)$. The solutions of the homogeneous wave equation are of the form

$$P_{\mathrm{a}}(\boldsymbol{y}, \omega) = A(e^{\pm ik_0 r}/r)$$

which are appropriate for propagating in free space, hence, the function

$$g(\boldsymbol{r}, \omega) = e^{\pm ik_0|\boldsymbol{r}-\boldsymbol{r}_0|}/4\pi|\boldsymbol{r} - \boldsymbol{r}_0|, \tag{2.112a}$$

a solution of

$$\nabla^2 g(\boldsymbol{r},\omega) + k_0^2 g(\boldsymbol{r},\omega) = -\delta(\boldsymbol{r}-\boldsymbol{r}_0) \tag{2.112b}$$

is called the "free space Green's function." The selection of $+i$ and $-i$ depends on the invoking of the radiation or absorption condition of propagating waves (Section 2.2). For outward-traveling waves, we select the convention $+i$ and express the traveling wave front as in constant phase moving to the right $(r>0)$ as t increases from a reference time. Thus, the retarded potential is given by

$$v\left(t-\frac{r}{c_0}\right) = \int_{-\infty}^{\infty} e^{-i\omega t} e^{+ik_0 r} V(\omega) d\omega \tag{2.113}$$

Substituting the inverse Fourier transform, Eq. (2.113), into Eq. (2.65) gives for this example of the Lighthill source

$$\begin{aligned} P_a(\boldsymbol{x},\omega) = &\iiint_{V_0} \frac{\partial^2 \tilde{T}_{ij}(\boldsymbol{y},\omega)}{\partial y_i \partial y_j} \frac{e^{+ik_0 r}}{4\pi r} dV(\boldsymbol{y}) \\ &+ \iint_{\Sigma} \left\{ \frac{e^{+ik_0 r}}{4\pi r} \frac{\partial P_a(\boldsymbol{y},\omega)}{\partial n} - P_a(\boldsymbol{y},\omega) \frac{\partial}{\partial n}\left(\frac{e^{+ik_0 r}}{4\pi r}\right) \right\} dS(\boldsymbol{y}) \end{aligned} \tag{2.114}$$

Here we have let $\tilde{T}_{ij}(\mathbf{y},\omega)$ be the Fourier transform of $T_{ij}(\mathbf{y},t)$, using the tilde and the changed independent variable to denote the transform. Eq. (2.114) is the Helmholtz integral equation. It could have been derived [6,7,12] from Eq. (2.111) by using the divergence theorem and the equation

$$\nabla_y^2 g(|\boldsymbol{x}-\boldsymbol{y}|,\omega) + k_0^2 g(|\boldsymbol{x}-\boldsymbol{y}|,\omega) = -\delta(\boldsymbol{x}-\boldsymbol{y}) \tag{2.115}$$

for the free space Green's function (Eq. (2.112a,b))

$$g(|\boldsymbol{x}-\boldsymbol{y}|) = e^{\pm ik_0|x-y|}/4\pi|\boldsymbol{x}-\boldsymbol{y}|$$

when ∇_y^2 denotes the Laplacian operation with respect to the variable y only.

We see, regarding Eq. (2.114) and Fig. 2.7, that if Σ is a control surface a distance R from a compact source region of volume V_0 then

$$P_a(\boldsymbol{x},\omega) = \iiint_V \frac{\partial^2 \tilde{T}_{ij}(\boldsymbol{y},\omega)}{\partial y_i \partial y_j} \frac{e^{+ik_0 r}}{4\pi r} dV(\boldsymbol{y}) \tag{2.116}$$

if

$$\lim_{R\to\infty}\left(-ik_0 R P_a(R,\omega) + R\frac{\partial P_a(R,\omega)}{\partial r}\right) = 0.$$

Recall that this last condition is just Sommerfeld's radiation condition (Section 2.2). Eq. (2.116) is the frequency-domain equivalent of Eq. (2.54).

A more general use of the Helmholtz integral equation lies in situations for which impedance boundaries or surfaces are present in the control volume. In these situations the free space Green's function of Eq. (2.112a,b) is replaced by $G(\mathbf{x}, \mathbf{y}, \omega)$, which is a solution of [6,7,10,12]

$$\nabla_y^2 G(\mathbf{x}, \mathbf{y}, \omega) + k_0^2 G(\mathbf{x}, \mathbf{y}, \omega) = -\delta(\mathbf{x} - \mathbf{y}). \tag{2.117}$$

In contrast to the free space Green's function, the function $G(\mathbf{x}, \mathbf{y}, \omega)$ is determined for the geometry under consideration and subject to certain boundary conditions. The Helmholtz integral equation corresponding to the reduced wave equation (2.111) is

$$P_a(\mathbf{x}, \omega) = \iiint_V \tilde{\sigma}(\mathbf{y}, \omega) G(\mathbf{x}, \mathbf{y}, \omega) dV(\mathbf{y}) + \iint_S \left\{ G(\mathbf{x}, \mathbf{y}, \omega) \frac{\partial P_a(\mathbf{y}, \omega)}{\partial n} - P_a(\mathbf{y}, \omega) \frac{\partial G(\mathbf{x}, \mathbf{y}, \omega)}{\partial n} \right\} dS(\mathbf{y}) \tag{2.118}$$

and similarly for the surface and volume source distributions in Eq. (2.103), where S is the surface of the boundary that is present, e.g., $S_1 + S_0$ in Fig. 2.8. The radiation condition has been invoked to eliminate the integral over the control surface S_2. Now, if $P_a(\mathbf{y}, \omega)$ is known on the surface, then the imposition of the boundary condition $G(\mathbf{x}, \mathbf{y}, \omega) = 0$ on S will put Eq. (2.118) into a form involving known functions. This boundary condition [4] is known as the Dirichlet boundary condition (see, e.g., Refs. [6,7,12,18–20,41]). Alternatively, if the normal gradient $\partial P_a(\mathbf{y}, \omega)/\partial n$ is known on S, then the boundary condition $\partial G(\mathbf{x}, \mathbf{y}, \omega)/\partial n = 0$ on S, known [6,7,12,18–29,41] as the Neumann boundary condition, puts Eq. (2.118) into a form that may be evaluated. The potency of the method is especially apparent if we consider the case of the rigid boundary. In this type of problem the velocity normal to the surface u_n is zero. Thus $\partial P_a(\mathbf{y}, \omega)/\partial n = 0$ so that the imposition of a Neumann boundary condition reduces Eq. (2.118) to the form

$$P_a(\mathbf{x}, \omega) = \iiint_V \tilde{\sigma}(\mathbf{y}, \omega) G(\mathbf{x}, \mathbf{y}, \omega) dV(\mathbf{y}). \tag{2.119}$$

The Green's function $G(\mathbf{x}, \mathbf{y}, \omega)$ now accounts for both the impedance and the geometry of the boundary.

Often the Green's functions may be separable, as in the case of the range $r = |\mathbf{x}|$ much larger than the extent $|\mathbf{y}|$ of the source region or in the case of enclosures of special geometry. Then

$$G(\mathbf{x}, \mathbf{y}, \omega) = G_x(\mathbf{x}, \omega) G_y(\mathbf{y}, \omega). \tag{2.120}$$

Accordingly if the boundaries are rigid, $G_y(\mathbf{y}, \omega)$ is selected so that

$$\partial G_y(\mathbf{y}, \omega)/\partial y_n = 0$$

on all boundaries S; then since $\partial P_a/\partial n = 0$ on rigid boundaries the Lighthill source term appearing in Eqs. (2.46) and (2.72) gives

$$P_a(\boldsymbol{x},\omega) = G_x(\boldsymbol{x},\omega)\iiint_V \tilde{T}_{i\,j}(\boldsymbol{y},\omega)\frac{\partial^2 G_y(\vec{y},\omega)}{\partial y_i \partial y_j}dV(\boldsymbol{y}) \qquad (2.121a)$$

since $\tilde{T}_{in}(\boldsymbol{y},\omega)$ elements of the full matrix also vanish on S as a result of u_n being zero there. In like manner for the dipole source of Eq. (2.93) situated near a rigid boundary, the appropriate form of Eq. (2.112a) is

Invoking the identities $\partial/\partial n = n_i\partial/\partial y_i = \vec{n}\bullet\nabla$,

$$P_a(\boldsymbol{x},\omega) = -Gx(\boldsymbol{x},\omega)\iiint_V F_i(\boldsymbol{y},\omega)\frac{\partial G_y(\boldsymbol{y},\omega)}{\partial y_i}dV(\boldsymbol{y}) \qquad (2.121b)$$

since

$$\iint_{S_s} F_n(\boldsymbol{y},\omega)G(\boldsymbol{y},\omega)dS(\boldsymbol{y}) = 0$$

where S_s is a surface surrounding the source region and $\nabla\cdot\mathbf{F}(\mathbf{y}, t)$ represents the right-hand side of Eq. (2.93).

The method was used by Ffowcs Williams and Hall [42] to determine the sound field from turbulence convected past a half plane. Some other aerodynamic noise problems have been attacked in this manner by Howe [35], Chase [43,44], Davies and Ffowcs Williams [45], and Crighton and Ffowcs Williams [46]; see Chapter 11, Noncavitating Lifting Sections.

A simplified example of the use of the Green's function for the Neumann boundary condition can be shown for the case of the plane boundary. A solution of Eq. (2.117) that is valid for the rigid plane boundary is given by

$$G(\boldsymbol{x},\boldsymbol{y},\omega) = \frac{e^{+ik_0 r_1}}{4\pi r_1} + \frac{e^{+ik_0 r_2}}{4\pi r_2} \qquad (2.122)$$

where $r_1^2 = (x_1-x_2)^2 + (y_1-y_2)^2 + (z_1-z_2)^2$ and $r_2^2 = (x_1-x_2)^2 + (y_1-y_2)^2 + (z_1+z_2)^2$. The ranges r_1, r_2 are the same as those shown for the primary and image source system in Fig. 2.2C, and they correspond to the r, r' in Fig. 2.9. The field point is $\mathbf{x} = (x_1, y_1, z_1)$, and the source point is $\mathbf{y} = (x_2, y_2, z_2)$. Also it can be easily shown that on the surface $z_2 = 0$

$$\left.\frac{\partial G(\boldsymbol{x},\boldsymbol{y},\omega)}{\partial n}\right|_{z_2=0} = \left.\frac{\partial G(\boldsymbol{x},\boldsymbol{y},\omega)}{\partial z_2}\right|_{z_2=0} = 0$$

and for $k_0 z_2 \ll 1$, Eq. (2.122) reduces to the functional form of Eq. (2.33). Substitution of Eq. (2.122) into Eq. (2.119) yields for the case of the Lighthill source term

$$P_a(\boldsymbol{x},\omega) = \iiint_{V_{0_1}+V_{0_2}} \frac{\partial^2 T_{ij}(\boldsymbol{y},\omega)}{\partial y_i \partial y_j}\frac{e^{+ik_0|\boldsymbol{x}-\boldsymbol{y}|}}{4\pi|\boldsymbol{x}-\boldsymbol{y}|}dV(\boldsymbol{y}) \qquad (2.123)$$

where integration now extends over both the physical source distribution and its image distribution. The mechanics of such an integration would have to account for the symmetric and nonsymmetric reflections of the T_{ij} about $y_2 = 0$, as discussed in Section 2.4.4. Eq. (2.123) is identical to Eq. (2.80).

Other known functions $G(\mathbf{x}, \mathbf{y}, \omega)$ for a wide variety of geometries have been given in books by Morse and Feshbach [41], Morse and Ingard [6], and Junger and Feit [10] as well as elsewhere in this book. Generally simple closed-form functions exist for circular cylinders, spheres, infinite planes. Analytically more complicated functions also exist for slits, half planes, and spheroidal bodies.

2.7.2 Generalized Transforms and Stochastic Variables

In Section 1.4.2 the fundamentals of correlation analysis were introduced, and in Section 2.3.3 we utilized the correlation function of the stress tensor to determine the time-averaged acoustic intensity, Eq. (2.59), far from a turbulent region. The correlation function was introduced in that analysis because the temporal and spatial variations of the velocity fluctuations are uncertain, yet occurring within certain limits of probability. For example, the velocity at any instant and location can be given by

$$u_i'(x,t) = U(x) + u_i(x,t)$$

where $U(\mathbf{x})$ is the time-averaged velocity, and $u_i(\mathbf{x}, t)$ is the stochastic velocity fluctuation with zero mean value, i.e.,

$$\overline{u_i(x,t)} = \lim_{T\to\infty} \frac{1}{T}\int_{-T/2}^{T/2} u_i(x,t)dt = \lim_{V\to\infty} \frac{1}{V}\iiint_V u_i(x,t)dV(x) \equiv 0. \quad (2.124)$$

where T is the time of averaging and V is the volume over which the velocity is instantaneously sampled. Fluid fields for which the equivalence of Eq. (2.124) holds meet one condition for the field to be statistically homogeneous. The time mean-square velocity fluctuation is

$$\begin{aligned}\overline{u_i'2(\mathbf{x},t)}^t &= \overline{(U(\mathbf{x})+u_i(\mathbf{x},t))^2}^t \\ &= \lim_{T\to\infty}\frac{1}{T}\int_{-T/2}^{T/2}(U(\mathbf{x})+u_i(\mathbf{x},t))^2 dt \\ &= \overline{U^2(\mathbf{x})} + \overline{u_i^2(\mathbf{x},t)}^t\end{aligned} \quad (2.125)$$

If the fluid region is truly homogeneous, then

$$\overline{u_i^2(\mathbf{x},t)}^t = \overline{u_i^2(\mathbf{x},t)}^x = \overline{u_i^2},$$

If it is also ergodic, it also equals the ensemble average

$$\overline{u_i^2(\mathbf{x},t)}^t = \langle u_i^2(\mathbf{x},t)\rangle$$

Note that the average over space of the fluctuating velocity at any instant in time will also be zero in the homogeneous turbulent field. Notice now that we have introduced the vincula $\overline{----}^{t}$ and $\overline{----}^{x}$ to formally distinguish between time and space averaging, respectively. Another condition of homogeneity is that the correlation function is independent of the datum of the **x**, **y**, or t variables and is dependent on the differences $\mathbf{y}-\mathbf{x}$ and τ. Thus a field with space–time homogeneity has a correlation function satisfying

$$\hat{R}_{uu}(\boldsymbol{y},\boldsymbol{x},\tau)=\hat{R}_{uu}(\boldsymbol{y}-\boldsymbol{x},\tau)=\hat{R}_{uu}(\boldsymbol{r},\tau)$$

where

$$\hat{R}_{uu}(r,\tau)=\lim_{T\to\infty}\frac{1}{2T}\int_{-T}^{T}u(\boldsymbol{x},t+\tau)u(\boldsymbol{x},t)dt$$

We now consider some general representations of the turbulent field and its resulting sound. Eq. (2.124) constitutes a boundedness on the integral of $u_i(\mathbf{x}, t)$ and it permits a definition of a generalized Fourier transform [40,47], which we shall write

$$\tilde{u}_i(\boldsymbol{x},\omega)=\frac{1}{2\pi}\int_{-\infty}^{\infty}e^{+i\omega t}u_i(\boldsymbol{x},t)dt \tag{2.126a}$$

and

$$u_i(\boldsymbol{x},t)=\int_{-\infty}^{\infty}e^{-i\omega t}\tilde{u}_i(\boldsymbol{x},\omega)d\omega \tag{2.126b}$$

The space–time *covariance* of the velocity fluctuations is given by (see, e.g., Batchelor [47], Lin [48], and Kinsman [49])

$$\hat{R}_{u_iu_j}(\boldsymbol{y},\boldsymbol{x},\tau)=\int_{-\infty}^{\infty}d\omega\int_{-\infty}^{\infty}d\omega'\left[\lim_{T\to\infty}\frac{1}{2T}\int_{-T}^{T}e^{i(\omega-\omega')t}dt\right]u_i(\boldsymbol{x},\omega)\overset{*}{u_j}(\boldsymbol{y},\omega)e^{-i\omega\tau} \tag{2.127}$$

where $\hat{R}_{u_iu_j}(\boldsymbol{x},\boldsymbol{x},0)=\hat{R}_{u_iu_j}(\boldsymbol{y},\boldsymbol{y},0)=\overline{u_i^2}>\hat{R}_{u_iu_j}(\boldsymbol{y},\boldsymbol{x},0)$. We have replaced the physical velocity fluctuation by the inverse transform of $\tilde{u}_i(\mathbf{x},\omega)$, using Eq. (2.126b).

The complicated integral can be cleared up by invoking the behavior of the Dirac delta function

$$\frac{1}{T}\int_{-T/2}^{T/2}e^{i(\omega-\omega')t}dt=\frac{1}{T}\frac{e^{i(\omega-\omega')T/2}-e^{-i(\omega-\omega')T/2}}{2i(\omega-\omega')}=\frac{\sin(\omega-\omega')T/2}{\frac{1}{2}T(\omega-\omega')}$$

As T increases, this function becomes more and more peaked near $\omega=\omega'$ so that if the integrals over all frequencies ($-\infty<\omega<\infty$) are to be equal we can write

$$\lim_{T\to\infty}\frac{\sin(\omega-\omega')T/2}{(\omega-\omega')T/2}=\frac{2\pi}{T}\delta(\omega-\omega') \tag{2.128}$$

The equivalence,

$$\delta(\omega-\omega')=\frac{1}{2\pi}\int_{-\infty}^{\infty} e^{\pm i(\omega-\omega')t}dt \tag{2.129}$$

can also be established by virtue of the definition of the Fourier transform of the Dirac delta function, Eq. (1.64) and its inverse.

This equivalence converts the Eq. (2.127) to the form (in the limit as $T\to\infty$)

$$\hat{R}_{u_iu_j}(\mathbf{y},\mathbf{x},\tau)=2\pi\int_{-\infty}^{\infty}\int_{-\infty}^{\infty}\frac{\delta(\omega-\omega')}{T}e^{-i\omega\tau}\tilde{u}_i^*(\mathbf{x},\omega')\tilde{u}_j(\mathbf{y},\omega)d\omega d\omega'$$

$$=\int_{-\infty}^{\infty} e^{-i\omega\tau}\left\{\frac{2\pi}{T}\tilde{u}_i^*(\mathbf{x},\omega)\tilde{u}_j(\mathbf{y},\omega)\right\}d\omega$$

This defines [12,28,47–51] the covariance function as an inverse Fourier transform of a function that we shall call the two-point *cross-spectral density* of the velocity fluctuation. We shall write this function

$$\lim_{T\to\infty}\left\{\frac{2\pi}{T}\tilde{u}_i^*(\mathbf{x},\omega)\tilde{u}_j(\mathbf{y},\omega')\right\}=\Phi_{u_iu_j}(\mathbf{y},\mathbf{x},\omega)\delta(\omega-\omega') \tag{2.130}$$

so that the space–time covariance $R(\mathbf{y}, \mathbf{x}, \tau)$ and the two-point cross-spectral density are Fourier transform pairs:

$$\Phi_{u_iu_j}(\mathbf{y},\mathbf{x},\omega)=\frac{1}{2\pi}\int_{-\infty}^{\infty} e^{i\omega\tau}\overline{u_i(\mathbf{y},t)u_j(\mathbf{x},t-\tau)}^t dt \tag{2.131}$$

and

$$\overline{u_i(\mathbf{y},t)u_j(\mathbf{x},t-\tau)}^t=\int_{-\infty}^{\infty} e^{-i\omega\tau}\Phi_{u_iu_j}(\mathbf{y},\mathbf{x},\omega)d\omega$$

In the following chapters the multidimensional space–time Fourier transform will be used. This is defined

$$u(k_1,\ldots,k_n,\omega)=\frac{1}{(2\pi)}\frac{1}{(2\pi)^n}\int_{-\infty}^{\infty}dy_1\ldots\int_{-\infty}^{\infty}dy_n\int_{-\infty}^{\infty}dt$$
$$\times\, u(y,t)e^{-i(k_1y_1+\cdots+k_ny_n)}\ldots e^{i\omega t}$$

where n varies from one to three space dimensions. The relationship equivalent to Eq. (2.130) for *spatially and temporally homogeneous turbulent fields* requires both space and time averaging. In the limit of $k_iL_i\to\infty$ and $\omega T\to\infty$

using Eq. (2.128) and its analogies for the spatial integrations we obtain the desired result

$$\Phi_{uu}(k_1,\ldots,k_n,\omega)\delta(\omega-\omega')\delta(k_1-k_1')\cdots\delta(k_n-k_n')$$
$$\equiv\frac{(2\pi)^n}{L_1\cdots L_n}\frac{2\pi}{T}\left\{u_1(k_1,\ldots,k_n,\omega)\overset{*}{u_2}(k_1',\ldots,k_n',\omega')\right\} \tag{2.132}$$

The function $\Phi_{uu}(k_1,\ldots,k_n,\omega)$ is called the wave number frequency spectral density of the disturbance u and the frequency spectrum $\Phi_{uu}(\omega)$ is the integral of $\Phi(k_1,\ldots, k_n, \omega)$ over all wave numbers. A generalization of Eqs. (2.130) and (2.132) in which the ensemble average is used is

$$\langle u_1(k_1,\ldots,k_n,\omega)u_2{}^*(k_1',\ldots,k_n',\omega')\rangle=\Phi_{u_1u_2}(\mathbf{k},\omega)\delta(\mathbf{k}-\mathbf{k}')\delta(\omega-\omega') \tag{2.133}$$

For the n-dimensional wave number frequency spectrum this relationship may be used whenever $u_1(\mathbf{x}, t)$ and $u_2(\mathbf{x}, t)$ are spatially and temporally homogeneous. Such homogeneity implies effectively infinite domains $\mathbf{x}$ and time duration t and formally rules out spatial edge effects and transients due to initial conditions. In a practical sense such homogeneity may often be assumed when the correlation length Λ, or correlation time θ, are much smaller than the spatial extent L, or the duration T. See Section 3.6, Review of Correlation and Spectrum Functions Used in Describing Turbulent Sources.

In the above derivation the variables $u_i(\mathbf{y}, t)$ and $u_j(\mathbf{x}, t)$ are used to define any two random variables of time and position. Actually the correlations may just as well be between any pairs of pressures, velocities, accelerations or any other physically measurable property of the fluid and surface motions.

There is no universally used equivalence between the spectrum function and the product of Fourier transforms. The above definition of equivalence has been used by Crandall [50] and by Goldstein [28], while others [47–49,51] have elected to use a slightly different representation of the ensemble average of the two quantities. Which definition is used in a given analysis is unimportant because the products of Fourier transforms will appear on both sides of an equal sign. In this book the definition introduced in Eq. (2.132) will be used because of its physical significance in the taking of the empirical time average.

We can now review the equivalence [47–49,51] between spatial and temporal averaging of statistically homogeneous random variables. An ensemble average, the time average, and the spatial average may be written (see also Section 3.5.2: Correlation Functions of Random Variables)

$$\langle u_i(y+\boldsymbol{r},t+\tau)u_jy(y,t)\rangle=\lim_{T\to\infty}\int_{-T/2}^{T/2}u_i(y+\boldsymbol{r},t+\tau)u_j(y,t)dt$$
$$=\lim_{L_i\to\infty}\frac{1}{L_i}\int_{-L_i/2}^{L_i/2}u_i(\mathbf{y}+\boldsymbol{r},t+\tau)u_j(\mathbf{y},t)dy_i$$

as long as the u_i field is temporally *and* spatially stationary. That is, the averaged products are dependent *only* on the separation variable **r** and τ and not on the averaging time or on the spatial locus of **y** for the spatial average. The ensemble average is ideally constructed of a large number of samples of the indicated product at an ensemble of **y** and t. The average is constructed for N samples in the limit as N is made large. When equivalence between the ensemble average and either of the other averages can be assumed, the process is said to be ergodic; see also Sections 1.4.2 and 3.5, Fundamentals of Correlation Analysis; The Stochastic Nature of Turbulence. In the statistics of steady fluid mechanics we deal mostly with time rather than space averages; homogeneity in space is only approximately attained in some special cases, but temporal stationarity is very often achieved. Examples where spatial stationarity may often be safely assumed are in the two dimensions in the plane of a fully developed turbulent boundary layer or in the one dimension along the span of a translating lifting surface or along the axis of a cylinder. In other circumstances, an approximation to spatial homogeneity may only be local and assumed to be maintained over distances that are only larger than the integral scale of the turbulent properties. Such approximations have limitations that must be clearly kept in mind. A specific example of this will be shown in Chapter 3, Shear Layer Instabilities, Flow Tones, and Jet Noise when we deal with turbulent jet noise.

We can also develop further the relationship between the correlation and spectrum function introduced in Section 1.4.2, Fundamentals of Correlation Analysis. The auto-spectrum $\Phi(\omega)$, the cross-spectrum $\Phi(\mathbf{r}, \omega)$, and the wave number frequency spectrum $\Phi(\mathbf{k}, \omega)$ are related to correlation functions. These relationships will be used extensively throughout this monograph. The temporal autocorrelation is

$$\langle u(\mathbf{y}, t)u(\mathbf{y}, t+\tau)\rangle = \hat{R}_{uu}(\tau) \tag{2.134}$$

so that the auto-spectrum function is

$$\Phi_{uu}(\omega) = \frac{1}{2\pi}\int_{-\infty}^{\infty} e^{i\omega t}\hat{R}_{uu}(\tau)d\tau \tag{2.135}$$

the variable u could represent any combination of physical variables, pressure, velocity, acceleration, displacement, etc.

The cross-spectral density and the wave number spectrum are related to the space–time correlation of two variables a and b for the spatially homogeneous field by Eq. (2.130)

$$\Phi_{ab}(\boldsymbol{r}, \omega) = \frac{1}{2\pi}\int_{-\infty}^{\infty} e^{i\omega t}\hat{R}_{ab}(\boldsymbol{r}, \tau)d\tau \tag{2.136}$$

For a spatially nonhomogeneous field, the equations for the correlations appearing prior to Eq. (2.126a,b) does not hold and accordingly the cross-correlation is not a function of the separation only; rather it is in general a function of the **y** and **y** + **r** variables separately. Occasionally, for simplicity the nonhomogeneity is handled by retaining separate dependence on

y and **y** + **r** in the mean-square variables and retaining $\hat{R}_{ab}$ (**y**, **y** + **r**, τ) locally as a function only of **r** and τ, i.e., say $\hat{R}_{ab}$ (**r**, τ; **y**). In this sense, the correlation function will have a similarity form that is independent of the location in the flow **y**, but its behavior with **r** may scale on a local integral correlation length that may itself depend on **y**.

The n-dimensional wave number spectrum is related to the space–time correlation function by

$$\Phi_{ab}(k_1,\ldots,k_n,\omega) = \frac{1}{(2\pi)^{n+1}}\int_{-\infty}^{\infty}\cdots\int_{-\infty}^{\infty} e^{i[\omega\tau-(k_1r_1+\cdots+k_nr_n)]} \times \hat{R}_{ab}(\boldsymbol{r},\tau)dr_1\cdots dr_n\,d\tau \tag{2.137}$$

where $n = 1$, 2, or 3 for n-dimensional **r**. The above convention allows us to denote the correlation function as homogeneous in y_1, y_3 and nonhomogeneous in $(y_2, y'_2) = (y_2, y_2 + r_2)$. The convention used in this monograph generally places the mean flow vector along the (1) axis with the lateral transverse direction along the (3) axis and the (2) direction being the stream-normal direction. In cylindrical flows the (3) direction is tangential, the (2) direction is radial, and, $dU/dy_3 = 0$, generally for circumferentially uniform flow. In the cross-stream direction, where usually U varies to produce shear (i.e., $dU/dy_2 \neq 0$), is given either the (2) axis or the radial (r) direction. It is especially in the y_2 or r direction that statistical homogeneity does not hold in a shear flow. Then Eq. (2.137) might only involve transforming only over r_1 and r_3 giving $\Phi_{ab}(k_1, \omega, y_2, y'_2)$ as the appropriate spectrum. Under these conditions the inverse transforms of Eq. (2.137) recover the cross-spectrum and the cross-correlation, i.e.,

$$\Phi_{ab}(\boldsymbol{r},\omega) = \int_{-\infty}^{\infty}\cdots\int_{-\infty}^{\infty}\Phi_{ab}(k_1,\ldots,k_n,\omega)e^{i(k_1r_1+\cdots+k_nr_n)}d^3\mathbf{k}$$

and

$$\hat{R}_{ab}(\boldsymbol{r},t) = \sqrt{\langle a^2\rangle\langle b^2\rangle}R_{ab}(\boldsymbol{r},t) = \int_{-\infty}^{\infty} e^{-i\omega\tau}\Phi_{ab}(\boldsymbol{r},\omega)d\omega$$

Correlation functions used in this monograph will generally have the normalizations on the mean square for statistically homogeneous variables. Therefore we define the normalized function

$$R_{ab}(\boldsymbol{r},\tau) = \langle a(\boldsymbol{y},t)b(\boldsymbol{y}+\boldsymbol{r},t+\tau)\rangle/[\langle a^2\rangle\langle b^2\rangle]^{1/2} \tag{2.138}$$

where the limits

$$\lim_{r\to 0} R_{ab}(\boldsymbol{r},\tau) = R_{ab}(0,\tau) = R_{ab}(\tau) \tag{2.139}$$

and

$$\lim_{\tau\to 0} R(\tau) = R(0) = 1.0 \tag{2.140}$$

apply as long as $\langle a^2(\boldsymbol{y},t)\rangle = \langle a^2(\boldsymbol{x},t)\rangle$ and similarly for $\langle b^2(\boldsymbol{x},t)\rangle$.

The spectral density functions for homogeneous statistics have integral values

$$\left[\langle a^2\rangle\langle b^2\rangle\right]^{1/2} = \int_{-\infty}^{\infty} d\omega \iiint_{\text{All } k} d^3\mathbf{k}\ \Phi_{ab}(\mathbf{k},\omega) \tag{2.141}$$

and

$$\Phi_{ab}(\omega) = \iiint_{\text{All } k} d^3\mathbf{k}\ \Phi_{ab}(\mathbf{k},\omega) \tag{2.142}$$

Occasionally throughout the text, spectrum functions will be normalized on $[\langle a^2\rangle\langle b^2\rangle]^{1/2}$ so that the integral (2.141) will be unity. In such cases the lowercase symbol ϕ rather than Φ will be used to designate the normalized spectrum functions and its normalization is

$$\int_{-\infty}^{\infty} \phi(\omega)d\omega = 1$$

2.7.3 Equivalent Integral Representation for the Acoustic Pressure

In this section the fundamental input–output relationships will be derived for structuring the problems of flow-induced noise as a random excitation of a linear deterministic acoustic or structural medium. This will be accomplished by continuing with our analysis of the sound field of free quadrupoles at subsonic speeds. As Eq. (2.118) shows, the acoustic pressure is a linear super-position of contributions of volume and surfaces distributions of sources. Often a Green's function can be selected so that even if surfaces are present (especially if the surfaces are rigid) the equation can be reduced to a simple integral over volume or surface. If surfaces are not present, of course, the integrals over S vanish. The sources are generally some combination of multipoles as depicted in Eq. (2.72).

Accordingly the acoustic pressure emanating from a region of sources can be represented by Eq. (2.119) in which the Green's function may be given by Eq. (2.112a,b), in the case of sources in free space, or by some other function if the sources are contiguous to a boundary. What Eq. (2.119) shows is that the sound pressure requires certain spatial matching, over the variable $\mathbf{y}$, between the characteristics of $G(\mathbf{x}, \mathbf{y}, \omega)$ and the characteristics of the sources. When, as is typically assumed, the sources are stochastic in time and space, then equations such as Eq. (2.119) must be handled using the representation of Section 2.7.2. We shall indicate the methods of treating such problems using Eq. (2.119) as an example, but they obviously apply to one- and two-dimensional source fields as well.

Thus the Fourier transform for the acoustic pressure in Eq. (2.119) must be considered in the same generalized sense as we have also done for the velocity. Therefore the acoustic pressure spectral density at the field point **x** is given by

$$\Phi_{pp}(\boldsymbol{x},\omega) = \lim_{T\to\infty}\frac{2\pi}{T}\{P_{\mathrm{a}}(\boldsymbol{x},\omega)P^{*}_{\mathrm{a}}(\boldsymbol{x},\omega)\} \tag{2.143}$$

and the integral of the spectral density over all frequencies is normalized according to

$$\int_{-\infty}^{\infty}\Phi_{pp}(\boldsymbol{x},\omega)d\omega = \overline{p_{\mathrm{a}}^{2}}$$

In the specific instance of radiated sound from a turbulent region, the pressure spectral density is related to the cross-spectral density of the source term,

$$\Phi_{\sigma\sigma}(\boldsymbol{y}_1,\boldsymbol{y}_2,\omega) = \lim_{T\to\infty}\frac{2\pi}{T}\{\tilde{\sigma}(\boldsymbol{y}_1,\omega)\tilde{\sigma}^{*}(\boldsymbol{y}_2,\omega)\}$$

by an integral relation that is analogous to the one for deterministic processes, Eq. (2.119),

$$\Phi_{pp}(\boldsymbol{x},\omega) = \iiint_V \cdot \iiint_V \Phi_{\sigma\sigma}(\boldsymbol{y}_1,\boldsymbol{y}_2,\omega)G^{*}(\boldsymbol{x},\boldsymbol{y}_1,\omega)G(\boldsymbol{x},\boldsymbol{y}_2,\omega)d^3\boldsymbol{y}_1\, d^3\boldsymbol{y}_2 \tag{2.144}$$

Eq. (2.144) is essentially the spectral representation of the acoustic pressure, and it is the spectral analog of combinations of integral functions in the time domain of which Eq. (2.59) is an example with Green's function being the one for free space. Actually integral equations in the form of Eq. (2.144) but involving surface integrals also appear regularly. (The equivalence can be appreciated by the reader by making the necessary substitutions into either of Eqs. (2.59) or (2.144).) An importance of Eq. (2.144) lies in the fact that the covariance and the cross-spectral density of the turbulence are (in theory) physically identifiable and measurable quantities, while the instantaneous source distribution is not necessarily a practical physical quantity to work with because it is a random variable of time and space. The utility of the spectral representation is found in the fact that it is often the acoustic intensities of specific frequencies, rather than the overall intensity, that is of importance in many applications. There are many forms of Eq. (2.144), which will be used in subsequent chapters, that involve some of the alternative source functions. For example, it may be suitable to invoke cross-spectra of any of the source terms appearing in Eq. (2.72) depending on the type of flow or flow–body interaction involved. Extensive use of Eq. (2.144) or of the methods to obtain it will be an underlying feature of the remainder of the chapters.

We shall make liberal use of the stochastic representations of this section in the remainder of this monograph. Most fluid dynamic processes that are unsteady are also turbulent. Their stochastic nature makes these, or similar time- (or space-) averaged quantities, the only useful means of representing the properties of the flow. Yet, as can be seen from a study of this chapter, the acoustic propagation characteristics are often deterministic. In these cases far-field acoustic power is just a convolution integral involving a measurable covariance or cross-spectral function and a geometrically influenced deterministic Green's function. In the case of acoustic reflection and refraction by local turbulent density and velocity fluctuations even the propagation characteristics must be considered in a stochastic sense.

Generalization of the stochastic representations to include both space and time is simple especially when we are concerned with the far-field acoustic spectrum. Restricting our attention to one frequency and a far-field distance much greater than the dimensions of the source volume the Green's function $G(\mathbf{x}, \mathbf{y}, \omega)$ can be separated into a product of functions of the source coordinate (expressing coupling of the source with the duct modes) and the field coordinate (expressing propagation away from the source zone), i.e., Eq. (2.120). Eq. (2.144) then becomes

$$\Phi_{pp}(\boldsymbol{x},\omega) = |G_x(\boldsymbol{x},\omega)|^2 \iiint_V \cdot \iiint_V \Phi_{\sigma\sigma}(\boldsymbol{y}_1,\boldsymbol{y}_2,\omega) G_y(\boldsymbol{y}_1,\omega) G_y(\boldsymbol{y}_2,\omega) d^3\boldsymbol{y}_1 d^3\boldsymbol{y}_2 \tag{2.145}$$

i.e., as long as the field point is in the far field $|\mathbf{x}| \gg |\mathbf{y}|$ and $k_0\ |\mathbf{x}| \to \infty$. We introduce the *spatial* Fourier transform pair of the source Green's function

$$\tilde{G}_y(\mathbf{k},\omega) = \iiint_{-\infty}^{\infty} G_y(\boldsymbol{y},\omega) e^{i\boldsymbol{k}\cdot\boldsymbol{y}} d^3\boldsymbol{y}$$

and

$$G_y(\boldsymbol{y},\omega) = \frac{1}{(2\pi)^3} \iiint_{-\infty}^{\infty} \tilde{G}_y(\boldsymbol{k},\omega) e^{-i\boldsymbol{k}\cdot\boldsymbol{y}} d^3\boldsymbol{k} \tag{2.146}$$

with a complex conjugate

$$G_y^*(\boldsymbol{y},\omega) = \frac{1}{(2\pi)^3} \iiint_{-\infty}^{\infty} \tilde{G}_y^*(\boldsymbol{k},\omega) e^{i\boldsymbol{k}\cdot\boldsymbol{y}} d^3\boldsymbol{k} \tag{2.147}$$

Substitution of Eqs. (2.120) and (2.146) into Eq. (2.144) yields

$$\begin{aligned} \Phi_{pp}(\boldsymbol{x},\omega) = {} & |G_x(\boldsymbol{x},\omega)|^2 \iiint_{-\infty}^{\infty} \cdot\, d^3\boldsymbol{k} \iiint_{-\infty}^{\infty} d^3\boldsymbol{\kappa} \\ & \times \left[\iiint_{-\infty}^{\infty} d^3\boldsymbol{y}_1 \iiint_{-\infty}^{\infty} d^3\boldsymbol{y}_2 \Phi_{\sigma\sigma}(\boldsymbol{y}_1,\boldsymbol{y}_2,\omega) e^{-i(\boldsymbol{k}\cdot\boldsymbol{y}_1 - \boldsymbol{\kappa}\cdot\boldsymbol{y}_2)} \right] \\ & \times \tilde{G}_y(\boldsymbol{k},\omega) \tilde{G}_y^*(\boldsymbol{\kappa},\omega) \end{aligned} \tag{2.148}$$

This integral expression for the far-field sound spectrum can be simplified considerably if the statistics of the turbulent source field are *spatially homogeneous*, i.e., that

$$\Phi_{\sigma\sigma}(y_1, y_2, \omega) = \Phi_{\sigma\sigma}(y_2, y_1, \omega) = \Phi_{\sigma\sigma}(y_2 - y_1, \omega) \tag{2.149}$$

The cross-spectral density of the source function is in this case a function only of the difference in the separation variables. Then, letting

$$\boldsymbol{y}_2 = \boldsymbol{y}_1 + \boldsymbol{r} \quad \text{and} \quad d^3\boldsymbol{y}_2 = d^3\boldsymbol{r}$$

Eq. (2.148) may be rewritten

$$\begin{aligned}\Phi_{pp}(\boldsymbol{x}, \omega) = |G_x(\boldsymbol{x}, \omega)|^2 \int_{-\infty}^{\infty} d^3\boldsymbol{k} \int_{-\infty}^{\infty} d^3\boldsymbol{\kappa}\tilde{G}_y(\boldsymbol{k}, \omega)\tilde{G}_y^*(\boldsymbol{\kappa}, \omega) \\ \times \frac{1}{(2\pi)^3}\int_{-\infty}^{\infty} \Phi_{\sigma\sigma}(\boldsymbol{r}, \omega)e^{i\boldsymbol{\kappa}\cdot\boldsymbol{r}}d^3\boldsymbol{r}\frac{1}{(2\pi)^3}\int_{-\infty}^{\infty} e^{i(\boldsymbol{k}-\boldsymbol{\kappa})\cdot\boldsymbol{y}_1}d^3\boldsymbol{y}_1\end{aligned} \tag{2.150}$$

The integral over $\mathbf{y}_1$, yields a delta function (see Section 1.6.4: Dirac Delta Function)

$$\frac{1}{(2\pi)^3}\iiint_{-\infty}^{\infty} e^{i(\boldsymbol{k}-\boldsymbol{\kappa})\cdot\boldsymbol{y}_1}d^3\boldsymbol{y}_1 = \delta(\boldsymbol{k} - \boldsymbol{\kappa}) \tag{2.151}$$

Now, introducing the wave number spectral density

$$\Phi_{\sigma\sigma}(\boldsymbol{k}, \omega) = \frac{1}{(2\pi)^3}\iiint_{-\infty}^{\infty} \Phi_{\sigma\sigma}(\boldsymbol{r}, \omega)e^{i\boldsymbol{k}\cdot\boldsymbol{r}}d^3\boldsymbol{r} \tag{2.152}$$

Eq. (2.150) reduces to the simple form

$$\Phi_{pp}(\boldsymbol{x}, \omega) = |G_x(\boldsymbol{x}, \omega)|^2 \iiint_{-\infty}^{\infty} \Phi_{\sigma\sigma}(\boldsymbol{k}, \omega)|\tilde{G}_y(\boldsymbol{k}, \omega)|^2 d^3\boldsymbol{k} \tag{2.153}$$

This equation is the spectrum analog of Eq. (2.144), and it applies to the acoustic far-field pressure of a spatially and temporally homogeneous source distribution. The equivalence between these two relationships, allowing of course for the insertion of Eq. (2.120) into Eq. (2.144), is a form of Parseval's theorem (see Refs. [7,51]).

In later chapters we shall be considering impedance relationships between the input and output spectrum functions of a linear system. For example, let $a(t)$ and $b(t)$ be the input and output variables, respectively, with $A(\omega)$ and $B(\omega)$ their generalized Fourier transforms. If $a(t)$ and $b(t)$ are linearly related

$$B(\omega) = Z(\omega)A(\omega)$$

where $Z(\omega)$ is an impedance function, then by invoking the above relationship Eq. (2.143), we find

$$\Phi_{BB}(\omega) = |Z(\omega)|^2\Phi_{AA}(\omega)$$

In this equation the spectral densities $\Phi(\omega)$ are related to the autocorrelation function by Eq. (2.135).

A useful combination of Eq. (2.179) and of Eq. (2.120) that is derived from the above for the Lighthill stress tensor is

$$P_a(\boldsymbol{x}, \omega) = G_x(\boldsymbol{x}, \omega) \iiint_{\infty}^{\infty} \left\{ \sum_{i\,j}(\boldsymbol{k}, \omega) \right\} G_y(\boldsymbol{k}, \omega) d^3\boldsymbol{k} \tag{2.154}$$

where

$$\left\{ \sum_{ij}(\boldsymbol{k}, \omega) \right\} = \frac{1}{(2\pi)^3} \iiint \frac{\partial^2 T_{\widetilde{ij}}(\boldsymbol{y}, \omega)}{\partial y_i \partial y_j} e^{i\boldsymbol{k}\cdot\boldsymbol{y}} d^3\boldsymbol{y} \tag{2.155}$$

is the generalized Fourier transform of the source density. Incorporation of the theorems of Section 2.7.2, under the assumption of homogeneity of the turbulence region will yield Eq. (2.153) directly.

Having now defined a sound *pressure* spectral density for the acoustic far field we can define the sound power spectral density following Eq. (2.16). Using the notation of Eq. (2.153)

$$\mathbb{P}_{\text{rad}}(\omega) = \iint_{S_0} \frac{\Phi_{pp}(\boldsymbol{x}, \omega)}{\rho_0 c_0} dS(\boldsymbol{x}) \tag{2.156}$$

where $\mathbb{P}_{\text{rad}}(\omega)$ is the far-field sound power spectral density.

Eqs. (2.145) and (2.153) are equivalent for spatially and temporally homogeneous stochastic fields. Their equivalence has been called Parseval's theorem [7,51]. They will be used interchangeably when an acoustic field or a structural member is linearly excited by a stochastic excitation field. Eq. (2.153) expresses that the response depends on both the spatial and temporal coupling of the turbulence and the responding medium or structure.

Eq. (2.145) is in general less restrictive than Eq. (2.153) because its application is in general not restricted to statistical homogeneity of the excitation field. These relations will be evoked for predictions of jet noise (Chapter 3: Shear Layer Instabilities, Flow Tones, and Jet Noise), eolian tone (Chapter 4: Dipole Sound From Cylinders), lifting surface noise (Chapters 11 and 12: Noncavitating Lifting Sections; Noise from Rotating Machinery), and boundary layer induced sound and vibration (Chapters 8 and 9: Essentials of Turbulent Wall-Pressure Fluctuations; Response of Arrays and Structures to Turbulent Wall Flow and Random Sound). The principal problem in *all* applications of flow-induced sound and vibration is to establish the behavior of the excitation function, exemplified by $\Phi_{\sigma\sigma}(\mathbf{k}, \omega)$, and of the response kernel, exemplified by $G_y(\mathbf{k}, \omega)$.

2.8 SOURCES IN DUCTS AND PIPES

When sources of sound are placed in ducts and pipes, the radiated sound power depends on the coupling of acoustic modes of the enclosure and the spatial directivity (or circumferential order) of the sources themselves. Three practical questions arise. First, what are the conditions under which an acoustic multipole source will generate propagating disturbances in the duct? Second, what is the relationship between propagating sound power in the duct and the source strength? Third, if the duct is terminated, what is the relationship between sound power radiated into the far field outside the duct opening and the acoustic power in the duct? Complete answers to the above questions are outside the scope of this book, but we shall examine the elementary answers to each question. More detailed examination of the acoustics of ducts are given by Morse and Ingard [6]; more practical considerations of absorption and propagation in ducts can be found in Beranek's book [52]. In this section we shall consider the fundamental behavior of sources in hard-walled ducts, adopting the context of the previously described modeling. We shall consider solutions of the inhomogeneous wave equation in the form, say, of Eq. (2.50), in which the source distribution $\sigma(\mathbf{y}, t)$ may assume any of the explicit forms given in Eqs. (2.48), (2.49), (2.72), (2.95), or (2.111); we seek a Green's function subject to the boundary conditions on the wall of zero velocity so that the acoustic pressure in the duct may be found by evaluating the appropriate form of Eq. (2.119). That is,

$$P_{\mathrm{a}}(\boldsymbol{x},\omega) = \iiint_V \tilde{\sigma}(\boldsymbol{y},\omega)G(\boldsymbol{x},\boldsymbol{y},\omega)dV(\boldsymbol{y}) \tag{2.157}$$

2.8.1 Elementary Duct Acoustics

When sources generate sound in acoustic enclosures, the radiated sound intensity depends on the coupling of the acoustic modes of the enclosure (represented by $G(\mathbf{x}, \mathbf{y}, \omega)$) and the spatial qualities of the sources (represented by $\tilde{\sigma}$ $(\mathbf{y}, \omega)$).

Acoustic sound propagation in ducts is affected by the finite cross section, which forces the sound to exist as standing waves across the duct cross section. Only certain of these modes produce propagating waves along the axis; pressures of other modes decay exponentially with axial distance from the source. To illustrate this effect, consider the sound pressure from monopole source that can exist inside the hard-walled duct of infinite axial extent illustrated in Fig. 2.14 for the case of a rectangular duct. The sound pressure at $\mathbf{x}$ of a wave of frequency ω is $p(\mathbf{x}, \omega)$, caused by a unit point source at $\mathbf{y}$ in the duct is just the Green's function satisfying Eq. (2.117) which is of the general form, where for example $\boldsymbol{x} = (x_1, x_2, x_3)$,

$$G(\boldsymbol{x},\boldsymbol{y},\omega) = \sum_{m,n} S_{mn}(x_1,x_2)S_{mn}(y_1,y_2)T_{mn}(x_3 - y_3) \tag{2.158}$$

where dependence on the source position is embedded in the functions. This Green's function satisfies the condition of zero velocity at the wall, i.e.,

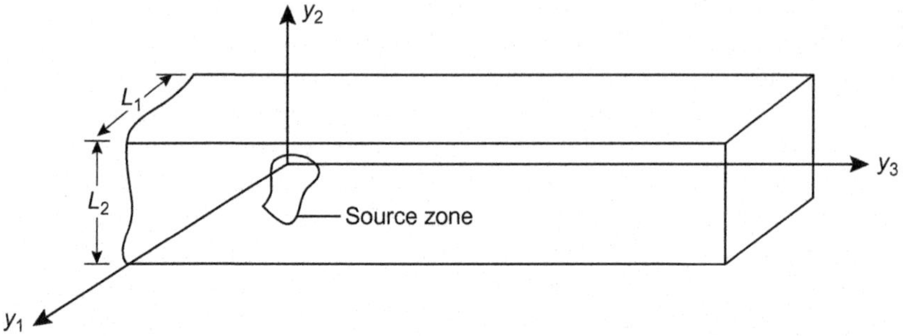

FIGURE 2.14 Coordinate system for the acoustic field in a rectangular duct.

$$\frac{\partial}{\partial x_n} = 0 \quad \text{on the inner surface of the duct} \tag{2.159}$$

The function $T_{mn}(x_3)$ is a modal transmission expressing propagation along the axis of the duct and $S_{mn}(x_1, x_2)$ are elliptic separable functions of the transverse cross section coordinates of the duct. Accordingly they satisfy the Laplacian

$$\nabla^2 S_{mn}(x_1, x_2) + k_{mn}^2 S_{mn}(x_1, x_2) = 0 \tag{2.160}$$

where the values of the wave parameter k_{mn}^2 are, as we shall see, modal constants determined by the boundary conditions, Eq. (2.159) for the specific cross section geometry.

The general form for the Green's function to be used in Eq. (2.118) will be found below. First, we note that all orders of $S_{mn}(x_1, x_2)$ are normalized over the area of the cross section to provide the orthogonality conditions, i.e.,

$$\iint_{A_D} S_{mn}(x_1, x_2) S_{pq}(x_1, x_2) dx_1, dx_2 = \begin{matrix} 0 \\ A_D \Lambda_{mn} \end{matrix} \quad \text{for} \quad \begin{matrix} m \neq p, \; n \neq q \\ m = p, \; n = q \end{matrix} \tag{2.161}$$

where A_D is the cross section area of the duct. The specific forms of $S_{mn}(x_1, x_2)$ and Λ_{mn} will be developed below and in Chapter 10, Sound Radiation from Pipe and Duct Systems for rectangular and circular ducts, respectively. The propagation function for the infinite duct, $T_{mn}(x_3)$, is independent of cross section geometry for internal potential fields expressible with separable functions and will be derived below. Substitution of Eq. (2.158) into Eq. (2.117) followed by expansion of the Laplacian and insertion of Eq. (2.160) gives

$$\sum_{m,n} S_{mn}(x_1, x_2) \left[\frac{\partial^2 T_{mn}(x_3)}{\partial x_3^2} + (k_0^2 - k_{mn}^2) \cdot T_{mn}(x_3) \right] = \delta(\boldsymbol{x} - \boldsymbol{x}_0)$$

Multiplying both sides by $S_{mn}(x_1, x_2)$, integrating over the cross section area of the duct, and invoking the orthogonality condition and normalization of Eq. (2.161) we have

$$\frac{\partial^2 T_{mn}(x_3)}{\partial x_3^2} + (k_0^2 - k_{mn}^2) \cdot T_{mn}(x_3) = \frac{S_{mn}(x_{10}, x_{20})}{A_D \Lambda_{mn}} \delta(x_3 - x_{30}) \tag{2.162}$$

$T_{mn}(x_3)$ is of the form $A_{mn} \exp\left\lfloor i\sqrt{k_0^2 - k_{mn}^2}|x_3 - x_{30}|\right\rfloor$ and

$$k_3 = \sqrt{k_0^2 - k_{mn}^2} \quad \text{with} \quad \sqrt{-1} = i \tag{2.163}$$

is the axial propagation wave number. The standard approach (e.g., [6]) for determining $A_{\rm mn}$ is to integrate Eq. (2.162) over $x_0 - \zeta$ to $x_0 + \zeta$ and in the result take the limit as ζ goes to zero, note being taken that the absolute value of the argument of the exponential creates a sign change in dT/dx_3 as x_3 passes through ζ. Accordingly one obtains

$$\lim_{|\xi| \to 0} \left[A_{mn} \left(\frac{dZ}{dx_3} \right) \right]_{-\xi}^{\xi} = A_{mn} 2ik_3 = \frac{S_{mn}(y_1, y_2)}{A_D \Lambda_{mn}}$$

This gives the monopole Green's function for the straight infinite duct with rigid wall boundaries,

$$G(\mathbf{x}, \mathbf{y}, \omega) = \sum_{m,n} \frac{S_{mn}(x_1, x_2) S_{mn}(y_1, y_2)}{i2A_D \Lambda_{mn} \sqrt{k_0^2 - k_{mn}^2}} \exp\left(\sqrt{k_0^2 - k_{mn}^2} |x_3 - y_3| \right) \tag{2.164}$$

The functions $S_{mn}(x_1, x_2)$ and the values of Λ_{mn} are specific to the cross section geometry of the duct, those for a rectangular cross section are given below.

For the rectangular cross sections of lengths L_1 and L_2,

$$S_{mn}(x_1, x_2) = \begin{pmatrix} \cos & & \cos & \\ & k_m x_1 & & k_n x_2 \\ \sin & & \sin & \end{pmatrix} \tag{2.165}$$

The boundary conditions yield

$$\begin{array}{llll} k_m = 2m\pi/L_1, & k_n = 2n\pi/L_2, & m, n = 0, 1, 2 \ldots & \text{for even modes} \\ k_m = m\pi/L_1, & k_n = n\pi/L_2, & m, n = 1, 3, 5 \ldots & \text{for odd modes} \end{array}$$

The cosine and sine functions describe "mode shapes" of the standing acoustic waves across the duct cross section. Even modes are those that are symmetric about the center lines $x_1 = 0$ and $x_2 = 0$ and so are described by the cosine functions; odd modes are antisymmetric and described by the sine functions.

Eq. (2.163) shows that as long as k_{mn} is less than k_0, the wave number k_3 is real and the waves propagate unattenuated. However, if k_0 is less than k_{mn}, then k_3 is imaginary so that

$$k_3 = \begin{cases} +i\sqrt{k_{mn}^2 - k_0^2}, & x_3 > 0 \\ -i\sqrt{k_{mn}^2 - k_0^2}, & x_3 < 0 \end{cases}$$

Thus for a given excitation wave number, say $(k_{mn})_e = \sqrt{(k_m)_e^2 + (k_n)_e^2}$, if the frequency is so low that $k_0 < (k_{m,n})_e$, then the sound will not propagate along the duct. The frequency ω_c for which

$$(k_{m,n})_e = k_0 = \omega_c / c_0 \tag{2.166}$$

is called the *cut-off* frequency of the mode. There is an exceptional mode, however, for which propagation always occurs. When $m = n = 0$, i.e., when the propagating wave is a plane wave across the duct, then $k_3 = k_0$ and no attenuation can occur. Thus plane waves always propagate in ducts.

The Green's function for a monopole source of unit strength set at position $\mathbf{y}$ in the duct of rectangular cross section is found by inserting the mode shape functions Eq. (2.165) into Eq. (2.161) to find

$$A_D \Lambda_{mn} = \frac{L_1 L_2}{4\varepsilon_m \varepsilon_n}$$

where

$$\begin{aligned} \varepsilon_m &= \varepsilon_n = 2, \quad m, n = 0 \\ \varepsilon_m &= \varepsilon_n = 1, \quad m, n \neq 0 \end{aligned}$$

so that the Green's function for the sound pressure radiated by the unit monopole source is given immediately by Eq. (2.164)

$$G_m(\boldsymbol{x}, \boldsymbol{y}, \omega) = \frac{2i}{L_1 L_2} \sum_{mn} \frac{S_{mn}(y_1, y_2) S_{mn}(x_1, x_2)}{\varepsilon_m \varepsilon_n \sqrt{k_0^2 - k_{mn}^2}} e^{i\sqrt{k_0^2 - k_{mn}^2}|x_3 - y_3|} \tag{2.167}$$

In like manner, for a distribution of dipoles aligned along the duct axis, say, of strength $\partial F_3/\partial x_3$, the Green's function is

$$G_{d_3}(\boldsymbol{x}, \boldsymbol{y}, \omega) = \frac{\partial}{\partial y_3}(G_m(\boldsymbol{x}, \boldsymbol{y}, \omega)) \tag{2.168a}$$

$$G_{d_3}(\boldsymbol{x}, \boldsymbol{y}, \omega) = \frac{-2}{L_1 L_2} \sum_{mn} \frac{S_{mn}(y_1, y_2) S_{mn}(x_1, x_2)}{\varepsilon_m \varepsilon_n} e^{i\sqrt{k_0^2 - k_{mn}^2}|x_3 - y_3|} \tag{2.168b}$$

and for a dipole oriented transversely to the axis of the duct (say -directed along "1" in Fig. 2.14)

$$G_{1_3}(\boldsymbol{x}, \boldsymbol{y}, \omega) = \frac{\partial}{\partial y_1}(G_m(\boldsymbol{x}, \boldsymbol{y}, \omega)) \tag{2.169a}$$

and

$$G_{d_1}(\boldsymbol{x}, \boldsymbol{y}, \omega) = \frac{2i}{L_1 L_2} \sum_{mn} \frac{\frac{\partial S_{mn}(y_1, y_2)}{\partial y_1} S_{mn}(x_1, x_2)}{\varepsilon_m \varepsilon_n \sqrt{k_0^2 - k_{mn}^2}} e^{i\sqrt{k_0^2 - k_{mn}^2}|x_3 - y_3|} \tag{2.169b}$$

These functions can now be inserted in the Helmholtz integral equation, Eq. (2.118). The propagation of sound therefore depends not only on the wave number of the mode excited relative to the acoustic wave number, but also on the local admittance of the duct, which is expressed by the mode shape evaluated at the source location: $S_{mn}(y_1, y_2)$.

2.8.2 Radiation From Multipoles in an Infinitely Long Pipe

When an axially-oriented dipole force $f_3(y_1, y_2, \omega)\ \delta(y_3)$ is concentrated on the axis of the duct, $y_3 = 0$; the sound pressure emitted by it is given by Eq. (2.121b) using the Green function given by Eqs. (2.168a,b); i.e.,

$$P_a(\boldsymbol{x}, \omega) = \iint_{A_s} G_{d_3}(\boldsymbol{x}, y_1, y_2, y_3 = 0, \omega) f_3(y_1, y_2, \omega) dy_1\, dy_2 \tag{2.170}$$

where $f_3(\mathbf{y}, \omega)$ is the component of force per unit length projected along the duct axis and where we will let A_s be the surface area of the dipole projected in the axial direction.

An important consequence of the duct at low frequencies ($k_0\sqrt{A_D} \ll 1$) is that for low orders of m and n, such that $k_0 < k_{mn}$ $(m, n > 0)$ propagation of higher-order modes is prevented because of the conditions leading to Eq. (2.160). Above this cut-off frequency only axially directed dipoles will radiate. Accordingly, when $m = n = 0$ in Eq. (2.152) and with the condition

$$\frac{\partial G_m}{\partial y_1} = \frac{\partial G_m}{\partial y_2} = 0$$

Eq. (2.170) gives

$$P_a(\boldsymbol{x}, \omega) = \frac{1}{2A_D} F_3 e^{\pm i k_0 x_3} \tag{2.171}$$

This describes plane waves of amplitude $F_3/2A_D$ propagating up and down the duct. Since $P_a(\mathbf{x}, \omega)$ is quadratically related to a sound pressure spectral density in the sense of Eq. (2.130), and $F_3(\omega)$ is related to a force spectral density $\Phi_{FF}(\omega)$, then the sound power spectral density in the duct radiated to one side of the source is, by Eq. (2.156)

$$[\mathbb{P}(\omega)]_{\text{Duct}} = \frac{\Phi_{FF}(\omega)}{4\rho_0 c_0 A_D} \tag{2.172}$$

where $k_0 L_1$ and $k_0 L_2$ are both much less than unity. The *total* sound power radiated to both sides is just twice this value.

The same force dipole in free space would generate a pressure amplitude deduced from Eq. (2.75)

$$P_a(\boldsymbol{x}, \omega) = \frac{k_0 F_3 \cos\theta}{4\pi r} e^{i k_0 r}$$

so that the sound power spectrum of the force in free space is, by Eq. (2.156) and Table 2.1

$$[\mathbb{P}(\omega)]_{\text{Free}} = \frac{1}{3}\left(\frac{k_0}{4\pi r}\right)^2 \frac{\Phi_{FF}(\omega)}{\rho_0 c_0} 4\pi r^2 = \frac{k_0^2 \Phi_{FF}(\omega)}{12\pi \rho_0 c_0} \tag{2.173}$$

The ratio of acoustic power radiated into free space by the dipoles to the total radiated into the duct at low frequencies is

$$\frac{[\mathbb{P}(\omega)]_{\text{Free}}}{[\mathbb{P}(\omega)]_{\text{In duct}}} = \frac{k_0^2 A_{\text{D}}}{3\pi} \tag{2.174}$$

for $\sqrt{k_0^2 A_{\text{D}}} \ll 1$. This result is independent of the geometry of the duct cross section or the distance of the source from the wall. It requires that the walls be hard, that the axis of the dipole coincide with the duct axis, and that standing waves are not developed between the source and the opening. In practice this might be achieved through slight absorption of sound by the walls. Eq. (2.174) is useful in converting source levels as deduced; say, by pressure transducers on the duct or pipe wall into equivalent far-field sound pressure levels.

The sound power in hard-walled pipe from enclosed quadrupoles at low frequencies ($k_0^2 A_{\text{D}} \ll 1$) also differs from that radiated by the same sources in the free-field sound. Using $G(\mathbf{x}, \mathbf{y}, \omega)$ given in Eq. (2.167), Davies and Ffowcs Williams [45] demonstrated that the only quadrupoles that radiate are axially oriented; i.e., the 3,3 combinations for which only $\Delta G = \partial\Delta/\partial y_3 \neq 0$ and $\partial^2 T_{33}/\partial y_3^2 \neq 0$. This is because only the plane wave propagation down the duct may occur at low frequencies. All other orientations of quadrupoles will not produce radiation far from the source. In the limit of vanishing Mach number the intensity of sound radiated down the pipe will be of the form (adopted previously in Eq. (2.62))

$$I(x_3) \sim \frac{V_0}{\rho_0 c_0^3}\left[\rho_0 U_c^2\right]^2 \left[\frac{U_c}{\Lambda}\right]^2 \Lambda \tag{2.175}$$

where Λ is the correlation length of the quadrupoles along the axis and Λ/U_c represents the time scale of the turbulence. This shows that the radiated sound power of longitudinal axial quadrupoles in a duct is enhanced over that in free space by an order $(\text{M}_c)^{-2}$ (for $M_c \ll 1$), but that radiation from other orientations is suppressed. At frequencies high enough that acoustic cross modes occur, the radiation from quadrupoles in a duct or pipe is essentially that of free space.

Although the above analysis has been worked out for ducts of rectangular cross section, circular ducts behave similarly as will be discussed in Chapters 10 and 12, Sound Radiation from Pipe and Duct Systems; Noise from Rotating

Machinery. For ducts with acoustically reactive walls, the formulation must be reconsidered along the lines of Eq. (2.118) in which a relationship can be drawn between $P_a(\mathbf{y}, \omega)$ and $\partial P_a(\mathbf{y}, \omega)/\partial n$ (see, e.g., Ref. [45]).

2.8.3 Radiation From the Opening of a Semi-Infinite Duct

This section will examine the acoustic power radiated down to the termination of a semi-infinite duct and subsequently radiated from the opening of the duct into a free space. The problem was considered by Heller and Widnall [53] and Heller et al. [54] in the limit of plane duct modes as illustrated in Fig. 2.15. When a source is placed in the duct, a distance x_3 from the open end, the power radiated down to the unterminated end is equal to the power that would be radiated to one side of an infinite pipe (as given by Eq. (2.172) in the case of an axial dipole) plus power reflected back from the open end. As long as the source is far enough inside the duct, say, $|x_3| > L_1L_2$, then the cross modes will be same as those responsible for the modal functions $S_{mn}(y_1, y_2)$ in Eq. (2.173). At the open end, $x_3 = L$, the acoustic waves face the impedance discontinuity so that

$$P_a(L, \omega) = Z_a u(L, \omega) \tag{2.176}$$

where Z_a is the impedance of the opening. If it is assumed that the duct is terminated in a baffle, as illustrated in Figs. 2.15 and 2.16 and if it is assumed that the frequencies are low enough that only plane waves propagate in the duct, then the fluid motion on the plane $x_3 = L$ represents a piston in a baffle for which the acoustic impedance is [27]

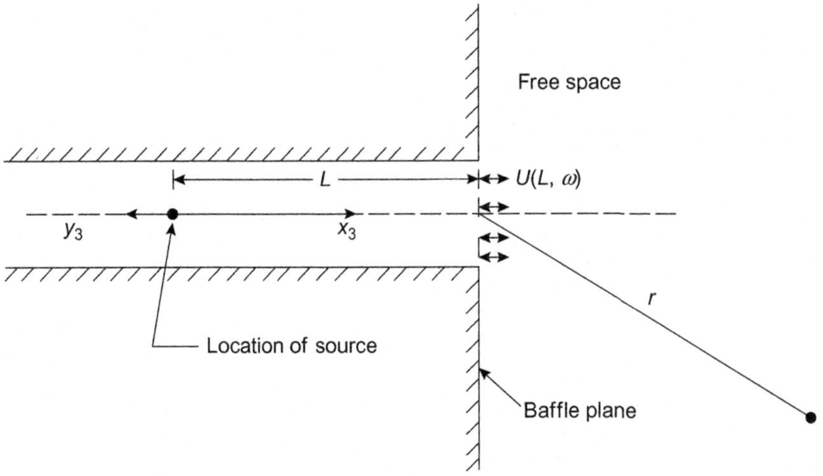

FIGURE 2.15 Cross section of a duct or pipe terminated in an infinite rigid baffle; y_3 denotes source coordinate, x_3 denotes field coordinate in duct.

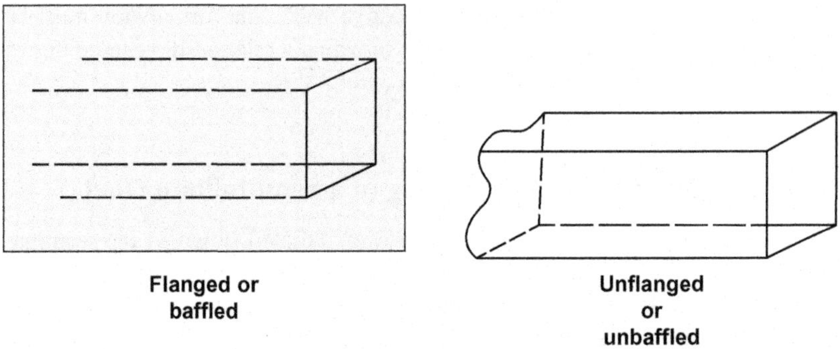

FIGURE 2.16 Flanged and unflanged duct exits.

$$Z_a = \rho_0 c_0 \left\{ \left(\frac{k_0^2 A_D}{2\pi} \right) + i \frac{8(k_0^2 A_D)^{1/2}}{3\sqrt{\pi^3}} \right\} \tag{2.177}$$

which can be derived from the methods presented later in Section 5.5, Essential Features of Structural Radiation. Since $\sqrt{k_0^2 A_D} \ll 1$, $(Z_a/\rho_0 c_0)$ is small enough to permit the imposition of a duct boundary condition of $P_a(x_3 = L, \omega)/[\rho_0 c_0 u(x_3 = L, \omega)] \approx 0$.

Accordingly for a source distributed over a distance y_3 situated a distance L from the end the low-frequency Green's function is

$$G_m(x_3, y_3, \omega) = \left(\frac{i}{2k_3 A_D} \right) \begin{cases} e^{ik_3(x_3 - y_3)} - e^{ik_3(x_3 - y_3 - 2L)} & x_3 \geq y_3 \\ e^{-ik_3(x_3 - y_3)} \left(1 - e^{2ik_3(L - y_3)} \right) & x_3 \leq y_3 \end{cases} \tag{2.178}$$

for $\sqrt{k_0^2 A_D} \ll 1$.

In a manner following that of the preceding section the acoustic pressure amplitude propagating toward the open end is

$$P_a(x_3, \omega) = \frac{F_3}{2A_D} \left(e^{ik_0 x_3} - e^{-ik_0(x_3 - 2L)} \right) \tag{2.179}$$

The second exponential function arises from the fact that all the power radiated by the source in the direction $x_3 > 0$ is reflected back down the duct. The radiated sound power that is transmitted down the infinite leg of the duct $(x_3 < 0)$ is just twice the value given by Eq. (2.172), i.e.,

$$[\mathbb{P}(\omega)]_{\text{In duct}} = \frac{\Phi_{FF}(\omega)}{2\rho_0 c_0 A_D}. \tag{2.180}$$

The acoustic power radiated from the baffled (flanged) open end of the pipe at $x_3 = L$ can be found from the relationship

$$[\mathbb{P}(\omega)]_{\text{Flanged}} = (Z_a)_{ac} \Phi_{uu}(L, \omega) A_p$$

where $(Z_a)_{ac}$ is the real part Eq. (2.177) and $\Phi_{uu}(L, \omega)$ is the spectral density of the acoustic particle velocity at the open end as obtained below. Since

$$U(x_3, \omega) = \frac{1}{i\omega\rho_0}\frac{\partial P_a(x_3, \omega)}{\partial x_3}$$

the velocity spectrum as found by the methods of Section 2.7 is

$$\Phi_{uu}(L, \omega) = \frac{\Phi_{FF}(\omega)}{A_D\rho_0 c_0}$$

and the power radiated from the flanged, or baffled, end due to a dipole well upstream of the duct opening is

$$[\mathbb{P}(\omega)]_{\text{Flanged}} = \frac{k_0^2\Phi_{FF}(\omega)}{2\pi\rho_0 c_0} \tag{2.181}$$

for $(k_0^2 A_D)^{1/2} \ll 1$, and ratio of sound power radiated from the opening of the flanged (baffled) duct to that generated by the dipole within the infinite duct (well upstream of the opening) is

$$\frac{[\mathbb{P}(\omega)]_{\text{Flanged}}}{[\mathbb{P}(\omega)]_{\text{In duct}}} = \frac{k_0^2 A_D}{\pi} \tag{2.182}$$

Eq. (2.181) is an important result because it shows that the radiation from a dipole in a flanged pipe will be six times (8 dB greater than) the radiation from a dipole of the same strength in free space. It will have the same frequency dependence, however. If the pipe is unflanged (Fig. 2.16) Levine and Schwinger [55] (see also Ref. [6] and Fig. 2.17) have shown that the acoustic power radiated out of the end is just one-half that radiated from the flanged pipe. Therefore the dipole radiation from the unflanged end will be

$$[\mathbb{P}(\omega)]_{\text{Unflanged}} = \frac{k_0^2\Phi_{FF}(\omega)}{4\pi\rho_0 c_0} \tag{2.183}$$

and

$$\frac{[\mathbb{P}(\omega)]_{\text{Unflanged}}}{[\mathbb{P}(\omega)]_{\text{In duct}}} = \frac{k_0^2 A_D}{2\pi} \tag{2.184}$$

This, at frequencies below the cut-off frequency of the duct, if the source is located farther than multiples of L_1 or L_2 of the end, then the acoustic power radiated from the pipe is expected to be bounded by Eq. (2.173), for the free dipole, and either Eq. (2.181) or (2.183) for the ducted dipole. In both of the cases of end-baffling, the ratio of sound power radiated from a duct opening by an enclosed axial dipole to that of the same dipole in free space is 3/2 and 3 for the flanged and unflanged duct, respectively.

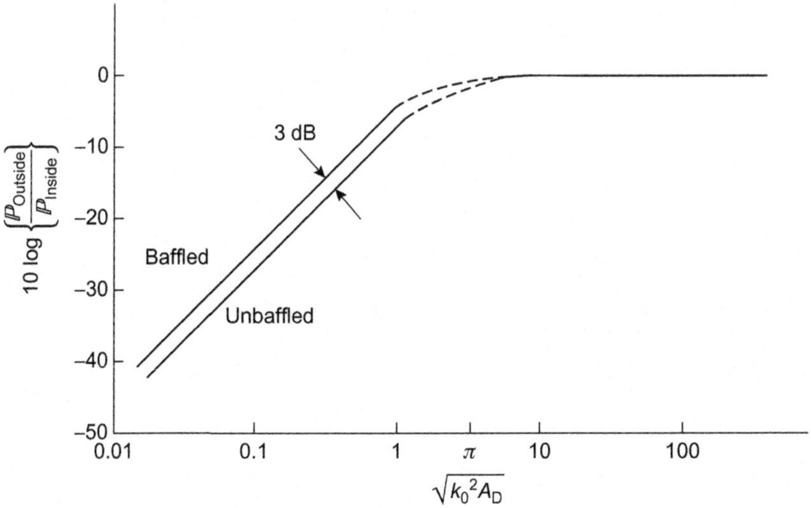

FIGURE 2.17 Sound power radiated from the open end of a semi-infinite duct relative to that developed inside the duct (Eq. (2.182) for baffled and Eq. (2.184) for unbaffled).

In the opposite limit of frequencies above the cut-off frequency, $\sqrt{k_0^2 A_D} \gg 1$, the acoustic impedance appearing in Eq. (2.176) is virtually $\rho_0 c_0$; therefore no internal reflections occur at the duct exit. Accordingly all the power that is radiated by the source in the direction of $x_3 > 0$ will (except for absorption) be radiated out the open end. As far as the acoustic power generated within the duct is concerned, that radiation will be governed by the Green's function given by Eq. (2.166). In this case the radiation is not confined only to those sources aligned with the axis of the duct because $\partial G_y/\partial y_1$ and $\partial G/\partial y_2$ are not necessarily zero (see Section 12.7). At very high frequencies such that the duct dimensions are much larger than an acoustic wavelength and the source dimensions are small compared with a duct dimension, the source excites many cross modes of the duct. In such cases, Davies and Ffowcs Williams [45] have shown the radiated sound power to be identical to that emitted in free space.

To summarize, Fig. 2.18 illustrates the power radiated from the open end by axial dipoles in a duct compared with that emanated by dipoles of the same strength in free space. Ratios at low frequencies and high frequencies will take the simple values shown. In an intermediate range of $\sqrt{k_0^2 A_D}$ there will be an interference pattern dependent on the location of the source relative to the mode shape of the duct. The sound power in this range of k_0 will therefore depend on the geometry of the duct. In the case of dipole sources in ducts radiating into free space, from a flanged opening according to the geometry illustrated in Fig. 2.14 the radiation to the outside depends on the orientation of the source within duct. As shown in Fig. 2.18 when sources are deep within the duct, $L/D \gg 1$, only those sources oriented with their axes

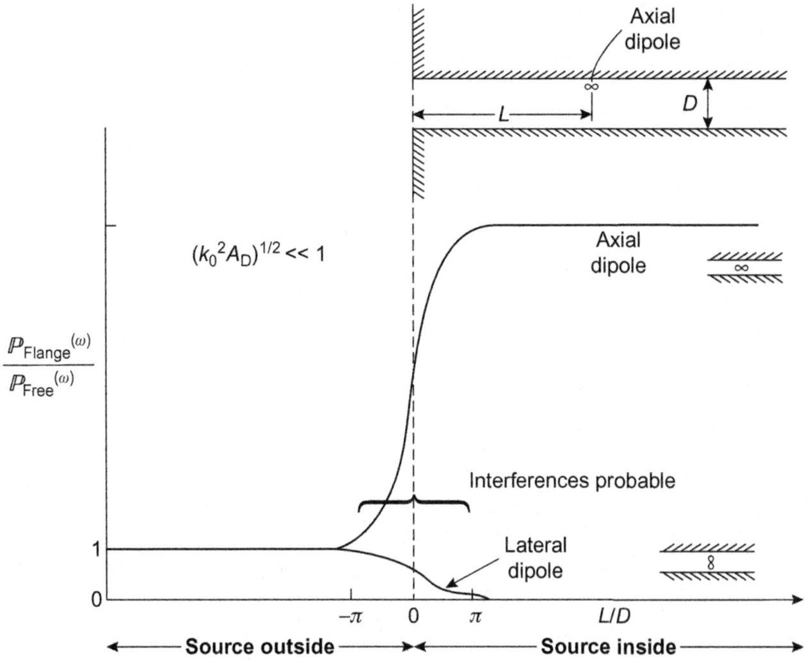

FIGURE 2.18 Sound power radiated into free space from the open end of a flanged pipe by dipole: inside the pipe compared to sound power radiated by dipoles of the same strength in an unbounded fluid.

coincident with that of the duct will radiate sound outside. The effects of the duct will be to enhance the radiation efficiency of the dipole. With the source near the opening, but still inside the pipe, the power radiated will be influenced by interferences due to scattering from the edges of the opening. When the source is just outside the entrance of the pipe, dipoles aligned with or lateral to the duct axis will radiate basically as if in the free field. These concepts have considerable practical engineering importance as discussed in Chapters 10 and 12, Sound Radiation from Pipe and Duct Systems; Noise from Rotating Machinery and in Ref. [56].

REFERENCES

[1] Milne-Thompson LM. Theoretical hydrodynamics. 4th ed. New York: Macmillan; 1960.

[2] Batchelor GK. An introduction to fluid dynamics. London and New York: Cambridge University Press; 2000.

[3] Sabersky RH, Acosta AJ, Hauptmann EG, Gates EM. Fluid flow. Upper Saddle River, N.J.: Prentice Hall; 1998.

[4] White FM. Fluid mechanics. 8th ed. New York: McGraw-Hill; 2015.

[5] Kundu PK, Cohen IM, Dowling D. Fluid mechanics. 6th ed. Waltham, Mass: Academic Press; 2015.

[6] Morse PM, Ingard KU. Theoretical acoustics. New York: McGraw-Hill; 1968.

[7] Pierce AD. Acoustics: an introduction to its physical principles and applications. New York: Acoustical Society of America; 1989.
[8] Kinsler LE, Frey AR, Coppens AB, Sanders JV. Fundamentals of acoustics. 4th ed. Hoboken, N.J.: Wiley; 2000.
[9] Fahey F, Gardonio P. Sound and structural vibration. 2nd ed. Waltham, Mass: Academic Press; 2007.
[10] Junger MC, Feit D. Sound structures and their interaction. Cambridge, MA: MIT Press; 1972.
[11] Sommerfeld A. "Partial differential equations in physics," lectures on theoretical physics, vol. 6. New York: Academic Press; 1964.
[12] Howe MS. Acoustics of fluid–structure interactions. Cambridge University Press; 1998.
[13] Lighthill MJ. On sound generated aerodynamically, I, General theory. Proc R Soc London, Ser A 1952;211:564–87.
[14] Lighthill MJ. On sound generated aerodynamically, II, Turbulence as a source of sound. Proc R Soc London, Ser A 1954;222:1–32.
[15] Lighthill MJ. Sound generated aerodynamically. Proc R Soc London, Ser A 1962;267: 147–82.
[16] Schlichling H. Boundary layer theory. New York: McGraw-Hill; 1979.
[17] Bateman H. Partial differential equations of mathematical physics. London and New York: Cambridge University Press; 1959.
[18] Stratton JA. Electromagnetic theory. New York: McGraw-Hill; 1941.
[19] Jackson JD. Classical electrodynamics. 3rd ed. New York: Wiley; 1999.
[20] Jones DS. The theory of electromagnetism. Oxford: Pergamon; 1964.
[21] Proudman I. The generation of noise by isotropic turbulence. Proc R Soc London, Ser A 1952;214:119–32.
[22] Crow SC. Aerodynamic sound emission as a singular perturbation problem. Stud Appl Math 1970;49:21–44.
[23] Curle N. The influence of solid boundaries upon aerodynamic sound. Proc R Soc London, Ser A 1955;231:505–14.
[24] Powell A. Aerodynamic noise and the plane boundary. J Acoust Soc Am 1960;32:982–90.
[25] Lamb H. Hydrodynamics. 6th ed. New York: Dover; 1945.
[26] Ffowcs Williams JE. Sound radiation from turbulent boundary layers formed on compliant surfaces. J Fluid Mech 1965;22:347–58.
[27] Gutin L. On the sound field of a rotating airscrew. Phys Z Sowjetunion A 1938;1:57 Translation NACA Tech Memo 1195, 1948.
[28] Goldstein ME. Aeroacoustics. New York: McGraw-Hill; 1976.
[29] Glegg, S., Devenport, W. The aeroacoustics of low Mach number flows: fundamentals, analysis, and measurements. Waltham, Mass: Academic Press; 2017.
[30] Ffowcs Williams JE, Hawkings DL. Sound generation by turbulence and surfaces in motion. Phil Trans R Soc London, Ser A 1969;264A(1151):321–42.
[31] Lowson MV. The sound field for singularities in motion. Proc R Soc, Ser A 1965;286: 559–72.
[32] Dowling A. Convective amplification of real simple sources. J Fluid Mech 1976;74(Part 2): 529–46.
[33] Crighton DG, Dowling AP, Ffowcs Williams JE, Heckl MFG. Modern methods in analytical acoustics lecture notes. Springer-Verlag; 1996.
[34] Powell A. Theory of vortex sound. J Acoust Soc Am 1964;36:177–95.

[35] Howe MS. Contributions to the theory of aerodynamic sound, with application to excess jet noise and the theory of the flute. J Fluid Mech 1975;71:625–73.

[36] Lauvstad VR. On nonuniform Mach number expansion of the Navier-Stokes equations and its relation to aerodynamically generated sound. J Sound Vib 1968;7:90–105.

[37] Obermeier F. Berechnung Aerodynamisch Erzeugter Schallfelder Mittels der Methode der "Matched asymptotic expansions" (L). Acustica 1967;18:238–9.

[38] Möhring WF, Müller E-A, Obermeier F. Schallerzeugung durch instationäre Strömung als singuläres Störungsproblem. Acustica 1969;21:184–8.

[39] Obermeier F. On a new representation of aeroacoustic source distribution, I. General theory, II. Two-dimensional model flows. Acustica 1979;42:58–71.

[40] Tichmarsh EC. Introduction to the theory of Fourier integrals. 2nd ed. London: Oxford University Press; 1948.

[41] Morse PM, Feshbach H. Methods of theoretical physics. New York: McGraw-Hill; 1953.

[42] Ffowcs Williams JE, Hall LH. Aerodynamic sound generation by turbulent flow in the vicinity of a scattering half plane. J Fluid Mech 1970;40:657–70.

[43] Chase DM. Sound radiated by turbulent flow off a rigid half-plane as obtained from a wave-vector spectrum of hydrodynamic pressure. J Acoust Soc Am 1971;52:1011–23.

[44] Chase DM. Noise radiated from an edge in turbulent flow. AIAA J 1975;13:1041–7.

[45] Davies HG, Ffowcs Williams JE. Aerodynamic sound generation in a pipe. J Fluid Mech 1968;32:765–78.

[46] Crighton DG, Ffowcs Williams JE. Real space-time Green's functions applied to plate vibration induced by turbulent flow. J Fluid Mech 1969;38:305–13.

[47] Batchelor GK. The theory of homogeneous turbulence. London and New York: Cambridge University Press; 1960.

[48] Lin YK. Probabilistic theory of structural dynamics. New York: McGraw-Hill; 1967.

[49] Kinsman B. Wind waves. Englewood Cliffs, NJ: Prentice Hall; 1965.

[50] Crandall SH. Random vibration, vol. I. Cambridge, MA: MIT Press; 1958.

[51] Lee YW. Statistical theory of communication. New York: Wiley; 1960.

[52] Beranek LL, editor. Noise and vibration control. New York: McGraw-Hill; 1971.

[53] Heller HH, Widnall SE. Sound radiation from rigid flow spoilers correlated with fluctuating forces. J Acoust Soc Am 1970;47:924–36.

[54] Heller HH, Widnall SE, Gordon CG. Correlation of fluctuating forces with the sound radiation from rigid flow-spoilers. Bolt Beranek and Newman Inc. Rep. 1734; 1968.

[55] Levine H, Schwinger J. On the radiation of sound from an unflanged circular pipe. Phys Rev 1948;73:383–406.

[56] Guerin S, Thomy E, Wright MCM. Aeroacoustics of automotive vents. J Sound Vib 2005;285:859–75.

Chapter 3

Shear Layer Instabilities, Flow Tones, and Jet Noise

3.1 INTRODUCTION

Whether or not a moving fluid is stable or unstable to some applied stimulation, say, an incident sound field, adjacent surface vibration, or buffeting from upstream turbulence, has largely to do with the spatial gradient and curvature of the mean velocity profile in the flow. A wide range of flow types, including jets, wakes, and flow-over cavities are not stable to applied disturbances. Often the time dependence of fluid motions in these types of flow is characterized by a predominant frequency that is dependent on a characteristic average velocity and a characteristic linear dimension of the flow region. As shown in the preceding chapter (e.g., Eqs. (2.46) and (2.86a)) sound is potentially produced whenever there is a disturbance-filled fluid region. Furthermore as shown in Chapter 2, Theory of Sound and its Generation by Flow the presence of surfaces complicates the sound field by providing not only acoustic reflections but also modifications in the primary hydrodynamic flow field that is responsible for the disturbance region. Therefore in this chapter, we shall consider in a basic fashion the unstable characteristics of flow that are required to create fluid disturbances, and relate those characteristics to the eventual breakdown into both regular and random vortex structures. We shall also introduce many of the analytical and experimental techniques that are used when the flow disturbances become irregular or turbulent.

As practical applications of the general theory of shear layer disturbances we shall develop rules for predicting the occurrence of various types of vortex-induced tones in holes, cavities, and obstructed jets. The part played by ambient turbulence in the basic flow and the influence of the Reynolds number on the vortex structures will be shown. Finally, some fundamental concepts in the similarity principles that govern noise from turbulent jets and some experimental approaches to validate those concepts will be introduced as a foundation to other flow types to be discussed in the body of this monograph.

Mechanics of Flow-Induced Sound and Vibration, Volume 1.
DOI: http://dx.doi.org/10.1016/B978-0-12-809273-6.00003-8

The disturbances in wakes behind cylinders and hydrofoils are, by themselves, so important that they will be dealt with separately in later chapters.

3.2 SHEAR FLOW INSTABILITIES AND THE GENERATION OF VORTICITY

Unstable flows are generally those that have gradients of mean velocity; the classical types, which have been extensively examined analytically and experimentally, are illustrated in Fig. 3.1. Profile A was the first to be theoretically examined by Helmholtz in 1868, see Rayleigh [1] or Lamb [2], who showed that the arrangement is unstable to disturbances of any frequency or

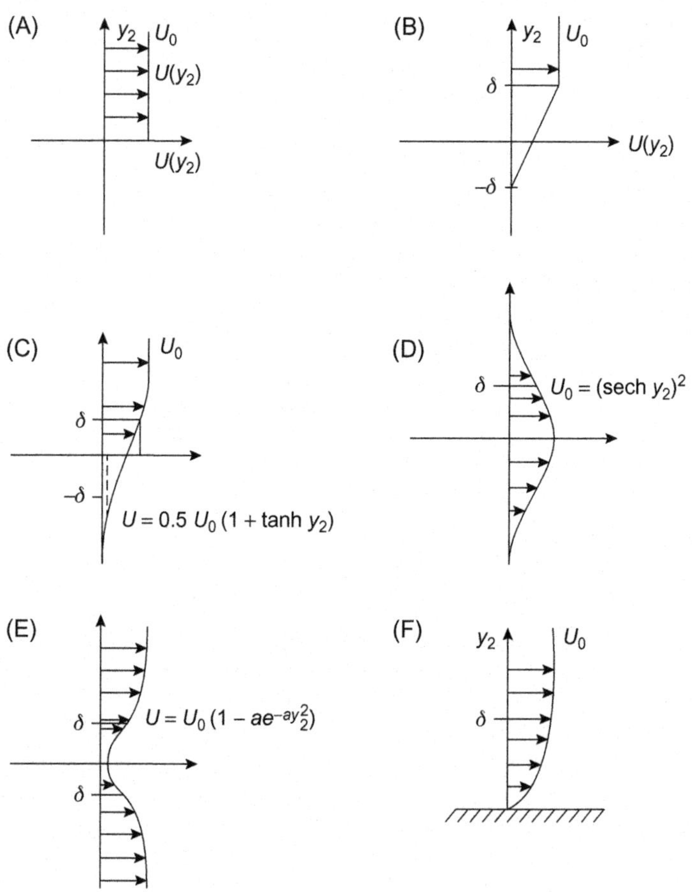

FIGURE 3.1 Classical types of shear flow: (A) discontinuous, (B) linear, (C) hyperbolic tangent, (D) jet, (E) Gaussian wake, and (F) Blasius laminar boundary layer. Maximum free-stream velocity has been normalized to unity.

wavelength. In this instance, the gradient of the velocity in y_2 is singular at the interface $y_2 = 0$ and the interface is therefore said to constitute a vortex sheet, i.e.,

$$\frac{\partial u_1(y_2)}{\partial y_2} = \lim_{\epsilon \to 0} \frac{U_1(y_2 = \varepsilon) - U_1(y_2 = -\varepsilon)}{2\varepsilon} = \omega_3 \delta(y_2)$$

which describes a distribution of vorticity $\omega_3 = \partial u_1 / \partial y_2$ that is zero everywhere but on the surface $y_2 = 0$. In more realistic circumstances the interface between the two moving fluids is less well defined, as in illustrations B and C. The former case of a linear velocity profile provides a constant region of vorticity $\omega_2 = U_1/\delta$ inside the region of $|y_2| < \delta$. This case has been examined by Rayleigh [1], Squire [3], and Esch [4], and the latter has been extensively examined by Michalke [5–7], Browand [8], Esch [4], Sato [9], Schade [10], and Tatsumi and Gotoh [11]; the hyperbolic tangent profile has been experimentally examined by Browand [8], and it has practical application in the production of cavity tones. The definition of δ is such that $d^3U(\delta)/dy^3 = 0$, i.e., that the curvature is a maximum. The jet profile D, approximated by a hyperbolic secant squared, has been examined by Sato and Sakao [12] and Sato [13]; the wake E, approximated by a Gaussian velocity profile, has been examined by Sato and Kuriki [14] both analytically and experimentally. Finally, the Blasius form of the boundary layer mean velocity profile has been exhaustively examined analytically and experimentally; extensive surveys include those of Lin [15], Betchov and Criminale [16], and Schlichting [17]. We shall reserve further comments on boundary layer waves and stability for Chapter 8, Essentials of Turbulent Wall-Pressure Fluctuations.

The traditional analyses of flow stability begin with an assumption that a small-magnitude disturbance is present in the flow and we are interested in the growth of that disturbance in time or space as it moves along with the remainder of the fluid, although see Tam and Morris [240] for an example of a more comprehensive treatment. Thus the total fluid velocity (mean plus fluctuating) is written in a two-dimensional mean flow field as

$$\boldsymbol{U} = U_1(y_2) + \mathbf{u}(y, y_2, t) \tag{3.1}$$

where the disturbance $\mathbf{u}(y_1, y_2, t)$ vanishes as $y_2 \to \pm\infty$ in an unbounded medium and simply as $y_2 \to \infty$ in a boundary layer. These problems are generally set in two-dimensional mean flow; then, according to Squires' theorem [3], the most unstable disturbance waves are those whose wave propagation directions are aligned with the flow direction. Now classically it is generally assumed further that the disturbances may be written as the gradient of a 2-dimensional stream function $\psi(y_1, y_2, t) = \Phi(y_2)e^{i\alpha(y_1 - ct)}$

and

$$\boldsymbol{u}(y_1, y_2, t) = (u(y_1, y_2, t), v(y_1, y_2, t)) = \nabla\psi(y_1, y_2, t)$$

so we can let $u(y_2) = \partial\Phi/\partial y_2$ giving

$$u(y_1, y_2, t) = u(y_2)e^{i\alpha(y_1 - ct)} \tag{3.2}$$

where the real wave number α is related to the frequency at which waves travel passed a fixed point of observation at the wave speed, c_r,

$$\omega/\alpha = c_r, \tag{3.3}$$

and c_r is the real part of the complex velocity

$$c = c_r + ic_i \tag{3.4}$$

Thus described, the disturbance amplitude has been modeled to grow exponentially in time at a rate

$$\frac{|u(y_1, y_2, t)|}{|u(y_2)|} = e^{\alpha c_i t} = e^{\alpha(c_i/c_r)y_1}$$

so that the least stable disturbances are those with the largest positive value of c_i.

In an alternative more contemporary formulation, the wave is assumed to grow exponentially in space; i.e., instead of Eq. (3.2) we have

$$u(y_1, y_2, t) = u(y_2)e^{i(\alpha y_1 - \omega t)}$$

where α is complex ($\alpha = \alpha_r - i\alpha_i$), ω is real, and a constant phase is maintained with $y_1 = c_r t$. The two points of view are not identical, as pointed out by Gaster [18], because in the spatial case the disturbance does not grow or decay with tome. However they are roughly equivalent as long as $c_i \ll c_r$ or $\alpha_i \ll \alpha_r$. Then the near equivalencies hold:

$$[\alpha_i]_{\text{Spatial}} \simeq [\alpha c_1/c_r]_{\text{Temporal}}$$

$$[\alpha_r]_{\text{Spatial}} \simeq [\alpha]_{\text{Temporal}}$$

Most early work on hydrodynamic analysis uses temporal growth rates, although for many shear flows the equivalency may break down since values of c_r and c_i are often similar.

When Eqs. (3.1) and (3.2) are substituted into Eqs. (2.40) (with $q = 0$) and (2.41) and all terms that include products of disturbance amplitudes are ignored in relation to others, the resulting equations retain only the linear first-order terms. It is called the classical Orr–Sommerfeld equation:

$$(U(y) - c)(\phi'' - \alpha^2\phi) - U''\phi + \frac{i}{\alpha R_\delta}(\phi^{iv} - 2\alpha^2\phi'' + \alpha^4\phi) = 0 \tag{3.5}$$

for small disturbances in an incompressible shear flow of a mean velocity distribution $U_1(y_2)$. In this equation we have expressed the fluctuating vertical velocity $u_2(y_1, y_2, t)$ in terms of a fluctuating stream function $\phi(y_2)$ of the linearized stream-normal disturbance; i.e. from Eq. (3.2)

$$u_2(y_1, y_2, t) = \hat{u}_2(y_2)e^{i\alpha(y_1 - ct)}$$

which may be used to introduce the stream function

$$u_2(y_1, y_2, t) = -i\alpha\phi(y_2)e^{i\alpha(y_1 - ct)} \tag{3.6}$$

The wave speed c and the wave number α are assumed to be independent of both y_1 and y_2. However, the relationship between c_r and c_i in the temporal instability, or between α_r and α_i in the spatial instability, will depend on the shape of $U_1(y_2)$ and on the Reynolds number

$$R_\delta = U_0\delta/\nu$$

where U_0 is the characteristic velocity of the flow. For a given type of flow, there is a critical value of R_δ above which c_i (or α_i) is positive and the disturbances grow. For the free shear flow types C–E this critical value can be as low as 30 (see also the stability diagram in Fig. 3.15), while for the Blasius layer bounded by the rigid wall it is of order 2500. Thus in relation to the wall layer, the free shear layer is less stable. Furthermore, when $R_\delta \gg (R_\delta)_{crit}$ then the dependence of c_i (or α_i) on the Reynolds number diminishes, while the dependence of c_i(or α_i) on wave number α remains dominant. When c_i (or α_i) is independent of Reynolds number, in the limit of large Reynolds number, the growth rates are as shown in Fig. 3.2 for all the free shear layers illustrated in Fig. 3.1. The relative instabilities of disturbances in the various shear layers at Reynolds numbers well above the critical values are given by substitution into eq 3.5 when $R\delta \rightarrow \infty$. Profile A is unstable to waves of all wavelengths, while the remainder of profiles are unstable to restricted ranges of wave number, generally greater than zero and less than $2\delta^{-1}$.

The large instability associated with the jets and wakes is caused by the pair of inflection points on either half of the shear layer, which makes these flow types very sensitive to acoustic and hydrodynamic stimulus. Furthermore, the characteristic length scale of the wavy motion is dependent on the shape of the velocity profile, and will be discussed later in Section 3.4.1.

Two- and three-dimensional jets are capable of two degrees of freedom, both of which have been observed in experimental environments (see Section 3.4 for details). The least stable mode is the wavy pattern diagrammed in Fig. 3.3A, while the more stable (symmetric) mode is diagrammed in Fig. 3.3B; both modes can often be reinforced by sound [13]. Jet instabilities can also involve dynamics of thin annular shear layers when the efflux contains a central region for which the mean velocity is constant (Fig. 3.3C). Such jets are generated by relatively short nozzles. In these cases an annular shear layer of thickness δ undergoes instabilities, much like those of single free shear layers. The waves are of shorter length, relative to the width or

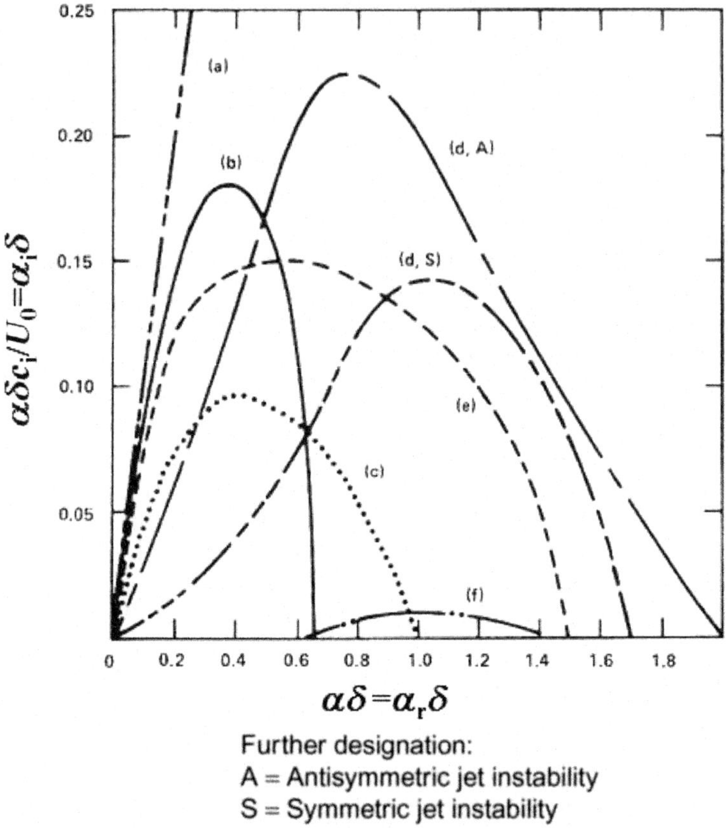

FIGURE 3.2 Theoretical growth rates in the limit of large Reynolds number based on spatial instabilities in the shear layers of Fig. 3.1.

diameter of the jet, than the waves shown in Fig. 3.3A and B; however, the characteristic length of the waves is δ rather than D. In the case of axisymmetric jets the growing waves, as shown in the excellent photographs of Brown [19] and later Becker and Massaro [20], begin to "crest" as sketched in Fig. 3.3C, causing a necking-down. In subsequent stages of development the necked-down regions separate the successive ballooned-out portions forming a street of rings or "puffs." Each of these puffs is a ring vortex. In the asymmetric mode, the later development results in a spiral vortex. The mode of instability determines the initial spatial scale for the vortices formed in later disturbances. This shall be seen more clearly below.

The development of a vortex structure from a particular mode of instability is an important concept in the generation of flow noise. It implies that a relationship ultimately exists between a flow type and the amount of noise produced. This relationship is made possible by the dependence of the sound

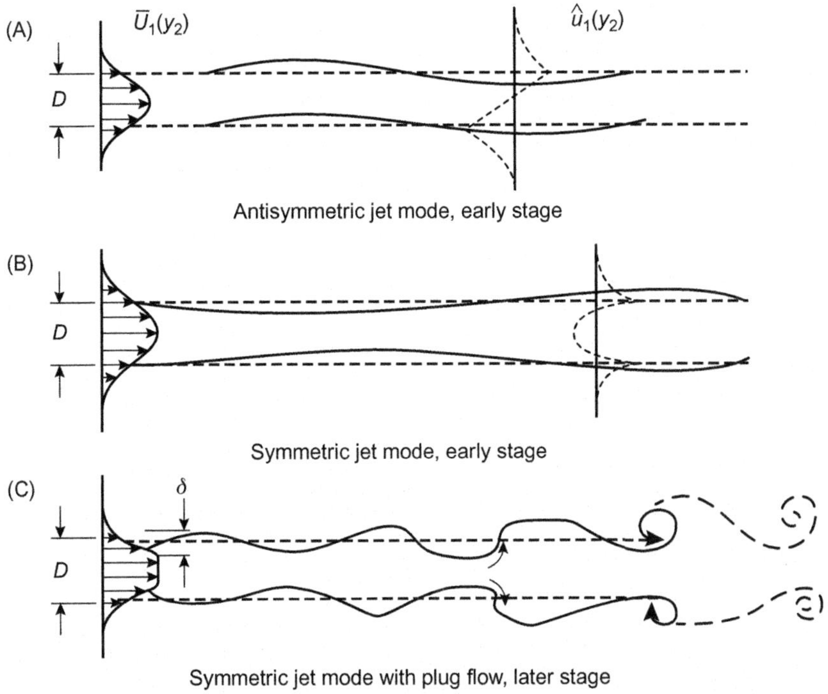

FIGURE 3.3 Illustrations of jet modes.

pressure on unsteady vorticity as expressed in Eq. (2.95). The less stable a type of flow, the more likely it is that the generation of vortex sound is possible. However, a formal mathematical connection between the mode of linear first-order instability of the type expressed in Fig. 3.2 and a vortex structure has been limited to only idealized planar shear layers. These shear layers consist of one or more parallel vortex sheets of the type A shown in Fig. 3.1. In this idealization the physical shear layer is concentrated into sheets. Rosenhead's [21] calculation for the single layer, Fig. 3.4, shows the gradual transition from a wavelike motion that involves a sheet of vorticity that is initially homogeneous in y_1 into a discrete set of point vortices as time increases. The distribution of vorticity in the sheet at $y_2 = 0$ may be expressed as

$$\omega(\mathbf{y}, t) = \gamma(y_1, 0, t)\delta(y_2)$$

Each wave steepens at the downstream side of a crest to form a single vortex ultimately. The character of flow changes from a crested sinusoid at $tU/\lambda = 0.30$ to a vortex at $tU/\lambda = 0.35$. The far-field disturbance caused by the redistribution of vorticity is given by Eqs. (2.104) or (2.106), and it is maximum when the local acceleration of momentum due to the redistribution of vorticity with time maximizes; this occurs in the interval between

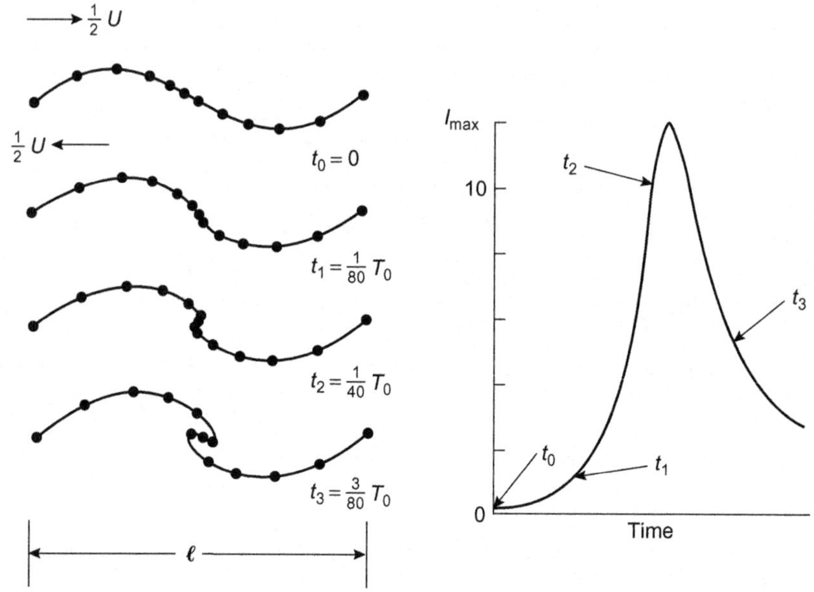

FIGURE 3.4 Shear layer represented by a set of point vorticities. The time t_0 is the period time $l/2U$. The maximum intensity of the resulting sound field is also shown: Corresponding points in the two diagrams are indicated by t_0, t_1, etc. *From Powell A. Flow noise: a perspective on some aspects of flow noise and of jet noise in particular. Noise Control Eng 1977;8:69–80, 108–119.*

$tU/\lambda = 0.30$ and 0.35. The dependence of the sound on time is also illustrated in Fig. 3.4, taken from Powell [22]. To quantify the illustration somewhat we combine Eqs. (2.99) and (2.104) to find the acoustic pressure as

$$p_a(\mathbf{x}, t) = \frac{\rho_0 \lambda \cos\theta}{4\pi c_0 r} \frac{\partial^2}{\partial t^2} \int \gamma\left(y_1, 0, t - \frac{|\mathbf{x} - y_1|}{c_0}\right) dy_1$$

where θ is the angle between the y_1 direction and the field location. When the flow lines begin to roll back on one another, the instantaneous *spatial distribution* of vorticity changes rapidly even though the total vorticity and circulation in the fluid remains constant. Therefore the phases of the induced motion of the fluid particles change with the development of the vortices so that the above integral is not instantaneously zero, but is time varying and its double derivative shown in Fig. 3.4. The most noise is generated at the moment when the change of this circulation distribution with time is greatest as illustrated. Other shear layer motions have been calculated by Michalke [5–7] for more realistic shear layers modeled by the hyperbolic tangent profile, and they show similar circulation regions, although far-field acoustic disturbances have not been calculated for such motions.

Instability modes have been calculated up to acoustically-relevant frequencies by Tam and Morris [240] for the free shear layer, (C) in Fig. 3.1 and include calculations of acoustic radiation. Deflections of streamlines, similar to those calculated by Rosenhead, have been calculated on a computer for a pair of parallel vortex sheets to resemble a wake by Abernathy and Kronauer [23] and Boldman et al. [24]. In the case of two vortex sheets each wavelength results in the formation of two concentrations of vorticity of opposite sign; see Chapters 4 and 11, Dipole Sound from Cylinders; Noncavitating Lifting Sections. More contemporary experimental examination of instability waves and transition in the wall flow, type (F) in Fig. 3.1, has been done by Eliahou et al [243] and Han et al [244] who examined the growth and breakdown of transverse (helical) waves in a pile flow.

3.3 THE FREE SHEAR LAYER AND CAVITY RESONANCE

3.3.1 General Considerations

An important application of the concept of flow instability occurs with the passage of flow past a slot in a wall, a gap between mating plates, or some other opening. This is an area of continuing interest, extensive reviews of developments being made available by Blake and Powell [25], Rockwell and Naudascher [26], Ahuja and Mendoza [27], and Rowley and Williams [28], which collectively focus on the fluid dynamics of instability, flow control, acoustic source mechanisms. Engineering applications are broad-based including automobile openings, e.g., Ma et al. [29], valves, e.g., Naudascher [30], and wheel wells and other fuselage openings, e.g., Ahuja and Mendoza [27] and Schmit et al. [31]. Fig. 3.5 uses three illustrations to collectively identify the various flow-acoustic mechanisms that control the existence or nonexistence of tones. The flow external to the opening may consist of a thin laminar boundary layer or a possibly thicker (relative to the opening dimension) turbulent boundary layer. Behind the opening is a closed cavity, a pipe, or some opening which may, or may not, be an acoustic resonator. In each of many situations of practical interest, it was disclosed in systematic experimental studies (e.g., Refs. [27,32]) that a somewhat different relationship will govern the frequency of the tone generated depending on whether the turbulent boundary layer upstream of the opening is laminar or turbulent. The fluid velocity, the opening size, the cavity volume, and its geometry all play a role as has been comprehensively studied in an analysis by Howe [33]. Based on the body of experimental and analytical work to date the creation of a cavity tone thus depends on the coupling of the fluid dynamics of the opening and the acoustic (or elastic) characteristics of the backing volume (e.g., Refs. [25,27,28,32–35]). Although a resonant cavity certainly enhances the tone generated, it is important to note that the cavity need not be resonant in order for a self-excited tone to be generated. The general view is that the sound external to the orifice, as illustrated in Fig. 3.5B, can be due to both a monopole and a dipole source mechanism in the opening. The former is caused by volume velocity changes due to fluid compressibility in the cavity, the latter due to shear layer impingement forces

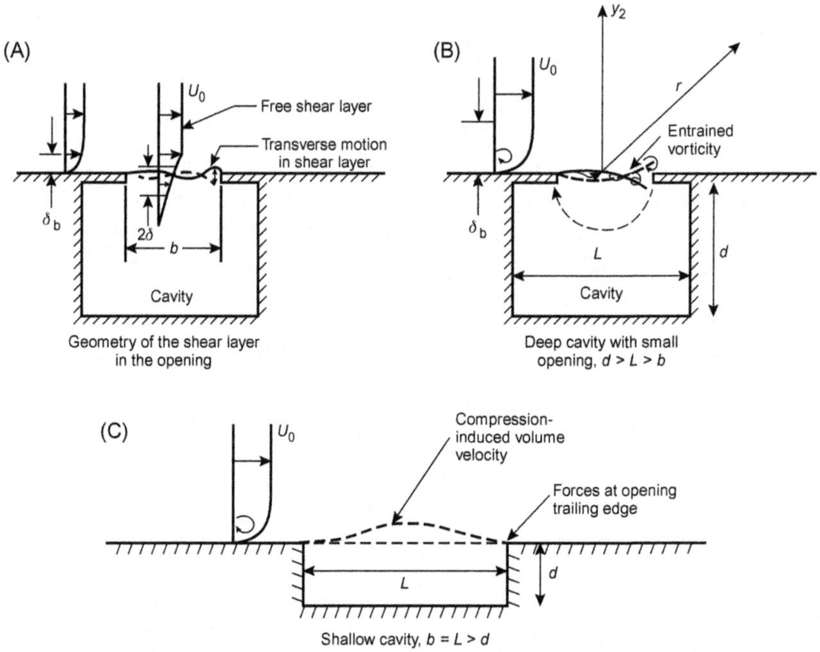

FIGURE 3.5 Idealizations of flow past cavities which are coupled to external boundary layers by slots, gaps, or orifices.

at the downstream edge of the opening as illustrated in Fig. 3.5C, and does not require compressibility of the fluid in the cavity.

The structure of the flow in the orifice that leads to self-sustained modes of the shear layer was first put forth by Rossiter [36], hence the term "Rossiter modes." To develop the argument for these modes, we consider the case of a laminar boundary layer flow over the opening shown in Fig. 3.5A for which the layer δ_b is presumed to be less than the streamwise dimension of the opening b. The passage of the fluid beyond the upstream edge and across the opening resembles the development of the classical free shear layer [27,29,32]. Idealistically, we presently assume that the shear layer does not change across the opening and remains thin compared with the length of the opening. Such a layer is well described by the hyperbolic tangent profile [5,7,8] (Fig. 3.1C), and it is theoretically least stable to disturbance wave numbers (Fig. 3.2) of magnitude

$$\alpha\delta = 0.42 \tag{3.7}$$

where α is given by Eq. (3.3). For the hyperbolic tangent velocity profile the momentum thickness, given by

$$\theta = \int_{-\infty}^{\infty} \frac{U}{U_0}\left(1 - \frac{U}{U_0}\right) dy_2$$

where U_0 is the local free stream velocity, is identically equal to $\delta/2$. Theoretically, the wave speed in the free shear layer is equal to the mean velocity at the inflection in the velocity profile, i.e.,

$$c_r = \tfrac{1}{2}U_0 \tag{3.8}$$

so that Eq. (3.7) gives the frequency at which the disturbances are propagated or convected downstream from the upstream edge as

$$f\theta/U_0 = 0.017 \tag{3.9}$$

In the simplified view that the shear layer does not spread across the width of the opening, θ and c are constants and Eq. (3.9) also gives the frequency at which the disturbances encounter the downstream edge.

3.3.2 The Flow in the Opening and the Strouhal Numbers of Tones

Now, the motion of this shear layer has associated with it a velocity, u_2 in the coordinate system of Fig. 3.5, which is, in the approximation of linear disturbances, nearly sinuous across the opening. At the upstream edge, the velocity transverse to the flat surface is zero so that separation is avoided, i.e., a Kutta condition exists [33,37], so that, as illustrated in Fig. 3.5, the flow grazes this edge. At the downstream edge of the aperture the transverse velocity in the shear layer is larger and it generates enhanced activity there. The impingement of the shear layer at this corner results in alternate inward and outward motions. This creates a small oscillating angle of attack at the downstream edge of the opening with accompanying vortex shedding in the outer flow. The details of the flow-edge interaction are dependent on the geometry of the edge, i.e., whether it is a corner, a thin plate, etc. In the cavity, the pumping at the downstream corner creates alternate pressure condensations and rarefactions in the cavity. These pulsations feed back to the upstream edge and synchronously drive the shear layer. Although the description of notional shear layer dynamics given above is a generally accepted idealization, the actual fluid mechanics is complicated by many factors and descriptions of the details of flow differ, e.g., DeMetz et al. [32], Ahuja and Mendoza [27], Ma et al. [29], Ronneberger [38], Rockwell et al. [39], Lucas [40], Oshkaia et al. [41], as shall be further discussed below. Yet, to fix ideas in an elementary way before further summarizing the literature, we consider a simple geometric model for the flow in the orifice that gives conditions for the behavior of the tones in general agreement with observation and widely-held viewpoints. The growing motion in the shear layer may be expressed following Eq. (3.2) as

$$u_2(y_1, y_2, t) = u_2(y_2)e^{\alpha_i y_1}e^{i(\alpha_r y_1 - \omega t)}, \quad 0 \le y_1 \le b \tag{3.10}$$

where $\alpha_i y_1$ expresses the spatially growing motion along the opening. Considering for the moment that the cavity is *not* a resonant volume then a rudimentary model of synchronization provides that whatever the small motion $u_2(y_1)$ is at the upstream edge, at the downstream edge it must be of the form

$$u_2(b) = u_2 e^{\alpha_i b} e^{i(\alpha_r b - \omega t)}$$

Now, the most optimum coupling will be such that when the maximum inflow occurs at the downstream edge (generating a relative pressure increase in the cavity), the transverse velocity at the upstream edge will be small and increasing. The phase $\alpha_r b$ of $u_2(b)$ with respect to $u_2(0)$ should thus be such that it is $\pi/2$ less than one complete cycle, i.e.,

$$\alpha_r b = 2n\pi - \pi/2, \quad n = 1, 2, 3, \ldots$$

for the most favorable synchronization. According to this simple model, since $\alpha_r = \omega/c_r$, then preferred frequencies, say $f_n = \omega_n/2\pi$, are given by

$$f_n b/c_r = \left(n - \frac{1}{4}\right), \quad n = 1, 2, 3, \ldots$$

which is a relationship that actually has considerable experimental support. The generation of the disturbance at the downstream edge will depend on the geometry of the edge and the real fluid mechanics of the shear layer [42] and its entrainment of fluid from the inner and outer media and the phase and strength of the feedback disturbance will depend on the nature of the cavity as well. Accordingly, a more general expression for the condition of synchronization is

$$\alpha_r b + \phi = 2\pi\left(n \pm \frac{1}{4}\right), \quad n = 1, 2, \ldots \tag{3.11}$$

where ϕ is an arbitrary phase angle that accounts for the possibility of a phase lag between the encounter of the disturbance with the edge and the response of the shear to this encounter and where the plus-or-minus sign allows for motions either into or out of the orifice. Flux into the cavity at downstream edge ($-1/4$ and $\phi = 0$) implies (in the nearly incompressible fluid limit for which $\omega b/c_0 \ll 1$ in the cavity used above) a forcing of the shear layer at the upstream edge directed out of the cavity due to an inward-directed velocity fluctuation at the downstream edge of the opening. Since $\alpha_r = 2\pi/\lambda$, where λ is the wavelength of the hydrodynamic instability mode across the opening width, we have the conditions appropriate to this hypothesis of reinforcement

$$b/\lambda = n - \tfrac{1}{4} - \phi/2\pi \tag{3.12}$$

The relationship between the wavelength and the phase can be deduced from Eq. (3.3) to give

$$\frac{f_n b}{U_0} = \frac{c_r}{U_0}\left(n - \frac{1}{4} - \phi/2\pi\right), \quad n = 1, 2, \ldots \tag{3.13a}$$

as the possible condition for the frequency of a standing wave of fluid disturbance across the opening; this condition, with $\phi = 0$, most generally applies to a shear layer in the mouth of an enclosure at low Mach numbers. This formula for the frequencies of shear layer modes has been attributed to Rossiter [36] and it has widespread use in the literature with various modifications. For finite Mach numbers, for example, a well-accepted modification due to Heller et al. [40] that accommodates fluid compressibility is

$$\frac{f_n b}{U_0} = \frac{c_r}{U_0} \frac{(n - 1/4 - \phi/2\pi)}{1 + \left(\frac{c_r}{U_0}\right) M \Big/ \sqrt{1 - \frac{\gamma - 1}{2} M^2}}, \quad n = 1, 2, \ldots \tag{3.13b}$$

where γ is the adiabatic gas constant for the fluid medium in the cavity. Of historical note, Rossiter [36] proposed a relationship identical to Eq. (3.13a) but with $\phi = 2\pi f b / c_0$ where c_0 is the speed of sound in the fluid to account for acoustic propagation back to the upstream edge.

The chain of events that leads to the establishment of a stable tone as described above and that accounts for frequency conditions of the type described by Eq. (3.13a,b) illustrate the essentials of flow-tone generation as first proposed by Powell [43,44] (see also Section 3.4.3). Such tones involve four processes that are necessary in the fluid–resonator interaction feedback loop that is common to all self-sustained tonal vibration or sound radiation. These loop gain elements are:

1. A fluid shear layer of the types depicted in Figs. 3.1 and 3.2 is driven by a disturbance at the upstream leading edge, in this case illustrated by Fig. 3.1B.
2. The disturbance is amplified as it is convected across the opening, as illustrated by Eq. (3.10), until it encounters a downstream edge (or other object). In general the upstream initiation surface and downstream impinging surfaces are parallel.
3. A second disturbance is initiated at the downstream edge in the form of a fluid momentum fluctuation. An unsteady pressure is launched from the downstream edge (surface). This behavior is provided by the alternating flux at the downstream edge and it is principally responsible for the generation of the sound in aerodynamic processes.
4. This pressure is propagated back upstream to the surface which launches the shear layer in the first place. The feedback path is in the cavity in the context of cavity tone generation, but the feedback path may not be associated with a resonating system of an acoustic, elastic, or hydrodynamic nature.

Depending on the phase of this fed-back disturbance with respect to the initiating disturbance, the initial disturbance will be reinforced. The condition set down in Eq. (3.13a,b) provides the appropriate type of relationship for most flow tone with $\phi = \pi/4$ for cavity tones.

To this point the discussion of shear layer dynamics has been in terms of instability modes of the shear layer. However, Dunham [45] proposed the existence of selected modes corresponding to integer numbers of vortices entrapped within the orifice. The later work of Oshkaia et al. [41] supports such considerations for larger gap openings. In such cases the vortex sheet breaks down to row of vortices and these replace the "waves" in the previous discussion. Analytical consideration of the shear layer dynamics (as a discontinuous vortex sheet) in the opening by Howe [46] agrees with Ronneberger's [38] observations and generally confirms the approximate validity of Eq. (3.13a,b) with $\phi = 0$ at least for $n = 1$ or 2 when a sheet of discrete vortices drives the opening.

Another rationalization that has relevance to the coupling of the shear layer with volumetric undulations of the cavity concerns the integrated disturbance across the opening. This class of cavity, relevant to acoustic monopole cavity tones was proposed by King et al. [47] and later extended by Martin et al. [48] and Rockwell [49]. Following Bilanin and Covert [50] the vertical displacement of the shear layer across the cavity, δ_2, is given by the proportionality

$$\delta_2(y_2, t) \propto \delta e^{\alpha_i y_i} \cos(\alpha_r y_1 - \omega t) \tag{3.14}$$

so that the instantaneous volume change per unit width imposed on the interior is

$$\delta V = \int_0^b \delta_2(y_1, t) dy_1 \tag{3.15}$$

This fluctuating volume must be absorbed in the elasticity of either the fluid or the cavity structure, but the important point is that a negative volume change will cause a positive pressure. This positive pressure reinforces a positive value of deflection at the origin of the shear layer, $\delta_2(0, t)$. Integration of Eq. (3.15) using Eq. (3.14) gives the volume change explicitly in terms of the parameters α_i, α_r, and c. Further, the condition that $\alpha_i b > 1$ provides a simple relationship just as Eq. (3.13a,b) with $\phi = 0$. Rockwell [49] has treated the relationships more exactly, replacing $\alpha_r y_1$ by integrated values across the opening. This procedure accounts for the fact that, for long cavities, the shear layer changes with y_1. For example, in Eq. (3.13a,b), the wave speed c is a function of distance from the leading edge of the opening. This point will be further examined when we discuss the jet tone phenomena in the next section.

Measurements of the frequencies of cavity tones have been made for a variety of external boundary layers, cavity lengths, Mach numbers, and resonance characteristics of the cavity. First, considering flows at low Mach number, collectively the measurements in Fig. 3.6A show a general decrease in Strouhal number as the boundary layer thickness increases in relation to the streamwise dimension of the opening. The values reported are for cases with either resonant or nonresonant acoustic cavities. Those by DeMetz and Farabee [32] were obtained on both circular openings and rectangular slots

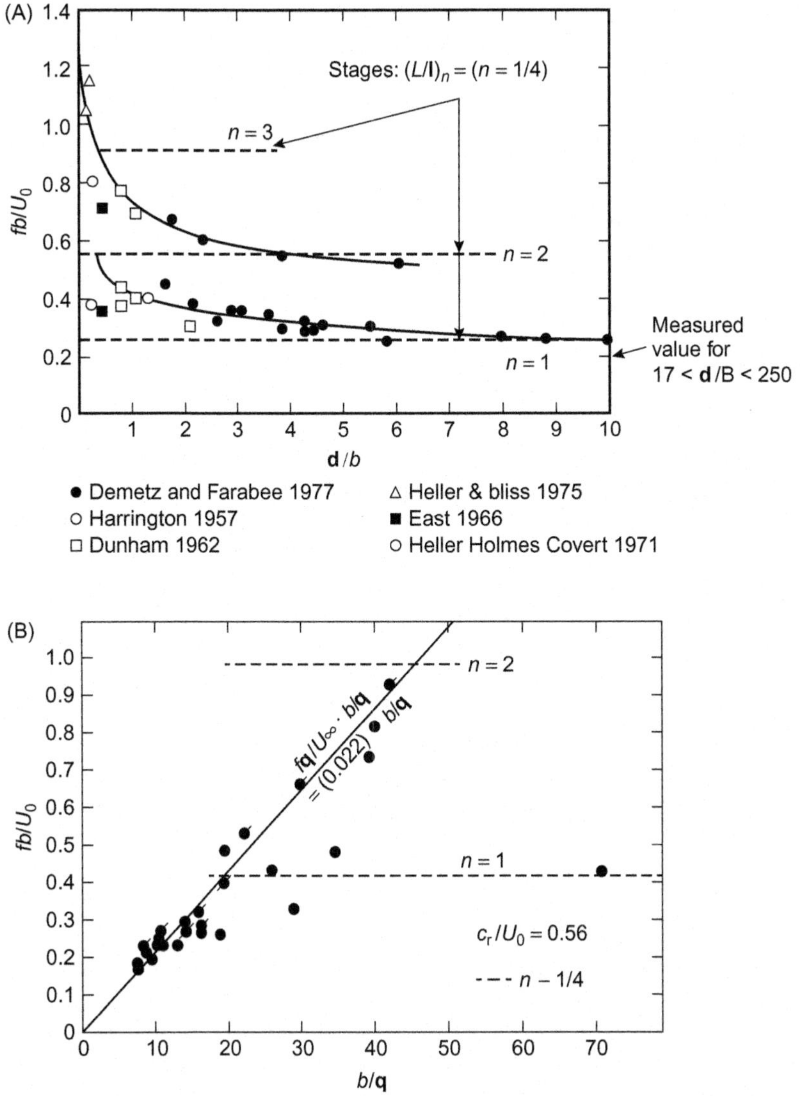

FIGURE 3.6 (A) Cavity resonance frequency scaling for turbulent boundary layer excitation from DeMetz and Farabee [32]. Predicted lines (dashed) using $c_r/U_0 \simeq 0.33$ and $fb/U_0 = c_r/U_0$; $n - \frac{1}{4}$. (B) Frequencies of cavity tones driven by laminar boundary layers. Data sources: DeMetz and Farabee [32], Harrington [51], Dunham [45], Heller and Bliss [52], East [53], and Heller, Holmes, and Covert [40]. *Copyright American Institute of Aeronautics and Astronautics; reprinted with permission.*

in air; Dunham's [45] results were obtained in air with slotted openings although a water cavity was used to visualize the shear layer dynamics. The tones of DeMetz and Farabee [32], Dunham [45], East [53], and Harrington [51] (all with resonant acoustical cavities) can all be explained from the flow visualizations of Dunham [45] and of Ma et al [29]. These frequencies correspond to the entrainment into the cavity mouth of either one or two vortices in the cavity; the vortices originate from the rapid breakdown of the shear layer instabilities downstream of the upstream lip as noted above. For small values of δ/b, corresponding to relatively long cavities, the reported values of fb/U_0 suggest a somewhat higher average convection velocity. This is clearly possible since more mixing of the inner stagnant and outer moving flow in the long gaps could very well make a larger mean flow across the gap. The apparent average convection velocity of these vortices across the mouth for moderate values of δ/b as reported by DeMetz and Farabee was on the order of $0.45U_0$ for $\delta/b \to 0$ and $0.33U_0$ for large δ/b. Each of the modes of oscillation, then, can be predicted from Eq. (3.12) with $0.33 < c_r/U_0 < 0.45$ and $\phi = 0$. The horizontal lines in Fig. 3.6A show the various modes for $n-1/4$ that generally agree with the observed Strouhal frequencies and that represent either one or two standing waves or vortices in the opening. The measured results of Heller and Bliss [52] and of Heller et al. [40] were obtained with very long shallow (not necessarily resonant) cavities and appear to correspond to higher modes of oscillation.

When the external boundary layer was laminar, DeMetz and Farabee [32] reported a less obvious quantizing of the Strouhal number. Instead, their results indicate a more uniform increase in frequency with velocity without the establishment of harmonics. This increase is described in a form similar to Eq. (3.9) with

$$f\theta/U_0 = 0.022$$

which is determined by the observed value of $c_r \simeq 0.56U_0$. However, some reinforcement by the opening feedback is evident about the $n = 1$ mode of condition Eq. (3.13a,b).

The measured Strouhal frequencies were all obtained at low Mach number. Earlier work, e.g., Refs. [34,40,50,52], did provide measurements over ranges of Mach number, but the later ones of Ahuja and Mendoza [27] provide a systematic broad-based examination of the effects of free stream Mach number. Fig. 3.7 shows the Strouhal frequencies of the first four Rossiter modes of shallow and deep (generally nonresonant) cavities as a function of Mach number. These results show frequencies which were relatively insensitive to temperature and boundary layer thickness, although the relative thickness of the turbulent boundary layer grazing the opening, δ/b, was small. It was found, however that the acoustic intensities of the tones were critically dependent on δ, however, reducing tens of decibels when $\delta/b = 0.038$ to 0.045.

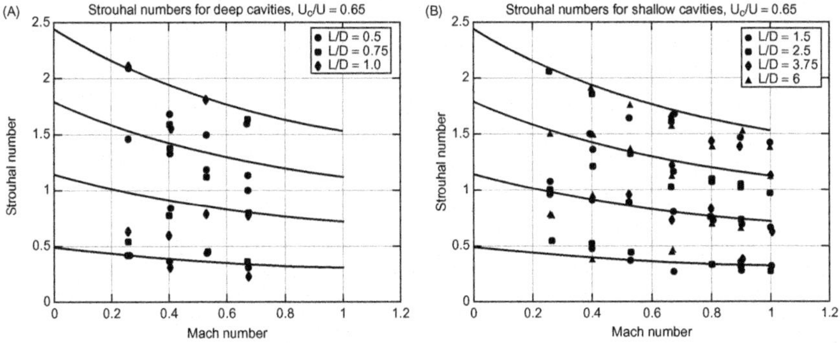

FIGURE 3.7 Cavity resonance frequency scaling deep (A) and shallow (B) rectangular cavities of the shape shown in Fig. 3.5C from Ahuja and Mendoza [27]. Predicted lines use $c_r/U_0 \simeq 0.65$ and Eq. (3.13b). In both cases $\delta/b = d/L = 0.038$.

Observations of the shear layer dynamics in the openings of orifices by Ronneberger [38] give insights into the limitations of the preceding hypotheses of fluxes in the opening. For Strouhal numbers $fb/U_0 > \sim 1$ and for a laminar grazing flow, dye streak visualizations of the shear layer dynamics disclosed the presence of motion of the type predicted by Eq. (3.14) across the opening. Inside the cavity a vortex structure could be observed coupled to the shear layer and fed by an influx at the trailing edge of the cavity. This system could be reinforced by forcing the cavity with an oscillating piston source as long as $f\theta/U_0 < \sim 0.024$ ($\omega\theta/U_0 < \sim 0.15$) and $fb/U_0 > \sim 0.16$ to ~ 1.11. The net flux in the cavity opening, however, was composed of nearly equal contributions of the net flux across the opening (as included above in Eq. (3.15)) as well as a flux local to the trailing edge of cavity. Thus for $fb/U_0 > \sim 1$ and $f\theta/U_0 < \sim 0.15$ the model suggested above is too superficial and ignores a trailing-edge flux. Self-excitation requiring coupling between the leading and trailing edges of the cavity by this flux was found to exist due to acoustic coupling between the shear layer motions at the upstream and downstream edges of the slot. Maintenance of a parallel flow (Kutta) condition at the leading edge of the opening and a finite value of displacement at the trailing edge are therefore essential to the modeling of self-sustained oscillations. The condition for self-sustained oscillation basically came down to an extraction of acoustic energy by the mean flow when fb/U_0 falls within fairly narrow limits that depend on the nature of the acoustic path, e.g., the presence or no presence of a cavity. At $f\theta/U_0 > 0.024$ with $b \gg \theta$ Ronneberger finds self-sustained oscillations impossible to generate in the opening. At these high Strouhal numbers the shear layer breaks down to a train of discrete vortices before it encounters the downstream edge, and the vortices decay downstream. This observation alone would seem to support the notion that self-excited oscillations that involve hydrodynamic instabilities and a feedback loop must occur before the hydrodynamic motion breaks down into a vortex structure. Ample support for this notion will be presented in sections below.

Thus far in this section, we have been concerned only with the shear-layer dynamics in the opening considering principally the influence of the cavity volume behind the aperture. The general agreement among investigators in the reported values of the Strouhal number attests to the relative first-order insensitivity of fb/U_0 to changes in cavity volume or cavity geometry. For rectangular U-shaped slots, Ethambabaoglu (reported by Rockwell [49]) reports a slight increase in fb/U_0 as the width exceeds the depth of the slot. The phase condition for cavity tone generation as well as Eq. (3.13a,b) has also been found to apply to axisymmetric cases, see Oshkaian et al. [41] mentioned above. Also Morel [54] found this Strouhal scaling to apply to the geometry of a round jet passing through a sealed enclosure. In particular, it was found that $fL/U_J \approx (n-0.25-fL/c_0)U_c/U_J$ where $U_c \approx 0.6U_J$, $n=1$, 2,. . ., and L is the length of the enclosure (i.e., the distance between the jet efflux plane and the exciting hole in the opposite wall). This formula is recognizably Rossiter's [36] form of Eq. (3.13a,b). The reader is also referred to the discussion of the hole tone in Section 3.4.

3.3.3 The External Radiated Sound

3.3.3.1 General Considerations

The physical acoustics of the external field has been given less systematic attention than has been given to the kinematics of the shear layer in the opening. The measurements of Block [34] and of DeMetz and Farabee [32] are among the first to characterize the acoustic intensities of the Rossiter mode tones as a function of external wind speed and mode order. In DeMetz and Farabee's case, since the tones were associated with acoustic cavity resonance, their assessments and those which followed by Elder [55,56] and Elder et al. [57] were focused on flow-cavity coupling dynamics and monopole source strengths. Most theoretical work in modeling the acoustic source prior to that of Howe [33] in 2004 only considered the monopole acoustic source embedded in a rigid wall. Hardin and Martin [58] modeled the acoustic emissions theoretically considering the radiation properties of cavities in terms of their entrained vortices. Ma et al [29] examine the interior pressure and model its levels on the details of flow structure in the cavity opening.

However, the most comprehensive analysis of far-field sound to date [33,37] formulates a rectangular rigid-walled cavity in a rigid wall as depicted in Fig. 3.5C with external coordinate system of Fig. 3.5B for the acoustic field above the wall. The internal cavity fluid and the external moving medium are separated by a shear layer in the cavity opening. Sound is produced by instability in this shear layer for which a Kutta condition at the opening upstream leading edge forces a tangential flow there. Analysis leads to the existence of both monopole and dipole sources at the opening, depending on whether the wall cavity sustains acoustic modes, or not. To achieve this, vorticity is shed into the opening, convects over the aperture, and impinges on the trailing edge. At the downstream edge this impingement induces a force that acts as a dipole, whether or not the medium within the cavity is compressible. Thus two sources emerge. One source is the force distribution applied to the fluid as a result of the vortex impingement, primarily of the form $\boldsymbol{\omega} \times \boldsymbol{u} \approx \omega_3 \frown \boldsymbol{k} \times \boldsymbol{u}$ contained within the opening. This time-varying force is in the flow direction and represents a fluctuating drag dipole moment. The second source occurs when the fluid within the wall cavity is compressible. Then, when there is an internal acoustic cavity mode for which a non-vanishing time-varying net volume velocity across the opening exists, a monopole source occurs.

In the notation of Fig. 3.5, the sound in the far, $r \gg b$, upper (flow) region above the orifice for an acoustically-compact cavity opening, $k_0 b \ll 1$, from the combined monopole and dipole sources is then of the frequency-domain form

$$P_{\text{rad}}(\omega, \boldsymbol{r}) = \frac{\rho(\pi b^2/4)\omega u_p(\omega)}{2\pi r} + \frac{\omega F_1(\omega) \sin\phi \cos\theta}{2\pi c_0 r}$$

for which (neglecting cross products) the spectral density is

$$\Phi_{\text{rad}}(\omega, \boldsymbol{r}) = \omega^2 \left\{ \left[\frac{\rho(\pi b^2/4))}{2\pi r}\right] \Phi_{uu}(\omega) + \left|\frac{\sin\phi \cos\theta}{2\pi c_0 r}\right|^2 \Phi_{11}(\omega) \right\} \quad (3.16)$$

The functions $\Phi_{uu}(\omega)$ and $\Phi_{11}(\omega)$, respectively, represent the spectral densities of the effective integrated particle velocity over the cavity opening and the unsteady drag force at the downstream edge of the opening. The volume velocity in the cavity opening is spectrally windowed around following a multi degree of freedom oscillator behavior for the acoustic cavity modes. Little is known of the characteristics of the unsteady drag on the opening although Howe [33] models it in terms of the pressure fluctuations on the wall surface downstream of the orifice. We will discuss this source again in Chapter 8, Essentials of Turbulent Wall-Pressure Fluctuations in the context of flow noise of gaps in walls.

An example calculation of the frequency spectrum of the radiated sound and of its directivity pattern is shown in Fig. 3.8 for a shallow cavity at Mach number $M = 0.1$. The spectrum levels are all relative to a common reference. For this case a coupled cavity resonance occurs near the expected Strouhal number for the second Rossiter mode, see Fig. 3.7. At other frequencies the directivity is notably aligned with the flow direction. Only when the resonance occurs does the monopole contribute to acoustic radiation normal to the wall. In the analysis, the strength of the drag dipole is attached to the magnitude of wall pressure on the wall just downstream of the trailing edge of the opening. The value of its spectrum is shown as the upper dashed line in Fig. 3.8A; the shape of the dipole sound follows closely

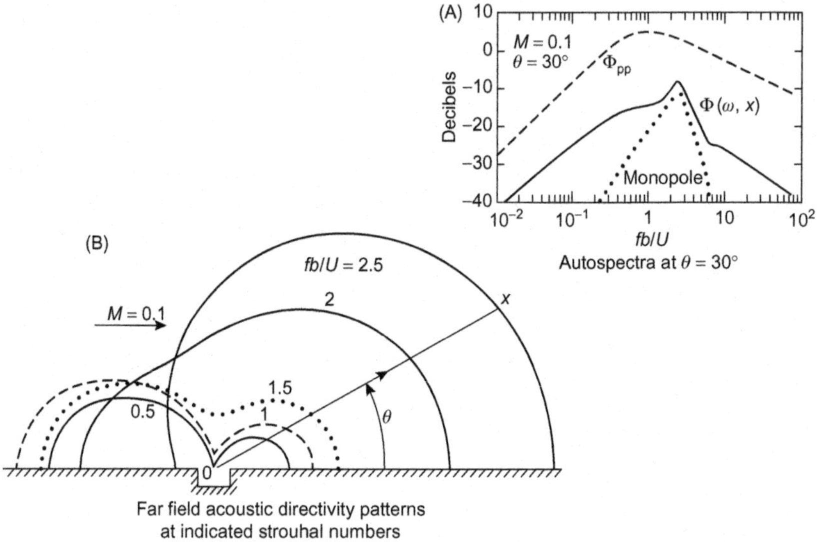

FIGURE 3.8 Example characteristics of spectra of radiated sound from a shallow cavity at low Mach number, Howe [33]. In (B) the numeric labels are values of fb/U; in (A) the dashed autospectrum is wall pressure, the solid and dotted lines are radiated sound levels from the dipole and monopole contributions, respectively at the indicated fly-over angle.

that spectrum shape since the cavity resonance is not relevant to this source. The monopole sound, on the other hand, occupies the expected narrow band of frequencies since it is associated with an acoustic cavity mode. This theory is not yet supported with measurements, except in a very general way. Ahuja and Mendoza [27] have provided a data set of spectra to accompany their reported observations of tone frequencies. That data shows a general broadband sound level underpinning the sequence of narrow band spectrum levels at the Rossiter mode frequencies. The speed dependence of their overall sound pressure levels was bounded by U^4 to U^6.

3.3.3.2 Monopole Sources With an Acoustically-Compact Cavity Width

The preponderance of analysis of the physics of acoustic radiation, therefore, has been concerned with the modeling of the monopole source and we will review some efforts to produce reasonable bounds to the monopole source levels in the case of deep cavities with compact cross section. Measurements of DeMetz and Farabee [32] were made with the openings of cylindrically shaped resonant (pipe-like) Helmholtz resonators with the center of the hole at the axis of the chamber. Relatively large pressure amplitudes were measured in the chamber at the resonance frequencies at the cavity resonance frequencies. The limiting maximum amplitude of pressure at the bottom of the cylindrical cavity occurring appears to be limited by

$$\overline{p_{\rm cav}^2}^{1/2}/q_0 < 1$$

where

$$q_0 = \tfrac{1}{2}\rho_0 U_0^2$$

At disturbance frequencies unequal to the cavity resonance frequency, the cavity pressures can be as low as $10^{-3}\ q_0$.

The pressure in the cavity has been related to the far-field acoustic pressure. When the dimension of the cavity is smaller than an acoustic wavelength, then, using the notation of Fig. 3.5B, the fluid displacement velocity, $u_{\rm p}$, in the opening represents a piston in a rigid baffle and accordingly causes a sound pressure

$$p_{\rm rad} \simeq \frac{i\rho_0\omega b^2 u_{\rm p}}{8\pi\, r} e^{ik_0 r} \tag{3.17}$$

where $k_0 b/2 \ll 1$ and the wall is essentially an infinite plane with an orifice of diameter, b. This result is easily derivable from Eqs. (2.118) and (2.122) for the otherwise source-free region ($\tilde{\sigma} = T_{ij} = 0$) external to the rigid plane. Therefore since on $y_2 = 0$

$$\frac{\partial P_{\rm a}}{\partial n} = \frac{\partial P_{\rm a}}{\partial y_2} = i\omega\rho_0 u_{\rm p}$$

only in the small opening and is zero on the surface otherwise, then Eq. (2.118) gives the result Eq. (3.17) directly, using the Green's function Eq. (2.122) with $r_1 = r_2 = r$.

Depending on the size of the cavity, the pressure in the cavity is related to $u_{\rm p}$ by either of two asymptotic forms. In either case, if the medium is an ideal gas, the pressure change for a fractional change in a unit volume is given by

$$\delta p = -\rho_0 c_0^2 \delta V/V$$

If the cavity is small compared to a wavelength, the $\delta\dot{V} = \pi a^2 u_{\rm p}$ and $V = V_{\rm cav}$ and, Eq. (3.17) gives ($|\delta\dot{V}| = \omega|\delta V|$, where ω is the resonance frequency of the cavity),

$$\frac{|p_{\rm rad}|}{|p_{\rm cav}|} \simeq \frac{1}{2\pi}\frac{a}{r}, \quad \lambda_0 >> \text{cavity dimensions} \tag{3.18}$$

when we make use of the Helmholtz frequency (given below) and assume that the hole radius is much larger than the depth of the hole that connects the inner and outer fluid. An alternative expression can be written for a cavity whose dimension exceeds an acoustic wavelength. If this cavity is represented as an organ pipe of radius $a_{\rm p}$, then the unit volume undergoing compression is determined by an acoustic wavelength. Thus the effective unit volume in the organ pipe behind the opening is

$$V = \lambda_0 \pi a_p$$

where λ_0 is the wavelength of sound. The pressure fluctuation in the pipe (or cavity) is, then,

$$\delta p_{cav} \simeq \rho_0 c_0 u_p$$

and we find

$$\frac{|p_{rad}|}{|p_{cav}|} = \frac{\omega A_0}{c_0 r} \tag{3.19}$$

where A_0 is the area of the pipe or cavity opening. This relationship, previously derived by Elder [55] for organ pipe excitation, shows an omnidirectional sound field with a sound pressure level inversely proportional to the sound speed. Thus the sound pressure has a monopole-like directivity, but a dipole-like wave number dependence. The factor ($p_{cav}A_0$) represents the time rate of change of the force exerted on the external fluid by the pressure in the cavity. Eq. (3.19) therefore includes the same parameters as the magnitude of Eq. (2.76) for the true dipole with the exception of the directivity factor. Howe [59] and Elder [56] have given more extensive treatments to the flow-excited Helmholtz resonator and other resonators.

The elastic character of the cavity structure has received little attention; however, in the case of rigid-walled cavities with fluid compressibility governing the cavity stiffness, relationships for the frequency have been given by Rayleigh [1]. The classical Helmholtz resonance frequency can be found in numerous texts [1,60]. For circular openings,

$$f_r = \frac{c_0}{2\pi}\sqrt{\frac{\pi}{V}\frac{a^2}{L + \Delta R}} \tag{3.20}$$

where $a = b/2$ is the radius of the opening, V is the volume of the cavity, L is the length of the opening, and ΔR is an end correction approximately equal to $1.64a$. This formula can be generalized to openings of other dimension by replacing πa^2 by the area of the opening S and ΔR by $\frac{1}{2}\sqrt{\pi S}$. Dunham [45] and later Covert [61], Ingard and Dean [62], Elder [55,56], and Elder et al. [57] have considered some general impedance characteristics of cavities and how these characteristics influence coupling of the cavity and shear layer dynamics. Miles and Watson [63] measured the flow-excited pressures of acoustic modes in a nearly cylindrical cavity with its axis set perpendicular to the flow direction and slotted along its length.

3.3.3.3 Mitigation of Sound and Cavity Pressure

The control of flow-induced cavity tones may be accomplished by interrupting the loop gain feature elements 1–4, enumerated in Section 3.3.2, at any point. A most effective passive means is to minimize or avoid the development of coherent shear layer disturbances, an example of which was examined by Ukeiley et al. [64]. This may be done through the use of opening gratings or leading-edge spoilers to disorder the flow. The second most effective means is to break or obstruct the feedback loop by eliminating the resonator, or by the addition of baffles or stiffening devices in the cavity. In supersonic flow, understanding the structure of the shock wave fronts in acoustically-large cavities, as in Ref. [65], can provide insights into suppression of cavity resonances.

Active control has been used to abate the internal cavity tones. For example, Kegerisea et al. [66,67] examined an approach to adaptively control the interior pressure of a cavity at acoustic resonance. In this case a section of the upstream surface of the wall that supported the (shallow) cavity and the upstream edge of the opening was oscillated with an actuator to impart a vertical, u_2, velocity to the shear layer. The control sensor was placed in the bottom wall of the cavity, although other sensor locations were considered. These locations may include a probe in the shear layer or a pressure transducer at the trailing edge of the cavity opening. Experimental results verified simultaneous control of the first three Rossiter modes (over the frequency range 500–1300 Hz) for three subsonic Mach numbers, 0.28–0.38.

3.4 SELF-EXCITATION OF JETS

3.4.1 The Essentials of Jet Tone Generation

The disturbances that occur in developing jets are now well understood to dependend on the character of the mean velocity profile of the efflux, and therefore somewhat on the type of nozzle used. For short, potential-flow nozzles (Fig. 3.9A), the efflux contains a modestly sized potential core with a thin (relative to the nozzle radius) annular laminar shear layer so that $2\delta/D \ll 1$, where δ is the shear layer thickness illustrated in Fig. 3.3C. When the nozzle is made many diameters longer, the flow in the tube is fully sheared so that velocity profiles with shapes generally similar to those shown in Figs. 3.1D and 3.3A,B occur (in these profiles the shear layer extends to the center line of the flow). The efflux will be laminar for Reynolds numbers (UD/ν) less than 1400 when the inlet to the nozzle is well formed to avoid separation of flow inside the nozzle. For jets ensuing from long square-edged orifices such as in Fig. 3.9C, the efflux is laminar only for $R_D < 600$ because vortices caused by separation of flow are generated at the inlet for larger Reynolds numbers. Finally, in the case of knife-edged orifices, the jets are disturbance sensitive at Reynolds numbers greater than 600 because of the rather thin shear layer in the efflux. The references cited in Fig. 3.9 are those for which extensive flow visualizations and quantitative measurements were obtained over a wide range of Reynolds number. The critical Reynolds numbers cited are the minimum values for which growing sinuous

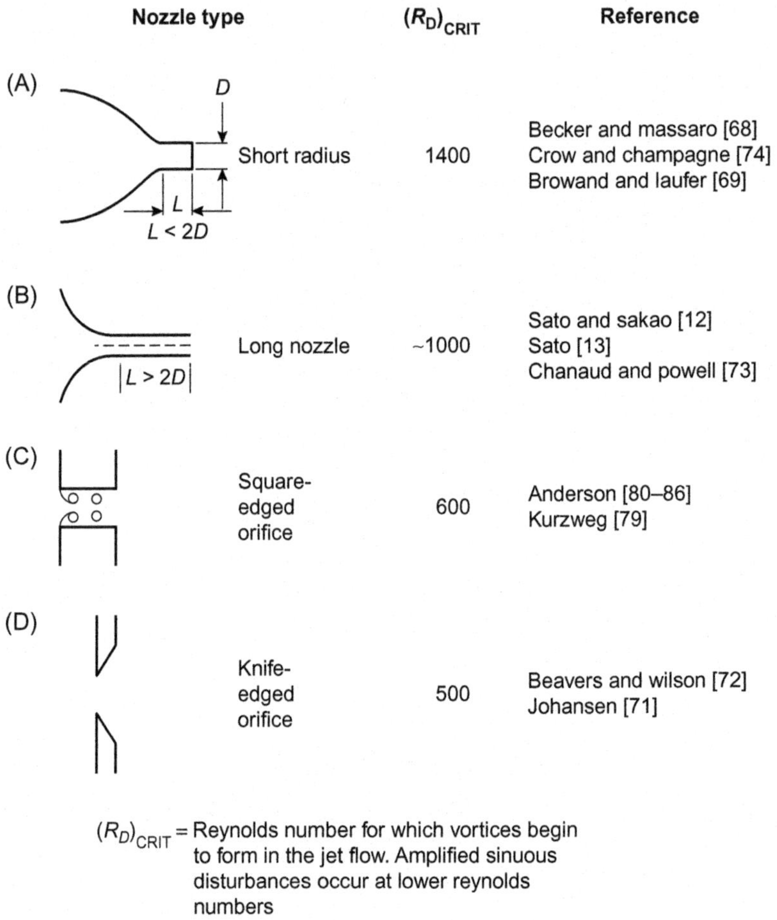
Nozzle type
$(R_D)_{CRIT}$
Reference
(A)
D
Short radius
1400
Becker and massaro [68]
Crow and champagne [74]
Browand and laufer [69]
L
L < 2D
(B)
Long nozzle
~1000
Sato and sakao [12]
Sato [13]
Chanaud and powell [73]
L > 2D
(C)
Square-edged orifice
600
Anderson [80–86]
Kurzweg [79]
(D)
Knife-edged orifice
500
Beavers and wilson [72]
Johansen [71]
$(R_D)_{CRIT}$ = Reynolds number for which vortices begin to form in the jet flow. Amplified sinuous disturbances occur at lower reynolds numbers

FIGURE 3.9 Illustrations of nozzles and orifices that produce jet tones.

disturbances give way to clearly defined vortex structures. This critical value of Reynolds number is not well defined since it is often influenced by the presence of extraneous disturbances and its identification also depends on experimental detail and on the manner of observation. As a general rule jets are disturbance sensitive at Reynolds numbers that are as low as 100.

The visualization of disturbances in jets has been a popular activity among researchers for well over 100 years, as outlined in a review article [25] and many others that deal with the evolution of larger scale coherent structures from instabilities, e.g. [242,245–247]. A good example of smoke visualization in a round jet is provided by Becker and Massaro [68] who have observed the axisymmetric disturbances in a jet formed with a short nozzle over a wide range of Reynolds numbers. In Fig. 3.10 many of their photographs are

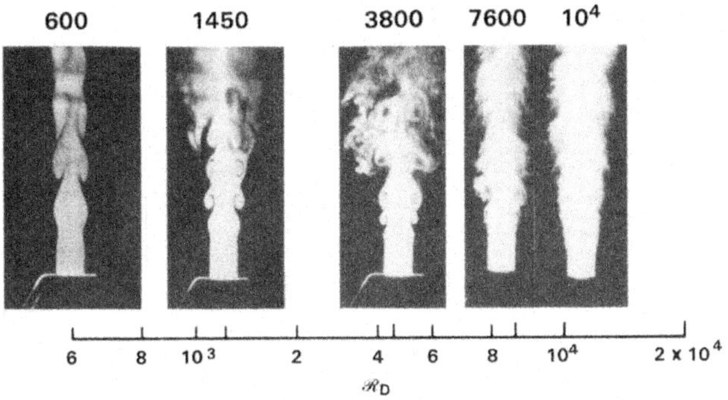

FIGURE 3.10 Photographs of smoke jets from laminar flow exiting an ASME short-radius circular nozzle. *Reprinted from Becker HA, Massaro TA. Vortex evolution in a round jet. J Fluid Mech 1968;31:435–48, by permission of Cambridge University Press.*

displayed on a Reynolds number scale. Their efflux was laminar at least until moderate values of R_0 and the vortex-growth process is clearly evident.

The propensity for jets (both round and rectangular) to develop such vortex structures, as shown in Fig. 3.10, makes them disturbance sensitive over such a wide range of Reynolds numbers that jet tones provide an immense practical importance. There are many ways of physically generating jet tones and they all may be described in the context of the elements of the flow-acoustic feedback enumerated in Section 3.3.2. The main ingredients are, then, the naturally unstable nature of jets and the generation and feedback of induced disturbances from some point in the flow generally, though not always, located near an impinging surface.

The class of jet tones which involve impinging surfaces are edge, hole, and ring tones. As illustrated in Fig. 3.11 these tones involve a jet incident on a wedge, a knife-edged hole in a baffle, or a ring set concentrically with the axis of a circular jet. Also included in this class are wall jets which might be regarded as "edge tones" with a 180° wedge. The hole and ring tones as shown involve axisymmetric modes of the jet, although a two-dimensional counterpart of the ring tone is easily envisaged. The feedback disturbances basically constitute force dipoles which are generated by vortex shedding from the ring. These dipoles may be axisymmetric or transverse to the axis of the jet, accordingly coupling into transverse of axisymmetric modes. These vortex formations and directivities are illustrated in Fig. 3.11. The fed-back disturbances reinforce axisymmetric modes of the jet in both the hole and ring tones. The edge tone involves nonsymmetrical oscillations of the jet, and the secondary disturbances are generated by vortex formation at the apex of the edge alternately on either side of the wedge. Jet tones which do not involve an impingement surface include the screech of high-speed choked jets as well as a nearly tonal phenomenon at higher Reynolds

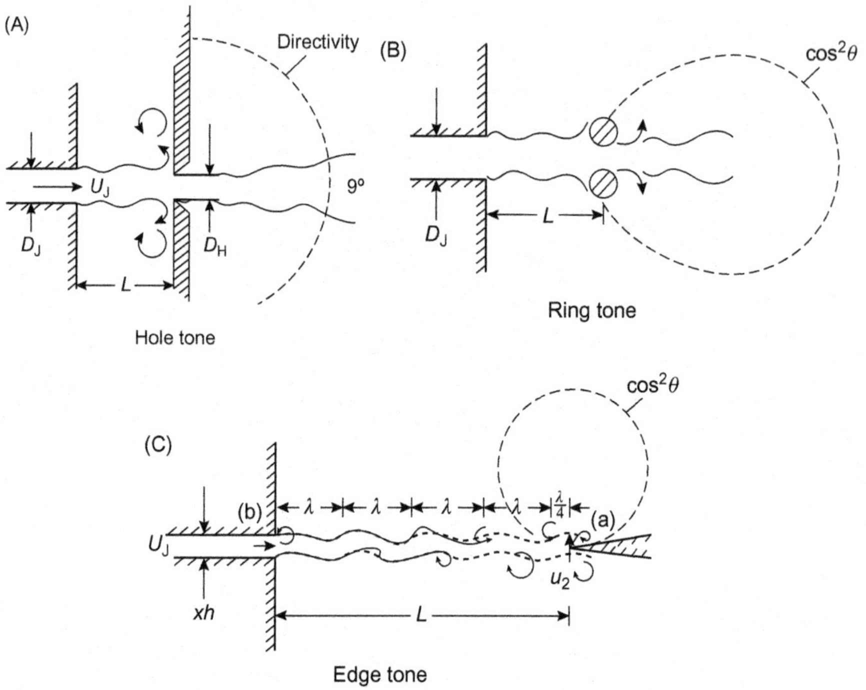

FIGURE 3.11 Schematics of the hole tone, ring tone, and edge tone systems.

numbers which involves flow separation at certain converging/diverging nozzles. Before dealing with these types of tones individually, the characteristic frequencies of jet disturbance modes will be reviewed since these frequencies are held in common collectively by jet tones.

3.4.2 Dimensionless Frequencies of Jet Tones

The bulk of, now classical, work has been conducted on the axisymmetric modes of circular jets, while work on the nonaxisymmetric modes has principally focused on the edge tones of rectangular jets of large width/height ratio. It is useful to recall the discussion at the beginning of Section 3.4.1 which delineated the differences between those effluxes with thin annular shear flows surrounding a laminar plug flow with little variation in mean velocity across the center of the jet and those effluxes with a fully sheared velocity profile across the entire diameter. These differences in profile provide different frequencies. In the class of fully sheared profiles are laminar flows at low Reynolds number that may generate tones and turbulent flows in long nozzles which will not generate tones.

We first consider flow instabilities of circular jets that involve an annular shear layer of thickness δ surrounding a zone of uniform velocity and diameter $D-2\delta$. We assume that $\delta \ll D$ in the discussion to follow. The frequencies of disturbances are determined by the thickness of the annular shear layer. The shear layer, therefore, is thought of as two parallel hyperbolic tangent profiles so that the unstable wave numbers of the annulus correspond to

$$\alpha_r \delta = \text{const}$$

Now, for an initially laminar jet, resulting from a laminar boundary layer on the wall of the nozzle, the shear layer thickness will depend on the Reynolds number approximately as (see Chapter 8: Essentials of Turbulent Wall-Pressure Fluctuations)

$$\frac{\delta}{D} \propto (R_D)^{-1/2}$$

Since we can write

$$\alpha_r = \frac{2\pi}{\lambda_c} = \frac{2\pi f}{c} \approx \frac{2\pi f}{U_J}$$

where U_J is the efflux velocity of the jet, the Strouhal number for these naturally growing jet instabilities can be written

$$S_D = \frac{fD}{U_J} \propto (R_D)^{1/2} \tag{3.21}$$

Visualizations [68–70] of these disturbances disclose short wavelength oscillations emanating from the nozzle lip. These oscillations merge as they move downstream such that pair-by-pair larger annular vortex structures evolve into a resultant stabilized pattern. Thus, the lip shear layer oscillations are embryos of larger vortex structures and they represent the shortest wavelength disturbances in the jet and form the upper bound of an envelope of possible Strouhal numbers as shown in Fig. 3.12. The reported flow visualizations show these wavelengths to be less than a diameter and to occur at Reynolds numbers more than 2000–3000. The Strouhal numbers of these disturbances follow Eq. (3.21) with the proportionality constant ranging from 0.012 to 0.02.

At Reynolds numbers less than 1000–2000 two features of the flow make the disturbances of circular jets with either fully sheared or plug flow profiles similar. First, the disturbance wavelengths for each type profile are at least on the order of a diameter or longer, second predicted Strouhal numbers of each are similar. At Reynolds numbers less than 6000–10,000, the axisymmetric modes of circular and two-dimensional jets with fully sheared velocity profiles may be accurately predicted using the theory of hydrodynamic stability and a profile similar to Fig. 3.1D. This profile is also somewhat representative of the more gradual shear layer occurring downstream in the higher Reynolds number turbulent jets where the larger scale vortex structures are observed to emerge. Laminar varicose motions of jets with fully sheared profiles are not often observed at Reynolds numbers above

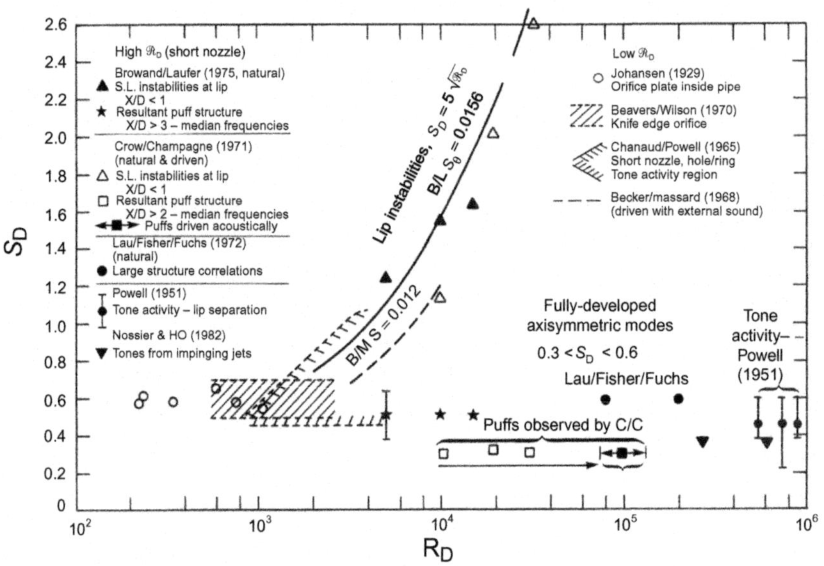

FIGURE 3.12 Strouhal numbers for vortex formation in circular jets with axisymmetric disturbances.

3000 because the nozzle flow that creates the jet undoubtedly goes turbulent before the jet ensues.

Visualizations at the low end of the Reynolds number range have been conducted with both externally driven jets and jets that undergo self-excited oscillations. For jets with shear layers that include most of the diameter, measurements of Sato and Sakao [12] on two-dimensional nozzles show frequencies of modes given by $S_{2h} \simeq 0.14$ for $2000 < R_{2h} < 10{,}000$, while at lower Reynolds number they found

$$S_{2h} \sim (7.7 \times 10^{-5})\sqrt{R_{2h}},$$

where $2h$ is the width of the jet. Frequencies of axisymmetric modes [13] follow $S_{2h} = 0.23$ for $1500 < R_{2h} < 8000$; while at values of R_{2h} of order 3×10^4, S_{2h} of order 1.25 were reported. Values of S_D for tones of circular knife-edged orifices are constant with R_D. Flow visualizations of Johansen [71] and of Beavers and Wilson [72] show that the responsible jet modes are axisymmetric. The hole tones observed by Chanaud and Powell [73] are bounded between the Strouhal numbers being either constant or proportional to $R_D^{1/2}$. Although the frequencies of hole tones depend on the ratio of orifice spacing L to diameter, L/D, in a manner to be examined subsequently, the tones are possible only because of the available range of wavelengths for jet instabilities to occur (Fig. 3.2). Thus the region enclosed within the branches in Fig. 3.12 describes the available Strouhal numbers of amplification.

As Reynolds numbers exceed 10^4, larger, clearly defined varicose growing waves are less apparent and axisymmetric vortex structures begin to dominate the jet dynamics. For Reynolds in excess of 4×10^4 well-defined annular lip oscillations have not been observed. In the measurements of Crow and Champagne [74] ($10^4 < R_0 < 4 \times 10^9$) wavelike disturbances initiated at the lip of the jet coalesce as they propagate downstream forming longer waves. After two stages of coalescence, the wavelike disturbances break down to vortices (or "puffs"). Strouhal numbers for the formation of waves and of vortices downstream of the nozzle (in the zone $y_1 > (1-2)D$) are both shown. The Strouhal number for vortex formation was of order 0.3; at this frequency the jet could be acoustically driven to larger-magnitude axisymmetric disturbances. Observations of Browand and Laufer [69] disclosed a similar breakdown of wavelike dynamics into vortex pairing. In both cases the observed Strouhal number of vortex formation was 0.5. At still larger values of R_D measurements by Lau et al. [75] of the frequency spectra of velocity and pressure fluctuations in the potential core were peaked about $fD/U_J \sim 0.5$–0.6. This frequency was later confirmed as related to a regular pattern of large vortices by the use of signal-conditioning techniques by Lau and Fisher [76]. The fluctuations sensed in the potential core are impressed by these large-scale axisymmetric vortices in the annular mixing layer. Alenus et al. [77] have conducted a high Reynolds number (200,000) large eddy simulation (LES) of a jet formed by a constriction in a circular duct finding Strouhal number of the principal tone emitted to be between 0.4 and 0.43 in agreement with measurements shown in Fig. 3.12 depending on the geometry of the constriction.

At Reynolds numbers approaching 10^6, tone activity or nearly tonal activity, has been reported by Powell [78] (see also Ref. [25]) in the sound field of a jet issuing from a diverging nozzle lip. Such a nozzle generated a separation zone at the lip which drove ordered sound-producing disturbances in the jet.

Tonal disturbances have also been observed to emanate from sharp-edged orifice plates by Kurzweg [79], and from squared-edged plates (Fig. 3.9C) by Anderson [80–86], Strouhal numbers based on the orifice diameter, D, and efflux velocity, U, are shown in Fig. 3.13 together with a schematic of the experimental arrangement used by Anderson. The sloped lines show that the numbers are functions of the ratio of thickness to diameter, t/D. For each value of the parameter

$$S_t = ft/U_J$$

the values of $S_D = f\,D/U$ generally range from 0.4 to 1.0 This range corresponds roughly to the observed range of numbers observed for free laminar jets in Fig. 3.12. The limit of $t/D = 0$ corresponds to the knife-edged orifices used by Beavers and Wilson [72] and Johansen [71]. Parametric dependence of S_D on t/D will be discussed at the end of the next subsection.

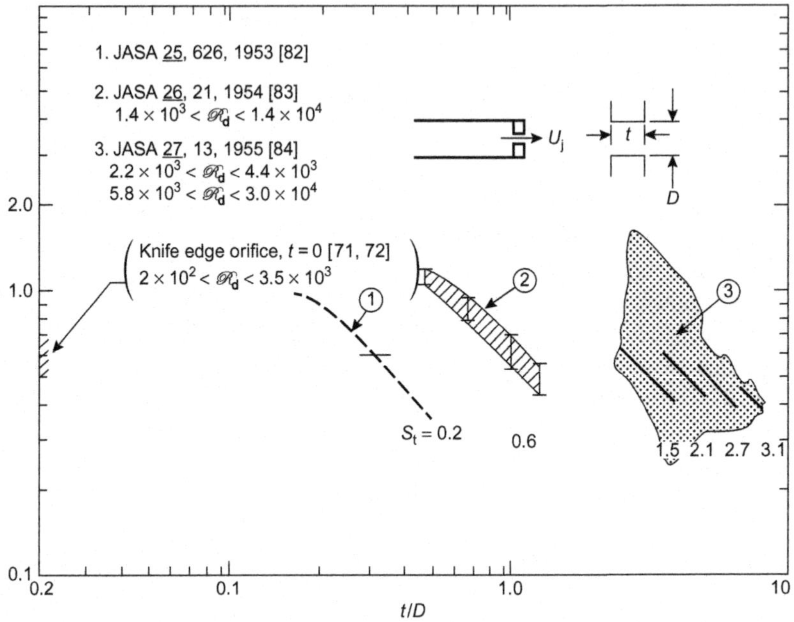

FIGURE 3.13 Tone frequencies for square-edge orifice plates terminating a pipe, from Anderson [82,84] compared with S_0 for knife-edged orifices from Johansen [71] and Beavers and Wilson [72].

3.4.3 Hole, Ring, and Edge Tones

The general sensitivity of laminar jets to external stimulation by sound, vibration, and by reflected hydrodynamic and acoustic disturbances has been recognized for well over a century as the primary cause of the many varieties of musical tones observed over the years. Note the variety of common-place observations of such instabilities afforded by acoustically excited flames (Tyndall [87] and Rayleigh [1]), bird calls, and whistling tea kettles (Rayleigh [1] and Chanaud and Powell [88]), human whistles (Wilson et al. [89]), Pfeifentone (pipe tone) [80–86] and a variety of musical instruments [1] (also Nyborg et al. [90], St. Helaire et al. [91], Elder [92], and Smith and Mercer [93] are some of the more recent and include interesting bibliographies). The fundamental feature of all these tones is that the jet as an oscillating system, with its available continuum of "resonance" frequencies, becomes dynamically coupled to another mechanical system as illustrated in Fig. 3.11.

The *hole tone*, previously discussed, is generated when a plate with a sharp-edged orifice is placed coaxially with the jet. Rayleigh [1] discusses in great detail the aspects of construction of this oscillator, which is the source of bird calls, tea kettle whistles, and human whistling. Axisymmetric disturbances are caused at the hole, which reinforces the initiating disturbances at the efflux. Sound radiation results from the pulsating efflux at the second plate. In the experiments of Chanaud and Powell [73], the second plate was larger than the wavelength of the sound emanated so that the radiation was omnidirectional. The *ring tone* [71] works in the same principle; however, the hole is replaced by a coaxial ring. Toroidal vortices shed from the ring cause alternating forces on the ring. These dipoles radiate in the direction of the jet axis. The *edge tone* (Fig. 3.11C) involves the interaction of the non-symmetrical modes of a jet with a rigid edge. The to and fro oscillation of the jet causes an alternating force on the edge. This force has a reaction at the jet efflux. Generally the edge tone has been examined for rectangular jets of relatively large width w to height h ratio. The edge tone has been given considerable attention in recent years because of its clear manifestation of the relationship between jet stability and geometric constraint. The relationship between the frequency of the tone and geometry that is generally accepted today has been given by Curle [94] and Powell [43] although arrived at by differing physical arrangements.

Because of its fundamental importance as a representative of jet tone behavior, the edge tone shall now be developed in some detail. Although given extensive attention over the years it was not until 1961 when Powell [44] provided the first comprehensive analysis of the phenomenon as a feedback-driven hydrodynamic jet instability. The essentials of this argument have already been introduced for the self-excited cavity tones in Section 3.3. The instability of the jet disturbance follows Eq. (3.10) which has the value at the edge downstream (see Fig. 3.11).

$$u_2(L, y_2, t) = \hat{u}_2(y_2) e^{\overline{\alpha}_i L} e^{i(\overline{\alpha}_r L - \omega t)} \tag{3.22}$$

where $\overline{\alpha}_r$ and $\overline{\alpha}_i$ are average eigenvalues across L. The dipole force $F(t)$ that is generated at the edge by the impinging jet is proportional to the transverse velocity $u_2(L, 0)$ at the jet centerline and it behaves as

$$F(t) = a_1 \rho_0 U_J w \delta u_2(L, 0, t)$$

where w and δ are the width and height of the jet at the point of impingement, respectively, and U_J is the centerline velocity of the jet at the edge location; $a_1 \approx 3$. This force induces a velocity field u' at distances away from the effective center of the dipole which depend on the distance between that centroid and the field point of interest. In cases where L is much less than an acoustic wavelength and the field point of the dipole is at the jet exit, the velocity induced is thus aerodynamic in nature and behaves as

$$u'(0, t) = a_2 \frac{F}{4\pi} \frac{e^{i\pi/2}}{\omega \rho_0 L^3} e^{-i\omega t}$$

where a_2 lies somewhere between 1 and 2 depending on the relationship between δ and L. This relationship is easily derived from the equations immediately preceding Eq. (2.31) which give the pressure and vertical (u_2) velocity in the acoustic near field ($k_0 \rightarrow 0$) of a dipole of strength $Q_i = Q_2$. By comparison with Eq. (2.75) the dipole strength and force are related by $Q_2 = -F_2/\rho_0$. Because of the tangential-flow boundary condition at the efflux tip, the reaction flow disturbance of the jet $u'_2(0, t)$ by this imposed external field is just

$$u'_2(0,t) + u'(0,t) = 0$$

so that the phase between $u'_2(0, t)$ and $u'(0, t)$ is π. Reinforcement with the initial disturbance $u_2(0, y_2, t)$ of Eq. (3.22) requires $u'_2(0, t)$ to be in phase with $u_2(0, y_2, t)$. Accordingly, the phase factor $\overline{\alpha}_r L$ must be governed by the sum of individual phases; i.e., the phase between $u'_2(0, t)$ and $u_2(0, y_2, t)$ is

$$2\pi n = \alpha_r L + 0 + \frac{1}{2}\pi + \pi, \quad \text{for} \quad n = 1, 2, 3$$

where, as before,

$$\overline{\alpha}_r L = 2\pi f L / \overline{c}_r$$

and $\overline{c}$ is the mean phase velocity of disturbances along L; it is roughly bounded [44] by $0.3 < \overline{c}/U_J < 0.5$. For a self-sustained oscillation, the loop gain, i.e., $u'_2(0, t)/u_2(0, y_2, t)$, must be greater than unity because the effect of all parts of the loop must result in a net increase in the disturbance at the efflux.

The allowed frequencies for self-sustained oscillation nondimensionalized as $f_s L / c_r$ are then

$$\frac{f_s L}{c_r} = \tfrac{1}{4}, 1\tfrac{1}{4}, 2\tfrac{1}{4}, 3\tfrac{1}{4}, \ldots$$

so that we have

$$\frac{f_s L}{U_J} = \frac{c_r}{U_J}\left(n + \frac{1}{4}\right) \quad \text{for} \quad n = 0, 1, 2, \ldots \tag{3.23}$$

The $n = 0$ mode has not been observed.

The most extensive data with which to validate Eq. (3.23) has been provided by Brown [95], Powell [43], and Curie [94] using rectangular jets with $w/h \gg 1$. Fig. 3.14 gives a sample of such data in both absolute and dimensionless form. The absolute frequencies from Brown [95] for tones emitted by a $h = 1$ mm slit with an $L = 7.5$ mm gap to the wedge have a jump discontinuity at about 17–19 m/s. For the parameters of the edge tone, Eq. (3.23) gives the loci of frequencies for each of the $n = 1, 2, 3$ stages. Collectively, the measurements show that in the regions of stage jumps there is a hysteretic behavior as illustrated in Fig. 3.14A by the dotted lines which signify the $f_s - U_J$ behavior for decreasing velocity.

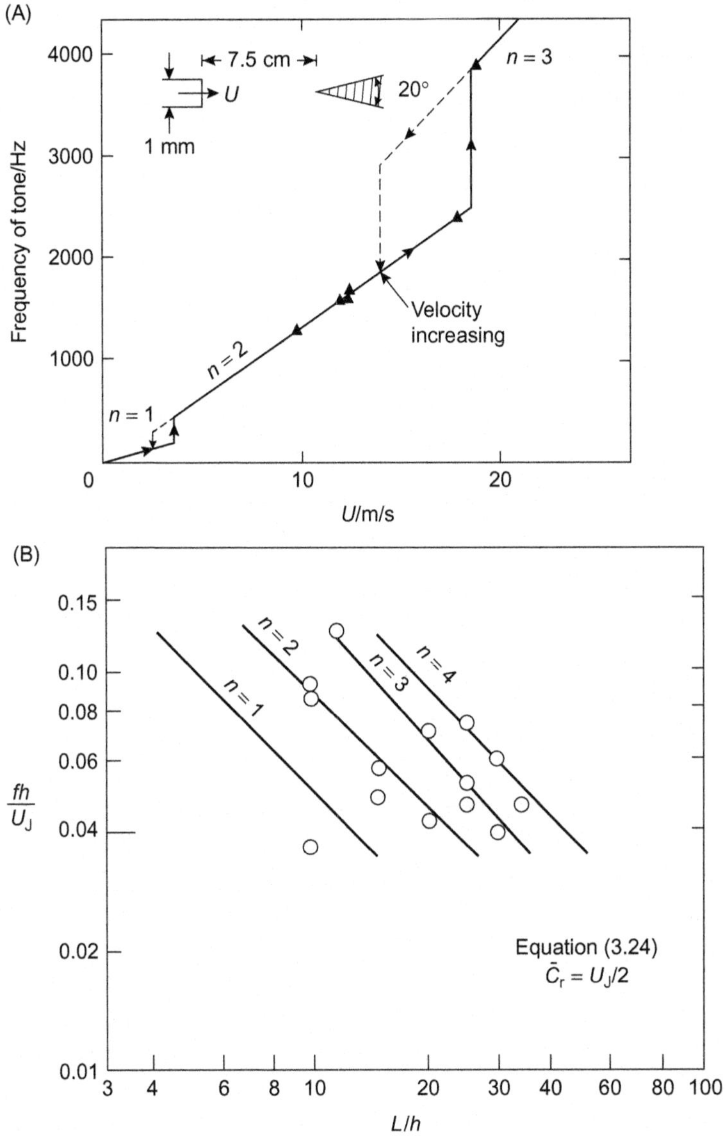

FIGURE 3.14 (A) Ranges of edge tone for a case examined by Brown [95]. Regions of hysteresis suggested by other data although not actually reported for this case, (B) Strouhal numbers for the first stages of Brown's [95] edge tone.

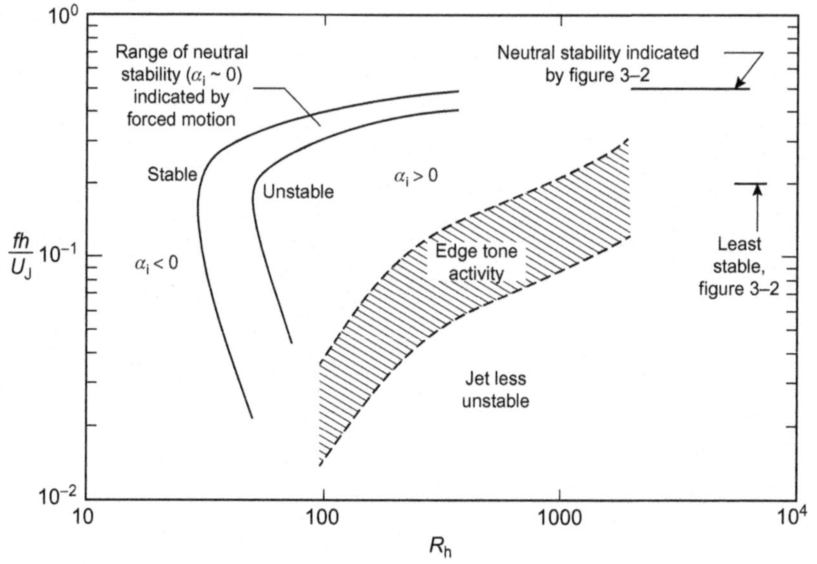

FIGURE 3.15 Stability diagram for two-dimensional jets showing the region of edge tone activity observed by Powell [43].

The tone frequencies may be cast in terms of the conventionally defined Strouhal number giving allowed frequencies as

$$\frac{f_s L}{U_J} = \frac{c_r}{U_J} \frac{h}{L} \left(n + \frac{1}{4} \right) \quad \text{for} \quad n = 0, 1, 2, \ldots \tag{3.24}$$

Fig. 3.14B shows a sample of Brown's measurements which are confined to the range

$$0.035 < fh/U_J < 0.15$$

Such a range is typical, as Powell's [43] measurements shown in Fig. 3.15 also disclose, and the range has some dependence on the Reynolds number. Within the range of S_D, various stages are observed as a function of the gap width L/h. For some values of L/h or of velocity, tones from two stages may exist simultaneously. This boundedness of the edge tone Strouhal numbers has been explained [43] on the basis of the hydrodynamic disturbance amplitude. Since the loop gain must be greater than unity the factor $\overline{\alpha}_i L$ appearing in Eq. (3.22) must be suitably large. This means that according to Fig. 3.2 the disturbance wave number $\overline{\alpha}_r L$ must also be bounded by some range due to the rather restricted region for which $\overline{\alpha}_i \overline{S}$ has a large positive value. There will thus be a corresponding region of the $\alpha_i - R_D$ domain as shown in Fig. 3.15 for which conditions are favorable for self-excitation.

Note that $\overline{\delta}$ is a mean value of the shear layer thickness or jet height which is nominally of the same order of the slit height h but not equal to it. Accordingly, since

$$\overline{\alpha}_r\overline{\delta} = 2\pi\frac{fh}{U_J}\left(\frac{U_J}{\overline{c}_r}\right)\left(\frac{\overline{\delta}}{h}\right) \tag{3.25}$$

a band of fh/U_J will be associated with a band of $\overline{\alpha}_r\overline{\delta}$ for which $\overline{\alpha}_i\overline{\delta}$ is larger than some threshold. For a given jet diameter and velocity, a stage will persist over a range of L/h such that $\alpha_i L$ lies in a band bounded by a value that is large enough for necessary loop gain and a value that is small enough that the orderly structure of a given mode in the jet does not break down before the disturbance reaches the edge. The persistence of a given stage over a restricted velocity range illustrated in Fig. 3.14A has also been explained [25] using a similar argument and Eq. (3.25). The instability characteristics scale on the shear layer width δ, not on the efflux height h; the local jet width decreases with increasing velocity since $\delta/h \sim R^{-1/2}$. Thus there will be a speed range ΔU_J over which $\overline{\alpha}_i L$ will be suitably large and that corresponds to a range of δ for which the conditions for favorable loop gain are maintained. On the basis of hydrodynamic stability, it would appear that edge tones would not exist for Reynolds numbers less than roughly 50 which is the lower limit for which α_i is greater than zero. In practice, the tones are generally (though not always) feeble outside the region $100 < R_h < 3000$ and for $2.5 < L/h < 4.5$.

The unsteady dipole force exerted on the edge by the jet has been shown both experimentally and analytically by Powell [43] to have an amplitude bounded by

$$|F| \lesssim \tfrac{20}{9}\rho_0 U_J^2 wh = \tfrac{40}{9} q_J wh \tag{3.26}$$

where w is the width and h is the thickness of the jet. This upper limit was deduced from measurements in air of the sound pressure radiated from laboratory edge tones, which showed that the force is actually a function of both L/h and R_h. The relationship between the force and the sound pressure is deduced from Eq. (2.74) or (2.75), as long as w, h, and L are all less than the wavelength of sound. Powell's measurements included independent evaluations of both the radiated sound and the aerodynamic force fluctuations which provided one of the first experimental confirmations of the validity of Eq. (2.75). Letting

$$f_i(t) = |F|e^{-i\omega t}$$

Eq. (2.75) gives

$$p_a(r,\omega) = \frac{-i\omega\cos\theta}{4\pi c_0 r}|F|e^{ik_0 r} \tag{3.27}$$

$\omega = 2\pi f$, and $\theta = \pi/2$ coincides with the vectorial direction of jet flow. This limit given by Eq. (3.26) has been supported [43] by kinematic considerations that take into account the strengths of vortices formed at the edge and their influence on the unsteady transverse momentum of the jet. Therefore this upper bound is expected to be generally applicable.

The hole tone of Chanaud and Powell [88] (Figs. 3.12 and 3.11A) can now be interpreted. For $L/D > (L/D)_{crit}$ the tone was emitted at one or more Strouhal numbers contained within the cross-hatched angular zone of Fig. 3.12 in the region $900 < R_D$ 2500. For a given value L/D the Strouhal number was constant over a range of R_D, and it would change to a second stage at some critical value of R_D, for example, at $L/D = 3$, $S_D \simeq 0.5$ for $1350 < R_D < 1900$, changing to $S_D \simeq 0.65$ at $R_D = 1900$ and continuing at this value until $R_D = 2500$. When the Reynolds number is reduced by reducing speed, this value of S_D was maintained down to $R_D = 1400$ when S_D reverted to ~ 0.5 giving further demonstration of hysteretic behavior that is common in self-sustained fluid oscillations. The conditions for stages of the hole tone can be summarized as

$$\frac{f_s h}{U_J} = \frac{c_r}{U_J}\left(n - \frac{1}{4}\right)$$

where $0.5 < c_r/U_J < 0.9U_j$ and for cross-hatched zone of Strouhal numbers shown in Fig. 3.12.

The behavior of the Strouhal numbers observed by Anderson, shown in Fig. 3.13, may also be interpreted in the above terms. Flow in sharp-edged orifices observed by Kurzweg [79] disclosed a train of axisymmetric ring vortices. Parallel to Eqs. (3.11) and (3.20), we write this condition

$$t = n\Lambda$$

where Λ is the vortex spacing, t is the thickness of the orifice plate, and n is the integral number of vortices in the orifice. Letting

$$f_s\Lambda / U_J = \Omega = \text{const}$$

where U_J is the velocity of the orifice flow, so that

$$S_t = \frac{f_s t}{U_J} = n\Omega, \quad n \geq 1$$

describes the Strouhal numbers for the various stages of tones. If we let the allowable range of S_D for the jet to be between $0.2 < S_D < 1$ (as suggested by Fig. 3.12) at the appropriate value of R_D, then the dependence of stages on d/t is given by

$$0.2 < S_D = (d/t)(S_t) < 1$$

which corresponds to the diagonal lines shown in Fig. 3.13 with the indicated values of S_t.

The conditions which favor the generation of the various jet tones of this section are summarized in Table 3.1 which includes supporting references

TABLE 3.1 Summary of Subsonic Jet Tone Arrangements

Mode Classification	$\frac{\phi_n}{2\pi}$	$\frac{U_c}{U_J}$	Reynolds Number Range	Arrangement	Reference
Hole and ring	$n-{}^1/_4$	0.5–0.9	1000–3000	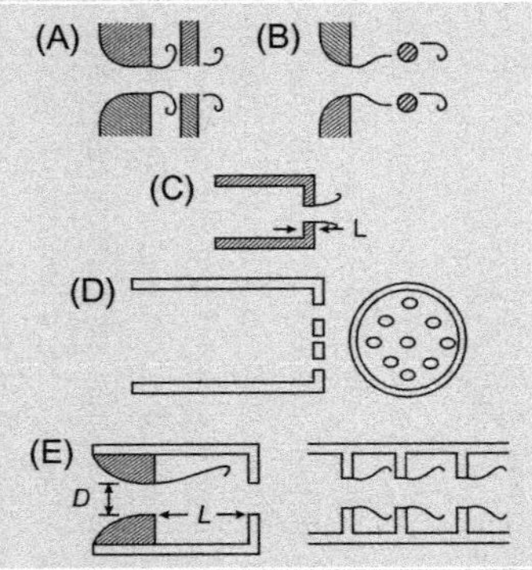	[88]
Axisymmetric modes $1<L/D<8$	Fig. 3.13	—	$10^3-3\times10^4$		[80–86,96,98]
	$n-{}^1/_4$		10^3-10^4		[54,71,89,97,98,100]
Wall jet: axisymmetric modes $2<L/D<8$	Integer closest to L/D	0.5	3×10^5–8×10^5 (roughly)	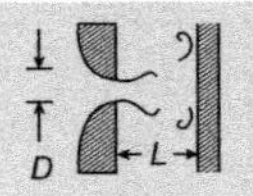	[103,106–109]
Edge tones:	$n+{}^1/_4$	0.3–0.5	100–7500	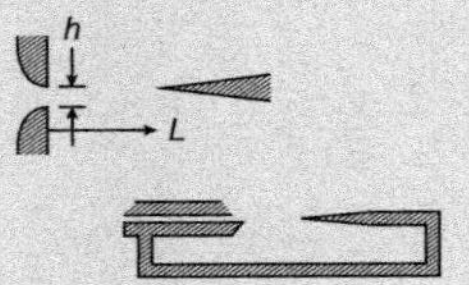	[44,63,101–104]
Nonsymmetric $2<L/h<12$			2×10^2–2×10^4		[29,63,105]

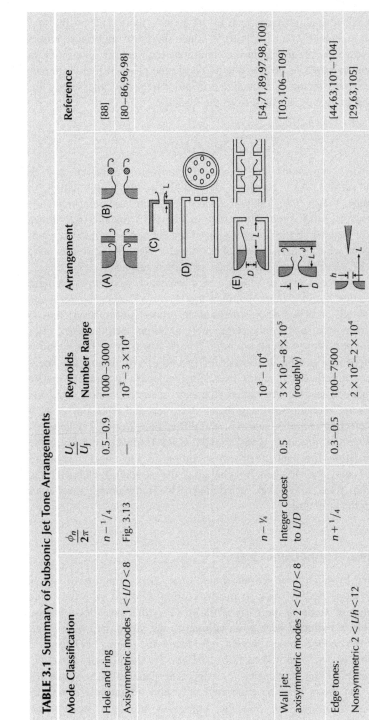

[95–107]. Each tone is characterized by its generic mode and by the basic geometric arrangement which supports it. Since the most important practical questions involve the prediction of the conditions required for tone onset, the Reynolds number range of observed activity and the phase velocity are tabulated. Also a phase factor ϕ_n is included in order to predict the stages. In a given case the frequency is found from the formula

$$\frac{f_s L}{U_J} = \frac{U_c}{U_J} \frac{\phi_n}{2\pi(1 + M_c)}$$

where L always represents the separation between the efflux and the impingement surface. As a check on the validity of any computation, the Strouhal number $= S_D = fD/U_J$ should always be within the bounds shown in Figs. 3.12 and 3.13 for circular jets and in Fig. 3.15 for slit (two-dimensional) jets. In general, at the lower region of the Reynolds number range the tones can occur without the support of a resonator. The generation of tones at $R_D > 3 \times 10^3$ generally, but not always, requires a resonator.

Control of jet tones may be effected in a few ways. For jets at moderate to high Reynolds number the conditions that favor a supporting resonance may be removed, or acoustic or structural absorption (dissipation) may be introduced. In most cases of either laminar or turbulent jets, the use of flow spoilers at the nozzle may break up ordered vortex structures. This might be done by roughening the inside surface of the nozzle or fluting the nozzle around its circumference. In cases involving enclosure resonances, baffles may be introduced to alter the modal character of the resonating fluid volume or to disrupt acoustic feedback paths. Also, jets and resonators may be detuned by selecting chamber sizes (L) and efflux sizes (h or D) outside of a parameter range that produces tones. Finally, if the feedback path involves a structural vibration mechanism (say a vibration of the nozzle wall), the reinforcement may be detuned by changing the structural characteristics involved. The tone may also be damped by introducing mechanical dissipation to the structure.

3.4.4 Tones in Supersonic Jets

Sound emitted by supersonic jets is an area of jet noise which goes beyond the subsonic scope of this book so in this section we shall give the reader an appreciation of some of the early work on tones that provides a physics overview. This is included here as an adjunct to the broad area of jet tones and to give the reader some relevant references and Tam [111] has provided a review of work published prior to 1995. As the velocity through a nozzle is increased the propensity for edge tone generation becomes less apparent so that tones may be generated only under relatively controlled situations. Accordingly, the acoustic emissions are less intense and

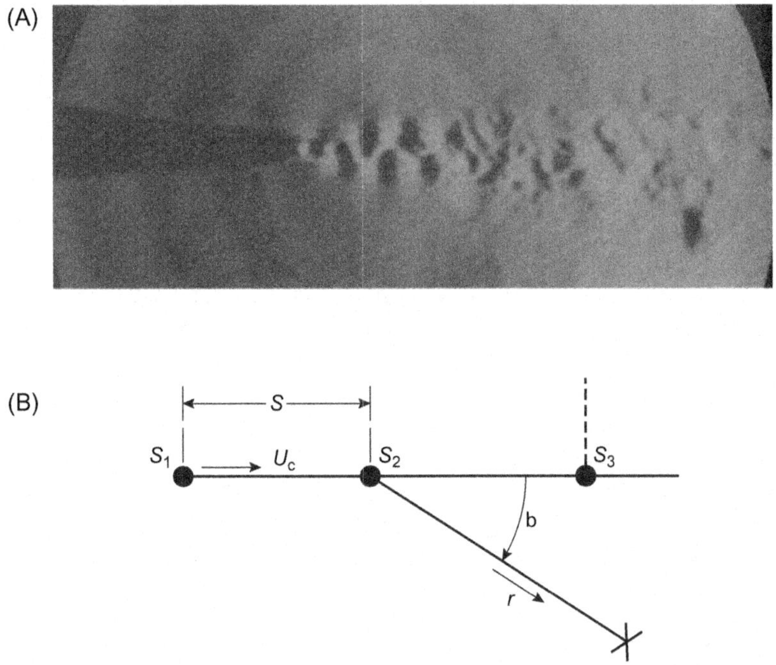

FIGURE 3.16 Shock cell patterns and source geometry for the sound field of a choked rectangular jet with breadth normal to page. (A) Schlieren photograph of choked jet and upstream-propagating soumd waves, $R = 3.05$, (B) Idealization of source geometry. *From Powell A. On the mechanism of choked jet noise. Phys Soc London 1953;B66:1039–57.*

broadband in nature. When the velocity in the nozzle becomes sonic, i.e., at and above the critical pressure, the characteristics of both noise and the jet structure change markedly. Superimposed on the broadband sound is an acoustic tone which is associated with the existence of a stable cellular pattern of expansion and compression zones along the jet axis. A schlieren photograph of such a pattern induced in a two-dimensional jet is shown in Fig. 3.16. These cells are formed from periodic compression and expansion zones spaced at intervals $(x_s)_i$ in the fluid as it passes down the jet axis; the downstream extent of this pattern is limited by dissipation to turbulence. The locations of cells in the pattern are diagrammed in the lower part of Fig. 3.16 as a series of points representing cell locations along the jet. The process is initiated at the nozzle exit as jet disturbances initiate an oscillation of the Mach lines originating there. Although generally studied in two-dimensional jets, the phenomenon also occurs in high-speed circular jets. Generally three to five cells occur until they are dispersed by downstream turbulent mixing. In the case of circular jets as many as 12 cells have been observed. The number of cells formed, and their axial dimension, is dependent on both pressure ratio and diameter.

The conditions required for the existence of these screech tones as well as an explanation of the directivity of the sound emitted were given by Powell [78,110,112,113] and follow the feedback mechanism enumerated in Section 3.3.2. The sound sources, being coincident with the shock cells, are stationary with respect to the jet exit and the source of oscillation arises from the convection of orderly vortex structures through the cells. Accordingly, the strongest tones are generated when the phases of the individual pressures radiated from each cell is such that the contributions best reinforce each other at the nozzle exit. These fed-back pressures drive the jet shear layer producing the aerodynamic disturbances which "pulse" the shock cells. The sound waves emitted are clearly visible as semicircular patterns of light and dark in the schlieren photograph of the slit jet in Fig. 3.16. The sound on opposite sides of the jet is 180° out of phase. The effective source center depends on conditions and may be situated three to four cells downstream. This location will be governed not only by geometrical aspects of acoustic reinforcement but also by the weighting of source strengths of adjacent cells down the jet axis. Near the nozzle the source strength will be relatively small, as the jet instabilities are still growing there, and a few cells downstream the strength will diminish because of the generation of disorder and decorrelation induced by turbulence. The most dominant acoustic sources are included between these extremes.

The directivity of the sound is determined by the acoustic wavelength, the cell dimension along the axis of the jet, and the convection speed of the jet vortices through the shock cells since this determines the relative phases of cell undulations as an acoustic line source. The primary directivity of sound, i.e., at the fundamental frequency, is upstream back to the jet exit. This directivity is required to determine the coupling of the sound to aerodynamic disturbances in the exit plane of the jet. The locations of these cells as acoustic sources were confirmed in a study that used an acoustically-directive mirror microphone by Glegg [248].

A fixed phase relationship between the emitted sound and the aerodynamic disturbances in the exit plane is required just as in the case of the subsonic edge tone. The stationary source distribution provides a fixed "acoustic center" at a distance x_s from the nozzle exit which is analogous to the slit-edge distance of the edge tone. In contrast to the edge tone however, the effective source location is a variable that depends on pressure ratio and diameter of the particular jet. Dealing with the case of two-dimensional jets, the condition analogous to Eq. (3.23) which is required for self-excitation of a tone of frequency f_s is (Powell [110,114,115])

$$\frac{f_s x_s}{U_J} = \frac{U_c}{U_e}\left(\frac{N+k}{1+M_c}\right)$$

The effective source location x_s for each jet must be deduced experimentally from schlieren photographs as the acoustic center of cylindrically spreading wave fronts. In cases reported by Powell [110] it appears to range typically from 1.7 to $2.3\lambda_c$, where λ_c is the effective eddy disturbance wavelength, but the generality of this relationship has not been established. Similarly, the convection velocity and the parameter k depend on the specific properties of the jet. Typically $U_c \simeq 0.6U_J$. The parameter k is a nonintegral measure of the number of eddy disturbance wavelengths along the axis which participate in the tone and the index N is number of complete acoustic wavelengths between the source and the jet exit. Thus the sum $N + k$ represents the total number of acoustic and aerodynamic cycles around the feedback loop. There are as yet no general rules for precisely determining x_s, K, or U_c/U_J. However for the jets examined by Powell [110], the parameter $N + k$ was a constant over a broad range of pressure ratios for a given jet and the frequency followed the rule

$$\frac{f_s h}{c_0} = \frac{0.2}{\sqrt{R - R_c}}$$

R is the ratio of the upstream stagnation pressure to the downstream ambient pressure; R_c is the critical pressure ratio ($R_c = 1.89$ for air); and h is the height of the jet. This rule was subsequently also observed by Krothapalli et al. [116] and it thus appears to be reasonably general.

The directivity of the sound is determined by the geometry of the cells in relation to the wavelength of sound. The dominant parameters are given in the bottom of Fig. 3.16. Each cell represents a source and the number of sources depends on the Reynolds number and turbulence level in the jet since turbulence will eventually determine the disintegration of the cells downstream. The cells are pulsed by eddy disturbances which convect at speed U_c; as with subsonic edge tones U_c is an effective value for a given jet exit velocity. The phase between adjacent sources is, thus $2\pi f_s s/U_c$, where s represents the mean of cell lengths, say $s = \langle (x_s)_{i+1} - (x_s)_i \rangle$ referring to Fig. 3.16. Although the number of participating cells and their relative strength is indefinite, the directivity pattern D_f of the main lobes is given in a representative fashion by only three sources of equal strength by which

$$D_f = \tfrac{1}{3} + \tfrac{2}{3}\cos\left[2\pi(s/\lambda_c)(1 - M_c\cos\beta)\right]$$

for the fundamental frequency, and

$$D_h = \tfrac{1}{3} + \tfrac{2}{3}\cos\left[4\pi(s/\lambda_c)(1 - M_c\cos\beta)\right]$$

for the first harmonic of $f = 2f_s$. Higher numbers of sources add additional factors of $1 - M_c \cos\beta$ that effect the side lobes, but not the main lobe directivity. Even though M_c may be a variable of the jet, the cell length and λ_c are given by the empirical relationship

$$s/h \simeq 1.9\sqrt{R - R_c}$$

and

$$\lambda_c/h = U_c/f_s d \approx 2.9\sqrt{R - R_c}$$

so that λ_c/s is a fixed number about equal to 0.66. For a range of typical values of M_c then the main lobe (greatest intensity of sound) will be nearly upstream (toward the jet) for the fundamental. The directivity of the first harmonic will be nearly broadside with much reduced intensity upstream. Accordingly this harmonic is incidental to the tone generation and its magnitude in relation to the fundamental is dependent on the details of the pulse shape of the eddy–shock interaction and the time stationarity of the flow. Thus, no further definition of characteristics of the fundamental and its harmonic is possible here.

Whether or not a tone of a given stage N is generated depends on principally the same factors which control the edge tone. The gain factors in the feedback loop analogous to the edge tone are continuously variable functions of frequency, but the frequency f_s pertains to the net disturbance at the jet exit from the transmission of pressures from the individual sources to the jet exit. Complete reinforcement, and therefore a maximum value of the fedback disturbance is determined [113] by the condition

$$(s/\lambda_c)(1 + M_c) = 1$$

Thus, reinforcement of the aerodynamic and acoustic disturbances will be facilitated within some band

$$\Delta[(s/\lambda_c)(1 + M_c)]$$

centered on $s(1 + M_c)/\lambda_c = 1$. The jet exit to flow feedback behavior is completely analogous to that occurring in subsonic cavity and jet tones thus Morris and McLaughlin [117] have found effective mitigation of the tones can be achieved using flow break-up corrugations in the jet exit.

In the case of the two-dimensional jet, the range of allowable tone frequencies is relatively large, with the frequency varying continuously with pressure ratio. In the case of the circular jet the observed dependence is discontinuous and hysteretic effects occur much as they do in the edge tone. Thus, the band $\Delta[s(1 + M_c)/\lambda_c]$ for circular jets appears to be much narrower than for two-dimensional jets. A prior prediction of the stages for the tones of circular jets is not yet possible, however, the frequencies observed generally discontinuously follow the rule

$$\frac{f_s D}{c_0} = \frac{0.33}{\sqrt{R - R_c}}$$

and involve a spiral mode of the jet disturbances. Follow-up work through the 1970s on sounds from choked jets further quantified some of these relationships and demonstrated that the feedback may be broken by the insertion of a shield forward of the exit plane, e.g., Hammitt [118] on two-dimensional jets, and Davies and Oldfield [119,120] on circular jets.

A topic of significant practical interest is concerned with the understanding and control of sound produced by supersonic jets impinging on flat plates, see, e.g., Powell [121,122], Tam and Ahuja [123], Krothapalli et al. [124]. When the shock wave cells are eliminated in the jet, the overall sound pressure behaves as the classical broadband character that will be discussed in Section 3.7 [125,126]. As shown by Maa et al. [126], the classical broadband character is maintained to pressure ratios (P_{stag}/P_{atm}) nearing 9.

3.5 THE STOCHASTIC NATURE OF TURBULENCE

3.5.1 Introduction

We have seen that in the case of jets and all other flows the disturbances become disordered, or random, at large values of Reynolds number. In order to deal with such disordered flows conveniently, it has been necessary to look toward statistical representations that usefully characterize the average, or expected, behavior of certain properties of the flow. These properties generally include fluctuating velocities in the turbulent region, fluctuating pressures, and density fluctuations. When radiated sound is the result of a random, or turbulent, process, then the far-field sound pressure will be a random variable. Therefore, in order to apply the deterministic relationships derived in Chapter 2, Theory of Sound and its Generation by Flow to realistic noise-producing flows, techniques have been developed for relating the statistics of flow fields to statistics of radiated sound. A suggestion of how the sound production from stochastic sources is represented using deterministic Green functions has already been given in converting Eq. (2.119) to Eq. (2.144). In this section we shall extend our analysis to cover alternative statistical formulations and apply them, in an elementary way, to the production of sound from turbulent jets.

The foundation of the classical statistical approaches to turbulence measurement and description is found in Taylor's series of papers on the statistics of isotropic turbulence [127] and on the spectrum of turbulence [128]. In these, the interrelationships between time and space through eddy convection were first realized, and methods of extracting measures of the largest and smallest eddies through correlation were outlined. In the years since, the importance of these statistical measures has become well recognized. Electronic instrumentation has also become more sophisticated, allowing observations of identifiable turbulence events through the merits of signal conditioning. Interpretation of correlations continually develop to meet the needs to describe physical behavior of transport phenomena and acoustics in virtually all forms of turbulent shear layers: jets, wakes, and boundary layers. Notable treatments of stochastic representations of many types of flows are those of Batchelor [129], Hinze [130], Townsend [131], Lumley [132], and

Pope [133]. The reader is referred to these sources for rigorous derivations and theorems of mathematical validity. Our discussions will deal with review and applications specifically as they apply to descriptions of flow-acoustic sources that we will use in the remainder of this book.

Noise and vibration are caused by fluidic and structural reactions to the contiguous turbulent field. In Chapter 2, Theory of Sound and its Generation by Flow linear cause-and-effect deterministic relationships were derived for the radiation process, and in Chapter 5, Fundamentals of Flow-Induced Vibration and Noise similar relationships will be derived for structural response as well. In either case the input is stochastic; although the transfer function is postulated as linear and deterministic so that the output of the linear system is also stochastic. As a historical note the fundamentals in the treatment of these systems were developed early in communication theory (see, e.g., Lee [134]), with subsequent development [135] of sampling criteria, frequency–time relationships, and nonstationarity. Treatises on the application of these techniques to physical systems were then developed for the generation of water waves [136] and the vibration of structures by random disturbances [137–139]. Essentially, this section continues our analysis of the random variable and its related functions that was introduced in Sections 1.4 and 2.6.2, Signal Analysis Tools of Vibration and Sound; Derivation of the Wave Equation With Vortical Sources.

3.5.2 Correlation Functions of Random Variables

In the treatment of random phenomena the variable in question, say, a vector component of velocity or vorticity, or a pressure, has a certain probability of attaining a value. We assume in the following that the flow is steady with invariant properties on which is superimposed relatively small turbulent characteristics. Giving the symbol u' to the random property of this variable and the symbol $P(u')$ to its probability of occurrence, then if the disturbance is to occur at all we have

$$\int_{-\infty}^{\infty} P(u')du' \equiv 1 \tag{3.28}$$

i.e., there is certainty that u will have some value between plus infinity and minus infinity. When u represents the streamwise velocity, then the average of u' or the expectation of u, is defined as the mean velocity

$$E(u') = \int_{-\infty}^{\infty} u'P(u')du' = \overline{U}' \tag{3.29}$$

Eq. (3.29) is also called the first moment of $P(u')$. In the hydroacoustic sense, Eq. (3.29) just defines the mean velocity, or in the case of pressures the static pressure, of the random quantity. A third property of the probability density is the mean square,

$$E(u'^2) = \int_{-\infty}^{\infty} u'^2 P(u')du' = \langle u'^2 \rangle \tag{3.30}$$

and the variance is defined as

$$\text{var}(u') = E(u'^2) - (E(u'))^2 \tag{3.31}$$

In fluid mechanical applications, the manipulations of relationships are simplified by treating the random variable as a variation about the mean. This permits us to decouple the mean and turbulent motions, finding first the mean, or steady, component of velocity or pressure, and then deduce the behavior of the random component to varying degrees of precision. In this way the random motion is viewed as superimposed on the mean flow. Thus we have been defining our random variable as a fluctuating plus mean

$$u' = \overline{U} + u \tag{3.32}$$

with a probability density of u given by $P(u)$ then it is easy to apply Eq. (3.28) through Eq. (3.31) to see that

$$E(u) = 0 \quad \text{and} \quad E(u^2) = \overline{u^2} = \text{var}(\overline{u^2})$$

which are simpler relationships to deal with. The expectation of $(u')^2$ is thus

$$E(u'^2) = \overline{U^2} + \overline{u^2}$$

To apply these notions to the problem of describing real turbulence quantities which are random in time and space let

$$u = u(\mathbf{y}, t)$$

where $\mathbf{y}$ is the space variable and t is time. The total flow variable is therefore expressed as a mean which is dependent on spatial coordinates and $u(\mathbf{y}, t)$.

$$u' = U(\mathbf{y}) + u(\mathbf{y}, t)$$

where now the over bar has been dropped from the time-averaged property. All operators on u' then reduce to a superposition of separate operations on the mean quantity $U(\mathbf{y})$ and the statistical quantity $u(\mathbf{y}, t)$. Often we shall be interested in the expectation of variables at two points and times accordingly, letting $u'_1 = u'(\mathbf{y}_1, t_1)$ and $u'_2 = u'(\mathbf{y}_2, t_2)$ then by the above definitions we illustrate this seperability of mean and fluctuating measures:

$$E(u_1', u_2') = U(\mathbf{y}_1)U(\mathbf{y}_2) + E(u_1, u_2)$$

where $E(u_1', u_2')$ is the joint expectation of u_1' and u_2'.

The joint expectation of the fluctuating velocities, u, at locations $\mathbf{y}_1$, and $\mathbf{y}_2$ and times t_1 and t_2 is a generalization of Eq. (3.30)

$$
\begin{aligned}
&E[u(\mathbf{y}_1,t_1)u(\mathbf{y}_2,t_2)] \\
&\quad = \langle u(\mathbf{y}_1,t_1)u(\mathbf{y}_2,t_2)\rangle \\
&\quad = \iint_{-\infty}^{\infty} u(\mathbf{y}_1,t_1)u(\mathbf{y}_2,t_2)P(u(\mathbf{y}_1,t_1))P(u(\mathbf{y}_2,t_2))du(\mathbf{y}_1,t_1)du(\mathbf{y}_2,t_2)
\end{aligned}
\tag{3.33}
$$

The angle brackets will henceforth denote the taking of an ensemble average in the formal sense. If the functions $P(u(\mathbf{y}, t))$ are independent of spatial position, then Batchelor [129] calls $u(\mathbf{y}, t)$ a *spatially homogeneous* random variable. The ensemble average $\langle u(\mathbf{y}_1,t)u(\mathbf{y}_2,t)\rangle$ is called a covariance or a correlation function. It is independent of the origin of $\mathbf{y}_1$, but it is dependent on the relative separation of $\mathbf{y}_1$ and $\mathbf{y}_2$. Alternatively, when $P(u)$ and therefore the ensemble average, is independent of the time, but dependent on time difference $\tau = t_2 - t_1$, then u is *temporally homogeneous*, or temporally stationary. This formalizes the introduction given in Section 1.4.2.2, Correlation Functions of Steady-State Signals.

To carry out the integration in Eq. (3.30) in practice, $u(\mathbf{y}_1, t_1)$ and $u(\mathbf{y}_2, t_2)$ would have to be sampled in a large number of experiments and then the integration performed over the complete ensemble of all possible values. This being impractical, we seek simpler alternative approaches in which the formal operation of Eq. (3.30) is replaced by a time or a space average. That is, we define the temporal average

$$
\begin{aligned}
\overline{u(\mathbf{y}_1,t_1)u(\mathbf{y}_2,t_2)}^{t} &= \overline{u(\mathbf{y}_1,t_2)u(\mathbf{y}_2,t+\tau)}^{t} \\
&= \lim_{T\to\infty}\frac{1}{T}\int_{-T/2}^{T/2} u(\mathbf{y}_1,t)u(\mathbf{y}_2,t+\tau)dt
\end{aligned}
\tag{3.34}
$$

and the spatial average

$$
\begin{aligned}
\overline{u(\mathbf{y}_1,t_1)u(\mathbf{y}_2,t_2)}^{y} &= \overline{u(\mathbf{y}_1,t_1)u(\mathbf{y}_1+r,t_2)}^{y_1} \\
&= \lim_{V\to\infty}\frac{1}{V}\iiint_V u(\mathbf{y}_1,t_1)u(\mathbf{y}_1+r,t_2)d\mathbf{y}_1
\end{aligned}
\tag{3.35}
$$

When

$$
\langle u(y_1,t_1)u(y_2,t_2)\rangle \equiv \overline{u(\mathbf{y}_1,t_1)u(\mathbf{y}_2,t_2)}^{t} \tag{3.36}
$$

$$
\langle u(y_1,t_1)u(y_2,t_2)\rangle \equiv \overline{u(\mathbf{y}_1,t_1)u(\mathbf{y}_2,t_2)}^{y} \tag{3.37}
$$

the process is said to be ergodic; there are certain formal requirements for ergodicity that are described for example by Lee [134], Bendat and Piersol [135], Kinsman [136], and Lin [137].

It will be assumed in this monograph that Eq. (3.36) holds unless stated otherwise and this assumption is often used in the analytical modeling of the fluid dynamics of turbulence. Relationship (3.37), however, does not generally hold for all types of flow. Specifically, the developing ordered disturbances in jets, wakes, and transitional boundary layers do not satisfy spatial homogeneity. Fully developed turbulent boundary layers also do not strictly satisfy Eq. (3.37); however, they are generally assumed to do so in planes that are parallel to the surface. This assumption is invoked in order to develop theorems for the description of the response of contiguous structures (Chapters 5 and 9: Fundamentals of Flow-Induced Vibration and Noise; Response of Arrays and Structures to Turbulent Wall Flow and Random Sound) in terms of boundary layer properties. Even though many flows do not formally satisfy Eq. (3.37), their correlation volumes are often small compared to the extent of the developing flow so that making the assumption leads to useful order-of-magnitude quantitative predictions. The relationships between the spatial correlations and wave number spectra that were developed in Section 2.7.2 apply to the above statistical quantities as shall be examined in some detail in the next section.

3.6 REVIEW OF CORRELATION AND SPECTRUM FUNCTIONS USED IN DESCRIBING TURBULENT SOURCES

3.6.1 Acoustically-Useful Representations for Homogeneous Turbulence

In this book we will consider turbulent flow sources whose properties exhibit a wide range of statistical characteristics. The statistical behavior of these properties, insofar as being characterized as "turbulent" is concerned, are generally identified by a threshold value of Reynolds number which is defined as the quotient $\rho U\delta/\mu$, where U and δ are velocity and length scales of the flow and ρ and μ are the density and viscosity of the fluid, respectively. When the Reynolds number of the flow is greater than some critical value the flow is said to be turbulent. In this section, we review the ways in which turbulence and turbulent acoustic sources are described analytically in order to calculate flow-induced sound and vibration. This will not be an exhaustive review, rather it will selectively draw on concepts that specifically apply to the description and modeling of flow-acoustic sources as used in this book; specifically acoustic source models of subsonic jet noise, turbulent-boundary-layer wall pressures, and turbulence in impinging wakes. Readers interested in further study of these concepts are referred to books that explore the subject in detail, e.g., by Batchelor [129], Hinze [130], Townsend [131], and Pope [133]. The turbulent flows of general interest to us will be temporally stationary in a statistical sense and will typically consist of flow-generated velocities and pressures. That is to say all time-averaged properties, e.g., temporal mean values, averages of products of velocities, space–time correlation tensors, and probability distributions of the variables, will all be invariant over time. No matter when these properties are measured, their time-averaged values will be the same as long as they are measured at the same locations in the flow. These temporally-averaged flow properties are expected to depend on the locations at which they are measured, but at these locations these averaged properties will be the same whenever they are measured. As discussed in Chapter 1, Introductory Concepts the principal of ergodicity will apply to these flows point-to-point so that temporal and ensemble averages are assumed to be interchangeable over long-enough averaging times or sample lengths. In the following pages, it is not meant to derive the required spectral quantities, they are derived elsewhere, but rather to cite certain key algebraic relationships and indicate their physical significance.

In general we will be assuming that the local mean velocity is much larger than the local fluctuating velocity, i.e.,

$$U_i(y_j) > u_j(y_j, t) \tag{3.38}$$

and that statistical quantities, such as a two-dimensional space–time correlation of velocities at two points in the flow satisfy an inequality that says their ensemble-averaged product is much smaller than the product of mean velocities taken at the two points:

$$\overline{u_i u_j} R_{ij}(x_k - y_j, \tau) = \frac{1}{2T}\int_{-T}^{T} u_i(y_j, t) u_j(x_k, t+\tau) dt = \overline{u_i(y_j, t) u_j(x_k, t+\tau)} \tag{3.39}$$

The wave number–frequency spectrum of the $u_i u_j$ velocity product tensor described by $R_{ij}(y_i - x_i, t)$ is

$$\Phi_{i,j}(\boldsymbol{k}, \omega) = \frac{1}{16\pi^4}\int_{-\infty}^{\infty}\int_{-\infty}^{\infty}\int_{-\infty}^{\infty} \overline{u_i u_j} R_{ij}(\boldsymbol{r}, \tau) \exp[i(\omega\tau - \boldsymbol{k} \bullet \boldsymbol{r})] d^3\boldsymbol{r} d\tau \tag{3.40a}$$

$$\Phi_{i,j}(\boldsymbol{k}, \omega) = \frac{1}{8\pi^3}\int_{-\infty}^{\infty}\int_{-\infty}^{\infty}\int_{-\infty}^{\infty} \Phi_{i,j}(\boldsymbol{r}, \omega) exp[-i\boldsymbol{k} \bullet \boldsymbol{r}] d^3\boldsymbol{r} \tag{3.40b}$$

Invoking Taylor's hypothesis of *frozen convection* [127,128,133] at velocity $\mathbf{U}_c$ we can express the wave number and frequency spectra in a separable form

$$\Phi_{i,j}(\boldsymbol{k},\omega) = \Phi_{i,j}(\boldsymbol{k})\phi_m(\omega - \boldsymbol{U}_c \cdot \boldsymbol{k}) \tag{3.41}$$

Where for frozen convection with a constant convection velocity

$$\phi_m(\omega - \boldsymbol{U}_c \cdot \boldsymbol{k}) = \delta(\omega - \boldsymbol{U}_c \cdot \boldsymbol{k}) \tag{3.42}$$

The dot product in the "moving axis" spectrum function simply aligns the wave vector with the mean convection direction. This geometric property of the turbulent field is important in aligning the flow field with surfaces since the flow–surface interaction encounter contributes to the frequency and magnitude characteristics of the generated sound.

For purposes of discussion here, we consider three generic classes of turbulent flows:

1. *Isotropic* turbulence for which the turbulence intensity is independent of direction, the correlation tensor is independent of spatial orientation, and the field statistics are independent of location in the field. Our ability to realize isotropic turbulence physically is very limited; in the laboratory conditions for isotropy are approximately met in flow generated downstream of a uniform grid in a uniform stream in wind tunnels. The conditions of isotropy require the wave number spectrum to be independent of coordinate rotation so that in isotropic turbulence, see Batchelor [129], Hinze [130], and Pope [133].
2. *Homogeneous turbulence* is that for which the mean velocity gradients and turbulent fluctuation statistics are independent of location in the mean field. Classes of nearly homogeneous turbulent fields which are of interest in flow-acoustic modeling are described by Townsend [131] as: nearly homogeneous and isotropic with uniform mean velocity; nearly homogeneous, but *anisotropic* turbulence with uniform mean velocity; nearly homogeneous with uniform gradient of mean velocity as is the case of some turbulent mixing layers. We shall see that conditions of homogeneity, or near homogeneity, are met approximately enough for acoustics modeling when the wave number of interest in a shear flow is larger than the reciprocal of a length scale of the shear flow mean velocity gradient. Isotropic turbulence is also homogeneous turbulence.
3. Inhomogeneous turbulence will be that for which the mean velocity and its *spatial gradients are not constant throughout the flow*, but are variable. These include such turbulent shear flows as jets, wakes, boundary layers on walls and within ducts. In practical cases, the turbulence may be viewed in a region of the flow for which the length scale of mean velocity, or mean velocity gradient, is δ and the minimum wave number of interest is, say $k > 2\pi\delta^{-1}$. In such cases we would consider the turbulence to follow the behavior of homogeneity within the zone defined by δ. The Kolmogrov theory of local isotropy [131] applied to these inhomogeneous flows asserts that the structures of these flows share in common a range of eddy scales: those which are small and do not dominantly affect the time mean energy of the flow and those of larger scale which do. The larger scales may or may not have scales approaching δ, but those whose scales approach or exceed δ cannot be considered without considering the characteristics of the whole fluid. If all the other eddies are also weak, in the context of Eqs. (3.38) and (3.39), then their statistics may be approximated by conditions on local homogeneity.

3.6.2 Spectrum Models for Homogeneous Turbulence

The condition of local isotropy stipulates that the two-point velocity product tensor, such as described by Eq. (3.39) is invariant under rotation of coordinate r_i. This leads to the definition of a correlation tensor of the exponential form [130,133]

$$R_{ij}(\boldsymbol{r}) = \left[(1 - r/(2\Lambda_f))\delta_{i,j} + \frac{1}{2}(r_i r_j)/(2\Lambda_f)\right]\exp(-r/\Lambda_f) \tag{3.43}$$

which characterizes the turbulence behind a uniform grid mesh quite well. Eq. (3.40) gives the equivalent spectrum function

$$\Phi_{ij}(\boldsymbol{k}) = \frac{2}{\pi^2}\overline{u^2}\Lambda_f^3 \frac{\left[(k\Lambda_f)^2\delta_{i\,j} - k_i k_j(\Lambda_f)^2\right]}{\left(1+(k\Lambda_f)^2\right)^3} \tag{3.44}$$

the length Λ_f is the integral length scale of the turbulence at that location of the shear layer:

$$2\Lambda_f = \int_{-\infty}^{\infty} e^{-r_1/\Lambda_f} dr_1 \tag{3.45}$$

or, more generally

$$2\Lambda_f = \int_{-\infty}^{\infty} R_{11}(r_1) dr_1 \tag{3.46a}$$

$$2\Lambda_f = \lim_{k_1 \to 0} \frac{\pi[\Phi_{11}(k_1)]}{\overline{u^2}} \tag{3.46b}$$

The wave number spectrum $\Phi_{ij}(\boldsymbol{k})$ is a key function with which we will be concerned and it can be characterized in different ways according to the assumptions made about the turbulence. In the strictly *isotropic* turbulent medium, this is [130,131,133]

$$\Phi_{ij}(\boldsymbol{k}) = \frac{E(k)}{4\pi k^4}\left[k^2\delta_{i\,j} - k_i k_j\right] \tag{3.47}$$

where

$$k^2 = k_1^2 + {k_2}^2 + {k_3}^2 \tag{3.48}$$

$E(k)$ is the scalar one dimensional energy spectrum of the turbulence which is related to the above spectrum function and to the turbulent kinetic energy in the flow by a series of integral relationships that begin with the longitudinal wave number spectrum of longitudinal velocity fluctuations (see e.g. Hinze [130]):

$$E_1(k_1) = 2\int_{-\infty}^{\infty}\int_{-\infty}^{\infty} \Phi_{11}(\boldsymbol{k})dk_2 dk_3 = \overline{u^2}\frac{1}{\pi}\int_0^{\infty} e^{ik_1 r_1} R_{1,1}(r_1, 0, 0)dr_1 = 2\Phi_{11}(k_1) \tag{3.49}$$

The kinetic energy, $\overline{q^2}$, the one-dimensional energy spectrum, $E(k)$, and the autospectrum of streamwise velocity fluctuations, $E_1(k_1)$, are all integrally related with $u^2 = u_1{}^2 = u_2{}^2 = u_3{}^2$ in isotropic flow and therefore:

$$\frac{1}{2}\overline{q^2} = \frac{3}{2}\overline{u^2} = \frac{3}{2}\int_0^\infty E_1(k_1)dk_1 \tag{3.50}$$

and

$$\overline{u_1^2} = \int_0^\infty E(k_1)dk_1 = \overline{u_2^2} = \overline{u_3^2}$$

There are numerous analytical function relationships for the one-dimensional energy spectrum function from the study of isotropic turbulence for the scalar energy density, a few of which derived in Hinze [130] and Pope [133] will be given below:

(A). Liepmann:

$$E(k) = \frac{8}{\pi}\overline{u^2}\Lambda_f \frac{(k\Lambda_f)^4}{(1+(k\Lambda_f)^2)^3} \tag{3.51}$$

This analytical function is the most commonly used and is easily used to give complementary analytical expressions for the turbulence spectra of individual directional components, as will be discussed below. This function closely resembles the behavior given by von Karman function provided below.

(B). Von Karman:

$$E(k) = \frac{55}{9}\frac{\Gamma(5/6)}{\sqrt{\pi}\Gamma(1/3)}\frac{\overline{u^2}}{k_e}\frac{(k/k_e)^4}{(1+(k/k_e)^2)^{17/6}} \tag{3.52a}$$

This spectrum function integrates using Eqs. (3.47) and (3.49) to obtain the one-dimensional spectrum of streamwise velocity

$$E_1(k_1) = \frac{2}{\sqrt{\pi}}\frac{\Gamma(5/6)}{\Gamma(1/3)}\frac{\overline{u^2}}{k_e}\frac{1}{(1+(k/k_e)^2)^{5/6}} \tag{3.52b}$$

with the inverse of the wave number k_e defining the average scale of energy containing eddies, $1/k_e$, and

$$k_e = \frac{\sqrt{\pi}}{\Lambda_f}\frac{\Gamma(5/6)}{\Gamma(1/3)} \approx \frac{3}{4\Lambda_f} \tag{3.52c}$$

Recall the identity Eq. (3.49)

$$E_1(k_1) = 2\Phi_{11}(k_1) \tag{3.52d}$$

This function is classically used in discussions of the dynamics of isotropic turbulence; the expression for k_e is also that of Dieste and

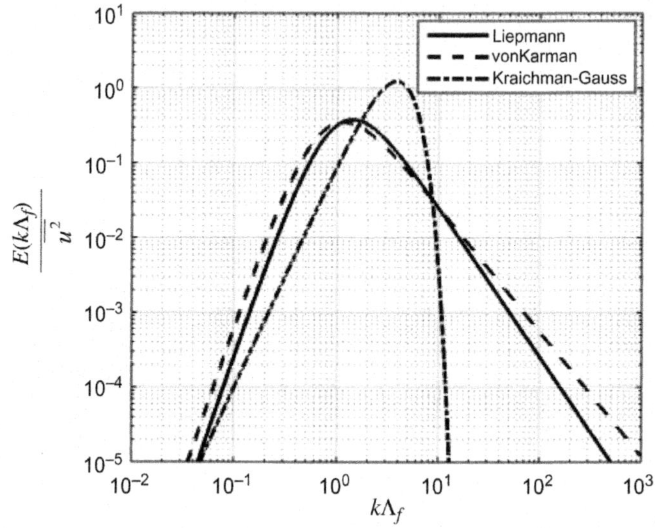

FIGURE 3.17 Energy spectra calculated using three analytical models: Leipmann, Eq. (3.51); von Karman, Eq. (3.52), and Kraichman-Gaussian, Eq. (3.54).

Gabard [140]. An empirical fit to the k_1 spectrum of the streamwise turbulence is the useful alternative to Eq. (3.52b):

$$E_1(k_1) = \frac{\overline{u^2}}{k_e} \frac{2}{\sqrt{\pi}} \frac{\Gamma(5/6)}{\Gamma(1/3)} \frac{1}{1 + \frac{9}{16}(k_1/k_e)^2} \tag{3.53a}$$

which substituting Eq. (3.52c) has an equivalent

$$E_1(k_1) = \overline{u^2} \frac{2}{\pi} \frac{\Lambda_f}{1 + (k_1 \Lambda_f)^2} \tag{3.53b}$$

This is derivable in Section 3.6.3 from the above cited Liepmann spectrum.

(C). Kraichman-Gaussian:

This function has been used in Euler solution methods apparently in offering numerical efficiency [140]. In comparison to those of Liepmann and von Karman this spectrum lacks high wave number content. Accordingly it may be viewed as providing a good representation of the low wave number energy-containing eddies in the turbulence.

$$E(k) = \frac{3}{\pi^2} \overline{u^2} \Lambda_f [k\Lambda_f]^3 \exp\left[-\left(\frac{k\Lambda_f}{\pi} \right)^2 \right] \tag{3.54}$$

This expression has the numerical coefficients given by Dieste and Gabard [140]. Fig. 3.17 gives a comparison of these three spectra.

3.6.3 Anisotropic Turbulence: Spectral Models Based on Stretched Coordinates

In considering the first relaxation of the assumption of isotropy, i.e., locally anisotropic turbulent fields, Ribner and Tucker [141] and later Batchelor and Proudman [142] developed a concept of selective stretching of integral scales in the three principal axes of strain. This concept, the corner stone of much conventional modeling of turbulent stresses, has been called "Rapid Distortion Theory." According to rapid distortion theory five conditions must be met: (1) the magnitude of turbulence fluctuation is linear with respect to the time mean velocity and pressure is linear, ensuring the inequality Eq. (3.38); (2) the displacement histories of the fluid particles are uniquely determined by the history of the mean distortion, thus the turbulence follows mean streamlines; (3) the influence of the turbulent motion on the mean streamlines is negligibly small; (4) the turbulence integral scale is small compared with the radius of curvature of the mean streamlines; (5) the turbulence distortion takes place in times small compared with decay times of the turbulence. In their view, the turbulence is seen as beginning upstream of a contracting stream tube with upstream conditions approaching isotropic. At the upstream condition the integral scales are all, say $\Lambda_1 = \Lambda_2 = \Lambda_3 = \Lambda_\phi$. At the downstream contracted condition the scales are changed so that

$$\begin{aligned}\Lambda_1 &= \alpha\Lambda_f \\ \Lambda_2 &= \beta\Lambda_f \\ \Lambda_3 &= \gamma\Lambda_f\end{aligned} \tag{3.55}$$

Since the volume of the fluid passing through contraction is assumed to be constant, due to the incompressibility condition assumed for the bulk flow, the coefficients satisfy the continuity equation ensuring

$$\Lambda_1 \cdot \Lambda_2 \cdot \Lambda_3 = \Lambda_f^3 \tag{3.56}$$

or

$$\alpha\beta\gamma = 1 \tag{3.57}$$

As corollaries to these relationships we derive a set of expressions that assert a similitude of the two-point correlation tensor, such as in Eq. (3.39) and such that at two points, a and b, along the contraction $\boldsymbol{x}_a$ and $\boldsymbol{x}_b$, such that $|\boldsymbol{x}_b - \boldsymbol{x}_\mathbf{a}| \gg \Lambda_f = \sqrt[3]{\Lambda_1\Lambda_2\Lambda_3}$, satisfying the local similarity form

$$R_{ij}(r_i/\Lambda_i(\boldsymbol{x}_\mathbf{a})) = R_{ij}(r_i/\Lambda_i(\boldsymbol{x}_\mathbf{b})) \tag{3.58}$$

In this case the associated three-dimensional wave number spectral density in an anisotropic turbulent field satisfies the similarity relationship for the wave number scaling

$$(k\Lambda_f(\boldsymbol{x}))^2 = (k_1\Lambda_1(\boldsymbol{x}))^2 + (k_2\Lambda_2(\boldsymbol{x}))^2 + (k_3\Lambda_3(\boldsymbol{x}))^2 \tag{3.59}$$

for the wave number spectrum tensor, e.g. for positions a and b

$$\begin{aligned}\Phi_{ij}(k_1\Lambda_1(\boldsymbol{x}_\mathbf{a}), k_2\Lambda_2(\boldsymbol{x}_\mathbf{a}), k_3\Lambda_3(\boldsymbol{x}_\mathbf{a}))\\ = \Phi_{ij}(k_1\Lambda_1(\boldsymbol{x}_\mathbf{b}), k_2\Lambda_2(\boldsymbol{x}_\mathbf{b}), k_3\Lambda_3(\boldsymbol{x}_\mathbf{b}))\end{aligned} \tag{3.60}$$

The dimensional wave number spectrum at a location, say, $\boldsymbol{x}_\mathbf{a}$ is obtained from the dimensionless spectrum by multiplying by the appropriate integral scales as in Eq. (1.60c):

$$\Phi_{ij}(k_1, k_2, k_3; \boldsymbol{x}_\mathbf{a}) = \Lambda_1(\boldsymbol{x}_\mathbf{a})\Lambda_2(\boldsymbol{x}_\mathbf{a})\Lambda_3(\boldsymbol{x}_\mathbf{a})\Phi_{ij}(k_1\Lambda_1(\boldsymbol{x}_\mathbf{a}), k_2\Lambda_2(\boldsymbol{x}_\mathbf{a}), k_3\Lambda_3(\boldsymbol{x}_\mathbf{a})) \tag{3.61}$$

We will continue to derive a series of wave number spectrum that, with those in Section 3.6.2, will be encountered throughout this book. Inserting the Leipmann energy spectra, Eq. (3.51) into Eq. (3.47) the spectrum function appearing in Eq. (3.61) is

$$\Phi_{ij}(\boldsymbol{k}) = \overline{u_i u_j}\frac{2}{\pi^2}\Lambda_f^3\frac{\left[(k\Lambda_f)^2\delta_{i,j} - k_i k_j(\Lambda_f)^2\right]}{(1+(k\Lambda_f)^2)^3} \tag{3.62}$$

$$(k\Lambda_f)^2 = [(\alpha k_1)^2 + (\beta k_2)^2 + (\gamma k_3)^2]\Lambda_f^2 \tag{3.63}$$

The three-dimensional wave number spectra for each of the three principal velocity components are automatically written as:

$$\Phi_{11}(\boldsymbol{k}) = \overline{u_1{}^2}\frac{2\alpha\beta\gamma}{\pi^2}\Lambda_f^3\frac{\left[(\beta k_2)^2 + (\gamma k_3)^2\right]\Lambda_f^2}{(1+(k\Lambda_f)^2)^3} \tag{3.64}$$

For the "1" direction aligned with the mean flow, and

$$\Phi_{22}(\boldsymbol{k}) = \overline{u_2{}^2}\frac{2\alpha\beta\gamma}{\pi^2}\Lambda_f^3\frac{\left[(\alpha k_1)^2 + (\gamma k_3)^2\right]\Lambda_f^2}{(1+(k\Lambda_f)^2)^3} \tag{3.65}$$

$$\Phi_{33}(\boldsymbol{k}) = \overline{u_3{}^2}\frac{2\alpha\beta\gamma}{\pi^2}\Lambda_f^3\frac{\left[(\alpha k_1)^2 + (\beta k_2)^2\right]\Lambda_f^2}{(1+(k\Lambda_f)^2)^3} \tag{3.66}$$

$$\Phi_{12}(\boldsymbol{k}) = \overline{u_1 u_2}\frac{2\alpha\beta\gamma}{\pi^2}\Lambda_f^3\frac{[-k_1k_2]\alpha\beta\Lambda_f^2}{(1+(k\Lambda_f)^2)^3} \tag{3.67}$$

For each velocity component u_i at location in the flow, at say, $\boldsymbol{x}_\mathbf{a}$, the mean square velocity is related to its spectrum function by:

$$\overline{u_i{}^2(\boldsymbol{x}_\mathbf{a})} = \int_{-\infty}^{\infty}\int_{-\infty}^{\infty}\int_{-\infty}^{\infty}\Phi_{ii}(\boldsymbol{k}, \boldsymbol{x}_\mathbf{a})dk_1 dk_2 dk_3 \tag{3.68}$$

Finally, we note that each of the integral scales is related to a corresponding diagonal correlation matrix component by

$$\Lambda_j = \int_{-\infty}^{\infty} R_{jj}(r_j)dr_j = \frac{1}{\overline{u_j^2}} \int_{-\infty}^{\infty} \int_{-\infty}^{\infty} \Phi_{jj}(k_j = 0, k_i, k_k)dk_i dk_k \tag{3.69}$$

Note that in isotropic turbulence the off-diagonal components of the Reynolds stress tensor are identically zero, $\overline{u_i u_j} = 0; i \neq j$ while those terms in anisotropic shear flow are proportional to the square of mean shear [131].

Models of anisotropy such as these have been used by Panton and Linebarger [143] who used a model of wave number-dependent streamwise anisotropy parameter to construct a turbulence spectrum in the boundary layer. Hunt and Graham [144], Graham [145] analytically, and Lynch et al. [146,147] experimentally have also examined the effects of anisotropy effected by two-dimensional cascades on incoming turbulence. In future chapters these spectral models will be invoked in modeling acoustic sources for turbulent boundary layers and lifting surfaces.

In the modeling of certain acoustic sources, e.g., turbulence-induced wall pressure fluctuations, a different approach to spectrum modeling is typically used. Specifically, the scalar wall pressure is most often measured on a rectangular grid pattern on the wall that is aligned with the direction of mean flow, accordingly it is described (see Chapter 8: Essentials of Turbulent Wall-Pressure Fluctuations) using correlation scales along orthogonal directions one of which is streamwise. As a matter of analytical convenience, this leads to a statistical description of space–time wall correlation that shares with Taylor's hypothesis of convection as well as, typically, a multiplicative function model of the form

$$R_{pp}(\boldsymbol{r}_{13}, \tau) = \overline{p^2} R_1(r_1) R_3(r_3) R_m(\tau - r_1/U_c) \tag{3.70}$$

which resembles Eq. (3.41) in employing Taylor's Convection Hypothesis to isolate the temporal and spatial behavior. A variation of this form has been used to characterize free stream turbulence for purposes of calculating sound from propellers in turbulent inflows, see Blake [148]. While functions of this type are not rooted in the kinematics of turbulence per se, they offer expediency in both characterizing measured spectral properties as curve fits and analytical simplicity.

3.6.4 Measured Turbulence in Plane Mixing Layers

Near the nozzle of the jet as discussed in Section 3.7.2.1, and in other flows such as open jet wind tunnels, the turbulent mixing layers that encompass a potential core of moving fluid appear locally as planar mixing layers. In such shear layers that are typical of many applications, the mean velocity of the flow transitions from zero outside the shear layer to a maximum value in the potential core or wind tunnel center notionally like that shown in Fig. 3.1C. In this transition layer there is a core region of mean shear that has a nearly

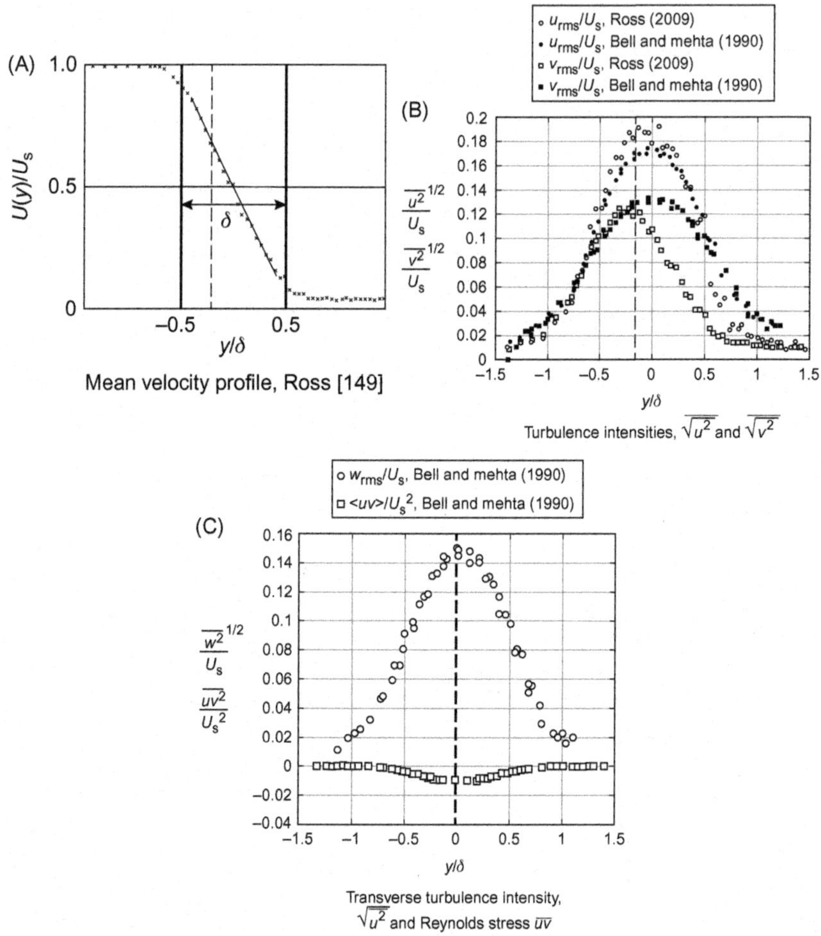

FIGURE 3.18 Profiles of mean velocity (A); rms turbulent u and v velocity (B); rms turbulent, w and Reynolds stress (C) at various points in two nearly-planar mixing layers of an open-jet wind tunnel. Data is Ross [149]; that of Bell and Mehta [150] was obtained with a layer formed by two co-planar turbulent boundary layers. Data v_{rms} of Ross was shifted in $y/\delta = 0.18$ to align maxima in turbulence intensities.

linear mean velocity profile, see Pope [133], as illustrated by measurements in Fig. 3.18A. This central nearly-linear region extends nearly over the shear layer thickness designated by δ which is defined between $U(y = -\delta/2) = 0.1U_s$ and $U(y = \delta/2) = 0.9U_s$ with y = 0 being the mid plane of the shear layer. In this region of the mixing layer both the mean shear and the turbulence reache a maximum (designated with the dashed line) near the center of the layer see Fig. 3.18b. The coordinate y is normal to the direction of mean flow and in the direction of mean velocity gradient; as is the velocity component v; the component u is in the flow direction. Here, we let the transverse direction across which the mean velocity is invariant for any given x,y

coordinate pair be z. For later use we shall invoke the Leipmann spectrum model and use Eqs. (3.64) and (3.65), to obtain the appropriate functions for the stream wise "u" direction For the flow-normal, "v", direction (Fig. 3.19)

$$\Phi_{uu}(k_1) = \int_{-\infty}^{\infty}\int_{-\infty}^{\infty} \Phi_{11}(k_x = k_1, k_2 = k_y, k_3 = k_z)dk_y dk_z$$

$$\Phi_{uu}(k_1) = \overline{u_1{}^2}\frac{\Lambda}{\pi}\frac{1}{1+(k_x\Lambda)^2} \tag{3.71a}$$

$$\Phi_{vv}(k_x) = \int_{-\infty}^{\infty}\int_{-\infty}^{\infty} \Phi_{22}(k_1 = k_x, k_2 = k_y, k_3 = k_z)dk_y dk_z$$

$$\Phi_{vv}(k_x) = \overline{u_2{}^2}\frac{\Lambda}{2\pi}\frac{1+3(k_x\Lambda)^2}{\left[1+(k_x\Lambda)^2\right]^2} \tag{3.71b}$$

Equation 3.71a was previously identified as the empirically-based Liepmann spectrum, Eq. 3.53b. These two functions are compared with the spectra measured by Ross [149] in Fig. 3.19 allowing for the Taylor's Hypothesis frozen turbulence interpretation for which $k_x = \omega/U(y_m)$. These spectra show evidence of structure in the spectra at low wave numbers. The turbulence integral scale of the stream-normal, v, component, determined as the $k_x \rightarrow 0$ asymptote of the v-component is about 0.26δ for this mixing layer. However, the bump in the v-component spectrum at $k_x\Lambda \sim 0.3$ represents an additional convected component with a spatial scale of about $\delta/4$. At this wave number scale the spanwise (z) correlation length defined below is a maximum as shown in Fig. 3.20. We consider this as another possibly vertical component with a spanwise correlation length to be found by combining Eqs. (3.41), (3.42), and (3.69). The spanwise correlation length of, say, the flow-normal turbulent velocity is therefore defined using the spectral functions for isotropic turbulence (see Lynch et al. [146,148]) an integrated cross-spectral density

$$\Phi_{vv}(\omega)\Lambda_z(\omega)\big|_V = \overline{u_2{}^2}\int_{-\infty}^{\infty} R_{vv}(r_z)dr_z = \int_{-\infty}^{\infty}\int_{-\infty}^{\infty} \Phi_{vv}(k_x, k_y, k_z = 0)\delta(k_x - \omega/U(y_m)dk_x dk_y \tag{3.72a}$$

The corresponding k_x-dependence of the spanwise, z, correlation length of the v component is

$$\Lambda_z(\omega)\big|_V = \frac{3\pi}{2}\frac{(k_x\Lambda)^2}{(1+(k_x\Lambda)^2)^{1/2}(1+3(k_x\Lambda)^2)} \tag{3.72b}$$

where $k_x = \omega/U(y_m)$, This is compared with experimental data of Ross [149] for the planar mixing layer in Fig. 3.20. The measurements were made with pairs of hot wire anemometers and show a maximum at the convection wave number corresponding with the maximum in the frequency spectrum that was identified above. We can see that the correlation length across the z coordinate for the shear layer exceeds that existing in isotropic turbulence.

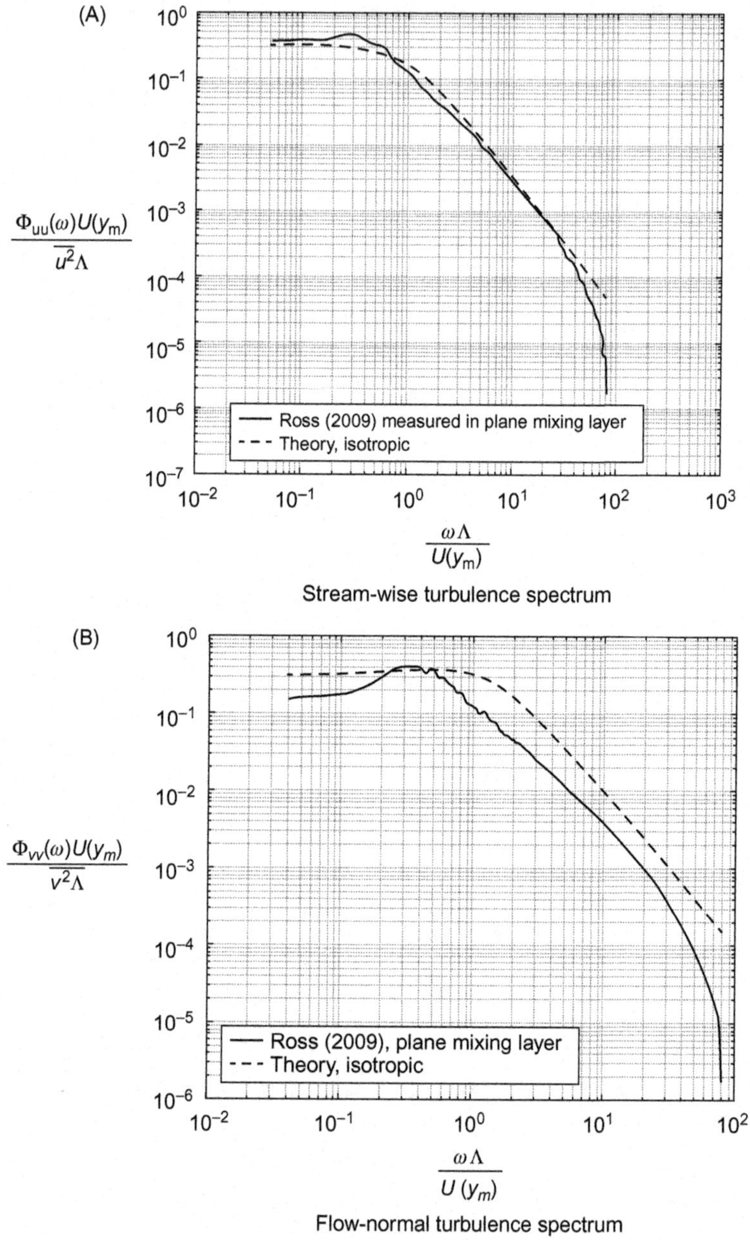

FIGURE 3.19 Spectral densities of the streamwise and flow-normal direction velocities in a planar mixing layer of an open-jet wind tunnel. (A) Streamwise turbulence spectrum and (B) flow-normal turbulence spectrum.

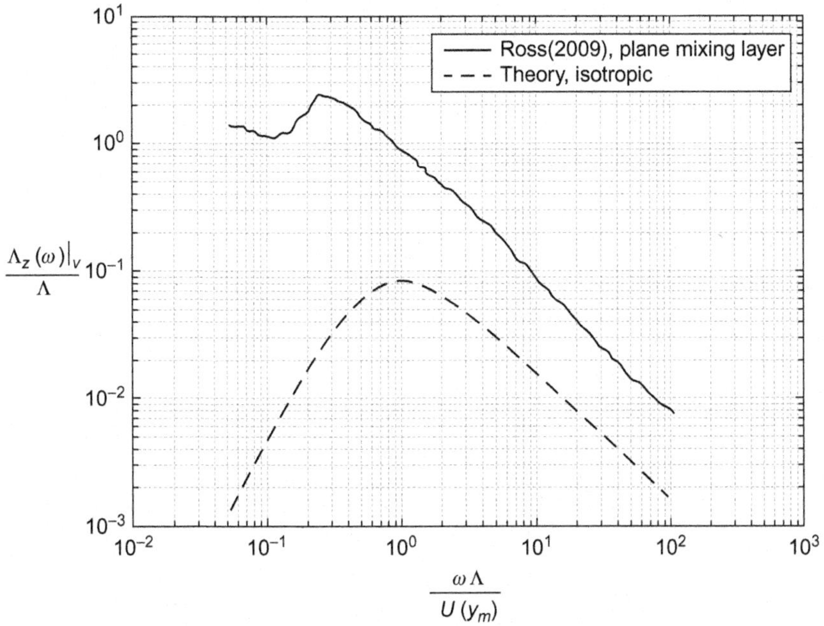

FIGURE 3.20 Correlation length of the flow-normal turbulence component over separation vectors in the transverse cross-stream direction determined by measurement (Ross [149]) and compared with the theory for isotropic turbulence (Lynch et al. [146,148]).

This example of an anisotropic turbulent flow is like those often encountered in practice because of the role played by orderly structure that is convected in a nearly frozen manner and has a length scale that is determined by the global flow dimension such as occurred with the large-scale structure in the jet. The departures from isotropic structure are seen to be most pronounced at relatively low wave numbers and affect both the spectral magnitude and the spatial correlation scales. We will return to discussion of this flow in Chapter 11, Noncavitating Lifting Sections when we discuss the acoustical implications of these structures on flow-induced sound from a lifting surface embedded in it. These expressions for wave number spectra and correlation length will be widely used in following chapters for modeling turbulent pressures and turbulence-induced unsteady forces.

3.7 FUNDAMENTALS OF NOISE FROM SUBSONIC TURBULENT JETS

In this section we examine some of the acoustically relevant characteristics of the turbulence in turbulent subsonic cold jets, the characteristics of the sound, scaling rules for jet noise, some common noise abatement methods, and recently formulated computational design prediction methods. Recent investigations, facilitated in part by the availability of accurate turbulent flow computer simulations, have radically altered the approach to this subject from largely empirical to first-principles-based prediction.

3.7.1 Application of Lighthill's Equation

3.7.1.1 Discussion of the Source Term

Noise from developing jets, being quadrupole in nature, is not generally a dominant source at the very low Mach numbers that are typical of hydroacoustic problems, but it is paramount in aeroacoustics. Jet noise is also of historical importance as a prime mover for Lighthill's [151,152] work and the some of the earliest investigations of turbulence statistics in relation to modeling acoustic sources. Subsonic jet noise is a mechanism of sound by convected quadrupoles in the absence of boundaries couple with the surrounding acoustic source-free medium to produce sound. The mathematical structure of this problem represents an application of the fundamental relationships such as Eqs. (2.54–2.57) or their alternatives, if another propagation function beyond the free space Green function is required. The approach used here is consistent with approaches used in other chapters for flow–structure interaction; i.e. the analytical or numerical combination of a statistical model of the turbulent sources with a deterministic model of a propagation function. Accordingly this section is meant to help unify the concepts. The traditional views of jet noise, motivated by investigations such as those of Goldstein [153], Crighton [156], Howe [154], Powell [22], Hubbard [251], Lilly [156], and Proudman [158] focused heavily on the production and statistics of turbulent sources and resulted in scaling laws that are still of interest, e.g., Fisher et al. [260], Tanna et al. [261], and Viswantha [262,263]. The more contemporary work uses results of computational fluid dynamics to realistically characterize the source dynamics and more rigorous propagation formulations than the simple free-space Green function with rectilinear source convection. Work in this area is exemplified by Tam and Auraiult [185], Depru Mohan et al. [183], Karabasov et al. [182], Khavaran et al. [257], Lieb and Goldstein [256], Goldstein and Leib [258], and Goldstein [254,259] as well as others referenced later in this section.

In pursuing the first and more classical form of analysis we implement the acoustic approximation Eq. (2.6) and insert Eq. (2.47) into Eq. (2.49) then subtract the time mean to yield the relationship

$$\frac{1}{c_0^2}\frac{\partial^2 p_a}{\partial t^2} - \nabla^2 p = \frac{\partial^2 \rho_0\left(U_iU_j - \overline{U_iU_j}\right)}{\partial y_i \partial y_j} \tag{3.73}$$

where $T_{ij} = \rho_0\left(U_iU_j - \overline{U_iU_j}\right)$ is the fluctuating Reynolds stress and p_{a} is the *fluctuating* pressure, both of which have zero time average. In considering subsonic cold jets we have approximated the source term in the turbulent field as including only incompressible Reynolds stresses. Fig. 3.21 shows a sketch of the arrangement of the jet. We have adopted the coordinate

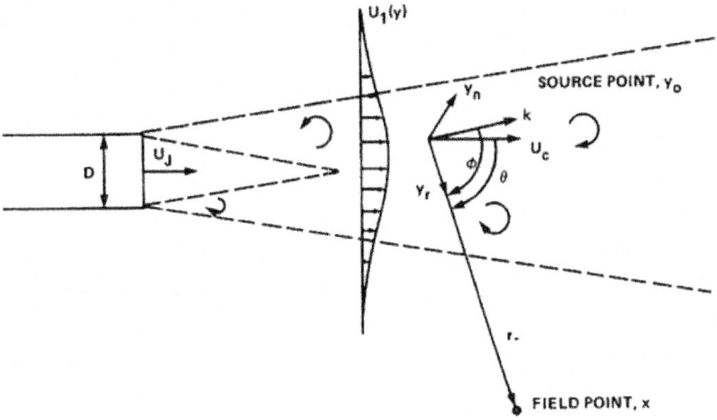

FIGURE 3.21 Geometry of convecting sources in a jet.

convention of "1" being the direction of mean flow and 2 the direction normal to the flow direction and in the direction of mean shear. We will use that convention here except that the "2" direction is radial in the round jet.

A split between mean and fluctuating velocities in this context is now classical and follows the early work of, e.g., Proudman [159], Jones [160], Ribner [161,163–166], and Goldstein et al. [167] giving

$$\boldsymbol{U}(\boldsymbol{y},t) = \boldsymbol{U}_J(\boldsymbol{y}) + \boldsymbol{u}(\boldsymbol{y},t)$$

where $\boldsymbol{U}_J(\boldsymbol{y})$ is the local mean velocity in the jet, assumed henceforth $= \mathbf{U}_1$.

An important component can be factored out of the above that is due only to the presence of a mean velocity profile, dU_1/dy_2. This contribution is loosely called the *mean-shear turbulence* interaction, Kraichman [164], *shear noise*, Ribner [153], or *shear-amplified* noise, Lilley [157]. It is of considerable classical importance in the theory of turbulent sources in turbulent boundary layers and it depends on the quantity $\boldsymbol{U}_I(\boldsymbol{y})\boldsymbol{u}_2(\boldsymbol{y})$ in the stress tensor, T_{ij}. For a planar compressible mixing layer the turbulence-mean shear "source" is exposed in Section 8.3.1 to be the first term on the right of

$$\frac{\partial^2\left(U_iU_j - \overline{U_iU_j}\right)}{\partial y_i\,\partial y_j} = 2\frac{\partial U_1}{\partial y_2}\frac{\partial u_2}{\partial y_1} + \frac{\partial^2\left(u_iu_j - \overline{u_iu_j}\right)}{\partial y_i\,\partial y_j} \tag{3.74}$$

The second term on the right is the purely turbulent quadrupole term; the first term is determined by the mean shear in free- and wall-bounded mixing layers. As discussed previously, Section 3.6.4, in this layer the turbulence intensity and Reynolds stresses attain their maximum values where the gradient $\partial U_1/\partial y_2$ is also maximum.

Even for subsonic cold jets other approaches of interest focused on examining alternative definitions of the sources on the right hand side of Eq. (3.73), notably the views of Lilley [156,157,244] and in Hubbard [251], in suggesting that the first term on the right hand side of Eq. (3.74) is not really an acoustic source at all, but represents a convection that is more properly included in the wave operator on the left. These views contributed to an extent to the current refinements in propagation modeling used in recent predictions [254,256]. Still, other views of, e.g., Michalke and Fuchs [168] and Morfey [174], examine various effects of mean convection, temperature gradients, time-dependent spatial gradients, modal decomposition of radiation, and forms of linearization. These early investigations showed that only low orders of circumferential variation of velocity modes, including axisymmetric, are indicated by correlations and source localization [e.g. 166–168, 214, 215, 253] and other references to be most relevant in the production of sound.

3.7.1.2 *Formal Analytical Relationships for Including Source Convection*

The expressions derived above include the effect of mean flow as a specific contributor to the source strength of the turbulent quadrupoles while other effects of mean shear flow are ignored in defining the sources. The basic Lighthill-Ffowcs Williams theory [151,152,176,241] does include convection as a simple Doppler shift of frequency, however, also giving rise to a substantial Mach number–dependent directivity. In its full rigor the details of the contemporary theory that involve otherwise-neglected convective propagation physics in shear flow are beyond the scope of this chapter, which is meant to focus on low speed flow and turbulence statistics. The reader is referred to the papers by Tam and Auriault [185] and of Goldstein [254] and Goldstein and Leib [260] for the discussions of more complete propagation functions. So, in considering low speed cold jets we will continue to use the free space Green function for the purpose of exposing the essentials of low speed jet noise that include the speed-power law and rudimentary effects of source convection on directivity. We continue to assume high Reynolds number jets, ignore refractive effects by assuming low subsonic flow, ignore acoustic scattering by turbulence, and neglect the viscous stress tensor by assuming a high enough Reynolds number flow. Accordingly to be clear the result equivalent to equation (2.57) is rewritten here for convenience to expose the salient fluctuating velocities

$$\frac{1}{c_0^2}\frac{\partial^2 p_a}{\partial t^2} - \nabla^2 p_a = \rho_0 \frac{\partial^2 (u_i u_j - \overline{u_i u_j})}{\partial y_i\, \partial y_j} + 2\rho_0 \frac{\partial U_1}{\partial y_2}\frac{\partial u_2}{\partial y_1}$$

$$\frac{1}{c_0^2}\frac{\partial^2 p_a}{\partial t^2} - \nabla^2 p_a = \frac{\partial^2 T_{ij}(\mathbf{y}, t)}{\partial y_i\, \partial y_j}$$

Following the methods of Chapter 2, Theory of Sound and Its Generation by Flow; the applicable integral equation solution for far field pressure is Eq. (2.58) with a time-dependent source to observer distance [see also, e.g., 166,176].

$$4\pi p_a(\boldsymbol{x},t)=\frac{1}{{c_0}^2}\int_{-\infty}^{\infty}\int_{-\infty}^{\infty}\int_{-\infty}^{\infty}\frac{(x_i-y_i(\tau))(x_j-y_j(\tau))}{[r(\tau)]^3}\left(\frac{\partial^2 T_{ij}(y,t-r(\tau)/c_0)}{\partial t^2}\right)dV(\mathbf{y}) \tag{3.75a}$$

where $\tau=t-r(\tau)/c_0$ is the retarded time at which the sound is emitted at the source.

In this and subsections 3.7.2.3 and 3.7.3 we shall rely heavily on Refs. [166,196,233,234], to form an expression for the autocorrelation of the far-field sound pressure using methods of Section 3.5.2 subject to the temporal stationarity restrictions of Sections 1.4.3.2 and 3.6.1. The autocorrelation of far field sound is

$$\begin{aligned} R_{pp}(\mathbf{x},\tau) &= \langle p_a(\mathbf{x},t)p_a(\mathbf{x},t+\tau)\rangle \\ &= \frac{1}{16\pi^2}\frac{1}{{c_0}^4}\int_{-\infty}^{\infty}dV(\mathbf{y}')\int_{-\infty}^{\infty}dV(\mathbf{y}'')\frac{(x_i-y_i{}')(x_j-y_j{}')(x_k-y_k{}'')(x_l-y_l{}'')}{r^6} \\ &\times\ldots\ldots\ldots\times\left\langle\frac{\partial^2 T_{ij}(y',t-|\boldsymbol{x}-\boldsymbol{y}'|/c_0)}{\partial t^2}\frac{\partial^2 T_{kl}(y'',t+\tau-|\boldsymbol{x}-\boldsymbol{y}''|/c_0)}{\partial t^2}\right\rangle \end{aligned} \tag{3.75b}$$

We interpret the various times and positions of sources in the above equation in the following way. The sources are taken to convect from a position at time τ_0, $\mathbf{y}'=\mathbf{y}(\tau_0)$, to another position at a later time $\tau_0+\tau$, i.e., $\mathbf{y}''=\mathbf{y}(\tau_0+\tau)$. The retarded time after $\tau=\tau_0$ is $\tau=t-|\boldsymbol{x}-\boldsymbol{y}(\tau_0)|/c_0$, which is the time of emission of the sound at the source at its initial point and and t is the time at the observer. The source convection therefore introduces an additional propagation-induced time delay in translating from $\boldsymbol{y}'$ to $\boldsymbol{y}''$. In the far field this acoustic time delay due to a change in source-receiver distance is approximated by

$$\frac{(|\boldsymbol{x}-\boldsymbol{y}''|-|\boldsymbol{x}-\boldsymbol{y}'|)}{c_0}\approx\frac{(y'-\boldsymbol{y}'')\cdot\boldsymbol{x}}{c_0|\boldsymbol{x}|} \tag{3.75c}$$

as can be seen by expanding the vector difference magnitudes using the law of cosines and rejecting small terms in the limit of $x \gg y$. We let η represent the convection-dependent separation vector, say

$$\eta=\zeta+c_0M_c\tau \tag{3.75d}$$

for which ζ is the starting value at $\tau=0$, then

$$\frac{(y'-\boldsymbol{y}'')\cdot\boldsymbol{x}}{c_0|\boldsymbol{x}|}=\frac{\eta\cdot\boldsymbol{x}}{c_0|\boldsymbol{x}|}$$

and therefore the separation vector includes a time-dependent component

$$\frac{\eta\cdot x}{c_0|x|}=\frac{(\zeta+c_0M_c\tau)\cdot x}{c_0|x|}=\frac{\zeta\cdot x}{c_0|x|}+|M_c|\cos\theta\cdot\tau$$

This result resembles the moving coordinate expression of Section 2.5. The first term on the right is the initial propagation delay that is associated with the initial separation vector, the second term on the right represents the convection-induced time delay. The variable ζ is a separation vector in the source region.

Now, the ensemble average of the Reynolds stress can be rearranged to identify separately the second and fourth order covariances.

$$\left\langle\frac{\partial^2T_{ij}(y',t-|x-y'|/c_0)}{\partial t^2}\frac{\partial^2T_{ij}(y'',t+\tau-|x-y''|/c_0)}{\partial t^2}\right\rangle$$
$$=\left\langle\frac{\partial^2T_{ij}'(y',t-|x-y'|/c_0)}{\partial t^2}\frac{\partial^2T_{ij}'(y'',t+\tau-|x-y''|/c_0)}{\partial t^2}\right\rangle$$
$$+[2\rho_0]^2\frac{\partial U_1(y')}{\partial y_2}\frac{\partial U_1(y'')}{\partial y_2}\left\langle\frac{\partial u_2(y',t-|x-y'|/c_0)}{\partial t}\frac{\partial u_2(y'',t+\tau-|x-y''|/c_0)}{\partial t}\right\rangle$$

where the primes on T_{ij}' denote strictly the nonlinear second order turbulent Reynolds stresses for which the covariance tensor is $\langle u_i(y,t)u_j(y,t)u_ku_l(y,t+\tau)u_l(y,t+\tau)\rangle$. This separation of linear and non linear sources has been used in early modeling of jet noise and it continues to be used in the modeling of sources in turbulent boundary layers. We set up an expression that will lead to a scaling law for Mach number by invoking a well-known theorem, see Hinze [130], which states that under the assumption of ergodicity for an isotropic turbulence the time derivatives may be replaced by derivatives of the covariance function with respect to time delay, i.e.,

$$\left\langle\frac{\partial u_i(y,t)}{\partial t}\frac{\partial u_i(y,t+\tau)}{\partial t}\right\rangle=\frac{\partial^2}{\partial\tau^2}\langle u_i(y,t)u_i(y,t+\tau)\rangle$$

and

$$\left\langle\frac{\partial^2u_i(y,t)}{\partial t^2}\frac{\partial^2u_i(y,t+\tau)}{\partial t^2}\right\rangle=\frac{\partial^4}{\partial\tau^4}\langle u_i(y,t)u_i(y,t+\tau)\rangle$$

Thus the correlation of the time derivatives of the stress tensor can be expanded to 2 terms involving both second and fourth-order correlations and their time derivatives

$$\left\langle\frac{\partial^2T_{ij}(y',t-|x-y'|/c_0)}{\partial t^2}\frac{\partial^2T_{ij}(y'',t+\tau-|x-y''|/c_0)}{\partial t^2}\right\rangle$$
$$=\rho_0{}^2\frac{\partial^4}{\partial\tau^4}\langle T_{ij}'(y',t-|x-y'|/c_0)T_{ij}'(y'',t+\tau-|x-y''|/c_0)\rangle$$
$$+4\rho_0{}^2\frac{\partial U_1(y')}{\partial y_2}\frac{\partial U_1(y'')}{\partial y_2}\frac{\partial^2}{\partial\tau^2}\langle u_2(y',t-|x-y'|/c_0)u_2(y'',t+\tau-|x-y''|/c_0)\rangle$$

Delaying further specifics on the structure of the Reynolds stress tensor, we will now apply these results to the general turbulence covariance in the integral for the autocorrelation of far field peessure. A general-purpose nondimensionalized fourth order covariance tensor of Eq. (3.75b) can be expanded

$$\left\langle \frac{\partial^2 T_{ij}(t-|\boldsymbol{x}-\boldsymbol{y}'|/c_o)}{\partial t^2}\frac{\partial^2 T_{kl}(t+\tau-|x-\boldsymbol{y}''|/c_0)}{\partial t^2}\right\rangle$$

$$= \rho_0 \overline{u_i u_j}(\boldsymbol{y}')\overline{u_k u_l}(\boldsymbol{y}'')\frac{\partial^4}{\partial \tau^4}R_{ij,kl}(\boldsymbol{y}'',\boldsymbol{y}',\tau-(|x-\boldsymbol{y}''|-|\boldsymbol{x}-\boldsymbol{y}'|)/c_0)$$

We will assume that the correlation length dimension of acoustic sources is no more than a small multiple of the jet diameter so that the above equation compresses to

$$\left[R_{pp}(\boldsymbol{x},\tau)\right]_{ijkl} = \frac{1}{16\pi^2}\frac{\rho_0{}^2}{c_0{}^4}\frac{(x_i x_j x_k x_l)}{r^6}\frac{\partial^4}{\partial\tau^4}\int_{-\infty}^{\infty} dV(\boldsymbol{y}')\int_{-\infty}^{\infty} dV(\boldsymbol{y})\times\cdots$$

$$\cdots\times\overline{u_i u_j}(\boldsymbol{y}')\overline{u_k u_l}(\boldsymbol{y}'')R_{ij,kl}(\boldsymbol{y}'',\boldsymbol{y}',\tau-(|x-\boldsymbol{y}''|-|\boldsymbol{x}-\boldsymbol{y}'|)/c_0)$$

It is assumed here and in the following that there is summation over all combinations of *ijkl.*

Since the observer is in the acoustic far-field we can now use the Eqs. (3.75a through 3.75e) for the convection-induced propagation time delay

$$\left[R_{pp}(\boldsymbol{x},\tau)\right]_{ijkl} = \frac{1}{16\pi^2}\frac{\rho_0{}^2}{c_0{}^4}\frac{(x_i x_j x_k x_l)}{r^6}\frac{\partial^4}{\partial\tau^4}\int_{-\infty}^{\infty} dV(\boldsymbol{y}')\int_{-\infty}^{\infty} dV(\boldsymbol{y}'')\times\cdots$$

$$\cdots\times\overline{u_i u_j}(\boldsymbol{y}')\overline{u_k u_l}(\boldsymbol{y}')R_{ij,kl}\left(\boldsymbol{y},\boldsymbol{y}',\tau-\frac{(\boldsymbol{y}''-\boldsymbol{y}')\cdot\boldsymbol{x}}{c_0|\boldsymbol{x}|}\right)$$

which gives

$$\left[R_{pp}(\boldsymbol{x},\tau)\right]_{ijkl} = \frac{1}{16\pi^2}\frac{\rho_0{}^2}{c_0{}^4}\frac{(x_i x_j x_k x_l)}{r^6}\frac{\partial^4}{\partial\tau^4}\int_{-\infty}^{\infty} dV(\boldsymbol{y}')\int_{-\infty}^{\infty} dV(\eta)\times\cdots$$

$$\cdots\times\overline{u_i u_j}(y')\overline{u_k u_l}(\eta)R_{ij,kl}\left(y',\eta,\tau+\frac{\eta\cdot\boldsymbol{x}}{c_0|\boldsymbol{x}|}\right) \quad (3.75e)$$

The full correlation of the far field sound pressure would be found as a summation over all *i, j, k,* and *l.*

We will use this result together with the characteristics of the space-time correlation function to develop scaling rules for jet noise in Sections 3.7.3. In preparation for this development, in the next section, we will examine the behavior of the space-time correlation function $R_{ij,kl}$ of the Reynolds stresses.

3.7.2 Qualities of Jet Turbulence Structure Relevant to Jet Noise

3.7.2.1 Flow Development in Turbulent Jets

Some of the important properties of turbulent jets at Reynolds numbers, R_D, greater than 10^5 are shown in Fig. 3.22. Within 4–6 diameters, the jet develops in an annular shear mixing layer that surrounds the so-called potential core. The core region has a mean velocity that is independent of radius and is therefore relatively disturbance free except for an unsteadiness that is imposed by large-scale vortex structure in the annular mixing zone, as shown by Lau et al. [75] and Ko and Davies [172]. The maximum turbulence intensities in the mixing zone measured by Davies et al. [171] are shown in the lower part of Fig. 3.22. The radial distribution of the turbulence in this region $\overline{u_1^2}^{1/2}$ satisfies the similarity rule,

$$\frac{\overline{u_1^2}^{1/2}_{(r,y_1)}}{\left(\overline{u_1^2}^{1/2}\right)_{\max}} = f_1\left(\frac{r - D/2}{y_1}\right) \tag{3.75f}$$

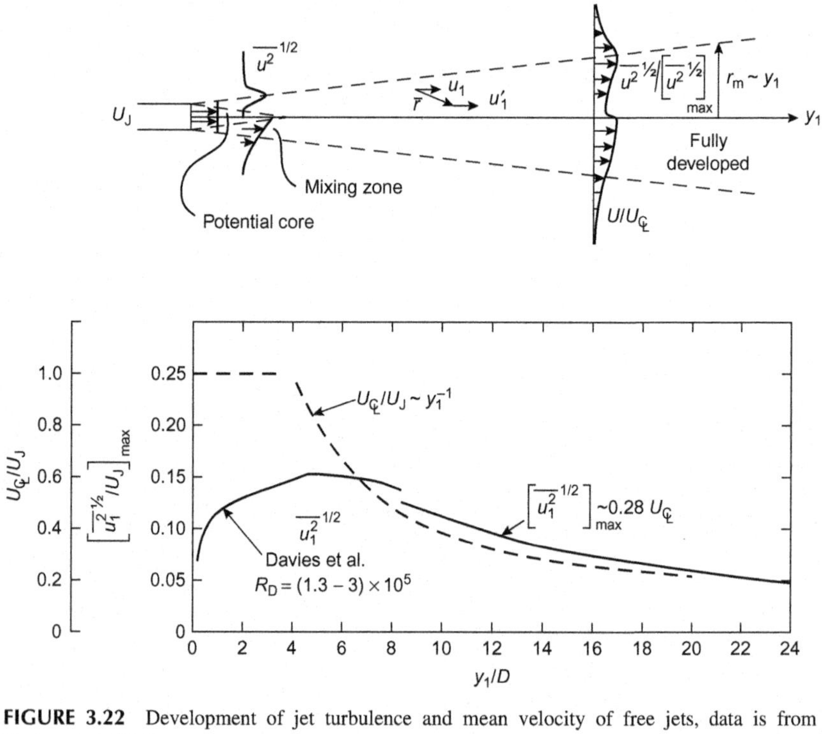

FIGURE 3.22 Development of jet turbulence and mean velocity of free jets, data is from Davies et al. [171].

where $\left(\overline{u_1^2}^{1/2}\right)_{max}$ is the value on the jet edge and where $f_1(0) = 1$. The mean velocity in the mixing region U_1 (r, y_1) behaves according to [171,172]

$$\frac{U_1(r,y_1)}{U_\mathbf{J}} = g_1\left(\frac{r - D/2}{y_1}\right) \tag{3.76}$$

These similarity functions reflect the fact that the length scale of the shear layer increases linearly with distance from the nozzle exit. This linear increase continues into the fully developed region downstream of $y_1/D = 4$ to 5. In the fully developed region the inner extremity of the annular shear layer has merged and, although the mean velocity is still maximum on $r = 0$, it decreases linearly with y_1. According to Forstall and Shapiro [173,174], the centerline velocity U_{cL} is given by

$$\frac{U_\mathbf{cL}}{U_\mathbf{J}} = \frac{(y_1)_c}{y_1} \tag{3.77}$$

where

$$\frac{(y_1)_c}{D} \sim (4 \text{ to } 5) \tag{3.78}$$

Replacing the two similarity forms (3.54) and (3.55) in the fully developed region, Hinze [130] and Townsend [131] give for the mean velocity

$$\frac{U_1(r,y_1)}{U_\mathbf{cL}} = f_2\left(\frac{r}{y_1 + (y_1)_\mathbf{c}}\right) \tag{3.79}$$

where $(y_1)_c$ is a constant measure of the effective datum of the jet given above and for the turbulent velocity

$$\frac{\overline{u_{1\,(r,y_1)}^2}^{1/2}}{U_{cL}} = g_2\left(\frac{r}{y_1 + (y_1)_\mathbf{c}}\right) \tag{3.80}$$

where g_2 is maximum near to or just off the centerline of the jet. Forstall and Shapiro [173,174] give an alternative functional dependence using as a length scale the variable jet radius $r_J(y_1)$

$$\frac{\overline{u_{1\,(r,y_1)}^2}^{1/2}}{U_{cL}} = f_3\left(\frac{r}{r_\mathbf{J}(y_1)}\right) \tag{3.81}$$

where $r_J(y_1)$ is determined by

$$\frac{U_1(r_\mathbf{J},y_1)}{U_\mathbf{cL}} = f_3(1) = \frac{1}{2} \tag{3.82}$$

and it depends on y_1 as

$$\frac{2r_\mathbf{J}(y_1)}{D} \sim \left(\frac{y_1}{(y_1)_\mathbf{c}}\right) \tag{3.83}$$

$(y_1)_c$ being the length of the potential core given by Eq. (3.77). These functions are illustrated in Fig. 3.22.

3.7.2.2 Space–Time Statistical Properties of the Turbulence in Circular Jets; the Second-Order Correlation Tensor

As introduced in Section 2.3.3, Acoustic Radiation From a Compact Region of Free Turbulence the mean-square sound pressure depends on the space–time covariance of the source tensor, we shall subsequently approximate the required 4-dimensional statistics of turbulent stresses using two-point space–time correlations of the velocity fluctuations (Sections 3.7.3.3 and 3.7.5.2). Ribner [163] and Goldstein and Rosenbaum [165] examined the covariance structures of the Reynolds stress covariance matrix, their relevance to acoustic intensity, and effects of anisotropy. Many of the methods of using measured statistics in acoustic modeling as described in this section are now classical and are still used in many models of flow-induced sound and vibration, e.g., boundary layer excitation, trailing edge, and rotor noise. The maturing of LES techniques (eg. [243]) for high Reynolds number has enabled some of these approaches to be used in exact solutions as surrogate "experiments" to extend physical capability. Accordingly the development here will be somewhat detailed to form a background for discussions of these methods later, Section 3.7.5 and Chapters 8–12.

Space–time correlations of the axial velocity fluctuations in the mixing zone obtained by Fisher and Davies [175] are shown in Fig. 3.23. These correlations, obtained as a function of time delay for axially-separated anemometer probes, are similar to others obtained more recently by Harper-Bourne [176] and Pokora and McGuirk [177]. These functions display maxima at a combination of r_1 and τ that defines the eddy convection velocity. An envelope R_m ($r_1 = U_c\,\tau$) can be drawn through the correlation values at these points and this envelope can be described in terms of the original correlation $R_{11}(r_1, \tau)$;

$$R_{11}(r_1, \tau) = R_{11}(r_1, \tau - r_1/U_c) \tag{3.84}$$

so that

$$R_{\mathbf{m}}(r_1) = R_{11}(r_1, \tau = r_1/U_c) \tag{3.85}$$

This moving-axis relationship between r_1 and τ not only defines a convection velocity ($U_c \simeq 0.6U_J$) of the disturbances as they move downstream from the nozzle, but also the moving-axis decorrelation of the velocity field. $R_m(r_1)$ can be called a "moving-axis correlation" [148], or a Laplacian correlation function, because the covariance is interpreted as a correlation over the distance r_1 traversed by the eddies in a frame of reference moving with the *average* speed of the eddy field. The deviation of $R_m(r_1)$ from unity is caused by a combination of turbulent mixing, in which eddies stretch and entrain fluid from the outer, undisturbed, environment and by viscous decay

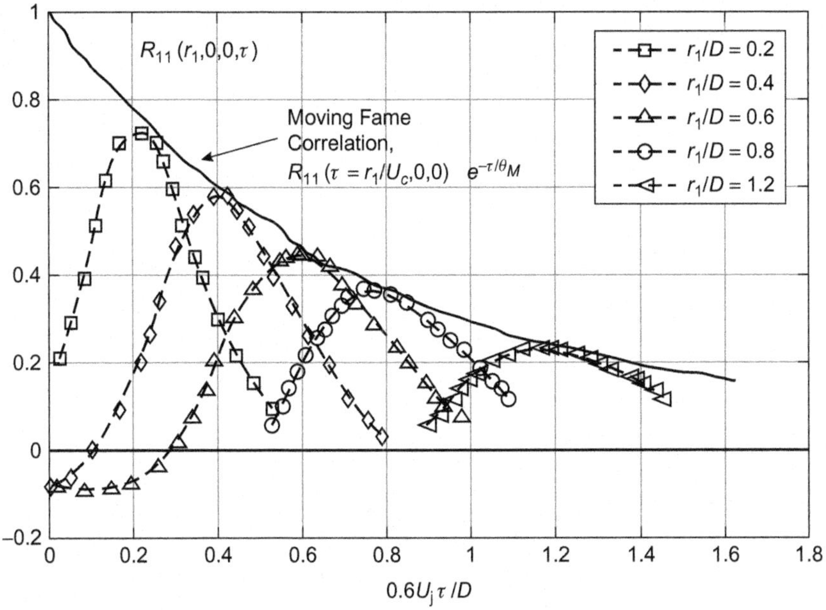

FIGURE 3.23 Examples of space–time correlations of axial velocity fluctuations (downstream separation). Fixed anemometer probe at $y_1/D = 1.5$, $y_2/D = 0.5$. $R_D = 1.2 \times 10^5$, $D = 2.54$ cm, $U_J = 74.7$ m/s. The moving-axis time scale θ_M is of order 0.8 D U_J. The data is from Fisher and Davies [175]. The indicated convection velocity is $0.6U_J$.

of smaller eddies as they are convected by the mean flow. In a low-speed jet it is the decorrelation caused by mixing and growth that allows the generation of sound! Without decorrelation, the convected eddy field would not have sources that couple with acoustic waves temporally and spatially.

The correlations shown in Fig. 3.23 are typical of those obtained in almost any turbulent field. In general the full covariance $R_{ij}(\mathbf{r}, \tau)$ depends on the location $\mathbf{y}$ of the separation vector $\mathbf{r}$. However, in some applications we place the $\mathbf{y}$ dependence in the intensities $\overline{u_1^2}$ and assume that the normalized covariance $R_{ij}(r_i, \tau, \mathbf{y})$ is a function of scaled position and time variables. The validity of this approximation is dependent on the type of flow and the conditions for validity will be made more apparent below. Accordingly, we shall here deal with the correlation and spectrum function derived from the variable $R_{ij}(\boldsymbol{r\tau})$, which in the following discussion shall be simplified $R(r_1, \tau)$.

The representation shown in Fig. 3.23 is only one perspective of the entire set of correlation functions $R_{11}(r_1, \tau)$ as it was obtained as a sequence of time delays at fixed values of separation. An alternative perspective would be to (if possible) obtain continuous functions of r_1 at fixed values of τ. Qualitatively it shows a ridge $r_1 = U_c\ \tau$, along which the correlations are locally maximum. For the space–time correlations generally observed in fluid mechanics the following relations hold, see also Section 2.7.2, Generalized Transforms and Stochastic Variables:

$$\begin{aligned} R(0,0) &= 1 \\ R(r_1, \tau = r_1/U_c) &\geq R(r_1, \tau) \\ R(0,0) &\geq R(r_1, 0) \simeq R(U_c\tau, 0) \end{aligned} \tag{3.86}$$

Correlations with separations along r_1, assumed here to be the flow direction, are qualitatively similar to a larger group of functions for which separations may be arbitrary relative to the flow direction. These correlation functions all follow the same convention used in Sections 2.7.2 and 3.6 in terms of normalization and definitions of integral scales. Denoting such correlations between two velocity components u_i and u_j as $R_{ij}(\mathbf{r}, \tau)$, we have also

$$R_{ij}(r_1, 0, 0, \tau) \geq R_{ij}(\mathbf{r}, \tau)$$

In the use of correlation functions for analytical developments as a simplification we sometimes will separate the complete function as

$$R_{ij}(\mathbf{r}, \tau) \simeq R_{ij}(r_1, \tau) R_{ij}(\mathbf{r}_n) \tag{3.87}$$

in which $R_{ij}(\mathbf{r}_n)$ describes spatial correlations in a plane perpendicular to the flow direction.

A separability of functions that is consistent with the presentation of data in Fig. 3.23 is given as a product of two time-domain correlation functions in Eq. (3.88)

$$R(r, \tau) \simeq R(r_1 - U_c\tau, \vec{\mathbf{r}}_n) R_m(\tau) \tag{3.88}$$

Alternatively, in view of Eq (3.85) we can have

$$R(r, \tau) \simeq R(r_1 - U_c\tau, \boldsymbol{r}_n) R_m(r_1) \tag{3.89}$$

If Taylor's hypothesis [127,128,133] holds, the two representations (3.88) and (3.89) are completely identical, otherwise they are not, as shall be apparent below.

To extend these considerations to spectrum functions we begin with Eq. (3.88) and write the cross-spectral density (which recall is normalized on the turbulence mean square) as a frequency transform (also normalized on the turbulence mean square), with $\mathbf{r}_n = 0$

$$\phi_{ij}(\omega, r_1) = \frac{1}{2\pi} \int_{-\infty}^{\infty} e^{i\omega\tau} R_{ij}(\tau - r_1/U_c) R_m(\tau) d\tau$$

and introduce the frequency spectrum of the temporal correlation $R_{ij}(\tau)$ as $\phi_{ij}(\omega)$. Then we obtain the result

$$\phi_{ij}(\omega, r_1) = \int_{-\infty}^{\infty} e^{i\omega' r_1/U_c} \phi_{ij}(\omega')\phi_{\rm m}(\omega - \omega')d\omega'$$
$$= \int_{-\infty}^{\infty} e^{i\omega' r_1/U_c} \left(\frac{1}{U_c}\right) \phi_{ij}(\frac{\omega'}{U_c})\phi_{\rm m}(\omega - \omega')d\omega'$$

In this expression the spectrum $\phi_{ij}\left(\frac{\omega'}{U_c}\right)$ represents a streamwise wave number spectrum of the turbulence; the second spectrum accounts for the decorrelation of turbulence along the moving axis. In frozen convection $R_{\rm m}(\tau) = 1$ and $\phi_{\rm m}(\omega - \omega') = \delta(\omega - \omega')$ so that the wave number spectrum $\phi_{ij}(k_1 = k_c)$ is preserved in the moving axis. $\phi_{ij}(k_c)$ and $\phi_{ij}(\omega)$are then simply related by

$$\left(\frac{1}{U_c}\right)\phi_{ij}(k_c) = \phi_{ij}(\omega)$$

For nonfrozen convection the correlation model of Eq. (3.89) leads to a frequency spectrum in the form of a convolution between ϕ_{ij} (ω) and $\phi_{im}(\omega - \omega')$. This model of the correlation is inconvenient for analysis because of the integration component. In one form or another we will use various of these expressions in applications throughout this book.

An alternative way of relating the frequency spectrum at a point to the general considerations presented in Section 3.6.1 is to consider the correlation function form of Eq. (3.89) and take the frequency-time Fourier transform to define a frequency-space cross-spectral density:

$$\phi_{ij}(\omega, \boldsymbol{r}) = \frac{1}{2\pi}\int_{-\infty}^{\infty} e^{i\omega\tau} R_{ij}(r_1 - U_c\tau, \boldsymbol{r}_n)R_{\rm m}(r_1)d\tau \qquad (3.90)$$

Assume also that the probes are aligned with the mean convection direction, as they were for the measurements shown in Fig. 3.23 with $\boldsymbol{r}_n = 0$, then consider for the moment the case of a single probe for which the temporal autocorrelation follows the following series of equalities

$$R_{ij}(\tau) = R_{ij}(0 - U_c\tau, \boldsymbol{r}_n = 0)R_{{\rm m},ij}(r_1 = 0) = R_{ij}(-U_c\tau) = R_{ij}(U_c\tau)$$

This application of Taylor's hypothesis describes the translation of the axial wave number spectral density obtained at zero time delay to the frequency spectrum obtained at a point through convection. Now in Eq. (3.89), we introduce a coordinate transformation $r_1 - U_c\tau = -\xi$ and introduce the wave number spectrum $\phi_{ij}(k_c)$ which is the spatial Fourier transform of $R_{ij}(r_1)$ evaluated at $k_1 = k_c$ to obtain

$$\phi_{ij}(\omega, \boldsymbol{r}) = \frac{1}{U_c}\left[\frac{1}{2\pi}\int_{-\infty}^{\infty} R_{ij}(\kappa_1)\ e^{i\omega\xi/U_c}d\xi\right] e^{-\,i\omega r_1/U_c}R_m(r_1)$$

or, together with the delta function operator produced by the central bracket integral (Eq. (1.60))

$$\phi_{ij}(\omega, \boldsymbol{r}) = \left(\frac{1}{U_c}\phi_{ij}\left(\frac{\omega}{U_c}\right)\right) R_m(r_1)e^{-i\kappa_1 r_1}$$

Separating out the frequency domain and spatial domain we have

$$\phi_{ij}(\omega, \boldsymbol{r}) = (\phi_{ij}(\omega))(R_m(r_1)e^{-i\kappa_1 r_1}) \tag{3.91}$$

which is the preferred analytical form in the literature. Because of the equivalence between the streamwise wave number and frequency spectra, and invoking Eq. (1.60c)

$$\frac{1}{U_c}\phi_{ij}\left(\frac{\omega}{U_c}\right)d\omega = \phi_{ij}\left(\frac{\omega}{U_c}\right)\frac{d\omega}{U_c} = \phi_{ij}(\omega)d\omega \tag{3.92}$$

The first bracketed term on the right of Eq. (3.91) is recognized as the streamwise frequency spectrum of the *i,j* velocity product in the context of Section 3.6, the second bracketed term of Eq. (3.91) captures the convective properties of the *i,j* product for which measurements will be shown below; $k_c = \omega/U_c$ is the convection wave number. The phase of the exponential in Eq. (3.91) can now be recognized as due to convection as defined by Eq. (3.88). Integral

$$\kappa_1 = \omega/U_c$$

An example frequency spectrum for u_1 velocity fluctuations of a subsonic jet is shown in Fig. 3.24. The autospectrum shown is related to the spectrum function above by

$$\Phi_{11}(\omega; \mathbf{y}) = \overline{u_1^2(\mathbf{y})}\phi_{11}(\omega)$$

and, more important, has the dimensionless, nearly universal form,

$$\Phi_{11}(\omega; \mathbf{y}) = \overline{u_1^2(y)}\frac{y_1}{U_\mathbf{J}}\phi_{11}\left(\frac{\omega y_1}{U_\mathbf{J}}\right)$$

in which we have now introduced the nondimensional spectrum function

$$\phi_{11}\left(\frac{\omega y_1}{U_\mathbf{J}}\right) = \phi_{11}(\omega)\frac{U_\mathbf{J}}{y_1}$$

The convection velocity is related to the jet exit velocity using Fig. 3.23 and Eqs. (3.46b) and Eq. (3.71a) and approximating $U_c \simeq 0.8U_J$ to derive the longitudinal integral scale, from the measurements as the limit of frequency (wave number) approaching zero

$$4\pi\frac{U_J}{U_c}\frac{\Lambda}{\pi y_1} = \frac{1}{2}\Lambda \quad \simeq 0.1y_1$$

$$\frac{\Lambda}{y_1} = \frac{1}{8}\frac{U_c}{U_J}$$

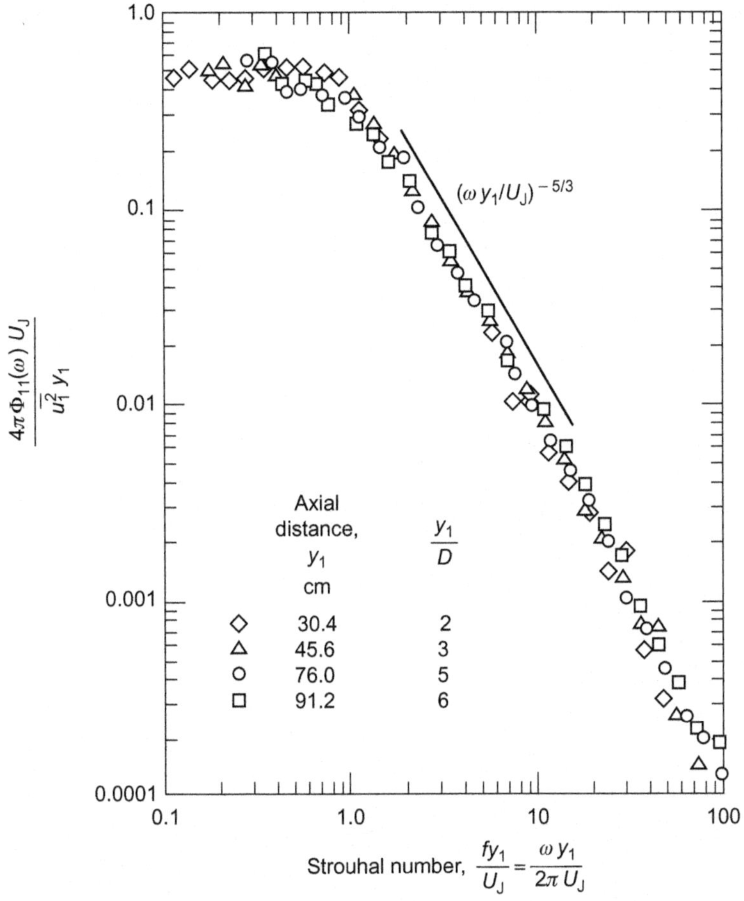

FIGURE 3.24 Turbulent axial velocity spectral density at lip axis. Mean nozzle exit velocity, 122 m/s; radial position, 7.6 cm, $R_D = 1.2 \times 10^6$. *From Karchmer AM, Dorsch GR, Friedman R. Acoustic tests of a 15.2-centimeter-diameter potential flow convergent nozzle. NASA Tech Memo. NASA TM X-2980, 1974.*

Thus the length scales of the eddies increase with y_1. Measurements of Λ_1 at Reynolds numbers near 10^5 by Davies et al. [171] and Laurence [178] provide corroboration, giving

$$\Lambda_1 \simeq 0.13 y_1$$

for $D < y_1 < 6D$, independent of radius. A radial integral scale, given by Laurence [178], is

$$\Lambda_r \simeq 0.05 y_1$$

Scales were measured by Jones [161] in a rectangular jet using both frequency-filtered signals and nonfiltered signals. With the 1, 2, 3 directions pertaining to the flow, vertical, and transverse directions, respectively, he found for broadband (unfiltered) velocities $\Lambda_2 \simeq \Lambda_3 \simeq 0.014(y_1 + h)$ and $\Lambda_1 \simeq 0.04(y_1 + h)$, where h is the height of the exit.

The frequency transform of Eq. (3.91) with retention of the form of Eq. (3.91) leads to a separable function for the cross-spectrum, that is, say

$$\Phi_{11}(\mathbf{r},\omega;y) = e^{i\omega r_1/U_c}\Phi_{11}(\omega)R(\mathbf{r}_n)R_m(r_1) \tag{3.93}$$

which is also experimentally available. The phase of the cross-spectral density arises from the convection of the turbulence field. Furthermore, the measurements of $\Phi_{11}(\mathbf{r},\omega;\mathbf{y})$ shown in Fig. 3.25 illustrate that due to moving-axis decay, the magnitude of the cross-spectral density will decrease with increasing r_1. Note that in this representation the spectrum functions Φ have not been normalized on the mean-square values of the turbulent velocities. Measurements of $\Phi_{11}(\mathbf{r},\ \omega;\ \mathbf{y})$ are shown in Fig. 3.25. The magnitude of the cross-spectrum has been normalized on the autospectrum,

$$\frac{|\Phi_{11}(r_1,\omega;\mathbf{y})|}{\Phi_{11}(\omega,\mathbf{y})} = A(r_1,\omega) \quad \sim e^{-\gamma_1\omega r_1/U_c} \tag{3.94}$$

where γ_1 is of order 0.2 to 0.3. We note that other functional representations that are not separable have been used. These may be quadratic or other forms that also may invoke scaling on local time mean variables. We shall encounter one of these in Section 3.7.2.3 and others when we discuss the turbulent boundary layer wall pressure fluctuations. The measurements in Fig. 3.25 were made by Kolpin [179] and Fisher and Davies [175], and in a round jet cover a range of frequencies and streamwise separations in the mixing zone, and they show clearly that the cross-spectral density with streamwise separation is a function of the variable $\omega r_1/U_J$. Recent measurements by Kerherve et al. [180] are similar. The convection velocities, determined from the filtered velocity signals over a range of frequencies, vary from $0.45U_J$ to $0.7U_J$ for the respective frequency range of fD/U_J from 2.08 to 0.52. For filtered signals Jones [161] in a rectangular jet found also that $\Lambda_2(\omega) \simeq \Lambda_3(\omega)$, but that all components of Λ_i/y_1 decreased as $\omega y_1/U_J$ increased, but not quite as rapidly as $(\omega y_1/U_J)^{-1}$.

The measurements shown in Figs. 3.23 and 3.25 give a partial picture of the structure of the second-order correlation, $\langle u_1u_1\rangle(\mathbf{y},r_1,\omega)$ of the jet turbulence as a function of frequency. The complete form of the cross-spectrum that is consistent with Eq. (3.93) is

$$\Phi_{11}(\mathbf{r}_1,\omega;\mathbf{y}) = \Phi_{11}(\omega)e^{ik_cr_1}A(k_cr_1)R(\mathbf{r}_n) \tag{3.95}$$

where $k_c = \omega/U_c$ has been inserted as the convection wave number. For these measurements $\mathbf{r}_n = 0$. This separable function has a parallel in the wave number domain (see Eq. 3.40)

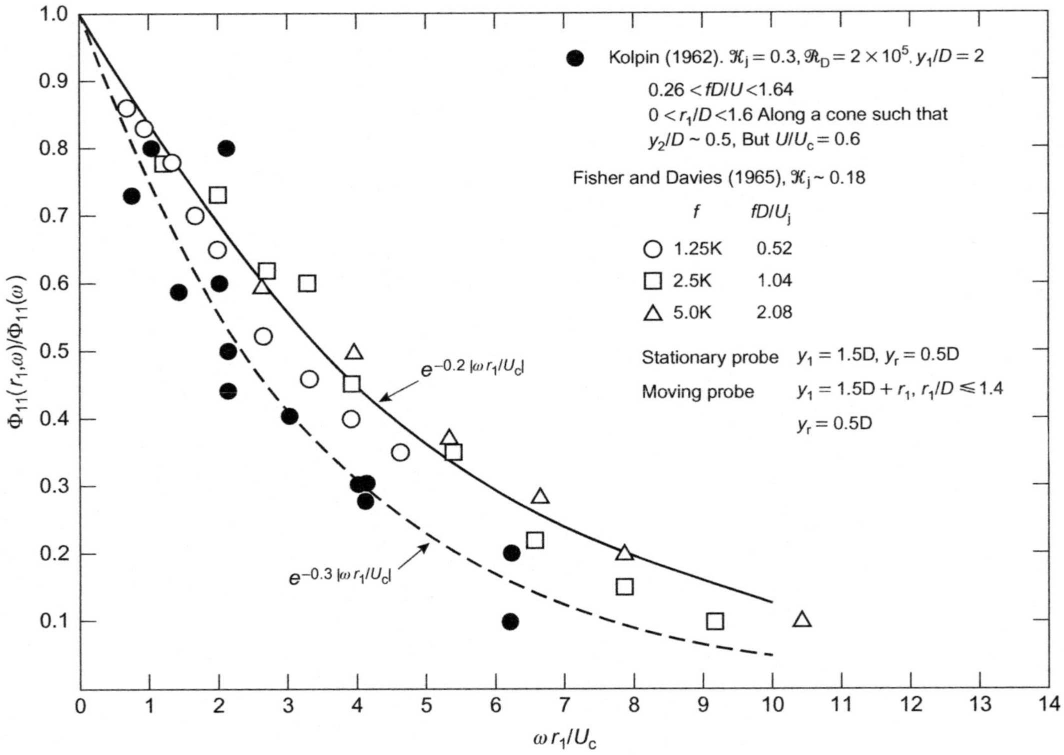

FIGURE 3.25 Normalized cross-spectral densities of jet turbulence at low Mach numbers. *Data sources: Kolpin MA. Flow in the mixing region of a jet. Aeroelastic Struct. Res. Lab. Rep. ASRL TR 92-3. Dep. Aeronaut. Astronaut., MIT, Cambridge, MA, 1962; Fisher MJ, Davies POAL. Correlation measurements in a non-frozen pattern of turbulence. J Fluid Mech 1965;18:97–116.*

$$\phi_{11}(r_1,\omega;\mathbf{y}) = \phi_{\mathrm{m}}(\omega - U_c k_1)\phi_1(k_1)\phi_n(\mathbf{k}_n) \tag{3.96}$$

3.7.2.3 Methods of Approximating the Fourth-Order Correlation Tensor

As an example of specific modeling that was used in early examples of predicting jet noise, we again take

$$\mathbf{U}(\mathbf{y},t) = \mathbf{U}_{\mathrm{J}}(\mathbf{y}) + \mathbf{u}(\mathbf{y},t)$$

where $\mathbf{U}_{\mathrm{J}}$ is the local mean velocity in the jet. When this is substituted into the correlation tensor the expected superposition of the mean shear and turbulence interaction terms emerge. For, say, the 1, 2 combination (where $\mathrm{U}_1 = |\mathbf{U}_J|$)

$$\begin{aligned}\langle T_{12}(\mathbf{y},t)T_{12}(\mathbf{y}+\mathbf{r},t+\tau)\rangle \simeq\, & U_{\mathrm{J}}(\mathbf{y})U_{\mathrm{J}}(\mathbf{y}+\mathbf{r})\langle u_2(\mathbf{y},t)u_2(\mathbf{y}+\mathbf{r},t+\tau)\rangle \\ & + \langle u_1(\mathbf{y},t)u_2(\mathbf{y},t)u_1(\mathbf{y}+\mathbf{r},t+\tau)u_2(\mathbf{y}+\mathbf{r},t+\tau)\rangle \\ & - \langle u_1(\mathbf{y},t)u_2(\mathbf{y},t)\rangle\langle u_1(\mathbf{y}+\mathbf{r},t)u_2(\mathbf{y}+\mathbf{r},t)\rangle \\ & + \text{third-order covariances}\end{aligned}$$

The early work in using this statistical model ignored values of third-order correlation as suggested by a hypothesis of nearly Gaussian (normal) statistics for the turbulence that provides for the vanishing of odd-order correlations, see e.g. Batchelor [129], and Goldstein and Rosenbaum [165]. Both of these correlations are therefore of the general dimensional form [164] and naturally involve the coexistence of both second- and fourth-order functions, i.e.

$$\langle T_{ij}(\mathbf{y},t)T_{kl}(\mathbf{y}+\mathbf{r},t+\tau)\rangle = \rho_0^2 U_1^4[\alpha_{jl}R_{jl}(\mathbf{r},\tau)\delta_{ij}\delta_{lk} + \beta_{ijkl}R_{ijkl}(\mathbf{r},\tau)] \tag{3.97a}$$

where α_{jl} and β_{ijkl} are second- and fourth-order intensity functions,

$$\alpha_{ijkl} = \frac{\overline{u_i u_j}}{U_1^2} \quad \beta_{ijkl} = \frac{\overline{u_i u_j}\cdot\overline{u_k u_l}}{U_1^4}$$

This separation is still widely used in modeling the near field pressures in a wall layer; see Chapter 8: Essentials of Turbulent Wall-Pressure Fluctuations. We note that $\delta_{ij} = 1$ for $i = j$ and $\delta_{lk} = 1$ for $k = l$, see Section 1.6.2, Differential Operators, and α_{jl} and β_{ijkl} are weighting factors. We hypothesize that both the second and fourth order correlation functions have qualitatively similar behavior so that the general features of $R_{jl}(r,\tau)$ are representative of those features of $R_{ijkl}(r,\tau)$. In fact for an assumed Gaussian statistics of the turbulence the fourth-order cross-correlation can be written as a summation of second-order correlations thus following Batchelor [129], Goldstein and Rosenbaum [165], and Goldstein and Howes [167]:

$$
\begin{aligned}
&\left\langle u_i(\boldsymbol{x},t)u_j(\boldsymbol{x},t)u_k(\boldsymbol{x}+\boldsymbol{r},t+\tau)u_l(\boldsymbol{x}+\boldsymbol{r},t+\tau)\right\rangle \\
&-\left\langle u_i(\boldsymbol{x},t)u_j(\boldsymbol{x},t)\right\rangle\left\langle u_k(\boldsymbol{x}+\boldsymbol{r},t+\tau)u_l(\boldsymbol{x}+\boldsymbol{r},t+\tau)\right\rangle = \ldots\ldots \\
&\ldots\ldots\ldots\left\langle u_i(\boldsymbol{x},t)u_k(\boldsymbol{x}+\boldsymbol{r},t+\tau)\right\rangle\left\langle u_j(\boldsymbol{x},t)u_l(\boldsymbol{x}+\boldsymbol{r},t+\tau)\right\rangle \\
&+\left\langle u_i(\boldsymbol{x},t)u_l(\boldsymbol{x}+\boldsymbol{r},t+\tau)\right\rangle\left\langle u_j(\boldsymbol{x},t)u_k(\boldsymbol{x}+\boldsymbol{r},t+\tau)\right\rangle
\end{aligned}
$$

or introducing the notation appropriate for the correlation in the form of $R_{ik}(\boldsymbol{r},\tau)$

$$
\begin{aligned}
&\left\langle u_i(\boldsymbol{x},t)u_j(\boldsymbol{x},t)u_k(\boldsymbol{x}+\boldsymbol{r},t+\tau)u_l(\boldsymbol{x}+\boldsymbol{r},t+\tau)\right\rangle \\
&-\left\langle u_i(\boldsymbol{x},t)u_j(\boldsymbol{x},t)\right\rangle\left\langle u_k(\boldsymbol{x}+\boldsymbol{r},t+\tau)u_l(\boldsymbol{x}+\boldsymbol{r},t+\tau)\right\rangle = \ldots\ldots \\
&\ldots\ldots\ldots\left\langle u_i(\boldsymbol{x},t)u_k(\boldsymbol{x},t)\right\rangle\left\langle u_j(\boldsymbol{x},t)u_l(\boldsymbol{x},t)\right\rangle R_{ik}(\boldsymbol{r},\tau)R_{jl}(\boldsymbol{r},\tau)+\ldots\ldots \\
&\ldots\ldots\ldots\left\langle u_i(\boldsymbol{x},t)u_l(\boldsymbol{x},t)\right\rangle\left\langle u_j(\boldsymbol{x},t)u_l(\boldsymbol{x},t)\right\rangle R_{il}(\boldsymbol{r},\tau)R_{jk}(\boldsymbol{r},\tau)
\end{aligned}
$$

The convection properties of R_{11} $(\boldsymbol{r}, \tau)$ already discussed (Figs. 3.23 and 3.25) can be used to approximate this more general function using an analytic function for $R_{ij}(\boldsymbol{r}, \tau) \approx R_{11}(\boldsymbol{r}, \tau)$. Then our quadrupole source covariance matrix may be populated with variables that are approximated by a representative measurement of the streamwise velocity correlation function.

$$
\left\langle T_{ij}(\mathbf{y},t)T_{kl}(\mathbf{y}+\mathbf{r},t+\tau)\right\rangle = \rho_0^2 U_1^4[\alpha_{jl}(\mathbf{y})R_{11}(\mathbf{y},\mathbf{r},\tau)\delta_{ij}\delta_{lk} + \beta_{ijkl}R_{11}(\mathbf{y},\mathbf{r},\tau)^2] \tag{3.98}
$$

The covariance matrix now depends on spatial integral scales using R_{11} and the mean values of Reynolds stresses, some of which might be known, others might be estimable based on published data for mixing layers.

Until recently, attempts at quantitative predictions of sound from jets were made on the basis of measured correlation functions which were limited to second-order tensors used to approximate the fourth-order source term as suggested above. The second-order tensors were also limited, generally to correlations of velocity as just described along principal axes. Assumptions of homogeneity and isotropy generally guided the analyses of, e.g., Proundman [160], Jones [161], and Ribner [158,159,162,163,181], Goldstein and Howes [167]. This contemporary propagation function therefore includes effects of mean shear in the Green function so that the sources only involve the nonlinear terms.

The use of super computers with LES has made possible "numerical experiments" that can make effective use of fourth order covariance matrices of the nonlinear sources to calculate radiated sound directly using Lighthill's equation and an implicit function for the propagation. Accordingly many of the above classical statistical model idealizations can be set aside to instead consider the 4th order correlations directly by either of two related methods. The first is the solution of the direct large eddy simulation that provides the full instantaneous u_iu_j, together with an implicit propagation function as described above. In the second, more approximate method, the large eddy simulation to provides a statistical model of the $R_{ij,kl}$ for the Lighthill equation formulation in a form analogous to Eq. (3.75b), but with the free space Green function replaced by the better propagation function. The investigators relied on a steady RANS calculation of the mean flow and the analytical model given below for the 4th order correlation function. This function relied on length scales of the shear layer that were produced by the RANS computation. These hybrid models have been developed for a performance study of noise abatement of a round turbulent jet at high Reynolds number; the models characterize the complete turbulent flow structure in the jet and provide all the required three-dimensional velocity statistics that go into computing the fourth-order space-time correlation tensor. These simulated "data" were used to generate the necessary "empirical" coefficients to populate the correlation function $R_{ij,kl}(y,\boldsymbol{r},\tau)$ [182–186].

$$R_{ijkl}(y,\boldsymbol{r},\tau)=\frac{\langle u_i(\boldsymbol{x},t)u_j(\boldsymbol{x},t)u_k(\boldsymbol{x}+\boldsymbol{r},t+\tau)u_l(\boldsymbol{x}+\boldsymbol{r},t+\tau)\rangle-\overline{u_iu_j(\boldsymbol{x})}\cdot\overline{u_ku_l(\boldsymbol{x}+\boldsymbol{r})}}{\overline{u_iu_j(\boldsymbol{x})}\cdot\overline{u_ku_l(\boldsymbol{x}+\boldsymbol{r})}} \tag{3.99}$$

where

$$R_{ij,kl}(\boldsymbol{y},\boldsymbol{r},\tau)=\exp\left\{-\left(\frac{r_z}{U_c(\boldsymbol{y})\tau_m(\boldsymbol{y})}\right)-(\ln 2)\left(\left(\frac{r_z-U_c(\boldsymbol{y})\tau}{\Lambda_z(\boldsymbol{y})}\right)^2+\left(\frac{r_\rho}{\Lambda_\rho(\boldsymbol{y})}\right)^2+\left(\frac{r_\phi}{\Lambda_\phi(\boldsymbol{y})}\right)^2\right)\right\} \tag{3.100}$$

All the variables τ_m , Λ_z, and Λ_ϕ are determined for by large eddy simulation results as are all combinations of Reynolds stresses. The covariance function $R_{11,11}(\boldsymbol{y},\boldsymbol{r},\tau)$ shown in Fig. 3.26 and again in Fig. 3.31, captures all the general space-time convection and decay characteristics that were given by the result of the LES. The numerical results disclosed that the integral over the source volume is dominated by the (11,11) combination, then with nearly equal magnitudes of covariances with combinations (12,12), (13,13), (22,22), (23,23), and (33,33). Other combinations contribute less than 1/2 of these to the net integrated sound level. In the reduced order model, all (ij,kl) correlations are weighted by products $<u_i^2><u_j^2>$. Further discussions of the computational methods will be in Sections 3.7.3 and 3.7.4.1 when we examine the explicit evaluation of the complete integral, Eq. (3.75e) or its analog with a convective propagation model.

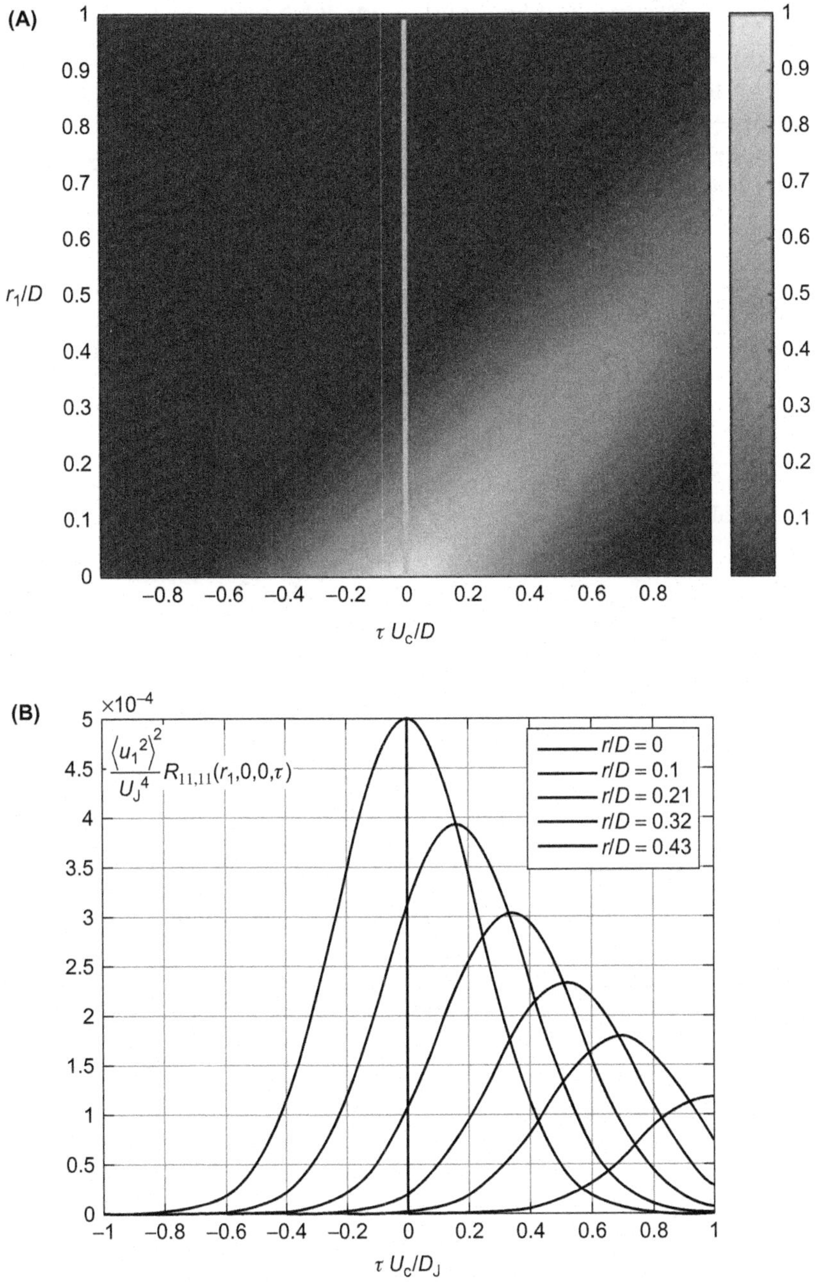

FIGURE 3.26 An example fourth-order space–time correlation function, $R_{11,11}(r_1,0,0,\tau)$, for $M_J = 0.75$ modeled using a Gaussian-style function, Eq. (3.100). Fixed probe position: $x/D = 4$, $\rho/D = 0.5$, $\theta = 0$.

3.7.3 Expressions and Scaling law for Jet Noise

We continue with the analysis of jet noise with an interest in developing scaling rules and use the early models for noise such as used by eg. Ffowcs Williams [176], Goldstein and Rosenbaum [167] and Lilley [157]. Consider the autospectral density of far field sound pressure which is found by invoking the Fourier transform, Eq. (1.35)

$$[\phi_{pp}(\mathbf{x},\omega)]_{ij,kl} = \frac{1}{2\pi}\int_{-\infty}^{\infty} e^{i\omega\tau}[R_{pp}(\mathbf{x},\tau)]_{ijkl}d\tau$$

This operation requires a spectrum function of the derivative of the correlation

$$\frac{1}{2\pi}\int_{-\infty}^{\infty} e^{i\omega\tau}\frac{\partial^4 R_{ij,kl}(\mathbf{x},\tau)}{\partial\tau^4}d\tau$$

Integration by parts 4 times and noting that $\lim R_{ij,kl}(|\tau|\to\infty)=0$ we obtain the equivalence

$$\frac{1}{2\pi}\int_{-\infty}^{\infty} e^{i\omega\tau}\frac{\partial^4 R_{ij,kl}(\mathbf{x},\tau)}{\partial\tau^4}d\tau = \frac{\omega^4}{2\pi}\int_{-\infty}^{\infty} e^{i\omega\tau}R_{ij,kl}(\mathbf{x},\tau)d\tau \tag{3.101a}$$

Using a change of variables in the case of a delayed correlation function, we have

$$\begin{aligned}\frac{1}{2\pi}\int_{-\infty}^{\infty} e^{i\omega\tau}\frac{\partial^4 R_{ij,kl}(\mathbf{x},\tau)}{\partial\tau^4}d\tau &= \frac{\omega^4}{2\pi}\int_{-\infty}^{\infty} e^{i\omega\tau}R_{ij,kl}(\mathbf{x},\tau+\tau_0)d\tau\\ &= \frac{\omega^4}{2\pi}\int_{-\infty}^{\infty} e^{i\omega(\tau-\tau_0)}R_{ij,kl}(\mathbf{x},\tau)d\tau\end{aligned}$$

which gives an operation that is appropriate to our autospectrum

$$\frac{1}{2\pi}\int_{-\infty}^{\infty} e^{i\omega\tau}\frac{\partial^4 R_{ij,kl}\left(\mathbf{y}',\boldsymbol{\eta},\tau+\frac{\eta\cdot x}{c_0|\boldsymbol{x}|}\right)}{\partial\tau^4}d\tau = \frac{\omega^4}{2\pi}\int_{-\infty}^{\infty} e^{i\omega\left(\tau-\frac{\boldsymbol{\eta}\cdot x}{c_0|\boldsymbol{x}|}\right)}R_{ij,kl}(\mathbf{y}',\boldsymbol{\eta},\tau)d\tau$$

Introducing the expression following Eq. (3.75e) the above equation for the convected time delay becomes

$$\begin{aligned}\frac{\omega^4}{2\pi}\int_{-\infty}^{\infty} e^{i\omega\left(\tau-\frac{\boldsymbol{\eta}\cdot x}{c_0|\boldsymbol{x}|}\right)}R_{ij,kl}(\mathbf{y}',\boldsymbol{\eta},\tau)d\tau &= \frac{\omega^4}{2\pi}\int_{-\infty}^{\infty} e^{i\omega\left(\tau-M_c\cos\theta\tau-\frac{\boldsymbol{\zeta}\cdot x}{c_0|\boldsymbol{x}|}\right)}R_{ij,kl}(\mathbf{y}',\boldsymbol{\zeta},\tau)d\tau\\ &= \omega^4 e^{-i\omega\left(\frac{\boldsymbol{\zeta}\cdot x}{c_0|\boldsymbol{x}|}\right)}\phi_{ij,kl}(\mathbf{x},\boldsymbol{\zeta},\omega(1-M\cos\theta))\end{aligned}$$

Finally the expression for the autospectral density of the sound pressure from the *ijkl* component that we shall use for dimensional analysis is

$$[\Phi_{pp}(\mathbf{x},\omega)]_{ijkl} = \frac{1}{16\pi^2}\frac{\rho_0{}^2}{c_0{}^4}\frac{(x_i x_j x_k x_l)}{r^6}\int_{-\infty}^{\infty} dV(\mathbf{y}')\int_{-\infty}^{\infty} dV(\varsigma)\times e^{-i\omega\left(\frac{\zeta\cdot x}{c_0|x|}\right)}\times\cdots$$

$$\cdots\times\overline{u_i u_j}(\mathbf{y})\overline{u_k u_l}(\zeta)\omega^4\phi_{ij,kl}(\mathbf{y}',\zeta,\omega(1-M\cos\theta)) \tag{3.101b}$$

The integration over ζ in Eq. (3.101b) extends over the separation variable of the covariance functions, that over $\mathbf{y}'$, extends over the jet volume and represents the reference point for the separation vectors. By analogy a frequency transform of the moving-axis decorrelation in R_{ijkl}, i.e., $\phi_{ij,kl}(\mathbf{y}',\varsigma,\omega)$, in Eq. (3.101b) creates similar effects expressed in the integral over ζ:

$$\int_{-\infty}^{\infty} e^{-i\omega\left(\frac{\varsigma\cdot x}{c_0|x|}\right)}\times\overline{u_k u_l}(\varsigma)\phi_{ij,kl}(\mathbf{y}',\varsigma,\omega)dV(\varsigma)$$

which is represents a wave number spectrum of the Reynolds stresses that is evaluated at the trace wave number, $k_0\cos\theta$.

This expression relates the autospectral density of far field sound pressure to the cross-spectral covariance functions of the Reynolds stress and was derived in a form similar to this by Goldstein [153] using essentially the same approach used here. Again the total sound pressure spectrum of all Reynolds stress components is found by the summing over all of *ij,kl*.

We could use this result in search of an explicit model for the sound as done in the early literature on jet noise. Instead we will use this theory to develop nondimensionalized spectrum functions for spectra of radiated sound as a starting point for deriving scaling rules. We will defer further discussion of explicit noise models to Section 3.7.4 when we examine the use of computational methods further. We will consider the autospectral density of sound as well as spectra that are measured in a one-third-octave band for which the bandwidth is proportional to frequency, i.e., $\Delta\omega\propto\omega$. Accounting for this bandwidth amounts to multiplying Eq. (3.101b) by an additional factor of ω making the power of frequency appearing outside the integral ω^5. Assume that the jets we are comparing are all turbulent for which the nozzle geometries and the flow in the jets are dynamically and geometrically similar. In Section 3.7.2.2 data was presented that showed the observed correlation lengths in the axial and radial coordinates to be small compared with distance from the nozzle lip; thus also they are smaller than, yet scale on, the diameter of the jet. Accordingly the spatial correlation, say turbulence integral scales of the separation variable in Eq. (3.100), are much smaller than the volume of the jet. Finally we note the frequency at the observer at rest is ω is related to that moving in the frame with the quadrupoles, Ω, by

$$\omega = \frac{\Omega}{1-M_c\cos\theta} \tag{3.102}$$

Equation (3.102) expresses the Doppler shift in frequency with Ω being the frequency at the source. We encountered this behavior already in Section 2.5.

Applying all these concepts to the volume integral in Eq. (3.101b) results in an algebraic expression that uses the integral scale volume, say $\delta V \approx \Lambda^3$, as the net correlation volume of the turbulence integrated over the jet shear layer. This is assumed to be proportional to D^3. V_J is the volume of the jet turbulence that contributes to the sound; this is also proportional to D^3, thus when we notionally ensemble-average all *ijkl* combinations of the tensor components we obtain a dimensionless scaling function for the net autospectrum of sound pressure

$$\Phi_{\text{rad}}(\omega,r,\theta)\cdot \propto \frac{1}{c_0^4 r^2}(1-M_c(\mathbf{y})\cos\theta)^2 \frac{(\rho_0 U_J{}^2)^2 \overline{\left\{\left[\frac{\Omega D}{U_J}\right]^4 [\Theta(\Omega)]\right\}}^{V_J}}{(1-M_c\cos\theta)^4} \frac{\Lambda^3 V_J U_J{}^4}{D^4}\left(\frac{D}{U_J}\right) \tag{3.103}$$

Now substituting for D^6 for the product of the correlation and geometric volumes, this becomes

$$\Phi_{\text{rad}}(\omega,r,\theta) \propto \left(\frac{U_J}{c_0}\right)^4 \left(\frac{D}{r}\right)^2 \frac{(\rho_0 U_J^2)^2}{(1-M_c\cos\theta)^4}\left(\frac{D}{U_J}\right) \overline{\left\{\left[\frac{\Omega D}{U_J}\right]^4 [\Theta(\Omega)]\right\}}^{V_J} \tag{3.104a}$$

with a matching representation for the mean square sound pressure in a proportional frequency band, say one-third octave:

$$\overline{p_{\text{rad}}^2}(\omega,r,\theta;\Delta\omega) \propto \left(\frac{U_J}{c_0}\right)^4 \left(\frac{D}{r}\right)^2 \frac{(\rho_0 U_J^2)^2}{(1-M_c\cos\theta)^5} \overline{\left\{\left[\frac{\Omega D}{U_J}\right]^4 [\Omega\Theta(\Omega)]\right\}}^{V_J} \tag{3.104b}$$

In both expressions we have inserted a dimensionless similarity function, either $[\Theta(\Omega)]$ or $[\Omega\Theta(\Omega)]$, that represents the proportional-band spectrum of Reynolds stresses in the moving Reynolds stress frame. The frequency in the moving quadrupole frame scales on jet diameter and velocity as provided in the similarity assumption. The $\{\bar{\ }\}$ symbol denotes a sort of ensemble average over the volume of turbulence in the jet of the dimensionless function that represents in the frequency domain the fourth order time derivative of the Reynolds stresses over the volume of the jet. This step allows placement of the jet volume explicitly in Eq. (3.103a). Note that the dimensions of $\overline{p_{\text{rad}}^2}(\omega,r,\theta;\Delta\omega)$ and $\Phi_{\text{rad}}(\omega,r,\theta)$ are pressure-squared, with time in the autospectrum being absorbed by multiplication by bandwidth. Finally the dimensionless spectral density of interest is

$$\frac{\Phi_{\text{rad}}(\omega,r,\theta)\dfrac{U_J}{D}}{(M)^4\left(\dfrac{D}{r}\right)^2 \dfrac{(\rho_o U_J{}^2)^2}{(1-M_c\cos\theta)^4}} = \Im_0\left(\frac{\omega D}{U_J}, \frac{\rho_0 U_j d}{\mu}\right) \tag{3.105a}$$

or, equivalently in a proportional frequency band, see Eqs. (1.44–1.46)

$$\frac{\overline{p_{\mathrm{rad}}^2}(f, r, \theta; \Delta f)}{(M)^4 \left(\frac{D}{r}\right)^2 \frac{(\rho_0 U_{\mathrm{J}}^2)^2}{(1-M_{\mathrm{c}}\cos\theta)^4}} = \mathfrak{I}\left(\frac{fD}{U_{\mathrm{J}}}, \frac{\rho_0 U_j d}{\mu}\right) \tag{3.105b}$$

The expressions on the left are dimensionless and are functions of Reynolds number and dimensionless frequency based on nozzle diameter and efflux velocity, all under the series of stipulations cited earlier that ensure similarity. They provide both spectral densities and mean square pressures (say in one-third octave band, $\Delta f \sim 0.23f$) for subsonic jet noise at different efflux velocities and it accounts for observer angle, Mach number and nozzle diameter. Eq. (3.105a) is the fundamental representation of jet noise and gives the $(1-M_{\mathrm{c}}\cos\theta)^5$ Doppler factor, which was first derived by Lighthill [151,152], Ffowcs Williams [170,241], Mani [188,189], and Goldstein [153]. A different form involving Mach number, specifically, $(1-M_{\mathrm{c}}\cos\theta)^3$ agrees well with some measurement [190,191,192,251].

Also since our development is in the frequency domain, the effective averaging volume of over which the quadrupoles are ensemble-averaged must also vary, increasing as frequency decreases because the frequency-dependent integral scale increases at lower frequencies. The factor $\mathfrak{I}\left(\frac{fD}{U_{\mathrm{J}}}, \frac{\rho_0 U_j d}{\mu}\right)$ represents nondimensional scaling of spectral density or proportional band levels that is dependent on jet efflux diameter and efflux velocity; all Doppler-related directivity is embodied in the term $(1-M_{\mathrm{c}}\cos\theta)^{-4}$ or $(1-M_{\mathrm{c}}\cos\theta)^{-5}$, which shows an augmentation for $\theta<\pi/2$ relative to the acoustic intensity at $\theta>\pi/2$. This behavior of enhanced sound in the direction of eddy convection is a well-known observation [e.g., 190–193] of sound from subsonic jets and it is the same as our result for the convecting single quadrupole obtained in Section 2.5 and plotted in Fig. 2.11. An additional attenuation of sound levels near the axis of the jet, i.e., $\theta<\pi/4$ occurs beyond that predicted by the Doppler effects as observed and widely accepted to be due in part to refraction [22,161,162,191]. However, recent first-principles propagation models, i.e., Tam and Aurialt [185], Goldstein and Leib [260], and Leib and Goldstein [256] are able to include refraction for high frequencies, $fD/c_0>0.3$ and apply to Mach numbers above 0.4. They also are capable of modeling supersonic and hot jets. Note that in some of the earlier measurements [194,195,196], high-frequency backward directed sound had been measured. This behavior was attributed to the existence of preferred quadrupole radiation from the highly sheared mixing region near the nozzle. However, the subsequent measurements, Mollo-Christensen et al. [195,196] and later had, perhaps incorrectly, indicated that this backward radiation was an artifact of experimental arrangement.

Fig. 3.27 shows the measured [191] and calculated angular dependence as a function of Mach number as appears with Eqs. (3.103–3.105).

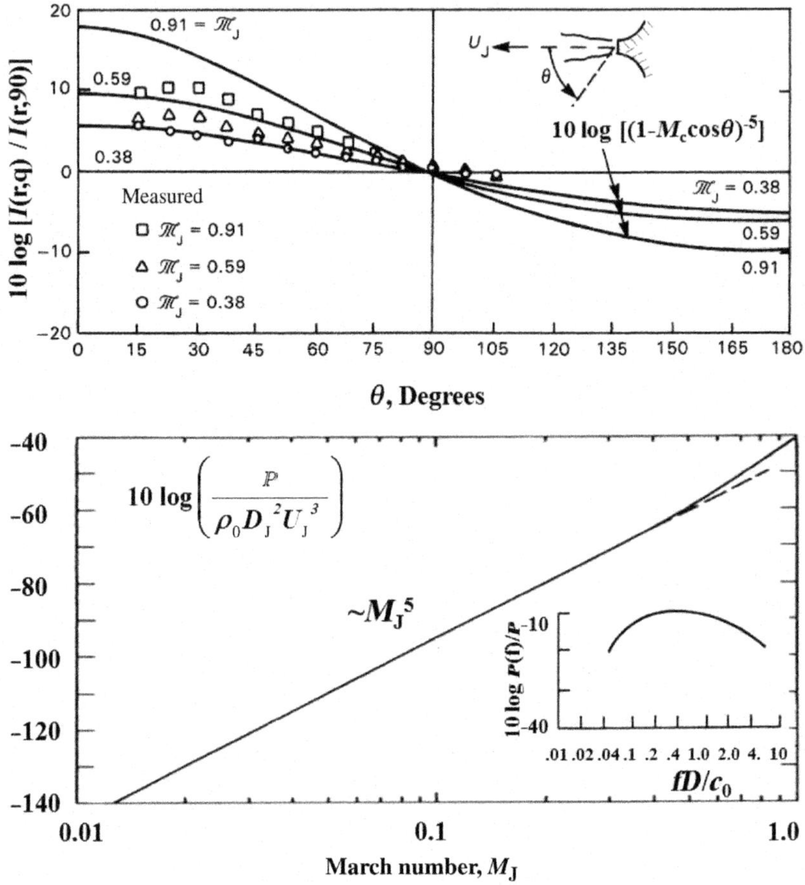

FIGURE 3.27 Directivity of overall intensity (Top) and overall sound power versus Mach number and its spectrum for jet noise (Bottom) using potential-flow nozzles, $M_c = 0.62$. M_j, $\sim 10^5 < R_D < \sim 10^6$. *Lush PA. Measurements of subsonic jet noise and comparison with theory. J Fluid Mech 1971;46:477–500.*

Integration over the full spherical surface surrounding the jet provides a simple formula for total radiated acoustic power [Eq. (2.17)] that is consistent with the directivity factor shown in Eq. (3.105) and illustrated in Fig. 3.27:

$$\mathbb{P} = \frac{\rho_0 U^8 D^2}{c_0^5} \frac{1 + M_c^2}{(1 - M_c^2)^4} F(R_D, M_c), \tag{3.106}$$

The function $F(R_D, M_c)$ denotes the scaling of an integral of overall frequency and angle and includes the dependence on the Reynolds and Mach numbers. The Mach number factor is the Lighthill-Ffowcs Williams [151,168,170,241] original value reflecting the quantity $[1-M\cos\theta]^5$

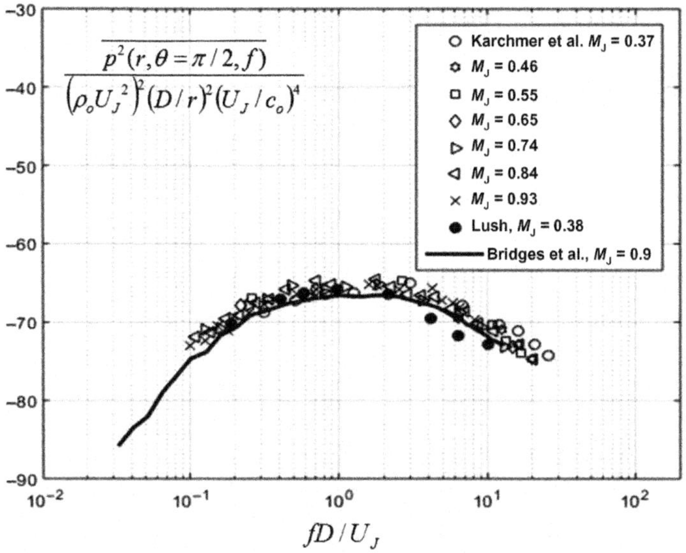

FIGURE 3.28 Normalized sound pressure levels in one-third octave frequency bands at a microphone angle of 90 degrees for 15.2-cm-diameter potential flow nozzle [190,191,198].

appearing in Eq. (3.105). Differences in the calculated effects of convection can be attributable to different assumptions made of quadrupole contributions dominate. Fig. 3.28 shows the total power nondimensionalized on $\rho_0 U_0^3 D^2$, as a function of the exit Mach number,

$$M_J = U_J/c_0$$

and $U_c \approx 0.63U_j$. This presentation represents the relative increase of acoustic power compared to jet power as the Mach number increases, and it is a classical result. Furthermore the data indicate that $F(R_D, M_c)$ is nearly constant valued between 0.3×10^{-4} and 1.2×10^{-4}.

The inset of Fig. 3.28 shows the one-third octave power spectrum nondimensionalized on the overall power as function of reduced frequency fD/c_o. The peak in the far-field intensity spectrum at any angle will change in frequency such that the peak will occur at constant values of $f\,(1-M_c \cos\theta)\,D/U_J$. When $\theta < \pi/4$, i.e., at angles near the jet axis, Lush [191] shows anomalously low values of noise at frequencies and M_c that are large enough that refraction is important. The spectrum of overall sound power therefore has a peak frequency that depends on D/c_0; the constancy of the dimensionless frequency of the peak in the *power* spectrum, incidentally, first observed by Fitzpatrick and Lee [197], is now well known. At $\theta = \pi/2$, the dependence of the sound pressure or sound intensity spectrum, Eq. (3.14) on M_c decreases and the spectrum becomes a function of fD/U_J as shown in Fig. 3.28. These data show clearly the scaling law for cold jets over a range

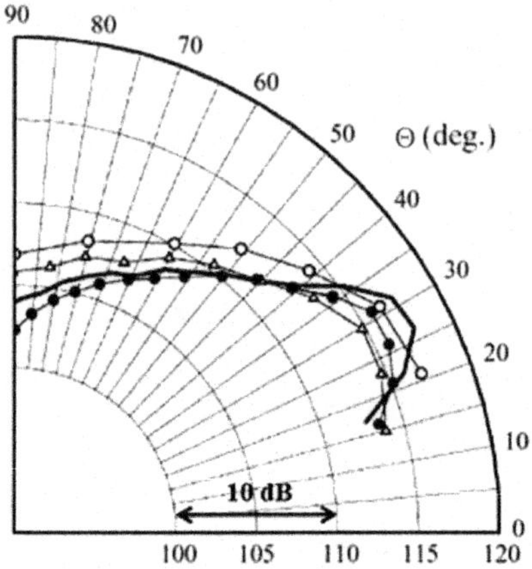

(A) Directivity of overall sound pressure dB re 1μPa, $\mathscr{R}_D$=3600 to 6x10[5]

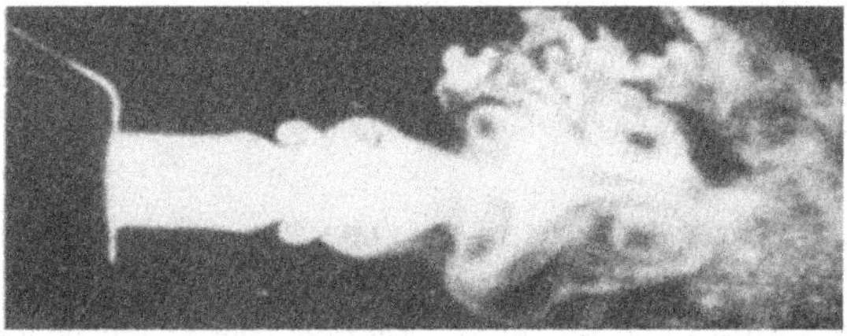

(B) Visualization, Becker and Massaro $\mathscr{R}_D$=3800 (1968)

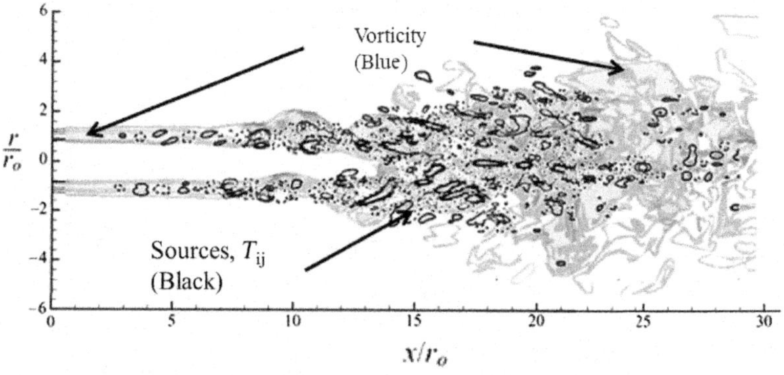

(C) Instantaneous vorticity and T_{ij} sources; Freund (2001); $\mathscr{R}_D$=3600 (DNS)

FIGURE 3.29 Comparison of the results of a compressible DNS of jet flow and radiated sound at $r = 30\ D_J$, $R_D = 3600$ with examples of data obtained with physical experiments: (A) overall sound pressure: ● Stromberg et al. [203], $R_D = 3600$; ○ Mollo-Christianson et al. [196], $R_D = 2 \times 10^5$; △ Lush[191], $R_D = 6\text{x}10^5$; —— Freund, [235], $R_D = 3600$; (B) Smoke visualization photograph of a jet at $t\ R_D = 3800$; Beavers and Wilson [72]; (C) DNS contours of instantaneous vorticity (gray), (contours are at constant values of $\omega D_J/2U_J$) and of the Reynolds stress source (black) $c_0{}^2\partial^2 T_{ij}/\partial y_i \partial y_j$

TABLE 3.2 Measurement Conditions Used for Fig. 3.28

Investigator	Diameter of Nozzle, cm	Approximate Efflux Reynolds Number
Lush (1971)	2.5	2×10^5
Karchmer et al. (1974)	15.2	1.3×10^6 to 3.3×10^6
Bridges et al. (2004)	50.8	2×10^7

of jet efflux Reynolds number. The empirical scaling laws of Fisher et al. [262], Tanna et al. [263], and Viswanatha [264,265], as they might apply to cold jets, are consistent with the formulas presented here. In their full application they [264,265] account for a temperature-dependent enthalpy source and atmospheric propagation Table. 3.2.

3.7.3.1 The Role of Specific Turbulent Structures in Jet Noise

As discussed above large-scale numerical simulation clearly and quantitatively interconnects the various events that take place in the development of subsonic jet structure and noise. Historically, though, experimental visualization of axisymmetric eddies in jets by Crow and Champagne [74] and Browand and Laufer was required to understand the roles these eddies in the development of disturbances and the role of vortex pairing as acoustic sources [199]. Furthermore, these eddies can be ultimately traced to the modes of axisymmetric instability, as can be deduced from the early studies of shear layers [21,23,24] previously discussed in Section 3.2 (Figs. 3.3 and 3.13). The existence of instability waves at high Reynolds number, although deduced by Crow and Champagne, has been analytically confirmed by Crighton and Gaster [200] using stability theory. Furthermore, a connection between the existence of forced shear layer instability and radiated sound has been experimentally investigated in the experiments of Mollo-Christensen [195], Moore [201], Bechert and Pfizenmaier [202], and Stromberg et al. [203], although more recently by many additional researchers, e.g., Wygnanski and coworkers [246,247,249,250]. See also the reviews appearing in Hubbard [251]. In these experiments the circular jet shear layer is often driven by a loudspeaker. The collective observations show pressure and velocity waves propagate along the center line from the nozzle and amplify the frequencies of those waves which are confined to a restricted frequency band width that coincides with the range of unstable axisymmetric mode frequencies of the undisturbed jet. These waves are determined by visualization to modulate the vortex formation and pairing in the mixing region. In Moore's [201] case with high-level excitation of the jet, the magnitude of induced disturbances at the driving frequencies did not increase in proportion to the excitation level, but rather the broadband center-line pressure and velocity levels were increased somewhat by the excitation. The broadband far-field sound pressure was also increased when driving the jet with a root-mean-square pressure of only 0.08% of the dynamic pressure of the center line $\left(\frac{1}{2}\rho_0 U_J^2\right)$. The results were typically interpreted to show that in the formation of ordered vorticity from the spatially growing shear layer vibration and entrainment of undisturbed fluid by the large eddies in the layer lead to increased turbulence that apparently increases the magnitude of the $\rho_0 u_i u_j$ acoustic sources. The wave motion of the shear layer itself cannot radiate sound unless the wave number spectrum of $\partial^2(\rho_0 u_i u_j)/\partial t^2$ has components in the wave number range of $|\boldsymbol{k}| \leq k_0$, i.e., unless this second time derivative is correlated over regions, which are comparable to acoustic wave lengths. The spatial growth and evolution of structure in the jet's shear layer make these components possible. It remains to be seen whether the conditions of vortex pairing [69,74,204,205,246,247,250] fully account for all the acoustic energy in subsonic jets by feeding the acoustic regions of the source wave number spectrum. That these motions may be plausibly considered as [206–208] radiators in supersonic jets is well-founded since their wave speeds U_c exceed the speed of sound, but this is a class of flow well beyond the scope of subsonic aeroacoustics and hydroacoustics. Other methods (under intensive development in the 1970s) pursued localizing the turbulent sources in the jet and led to the clarification of the mixing region used array-based correlation techniques in the acoustic near and far field with pressure–pressure and anemometer–sound pressure correlation methods, see, e.g., Refs. [209–219].

3.7.4 Emergence of Computer-aided Design and Suppression of Broadband Jet Noise

3.7.4.1 Enablers for Computational Methods in Parametric Studies

The large eddy simulations and direct numerical simulations of recent years have complemented and extended physical experiments in design studies. LES approaches have been used, e.g., Bogey et al. [220], Karabasov, et al. [182], Depru Mohan et al. [183,184], Goldstiein and Leib [260], Leib and Goldstein [256], and Khavaran et al. [257] in combination with analytical propagation models. Those propagation models account for source convection and other influences on the acoustic impedance of the medium in the jet, see Tam and Auriault [185] and Goldstein [254] for accounts of these propagation models. Collectively these simulations agree well with the measurement at Mach numbers of 0.5–1.4 in the vicinity of Reynolds number ($\sim 10^6$). Direct numerical simulation also has been used at lower Reynolds number (500–3600, say) research studies by, Colonius et al. [221], Jaing et al. [222], Gohill et al. [223], and Freund [235].

The compressible flow DNS simulation used by Freund [235] provided results for the flow structure and the radiated sound that compare well with the collective body of knowledge. Fig. 3.29 summarizes these findings. The calculated directivity of the overall far field sound pressure level of the sound agrees well with the prior results of Mollo-Chirstianson et al. [195], Lush [191] (also shown in Figs. 3.27 and 3.28), and Stromberg et al. [203]. The calculated instantaneous structure of the turbulence as well as the stress tensor sources align well with the photographed large-scale structure development of Beavers and Wilson [72] as shown in the center of the figure. The lower part of the figure illustrates both spatial distributions of instantaneous vorticity and acoustic source strength calculated by the simulation. The observed smoke pattern, taken from Fig. 3.10 keys the visualized development of large scale to the simulated sources.

Engineering application depends on the use of large eddy simulations of the turbulence, as referenced repeatedly previously. In the notation of this chapter the relevant equation to be evaluated can be represented as an autospectrum of sound pressure

$$\Phi_{\text{rad}}(\omega, r, \theta) = {\rho_0}^2 \iiint_{-\infty}^{\infty} dV_{\text{J}}(\mathbf{y}) \iiint_{-\infty}^{\infty} dV_{\text{J}}(\mathbf{r}) G_{ij}(\mathbf{x}, \mathbf{y}, \omega((1 - M_{\text{c}}(\mathbf{y})\cos\theta(\mathbf{y})))$$
$$\times G_{kl}^{*}(\mathbf{x}, \mathbf{y} + \mathbf{r}, \omega(1 - M_{\text{c}}(\mathbf{y} + \mathbf{r})\cos\theta(\mathbf{y} + \mathbf{r})) R_{ij,kl}(\mathbf{y}, \mathbf{r}, (\omega(1 - M_{\text{c}}\cos\theta)) \tag{3.107}$$

where $G_{ij}(\mathbf{x}, \mathbf{y}, \omega)$ represents the propagation Green function for the quadrupoles and the single Doppler factor $(1 - M_{\text{c}}\cos\theta)$ represents an average over the separation vector in the correlation function. For example, the free space Green function that is included in Eq. (3.75a) is a the simple one appropriate for cold subsonic jets ignores that refraction

$$G_{kl}(\mathbf{x}, \mathbf{y} + \mathbf{r}, \omega) = \frac{1}{4\pi c_0^2 |\mathbf{x} - \mathbf{y} - r|} \frac{\omega^2}{} \left(\frac{(x_k - y_k - r_k)(x_l - y_l - r_l)}{|\mathbf{x} - \mathbf{y} - r|^2} \right) \times \cdots$$
$$\cdots \times e^{i\omega|\mathbf{x}-\mathbf{y}-\mathbf{r}|\cos\theta(\mathbf{y}+\mathbf{r})/c_0} \tag{3.108}$$

Lew et al. [266] used this free space Green function together with LES results for the Reynolds stresses to examine the relative importance of the mean shear–turbulence and the turbulence–turbulence interaction contributions to the sound. Karabasov et al. [182], Depru et al. [183,184], Goldstein and Leib [260], Leib and Goldstein [256], and Khavaran et al. [257] have used the LES to populate the correlation model, Eq. (3.100) for use with their particular implicit propagation Green functions to effectively evaluate Eq. (3.107) providing the autospectra of radiated sound by jets. The more general approaches for propagation use systems of functions that replace the simple one-equation form of Eq. (3.108). The propagation functions used in the cited references require information about the mean velocity distributions in the jet that may be provided by provided by appropriate simulation.

In concept design studies that require evaluation of a series of design variants it is attractive to replace the full reliance on an LES solution with an algebraic form like Eq. (3.100) as discussed previously in that section. Then the spatial distribution of time-mean Reynolds stresses and mean velocities are provided by a RANS solution, all the length scales appearing in Eq. (3.100) are provided by a baseline LES calculation the statistical functions which can be scaled to the other cases of interest by using the turbulence quantities provided by the RANS solution. A $\kappa - \varepsilon$ RANS algorithm is an attractive choice in this regard because it directly provides the turbulent kinetic energy, κ , with dimensions $(L/T)^{2}$, and dissipation rate, ε, with dimensions (L^2/T^3). Other turbulence models may be used, however, see Ref. [182] as long as these scales can be determined from the parameters of the model. We define a length scale (see Pope [133]) that characterizes sizes of the largest eddies and which is related to the Kolmogoroff integral scale

$$L = \frac{\kappa^{3/2}}{\varepsilon}$$

The length scales appearing in an analytic simulation of the correlation matrix, Eq. (3.100) are assumed to be proportional to this L, i.e., for the three coordinate directions we have $\Lambda_i = c_i L$ in Eq. (3.100) and similarly the time scale is $\tau_m = c_m L/\kappa^{1/2}$. When a Reynolds number based on the microscale of turbulence, R_{λ_g}, exceeds 200, according to Pope [133], $L \approx 0.45\,\Lambda_f$, with Eq. (3.46) defining the integral scale, Λ. In terms related to the factors given in a computed result for dissipation rate, this works out to be $R_{\lambda_g} = {u_1}^2 \left(\frac{15}{\nu\varepsilon}\right)^{1/2} > 200$. The proportionality constants c_i, where $i = 1{:}3$, and c_m are not universally accepted, however, as the accompanying table shows. As examples, Depru Mohan et al. [183] provide several values that are repeated below Table. 3.3.

The values of the scale factors that were used for the calculation of sound shown in Fig. 3.31, and for covariance functions in Figs. (3.26) and (3.30) (which were also used by Karabasov et al. [182]) corresponded to $\tau_m = 0.00015$ second and $\Lambda_1 = 0.00085$ m at $x = 4\text{D}$. Note that in a jet, the length scales are determined by the spreading rate of the jet and as such increase linearly with distance from the efflux (as can be gleaned from inspection of Figs. 3.10 and 3.29). Thus the values of τ_m and Λ_i just given are good for one axial position only. For other axial positions, one requires the examination of fully populated spatial distribution of kinetic energy and dissipation rates in the jet to calculate the scales τ_m and Λ_i everywhere using the coefficients like those in the table. These values then give a distribution of scaled covariance functions for use in the integral Eq. (3.107). Finally the magnitudes of the Reynolds stresses that multiply the Gaussian function of Eq. (3.100),

TABLE 3.3 Comparison of the Proportionality Constants for Amplitude, Length, and Time Scales

Source	Mach Number	Length Scale	Time Scale	Amplitude Scale
		c_l	c_m	$\sqrt{C_{1111}}$
Tam and Auriault [185]	0.9	0.13	0.31	0.26
Morris and Farassat [187]	0.91	0.78	1	0.26
Karabasov et al. [182]	0.75	0.37	0.36	0.25
Karabasov et al. [183]	0.9	0.26, 0.08, 0.08	0.14	0.25

i.e., $\langle u_i u_j(y)\rangle\langle u_k u_l(y+r)\rangle$ deserve comment. It turns out that $\left|\langle u_i u_j(y)\rangle\langle u_k u_l(y+r)\rangle\right| = C_{ii,jj}(2\kappa)^2$ and the C_{1111} of all investigators in Table 3.3 nearly equal 0.25. Refs. [182–184] reported that $R_{11,11}$, $R_{22,22}$, and $R_{33,33}$ dominate the magnitudes of all other tensor entries. Also, in a round jet $\langle uw\rangle = \langle wu\rangle = \langle wv\rangle = \langle vw\rangle = 0$ due to axisymmetry (Pope [133]). Goldstein and Leib [260] find some variation of these coefficients with ij,kl combination so no universal rule appears to apply.

Fig. 3.30 shows comparisons with data of examples of $R_{11,11}(r_1,0,0,\tau)$ given by ensemble-averaged LES covariances used in (Eq. 3.100). The hybrid method that relied on these correlation functions was used to create the calculations of sound shown in Figs. 3.31 and 3.32. Comparison with the directly calculated sound that used the LES results directly in the Ffowcs Williams and Hawkings integral equation were within 2 dB over frequencies fD_J/U_J between 0.05 and 5 [183].

Goldstein and Leib [260] and Leib and Goldstein [256] applied different values of scales for some of the coefficients $C_{ij,kl}$ and achieved similar comparisons with data shown in Fig. 3.32. Their coefficients did not rely on a large eddy simulation, rather they were based on a set of measured data [260]. The example shown in Fig. 3.32 is the acoustic intensity in 1 Hz bandwidth at three observation angles and three Mach numbers at a range $r/D_J = 100$. The nozzle diameter is $D_J = 0.051$ m (2 in.). The data is that of Khararan et al. [257], that was taken as part of a larger test program as reported by Kharavan et al. [257], and Brown and Bridges [259] the calculation was by Leib et al. [257]. The agreement between measurement and computation is close at all Mach numbers and observation angles. Further the progression of narrowband spectral values (relative to 90 degrees) at $fD_J/U_J = 0.3$ (the spectral peak) for angles $\theta = 90$, 60, and 30 degrees were measured to be 0, 3, and 5.5 dB; calculated values of the directivity factor $10\log[1 - M_J\cos(\theta)]^{-4}$ for the *spectral density* (Eq. 3.104a) are 0, 4, and 8 dB, respectively. Further discussion on computational methods is in Section 3.7.4.

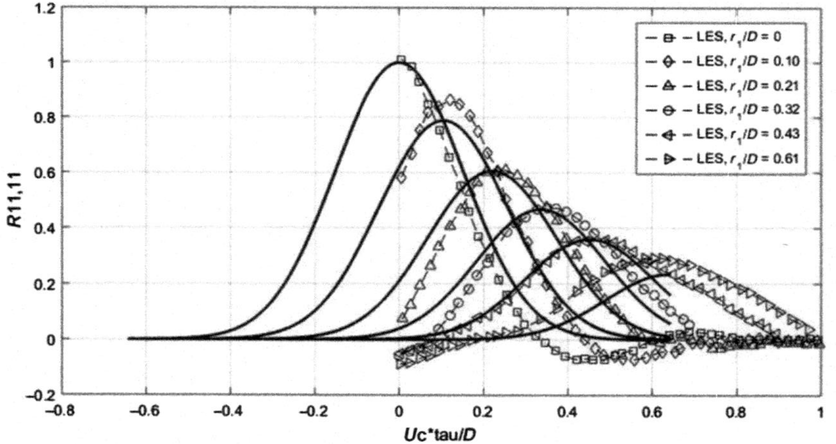

FIGURE 3.30 Comparison of the $R_{1111}(r_1,0,0,\tau)$ covariance function obtained with postprocessing LES velocities that obtained with Eq. (3.100). The points are from LES solutions directly, the lines (Eq. 3.100) with $\tau_m = 0.00015$ s and $\lambda_1 = 0.00085$ m at $x_1 = 4D$.

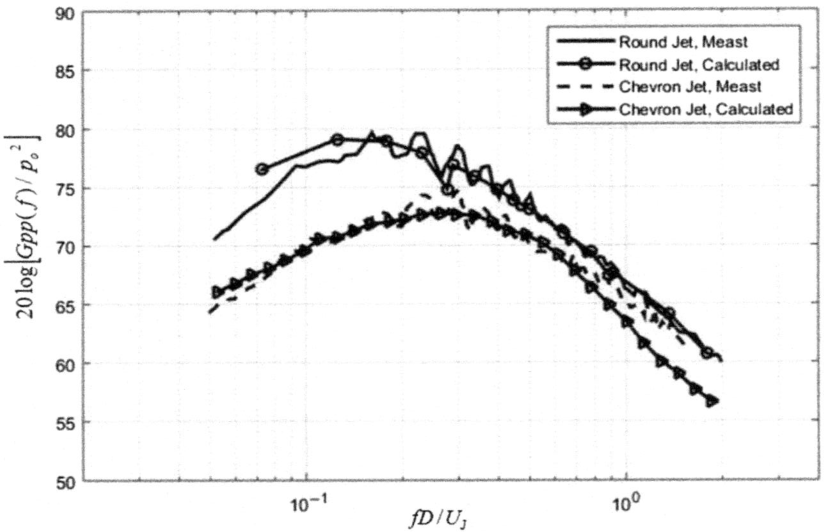

FIGURE 3.31 Comparison of calculated and measured spectra of sound at 30 degrees to the axis of the jet. Conditions are those of Fig. 3.25 and the calculations use the hybrid RANS-Green function approach that is populated with statistical scales provided by an LES simulation.

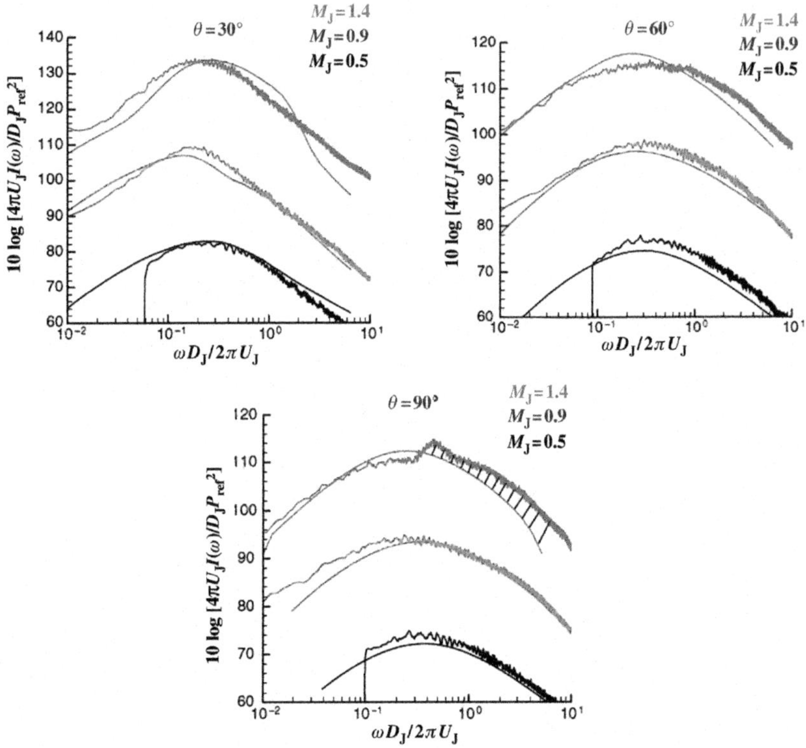

FIGURE 3.32 Comparisons of predictions based on Eq. (6.24) of Goldstein and Leib [260] and the present hybrid source model with the acoustic data of Khavaran et al. [257].

The calculations of radiated sound collectively by Karabasov et al. [182], Depru Mohan et al. [183,184], using the hybrid method in combination with analytical propagation model of Tam and Auriault [185] are shown in Figs. 3.31 and 3.33. These were made at 30 degrees and 90 degrees to the jet axis, respectively. The calculation of sound for the straight large eddy simulation using the same analytical propagation model agreed with measurement at both these angles to within a decibel for Strouhal numbers less than 3 [183]. Above that frequency the LES solution under predicted possibly due to unresolved sources.

The approach of using RANS solutions as a basis of scaling length, time, and turbulence intensity factors to provide scales such as integral scales in analytical spectrum and correlation functions has been used for turbulent boundary layer pressure fluctuations (Chapter 2 of Volume 2), trailing edge noise (Chapter 5 of Volume 2), and turbulence ingestion noise (Chapters 5 and 6 of Volume 2). The methods are given names such as "Hybrid RANS-LES" or "RANS-Statistical," but in essence they are all based on the use of steady flow RANS solvers, analytical functions for the turbulence correlation functions, and a Green function for the acoustic propagation or structural response.

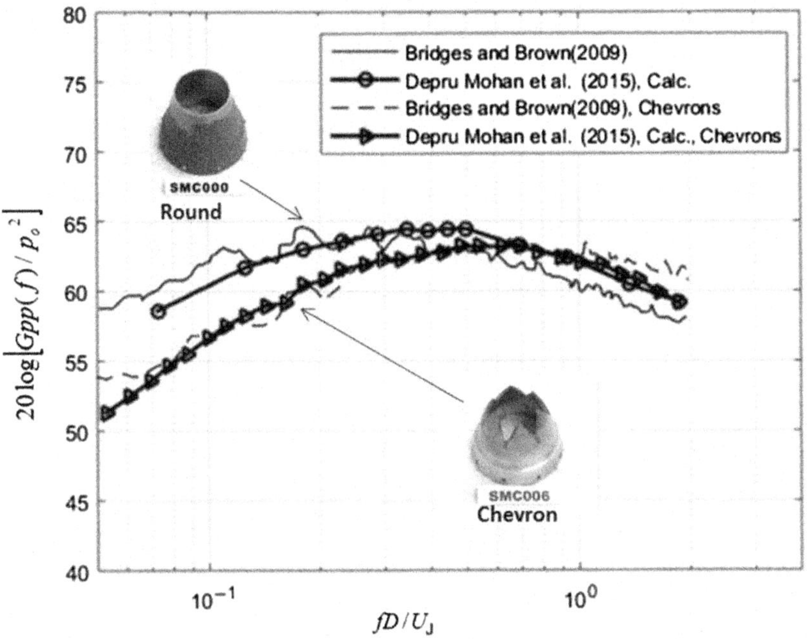

FIGURE 3.33 Spectra of sound in 1 Hz bands at 90 degrees from the jet axis from jets produced by two NASA nozzle concepts. The round nozzle produces the upper curve and chevron nozzle produces the lower curve. Two spectra are shown for each nozzle, a measurement and a result of the hybrid RANS-Green function approach. The measurement range was 40D, $p_0 = 20 \times 10^{-6}$ Pa.

3.7.4.2 Noise Suppression

This is a broad topic for which space allows only a few concepts. The similarity considerations presented in this section suggest that the simplest technique of noise reduction is to reduce jet diameter or efflux velocity, while maintaining equality in mean momentum. By Eq. (3.106) the overall broadband sound pressure level, averaged over solid angle, as illustrated in Fig. 3.27, is

$$L_s = 80 + 80 \log M_\mathrm{J} + 20 \log D$$

where $L_S = 10\ \log(\overline{p^2}/p_0^2)$ with $p_0 = 20\ \mu P_\mathrm{a}$ and D referred to 1 mm is the diameter of the jet. This equation shows that reduced speed and smaller diameters will decrease the overall sound pressure level. Increasing the efflux area for a fixed efflux momentum will reduce M_J and therefore overall sound by $60 \log M_\mathrm{J} \sim 60 \log(1/D)$. Replacing a single jet by multiple jets with the same total efflux area cannot result in sound power reductions unless the jets interact and break up the large energy producing eddies. This can be seen by observing that the overall sound pressure level depends on D^2 or, rather, on the total cross-sectional area of jet efflux. In the case of multiple jets, then, the total sound power can be reduced only when the constituent jets interact to alter their mixing regions. Middleton [224] has shown, e.g., that clustering of multiple jets can affect overall sound power reductions of from 3 to 5 dB with efflux solidities of order 0.5. Solidity here is defined as the enclosed jet area divided by the minimum rectilinear area circumscribing the group. For solidities less than 0.2 there are no benefits of clustering to achieve reduction in overall sound level. For a 3 × 3 array of identical constituent nozzles this translates to a ratio of center-to-center spacing (b) to nozzle diameter (D) of $b/D = 2.5$ or less for noise reduction of 0–5 dB and of $b/D = 1.4$ for reductions of 5 dB or more. Other measures for abatement techniques motivated by the breakup of turbulence scales are reviewed by Jordan et al. [225].

Replacing a single jet with an array of smaller jets has been shown by Maa et al. [126] to reduce the physiological impact of the noise. As indicated by the nondimensional spectra shown in Fig. 3.28, smaller jets will generate sound power levels with a center frequency changing with diameter as U_J/D. The magnitude of the sound power spectrum will, however, remain unchanged unless b/D is less than 2.5 as described previously. Thus the sound power spectrum just shifts to higher frequencies as D is reduced without much alteration in the integrated total acoustic power. The physiological impact of the noise, however, as indicated by A-weighted overall sound levels, may be markedly reduced by decreasing the diameter of the nozzles by a factor of 3 while maintaining constant efflux area.

Another reduction in sound level may be achieved by reducing the turbulence intensity in the mixing region of the jet. This may be accomplished by enhancing entrainment of ambient still fluid by, say, perforating the nozzle near the efflux. This mixing broadens the shear layer, reducing the mean velocity gradient and therefore the turbulence intensity. Other means of altering nozzle geometry for reduced noise have been reviewed by Elvhammar and Moss [226].

The use of chevron nozzles has been found advantageous in reducing the sound at low frequencies with a slight noise penalty at high frequencies. Figs. 3.31 and 3.33 show two examples of performance taken from a recent parametric study. These figures have attached to them multiple further discussion points. The first is that the application of our knowledge that large spatial scales of turbulence are more efficient radiators than smaller ones and that the use of chevrons is a noise-reduction technology with minimum other performance penalty, the second is that the concept itself is marked by a large number of design features which can affect turbulent structure in subtle ways that stretch the understanding of how turbulence structure affects noise. Thus computer-aided optimization facilitates concept tradeoffs. These optimizations, as we shall discuss, are now facilitated by hybrid computational fluid dynamics and acoustical methods.

Examining the performance of the chevron concept itself, Bridges and Brown [198] covered seven proposed alternatives which at the time could be best evaluated empirically. The concept shown in Figs. 3.31 and 3.33 was one of the better ones in terms of momentum and acoustic requirements, Saiyed et al. [227]. The efficacy of chevron nozzles lies in the ability of the chevrons, which protrude radially into the flow, to break up large eddies in the jet. As noted in the last subsection, these eddies promote energy production in the flow and large turbulent structures that generate sound. By breaking up the eddies, the acoustically relevant correlation volumes are reduced. Acoustic performance of the nozzle appears to be affected by the pitch, length, and protrusion angle of the chevron.

3.8 NOISE FROM UNSTEADY MASS INJECTION

This section will deal with certain general characteristics of noise from a jet efflux that is either turbulent or bubbly. The latter case has to do with unsteady two-phase motions (liquid and gas), although the complete subject will be examined in Chapters 6 and 7, Introduction to Bubble Dynamics and Cavitation; Hydrodynamically Induced Cavitation and Bubble Noise. Unsteadiness of exiting flow brings about a U_J^4 or U_J^6 speed dependence, depending on the mechanism, and this dependence contrasts sharply with the classical U_J^8 dependence that is known for the developing free jet. Furthermore, in the case of two-phase turbulent jets one would also expect a U_J^8 speed dependence, but amplified by a factor which is the fourth power of the ratio of the speed of sound in the acoustic medium to the speed of sound in the two-phase jet.

3.8.1 Sound From Efflux Inhomogeneities

When an inhomogeneous fluid is exhausted into a homogeneous medium, as illustrated in Fig. 3.33, sound can be created as a result of volume–velocity fluctuations induced at the orifice and turbulence in the ensuing two-phase jet. The topic of the unsteady orifice flow was given attention by Ffowcs Williams and Gordon [229] and Ffowcs Williams [230], who examined the noise from low-speed turbulent exhaust. Ffowcs Williams [230] deduced a U^4 speed dependence for this noise, but the result is now superseded by a later paper (Ffowcs Williams and Howe [231]), which treats the matter in more detail. Apparently, in the earlier paper, the compressibility of the fluid in the nozzle was overlooked. It should be noted that edge dipole sound (see Chapter 11: Noncavitating Lifting Sections) is generated by low-subsonic turbulent effluxes and gives rise to a U^6 dependence. Such deviations of jet noise from the U^8 dependence at low Mach numbers have been called "excess noise."

A U^4 speed dependence applies to the case of the unsteady pumping of an incompressible fluid from an enclosed plenum into a free space. In this case the pipe length from the pump to the exit must be shorter than an acoustic wavelength. There is then a time-varying rate of volume injection q so that the far-field sound pressure from the unbaffled nozzle is Eq. (2.24a), but see Eq. (3.17) for the baffled orifice.

$$p(r,t) = \frac{\rho_0[\dot{q}]_{t-r/c_0}}{4\pi r}.$$

Now, $|\dot{q}| = \omega_v u A_n$, where ω_v is the pulsation frequency of the pump, u is the amplitude of the velocity fluctuation at the nozzle, and A_n is the nozzle area. The pulsation frequency ω_v will be proportional to the shaft speed Ω; the velocity fluctuation will also increase in proportion to ΩD (where D is pump diameter) with the proportionality dependent on the pump type. Accordingly for a given pump the mean-square sound pressure should increase as

$$\overline{p^2} \sim \rho_0^2 \frac{\Omega^4}{r^2} A_n^2 D^2$$

which is the fourth power of shaft speed.

We shall now consider cases in which compressibility enters the solution. In the first case, consider the noise produced by a density inhomogeneity in the efflux. In this case the fluid issuing from the pipe is not always of the same density and compressibility; rather it consists of intermittent ejections of different fluids which often manifests itself as a surging. This behavior will be modeled as a slug of differing fluid density shown generally in Fig. 3.33 and in detail in Fig. 3.34, which passes through a nozzle into the

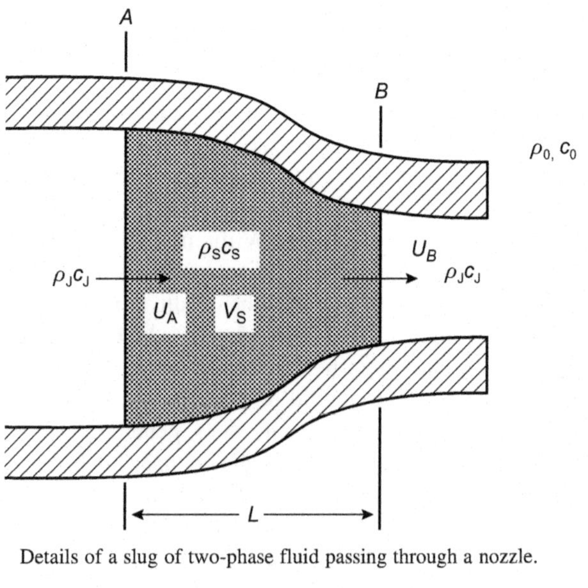

FIGURE 3.34 Details of a slug of two-phase fluid passing through a nozzle.

outer fluid. The length of the fluid slug, the length of the nozzle, and the diameter of the discharge nozzle will be considered small compared to an acoustic wavelength. This problem was treated generally and elegantly by Ffowcs Williams and Howe [231], but here we shall offer a few physical arguments (retrospectively) in order to illustrate the fundamentals.

The average flow through the nozzle is essentially one dimensional. The noise comes from two sources. The first occurs as each of the slug interfaces A and B are ejected; a pressure surge occurs in the nozzle that is equal to the change dynamic pressure (due to the different densities). This pressure surge induces a piston-like particle motion across the orifice. The second is due to a volume change in the slug as it passes through the pressure field of the contraction. The sound from the first source is derived from the one-dimensional momentum Eq. (2.2) integrated across each interface:

$$\rho\frac{\partial u_1}{\partial t}+\frac{\partial}{\partial x_1}\left(P+\rho\frac{u_1^2}{2}\right)=0$$

so the integral across interface A is

$$\int\rho\dot{u}_A dx_1=-\frac{\rho_s-\rho_J}{2}U_A^2$$

and across interface B it is

$$\int\rho\dot{u}_B dx_1=-\frac{\rho_J-\rho_s}{2}U_B^2.$$

The left-hand sides represent the potential jumps across the interface. The net jump across the slug represents a pressure fluctuation exerted on the nozzle orifice p_0, where

$$\int\rho\dot{u}_A\, dx_1-\int\rho\dot{u}_B\, dx_1=\frac{\rho_s-\rho_J}{\rho_s}(P_A-P_B)=p_0 \tag{3.109}$$

where we have used Bernoulli's equation along the slug:

$$\frac{U_B^2}{2}+\frac{p_B}{\rho_J}=\frac{U_A^2}{2}+\frac{p_A}{\rho_J}.$$

The acoustic particle velocity in the exit plane u_p is

$$u_p=p_0/\rho_J c_J \tag{3.110}$$

as long as the discharge pipe is longer than an acoustic wavelength. The far-field sound pressure at $\mathbf{x}$ (from an unbaffled nozzle) is

$$p_{\text{rad}_d}(\boldsymbol{x},t)=\frac{\rho_0 A_0}{4\pi r}\left[\dot{u}_p\right]\simeq-\frac{A_0}{4\pi r c_J}\frac{\rho_0}{\rho_J}\left(\frac{\rho_s-\rho_J}{\rho_J}\right)\left[\frac{\partial}{\partial t}(P_A-P_B)\right] \tag{3.111}$$

where the brackets denote evaluation at the retarded time $t-r/c_0$. This sound pressure could be considered dipole like, since it depends on the speed of sound of the jet liquid, or it could be considered monopole like because it is omnidirectional.

Before discussing Eq. (3.111) further, we shall consider the second source of radiation due to the volume pulsation of the slug passing through the contraction. This pulsation occurs because of the relative compressibility of the slug. By the equation preceding Eq. (3.18), the time rate of change of the pulsating volume $\dot{V}_s$ is related to the pressure change in the slug as

$$\dot{p} \simeq -\rho_s c_s \frac{\dot{V}_s}{V_s}$$

where V_s is the equilibrium volume of the slug, and the time change is referred to the moving slug.

The pressure fluctuation in the slug is

$$P_s = P_A - P_B$$

therefore

$$\dot{V}_s = -\frac{V_s}{\rho_s c_s^2 \partial t}(P_A - P_B)$$

as the slug moves through the nozzle the volumetric acceleration of the issued slug is

$$\ddot{V}_s = -\frac{V_s}{\rho_s c_s^2} \frac{\partial^2}{\partial t^2}(P_A - P_B)$$

or approximately since $\partial/\partial t \sim U/L$, where U is the average forward velocity of the slug

$$\ddot{V}_s \simeq -\frac{V_s}{\rho_s c_s^2} \frac{U}{L}\left(\frac{\partial}{\partial t}(P_A - P_B)\right)$$

where L is the length of the nozzle. The amplitude of the acoustic radiation from the unbaffled pipe is then given by Eq. (2.27), in a form that may be compared with its counterpart Eq. (3.111),

$$p_{\mathrm{rad_m}}(\boldsymbol{x}, t) \simeq \sim \frac{V_s}{4\pi r L} \frac{\rho_0}{\rho_s} \left(\frac{U}{c_s}\right)^2 \frac{1}{U} \left[\frac{\partial}{\partial t}(P_A - P_B)\right] \tag{3.112}$$

where the brackets denote evaluation at the retarded time. This radiation is monopole like owing to its direct dependence on the pulsation of the volume V_s.

The two components $p_{\mathrm{rad_m}}$ and $p_{\mathrm{rad_d}}$, however, depend on speeds of sound in such a way that the *dipole* component overwhelms the monopole component whenever

$$\frac{U}{c_J}\left(\frac{c_s}{U}\right)^2 >> 1$$

or when

$$\left(\frac{c_s}{c_J}\right)^2 < < \frac{U}{c_J} \tag{3.113}$$

this is generally the case unless speed of sound in the slug is very low and the discharge velocity very large. One such instance could occur if the slug is a bubble slurry in an otherwise homogeneous jet liquid. In this case (see Chapter 6: Introduction to Bubble Dynamics and Cavitation) the speed of sound in the slug could be quite small relative to c.

Note that the two contributions to the radiation have different dependences on speed. Introducing the pressure coefficient, let

$$\frac{\partial P}{\partial t} = -U\frac{\partial P}{\partial x} = -\frac{U}{L}\left(\frac{1}{2}\rho_s U^2\right)\frac{\partial c_p}{\partial(x/L)}$$

we find, since $\partial c_p/\partial(x/L)$ is nondimensional and constant for a given nozzle geometry,

$$p_{\text{rad}_d} \propto \frac{-A_0}{8\pi r L}\frac{\rho_0}{\rho_J}\left(\frac{\rho_s-\rho_J}{\rho_J}\right)\left(\frac{U}{c_J}\right)^3 \rho_J c_J^2 \tag{3.114}$$

while

$$p_{\text{rad}_m} \propto \frac{-V_s}{8\pi r L^2}\frac{\rho_0}{\rho_s}\left(\frac{U}{c_J}\right)^4 \rho_J c_J \tag{3.115}$$

Therefore the dipole sound power increases as U^6, while the monopole sound power increases as U^8, reflecting the inequality (3.113) that the monopole source is essentially a high-speed source.

Referring, again, to radiation caused by a bubble slurry and considering now the dipole component only, we write the density of the slug in terms of the jet liquid density ρ_J, the gas density, ρ_g, and the volume fraction of gas, β. The density is

$$\rho_s = \beta\rho_g + (1-\beta)\rho_J$$

Therefore in Eq. (3.114)

$$\frac{\rho_s-\rho_J}{\rho_J} \approx \frac{\beta(\rho_g-\rho_J)}{\rho_J} \simeq \beta$$

if $\rho_g \ll \rho_J$. The radiated sound pressure can now be written in an alternative fashion to Eq. (3.114):

$$p_{\text{rad}_d} \approx \frac{-A_0}{8\pi r L}\beta\frac{\rho_0}{\rho_J}\left(\frac{U}{c_J}\right)^3 \rho_J c_J^2 \tag{3.116}$$

The sound pressure will increase linearly with the volume fraction of gas and as efflux velocity cubed. Lengthening the nozzle (increasing L) reduces the noise. For a given volumetric flow rate (A_0 U) the sound pressure increases as the orifice area is decreased, i.e., as A_0^{-2}.

A similar result holds in the case of a turbulent efflux of a compressible fluid. To see this we find the average velocity-induced pressure fluctuation across the orifice, $\rho_J u_0 U$, which drives the fluid in the orifice. This causes an acoustic particle velocity in the pipe orifice given by Eq. (3.110),

$$u_p \approx \frac{\rho_J u_0 U}{\rho_J c_J} \approx u_0 \frac{U}{c_J}$$

so that the radiated sound pressure in the free space is

$$p_{rad} \approx \frac{\rho_0 A_0}{4\pi r}[\dot{u}_p]$$

$$\approx \frac{\rho_0 A_0}{4\pi r}\frac{U}{c_J}\frac{U}{\Lambda_1}u_0$$

where we have replaced $\partial/\partial t$ by U/Λ_1, where Λ_1 is the integral scale of the turbulence in the axial direction. Since $u_0 \sim U$ we can write

$$p_{rad} \approx \frac{A_0}{4\pi r\,\Lambda_1}\rho_0 c_J^2 \left(\frac{U}{c_J}\right)^3 \tag{3.117}$$

which shows the sound power increasing as U^6.

Other problems of this sort have been examined by Ffowcs Williams and Howe [231] who treated the effluxes consisting of occasional slugs of fluid with density varying from the main fluid; Whitfield and Howe [232] who considered noises from volumetric pulsations of bubbles passing from the nozzle into an ambient pressure field; and Plett and Summerfield [233] who considered the density and velocity inhomogeneities analogous to the above.

3.8.2 Inhomogeneities in the Free-Turbulent Field

Another problem related to two-phase jets has been treated by Crighton and Ffowcs Williams [234] and Crighton [154]. Basically, the analysis considers a bubbly turbulent region of gas concentration β which is surrounded by the ambient fluid with properties ρ_0, c_0. In the absence of any net mass injection or forces on the fluid, the appropriate form of the stress tensor becomes [236]

$$T_{ij} = (1-\beta)\rho_0 u_i u_j + (p - c_0^2\rho)\delta_{ij} + \tau_{ij} \tag{3.118}$$

instead of Eq. (2.47), where p is the dilatational pressure fluctuations in the turbulent two-phase medium. Eq. (2.57) is written

$$\begin{aligned} p_a(\mathbf{x},t) &= \frac{1}{4\pi}\frac{\partial^2}{\partial x_i \partial x_j}\iiint \frac{[T_{ij}]}{r}dV(\mathbf{y}) \\ &= \frac{1}{4\pi c_0^2}\frac{x_i x_j}{r^2}\frac{\partial^2}{\partial t^2}\iiint \frac{[T_{ij}]}{r}dV(\mathbf{y}) \end{aligned} \tag{3.119}$$

which becomes [155]

$$p_a(\boldsymbol{x},\ t) \simeq \frac{1}{4\pi c_0^2}\ \frac{1}{r}\ \frac{\partial^2}{\partial t^2} \iiint [p - \rho c_0^2] dV(\mathbf{y}) \tag{3.120}$$

Adiabatic pressure fluctuations in the two-phase region are related to density fluctuations in that region $\delta p = c_m^2\ \delta\rho$ so that the time-varying part of the integrand in Eq. (3.120) becomes $\delta p - c_0^2(\delta p/c_m^2)$. When the external fluid is water and the bubbly region contains gas $c_m < c_0$, then

$$p(\boldsymbol{x},t) \simeq \frac{-1}{4\pi c_0^2 r} \iiint \frac{c_0^2}{c_m^2} \frac{\partial^2 p}{\partial t^2} dV(\mathbf{y}) \tag{3.121}$$

Now if disturbances in the region are characterized by velocity u, then the pressure is given by $p \sim \rho u^2 \sim \rho_0 u^2$ and $\partial/\partial t \sim u/l$, where l is the macroscale of the disturbance. Therefore with the elemental volume of the disturbances $\delta V l^3$, Eq. (3.120) gives

$$p(\mathbf{x},t) \sim \left(\frac{\rho_0 l}{c_0^2 r}\right)\left(\frac{c_0^2}{c_m^2}\right) u^4 \tag{3.122}$$

which dominates the usual quadrupole radiation

$$p(\boldsymbol{x},\ t) \sim \left(\frac{\rho_0 l}{c_0^2 r}\right) u^4$$

by the factor $c_0^2/c_m^2 > 1$. Also the field has an omnidirectional radiation rather than of the typical quadrupole nature; compare Eqs. (3.119) and (3.120).

At frequencies that are well below the resonance frequencies of the bubbles

$$c_m = \frac{R_b \omega_b}{\sqrt{3\beta}} \quad \text{or} \quad \left(\frac{c_0}{c_m}\right)^2 \simeq \left[1 + \beta\frac{\beta_0 c_0^2}{\rho_g c_g^2}\right] \simeq \beta\frac{\rho_0}{\rho_g}\frac{c_0^2}{c_g^2}$$

where ω_b is the resonance frequency of the bubbles of radius R_b (see Section 6.1.2). The enhancement of radiated noise intensity brought about by the factor $(c_0/c_m)^4$, as pointed out by Crighton [154] will be as much as 50 dB over that of single-phase jets for volumetric concentrations β of order 1%. This and the previous analysis of unsteady mass flux suggest important hydroacoustical sources for which experimental confirmation is totally lacking.

REFERENCES

[1] Rayleigh JWS. The theory of sound, vol. 2. New York: Dover; 1945.

[2] Lamb H. Hydrodynamics. New York: Dover; 1945.

[3] Squire HB. On the stability of the three dimensional disturbances of viscous flow between parallel walls. Proc R Soc 1933;A142:621–8.

[4] Esch RH. The instability of a shear layer between two parallel streams. J Fluid Mech 1957;3:289–303.
[5] Michalke A. On the inviscid instability of the hyperbolic tangent velocity profile. J Fluid Mech 1964;19:543–56.
[6] Michalke A. Vortex formation in a free boundary layer according to stability theory. J Fluid Mech 1965;22:371–83.
[7] Michalke A. On spatially growing disturbances in an inviscid shear layer. J Fluid Mech 1965;23:521–44.
[8] Browand FK. An experimental investigation of the instability of an incompressible, separated shear layer. J Fluid Mech 1966;26:281–307.
[9] Sato H. Experimental investigation on the transition of laminar separated layer. J Phys Soc Jpn 1956;11:702–9.
[10] Schade H. Contribution to the non-linear stability theory of inviscid shear layers. Phys Fluids 1964;1:623–8.
[11] Tatsumi T, Gotoh K. The stability of free boundary layers between two uniform streams. J Fluid Mech 1960;2:433–41.
[12] Sato H, Sakao F. An experimental investigation of the instability of a two-dimensional jet at low Reynolds numbers. J Fluid Mech 1964;20:337–52.
[13] Sato H. The stability and transition of a two-dimensional jet. J Fluid Mech 1960;7:53–80.
[14] Sato H, Kuriki K. The mechanism of transition in the wake of a thin flat plate placed parallel to a uniform flow. J Fluid Mech 1961;11:321–52.
[15] Lin CC. The theory of hydrodynamic stability. London and New York: Cambridge University Press; 1966.
[16] Betchov R, Criminale WO. Stability of parallel shear flows. Waltham, Ma: Academic Press; 1967.
[17] Schlichting H. Boundary-layer theory. New York: McGraw-Hill; 1960.
[18] Gaster M. A note on the relation between temporally-increasing and spatially-increasing disturbances in hydrodynamic stability. J Fluid Mech 1964;14:222–4.
[19] Brown GB. On vortex motion in gaseous jets and the origin of their sensitivity to sound. Proc Phys Soc London 1935;47:703–32.
[20] Becker HA, Massaro TA. Vortex evolution in a round jets. J Fluid Mech 1968;31: 435–48.
[21] Rosenhead L. The formation of vortices from a surface of discontinuity. Proc R Soc London Ser A 1931;134:170–93.
[22] Powell A. Flow noise: a perspective on some aspects of flow noise and of jet noise in particular. Noise Control Eng 1977;8:69–80 108–119.
[23] Abernathy FH, Kronauer RE. The formation of vortex streets. J Fluid Mech 1962;13: 1–20.
[24] Boldman DR, Brinich PF, Goldstein ME. Vortex shedding from a blunt trailing edge with equal and unequal external mean velocities. J Fluid Mech 1976;75:721–35.
[25] Blake WK, Powell A. The development of contemporary views of flow-tone generation. International symposium on recent advances in aerodynamics and acoustics. Berlin: Springer-Verlag; 1985.
[26] Rockwell D, Naudascher E. Self-sustained oscillations of impinging free shear layers. Ann Rev Fluid Mech 1979;11:67–94.
[27] Ahuja K.K., Mendoza J. Effects of cavity dimensions, boundary layer, and temperature on cavity noise with emphasis on benchmark data to validate computational aeroacoustic codes. NASA CR 4653. April 1995.

[28] Rowley CW, Williams DR. Dynamics and control of high-Reynolds-number flow over open cavities. Ann Rev Fluid Mech 2006;38:251–76.
[29] Ma R, Slaboch PE, Morris SC. Fluid mechanics of the flow-excited Helmholtz resonator. J Fluid Mech 2009;623:1–26.
[30] Naudascher E. From flow-instability to flow-induced excitation. J Hydraul Div Am Soc Civ Eng 1967;93:15–40.
[31] Schmit RF, Grove JE, Semmelmayer F, Haverkamp M. Nonlinear feedback mechanisms inside a rectangular cavity. AIAA J 2014;52:2127–42.
[32] DeMetz FC, Farabee TM. Laminar and turbulent shear flow induced cavity resonances. AIAA Paper 1977;77–1293.
[33] Howe MS. Mechanism of sound generation by low Mach number flow over a cavity. J Sound Vib 2004;273:103–23.
[34] Block PJW, Noise response of cavities of varying dimensions at subsonic speeds. NASA TN D-8351. 1976.
[35] Tam KW, Block PJW. On the tones and pressure oscillations induced by flow over rectangular cavities. J Fluid Mech 2004;89:373–99.
[36] Rossiter JE. Wind tunnel experiments in the flow over rectangular cavities at subsonic and transonic speeds. Rep. & Memo. No. 3438. Aeronaut. Res. Council, London 1966.
[37] Howe MS. Edge, cavity and aperture tones at very low Mach numbers. J Fluid Mech 1997;330:61–84.
[38] Ronneberger D. The dynamics of shearing flow over a cavity—a visual study related to the acoustic impedance of small orifices. J Sound Vib 1980;71:565–81.
[39] Lucas M, Rockwell D. Self-excited jet: upstream modulation and multiple frequencies. J Fluid Mech 1984;147:333–52.
[40] Heller HH, Holmes DG, Covert EE. Flow-induced pressure oscillations in shallow cavities. J Sound Vib 1971;18:545–53.
[41] Oshkaia P, Rockwella D, Pollack M. Shallow cavity flow tones: transformation from large- to small-scale modes. J Sound Vib 2005;280:777–813.
[42] Rockwell D, Knisely C. The organized nature of flow impingement upon a corner. J Fluid Mech 1979;93:413–32.
[43] Powell A. On edge-tones and associated phenomena. Acustica 1953;3:233–43.
[44] Powell A. On the edge tone. J Acoust Soc Am 1961;33:395–409.
[45] Dunham WH. Flow-induced cavity resonance in viscous compressible and incompressible fluids. *Symp. Nav. Hydromech., 4th*. Washington, DC: 1962. p. 1057–1081 (1962).
[46] Howe MS. On the theory of unsteady shearing flow over a slot. Phil Trans R Soc London Ser A 1981;303:151–80.
[47] King JL, Doyle P, Ogle JB. Instability in slotted wall tunnels. J Fluid Mech 1958;4: 283–305.
[48] Martin WW, Naudascher E, Padmanabhan M. Fluid dynamic excitation involving flow instability. J Hydraul Div Am Soc Civ Eng 1975;101:681–98. No. HY6.
[49] Rockwell D. Prediction of oscillation frequencies for unstable flow past cavities. J Fluids Eng 1977;99:294–300.
[50] Bilanin AJ, Covert EE. Estimation of possible excitation frequencies for shallow rectangular cavities. AIAA J 1973;11:347–51.
[51] Harrington MC. Excitation of cavity resonance by air flow. J Acoust Soc Am 1957;29:187.
[52] Heller HH, Bliss DB. Aerodynamically induced pressure oscillations in cavities–physical mechanisms and suppression concepts. Tech. Rep. AFFDL-TR-74-133. AF Flight Dyn. Lab. Wright Patterson Air Force Base, Dayton, OH, 1975.

[53] East LF. Aerodynamically induced resonance in rectangular cavities. J Sound Vib 1966;3:277–87.

[54] Morel T. Experimental study of a jet-driven Helmholtz oscillator. Am Soc Mech Eng Pap 1978; 78-WA/FE-16.

[55] Elder, S.A. A root locus solution of the cavity resonator problem. Rep. No. E7801. Michelson Lab., U.S. Nav. Acad., Annapolis, MD, 1978.

[56] Elder SA. Self-excited depth-mode resonance for a wall-mounted cavity in turbulent flow. J Acoust Soc Am 1978;64:877–90.

[57] Elder SA, Farabee TM, DeMetz FC. Mechanisms of flow-excited cavity tones at low Mach number. J Acoust Soc Am 1982;72:532–49.

[58] Hardin JC, Martin JP. Broadband noise generation by a vortex model of cavity flow. AIAA J 1977;15:632–7.

[59] Howe MS. On the Helmholtz resonator. J Sound Vib 1976;45:427–40.

[60] Kinsler LE, Frey AR. Fundamentals of acoustics. 2nd ed. New York: Wiley; 1962.

[61] Covert EE. An approximate calculation of the onset velocity of cavity oscillations. AIAA J 1970;8:2189–94.

[62] Ingard U, Dean LW III. Excitation of acoustic resonators by flow. *Symp. Nav. Hydrodyn., 4th*. Washington, DC: 1962.

[63] Miles JB, Watson GH. Pressure waves for flow-induced acoustic resonance in cavities. AIAA J 1971;9:1402–4.

[64] Ukeiley LS, Ponton MK, Seiner JM, Jansen B. Suppression of pressure loads in cavity flows. AIAA J 2004;42:70–9.

[65] Raman G, Envia E, Bencic TJ. Jet-cavity interaction tones. AIAA J 2002;40.

[66] Kegerisea MA, Cabellb RH, Cattafesta III LN. Real-time feedback control of flow-induced cavity tones—Part 1: fixed-gain control. J Sound Vib 2007;307:906–23.

[67] Kegerisea MA, Cabellb RH, Cattafesta III LN. Real-time feedback control of flow-induced cavity tones—Part 2: adaptive control. J Sound Vib 2007;307:924–40.

[68] Becker HA, Massaro TA. Vortex evolution in a round jet. J Fluid Mech 1968;31:435–48.

[69] Browand FK, Laufer J. The role of large scale structures in the initial development of circular jets. Turbul Liq 1975;4:333–4.

[70] Schade H, Michalke A. Zur Entstehung von Wirbeln in einer freien Grenzschicht. Z Flugwiss 1962;10:147–54.

[71] Johansen FC. Flow through pipe orifices at low Reynolds numbers. Proc R Soc London Ser A 1929;216:231–45.

[72] Beavers GS, Wilson TA. Vortex growth in jets. J Fluid Mech 1970;44:97–112.

[73] Chanaud RC, Powell A. Experiments concerning the sound-sensitive jet. J Acoust Soc Am 1962;34:907–15.

[74] Crow SC, Champagne FH. Orderly structure in jet turbulence. J Fluid Mech 1971;48:547–91.

[75] Lau JC, Fisher MJ, Fuchs HV. The intrinsic structure of turbulent jets. J Sound Vib 1972;22:379–406.

[76] Lau JC, Fisher MJ. The vortex-street structure of turbulent jets, Part 1. J Fluid Mech 1975;67:299–337.

[77] Alenus E, Abom M, Fuchs L. Large eddy simulations of acoustic flow at an orifice plate. J Sound Vib 2015;345:162–77.

[78] Powell A. A schlieren study of small scale air jets and some noise measurements in two-inch diameter air jets. ARC 14726 FM. 1694.

[79] Kurzweg H. Neue Unterschungen über die Entstehung der turbulenten Rohrströmung. Ann Phys (Leipzig) 1933;18:193–216.

[80] Anderson ABC. Dependence of Pfeifenton (pipe tone) frequency on pipe length, orifice diameter, and gas discharge pressure. J Acoust Soc Am 1952;24:675–81.

[81] Anderson ABC. Dependence of the primary Pfeifenton (pipe tone) frequency on pipe-orifice geometry. J Acoust Soc Am 1953;25:541–5.

[82] Anderson ABC. A circular-orifice number describing dependency of primary Pfeifenton frequency on differential pressure, gas density, and orifice geometry. J Acoust Soc Am 1953;25:626–31.

[83] Anderson ABC. A jet-tone orifice number for orifices of small thickness-diameter ratio. J Acoust Soc Am 1954;26:21–5.

[84] Anderson ABC. Metastable jet-tone states of jets from sharp-edged, circular, pipe-like orifices. J Acoust Soc Am 1955;27:13–21.

[85] Anderson ABC. Structure and velocity of the periodic vortex-ring flow pattern of a primary Pfeifenton (pipe tone) jet. J Acoust Soc Am 1955;27:1048–953.

[86] Anderson ABC. Vortex ring structure-transition in a jet emitting discrete acoustic frequencies. J Acoust Soc Am 1956;28:914–21.

[87] Tyndall J. The science of sound. 1875; reprinted by Citadel Press, New York, 1964.

[88] Chanaud PC, Powell A. Some experiments concerning the hole and ring tone. J Acoust Soc Am 1965;37:902–11.

[89] Wilson TA, Beavers GS, DeCoster MA, Holger DK, Regenfuss MD. Experiments on the fluid mechanics of whistling. J Acoust Soc Am 1971;50:366–72.

[90] Nyborg WL, Woodbridge CL, Schitting HK. Characteristics of jet-edge-resonator whistles. J Acoust Soc Am 1953;25:138–46.

[91] St. Helaire A, Wilson TA, Beavers GS. Aerodynamic excitation of the harmonium reed. J Fluid Mech 1971;49:803–16.

[92] Elder SA. On the mechanism of sound production in organ pipes. J Acoust Soc Am 1973;54:1554–64.

[93] Smith RA, Mercer DMA. Possible causes of wood wind tone color. J Sound Vib 1974;32:347–58.

[94] Curle N. The mechanics of edge-tones. Proc R Soc London Ser A 1953;216:412–24.

[95] Brown GB. Vortex motion causing edge tones. Proc Phys Soc London 1937;49:493–507 520.

[96] Shachenmann A, Rockwell D. Self-sustained oscillations of turbulent pipe flow terminated by an axisymmetric cavity. J Sound Vib 1980;73:61–72.

[97] Rebuffet P, Guedel A. Model based study of various configuration of jet crossing a cavity. Rech Aerospatiale 1982;1:11–22.

[98] DeMetz FC, Mates MF, Langley RS, Wilson JL. Noise radiation from subsonic airflow through simple and multiholed orifice plates. DTNSRDC Rep. SAD 237E-1942. 1979.

[99] Stull FD, Curran ET, Velkoff HR. Investigation of two-dimensional cavity diffusers. AIAA Paper No. 73, 685. 1973.

[100] Isaacson LK, Marshall AG. Acoustic oscillations in internal cavity flows: nonlinear resonant interactions. AIAA J 1982;20:152–4.

[101] Rockwell DO. Transverse oscillations of a jet in a jet-splitter system. ASME J Basic Eng, 94. 1972. p. 675–81.

[102] Stegun GR, Karamcheti K. Multiple tone operation of edgetone. J Sound Vib 1970;12:281–4.

[103] Powell A, Unfried HH. An experimental study of low speed edgetones. Department of Engineering, University of California; 1964.

[104] Brown GB. On vortex motion in gaseous jets and the origin of their sensitivity to sound. Proc Phys Soc (London) 1935;41:703–32.

[105] Powell A. Mechanism of aerodynamic sound production. AGARD Rep 1963;466.

[106] Ho CM, Nosseir NS. Dynamics of an impinging jet. Part 1. The feedback phenomenon. J Fluid Mech 1981;105:119–42.

[107] Nosseir NS, Ho CM. Dynamics of an impinging jet. Part 2. The noise generation. J Fluid Mech 1982;116:379–91.

[108] Quick AW. Zum Schall-und Stromungsfeld eines axialsymmetrischen Freistrahls beim Auftreffen auf eine Wand. Zeit Flugwissenschaftern 1971;19:30–44.

[109] Marsh AH. Noise measurements around a subsonic air jet impinging on a plane rigid surface. J Acoust Soc Am 1961;33:1065–6.

[110] Powell A. On the noise emanating from a two-dimensional jet above the critical pressure. Aero Quart 1953;4:103–22.

[111] Tam CKW. Supersonic jet noise. Ann Rev Fluid Mech 1995;27:17–43.

[112] Powell A. The noise of choked jets. J Acoust Soc Am 1953;25:385–9.

[113] Powell A. On the mechanism of choked jet noise. Phys Soc London 1953;B66:1039–57.

[114] Powell A. Nature of the feedback mechanism on some fluid flows producing sound. *Int. Congr. on Acoustics, 4th.* Copenhagen: 1962.

[115] Powell A. Vortex action in edgetones. J Acoust Soc Am 1962;34:163–6.

[116] Krothapalli A, Baganoff D, Hsia Y. On the mechanism of screech tone generation in underexpanded rectangular jets. AIAA-83-0727. 1983.

[117] Morris PJ, McLaughlin DK, Kuo C-W. Noise reduction in supersonic jets by nozzle fluidic inserts. J Sound Vib 2013;332:3992–4003.

[118] Hammitt AG. The oscillation and noise of an overpressure sonic jet. J Aero Sci 1961;28:673–80.

[119] Davies MG, Oldfield DES. Tones from a choked axisymmetric jet. I. Cell structure, eddy velocity and source locations. Acustica 1962;12:257–67.

[120] Davies MG, Oldfield DES. Tones from a choked axisymmetric jet. II. The self-excited loop and mode of oscillation. Acustica 1962;12:267–77.

[121] Powell A. The sound-producing oscillations of round underexpanded jets. J Acoust Soc Am 1988;83:515–33.

[122] Powell A. Experiments concerning tones produced by an axisymmetric choked jet impinging on flat plates. J Sound Vib 1993;168:307–26.

[123] Tam CKW, Ahuja KK. Theoretical model of discrete tone generation by impinging jets. J Fluid Mech 1990;214:67–87.

[124] Krothapalli A, Rajkuperan E, Alvi F, Lourenco L. Flow field and noise characteristics of a supersonic impinging jet. J Fluid Mech 1990;214:67–87.

[125] Dosanjh DS, Yu JC, Abdelhamed AN. Reduction of noise from supersonic jet flows. AIAA J 1971;9:2346–53.

[126] Maa D-Y, Li P-z, Lai G-H, Wang H-Y. Microjet noise and micropore diffuser-muffler. Sci Sin (Engl Ed), 20. 1977. p. 569–82.

[127] Taylor GI. Statistical theory of turbulence, Parts I–IV. Proc R Soc London Ser A 1935;151:421–78.

[128] Taylor GI. The spectrum of turbulence. Proc R Soc London Ser A 1938;164:476–90.

[129] Batchelor GK. Homogeneous turbulence. London and New York: Cambridge University Press; 1960.

[130] Hinze JO. Turbulence. 2nd ed. New York: McGraw-Hill; 1975.

[131] Townsend AA. Structure of turbulent shear flow. London and New York: Cambridge University Press; 1976.
[132] Lumley JL. Stochastic tools in turbulence. New York: Academic Press; 1970.
[133] Pope SB. Turbulent flows. London, New York: Cambridge University Press; 2000.
[134] Lee YW. Statistical theory of communication. New York: Wiley; 1964.
[135] Bendat JS, Piersol AG. Random data analysis and measurement procedures. 4th ed. New York: Wiley; 2010.
[136] Kinsman B. Wind waves: their generation and propagation on the ocean surface. Englewood Cliffs, NJ: Prentice-Hall; 1965.
[137] Lin YK. Probabilistic theory of structural dynamics. New York: McGraw-Hill; 1967.
[138] Crandall S. Random vibration, vol. 1. Cambridge, MA: MIT Press; 1958.
[139] Crandall S. Random Vibration, vol. 2. Cambridge, MA: MIT Press; 1963.
[140] M. Dieste, G. Gabard. Broadband interaction noise simulations using synthetic turbulence. 16th international congress on sound and vibration, Krakow, 5–9 July 2009.
[141] Ribner HS, Tucker M, Spectrum of turbulence in a contracting stream. NACA Tech. Note 2606, 1952.
[142] Batchelor GK, Proudman I. The effect of rapid distortion of a fluid in turbulent motion. Quart J Mech Appl Math 1954;7(1):83–103.
[143] Panton RL, Linebarger JH. Wall pressure spectra calculations for equilibrium boundary layers, *Journal of Fluid Mechanics*, **65**, 1974;261–87.
[144] Hunt JCR, Graham JMR. Free stream turbulence near plane boundaries. J Fluid Mech 1974;84:209–35.
[145] Graham JMR. The effect of a two-dimensional cascade of thin streamwise plates on homogeneous turbulence. J Fluid Mech 1994;356:125–47.
[146] Lynch DA, Blake WK, Mueller TJ. Turbulence correlation length-scale relationships for the prediction of aeroacoustic response. AIAA J 2005;43(6):1187–97.
[147] Lynch DA, Blake WK, Mueller TJ. Turbulent flow downstream of a propeller, Part 2: ingested propeller-modified turbulence. AIAA J 2005;43(6):1211–20.
[148] Blake WK. Mechanisms of flow induced sound and vibration, vol. 2 complex flow-structure interactions, Ed 1. Academic Press; 1986.
[149] Ross MH. Radiated sound generated by airfoils in a single stream shear layer. MS thesis Department of Aerospace and Mechanical Engineering. University of Notre Dame; 2009.
[150] Bell JH, Mehta RD. Development of a two stream mixing layer from tripped and untripped boundary layers. AIAA J 1990;28. 2034–2042.
[151] Lighthill MJ. On sound generated aerodynamically, I General theory. Proc R Soc London Ser A 1952;211:564–87.
[152] Lighthill MJ. On sound generated aerodynamically, II Turbulence as a source of sound. Proc R Soc London Ser A 1954;222:1–32.
[153] Goldstein ME. Aeroacoustics. New York: McGraw-Hill; 1976.
[154] Crighton DG. Basic principles of aerodynamic noise generation. Prog Aerosp Sci 1975;16:31–96.
[155] Howe MS. Acoustics of fluid–structure interactions. Cambridge University Press; 1998.
[156] Lilley GM. On the noise from air jets. Aeronautical Research Council, Report ARC-30276 1958.
[157] Lilley GM. The radiated noise from isotropic turbulence with applications to the theory of jet noise. J Sound Vib 1996;190:463–76.
[158] Kraichman RH. Pressure fluctuations in turbulent flow over a flat plate. J Acoust Soc Am 1956;28:378–90.

[159] Ribner HS. Theory of two-point correlations of jet noise. NASA Tech Note NASA TN D-8330 1976.
[160] Proudman I. The generation of noise by isotropic turbulence. Proc R Soc London Ser A 1952;214:119–32.
[161] Jones IS. Fluctuating turbulent stresses in the noise-producing region of a jet. J Fluid Mech 1969;36:529–43.
[162] Ribner HS. The generation of sound by turbulent jets. Adv Appl Mech 1964;8:103–82.
[163] Ribner HS. Quadrupole correlations governing the pattern of jet noise. J Fluid Mech 1969;38:1–24.
[164] Ribner HS. Perspectives on jet noise. AIAA J 1981;19:1513–26.
[165] Goldstein ME, Rosenbaum B. The effect of anisotropic turbulence on aerodynamic noise. J Acoust Soc Am 1973;54:630–45.
[166] Michalke A, Fuchs HV. On turbulence and noise of an axisymmetric shear flow. J Fluid Mech 1975;70:179–205.
[167] Goldstein ME, Howes WL, New aspects of subsonic aerodynamic noise theory. NACA TN D-7158, 1973.
[168] Lighthill MJ. The Bakerian Lecture, "Sound Generated Aerodynamically". Proc R Soc London Ser A 1961;267:147–71.
[169] Morfey CL. Amplification of aerodynamic noise by convected flow inhomogeneities. J Sound Vib 1973;31:391–7.
[170] Ffowcs Williams JE. Sound production at the edge of a steady flow. J Fluid Mech 1974;66:791–816.
[171] Davies POAL, Fisher MJ, Barratt MJ. The characteristics in the mixing region of a round jet. J Fluid Mech 1963;15:337–67.
[172] Ko NWM, Davies POAL. Some covariance measurements in a subsonic jet. J Sound Vib 1975;41:347–58.
[173] Forstall Jr. W, Shapiro AH. Momentum and mass transfer is co-axial gas jets. J Appl Mech 1950;72:399–408.
[174] Forstall Jr. W, Shapiro AH. Momentum and mass transfer in co-axial gas jets, Discussion. J Appl Mech 1951;73:219–20.
[175] Fisher MJ, Davies POAL. Correlation measurements in a non-frozen pattern of turbulence. J Fluid Mech 1965;18:97–116.
[176] Harper-Bourne M. Jet noise turbulence measurements. 9th AIAA/CEAS Aeroacoustics Conference, AIAA Paper 2003-3214, 2003.
[177] Pokora C, McGuirk JJ. Spatio-temporal turbulence correlations using high speed PIV in an axisymmetric jet. 14th AIAA/CEAS aeroacoustics conference, AIAA Paper 2008-3028, May 2008.
[178] Laurence JC. Intensity, scale, and spectra of turbulence in mixing region of free subsonic jet No. 1292 Natl Advis Comm Aeronaut Rep. 1956
[179] Kolpin MA. Flow in the mixing region of a jet. Aeroelastic Struct. Res. Lab. Rep. ASRL TR 92-3. Cambridge, MA: Dep. Aeronaut. Astronaut., MIT; 1962.
[180] Kerherve F, Fitzpatrick J, Jordan P. The frequency dependence of jet turbulence for noise source modeling. J Sound Vib 2006;296:209–25.
[181] Amiet RK. Refraction of sound by a shear layer. J Sound Vib 1978;58:467–82.
[182] Karabasov SA, Afsar MZ, Hynes TP, Dowling AP, McMullan WA, Pokora CD, et al. Jet noise: acoustic analogy informed by large eddy simulation. AIAA J 2010;48:1312–25.

[183] Depru Mohan NK, Dowling AP, Karabasov SA, Xia H, Graham O, Hynes TP, Tucker PG. Acoustic sources and far-field noise of chevron and round jets. AIAA J 2015;53:2421–36.

[184] Depru Mohan NK, Karabasov SA, Xia H, Graham O, Dowling AP, Hynes TP, et al. Reduced-order jet noise modelling for chevrons. 18th AIAA/CEAS aeroacoustics conference, AIAA Paper 2012-2083, May 2012.

[185] Tam CKW, Auriault L. Mean flow refraction effects on sound radiated from localized sources in a jet. J Fluid Mech 1998;370:149–74.

[186] Tam CKW, Aurialt L. Jet mixing noise from fine scalar turbulence. AIAA J 1999;37:145–53.

[187] Morris P, Farassat F. Acoustic analogy and alternate theories for jet noise prediction. AIAA J 2002;40:671–80.

[188] Mani R. The influence of jet flow on jet noise, Part I—The noise of unheated jets. J Fluid Mech 1976;73:753–78.

[189] Mani R. The influence of jet flow on jet noise, Part 2—The noise of heated jets. J Fluid Mech 1976;73:779–93.

[190] Karchmer AM, Dorsch GR, Friedman R. Acoustic tests of a 15.2-centimeter-diameter potential flow convergent nozzle. NASA Tech Memo 1974; NASA TM X-2980.

[191] Lush PA. Measurements of subsonic jet noise and comparison with theory. J Fluid Mech 1971;46:477–500.

[192] Olsen WA, Gutierrez OA, Dorsch RG. The effect of nozzle inlet shape; lip thickness, and exit shape and size on subsonic jet noise. NASA Tech Memo 1973; NASA TM X-68182.

[193] Olsen WA, Friedman R. Jet noise from co-axial nozzles over a wide range of geometric and flow parameters. NASA Tech Memo 1974; NASA TM X-71503 (1974); also *AIAA Pap.* 74–43.

[194] Gerrard JH. An investigation of noise produced by a subsonic air jet. J Aerosp Sci 1956;23:855–66.

[195] Mollo-Christensen E, Kolpin MA, Martuccelli JR. Experiments on jet flows and jet noise far-field spectra and directivity patterns. J Fluid Mech 1964;18:285–301.

[196] Mollo-Christensen E. Jet noise and shear flow instability seen from an experimenter's viewpoint. J Appl Mech 1967;34:1–7.

[197] Fitzpatrick H, Lee R. Measurements of noise radiated by subsonic air jets. DTBM Rep. 835, David Taylor Model Basin (David Taylor Naval Ship Res. and Dev. Ctr.). Washington, DC: 1952.

[198] Bridges J, Brown CA. Parametric testing of chevrons on single flow hot jets. NASA/TM-2004-213107, Sept. 2004.

[199] Bradshaw P, Ferris DH, Johnson RF. Turbulence in the noise-producing region of a circular jet. J Fluid Mech 1964;19:591–624.

[200] Crighton DG, Gaster M. Stability of slowly diverging jet flow. J Fluid Mech 1976;77:397–413.

[201] Moore CJ. The role of shear-layer instability waves in jet exhaust noise. J Fluid Mech 1977;80:321–67.

[202] Bechert D, Pfizenmaier E. On the amplification of broadband jet noise by a pure tone excitation. J Sound Vib 1975;43:581–7.

[203] Stromberg JL, McLaughlin DK, Troutt TR. Flow field and acoustic properties of a low Mach number jet at a low Reynolds number. J Sound Vib 1980;72:159–76.

[204] Winant CD, Browand FK. Vortex pairing—the mechanism of turbulent mixing-layer growth at moderate Reynolds number. J Fluid Mech 1974;63:237–55.

[205] Acton E. The modeling of large eddies in a two-dimensional shear layer. J Fluid Mech 1976;76:561–92.

[206] Merkine L, Liu JTC. On the development of noise producing large-scale wave like eddies in a plane turbulent jet. J Fluid Mech 1975;70:353–68.

[207] McLaughlin DK, Morrison GL, Troutt TR. Experiments in the instability waves in a supersonic jet and their acoustic radiation. J Fluid Mech 1975;69:73–95.

[208] McLaughlin DK, Morrison GL, Troutt TR. Reynolds number dependence in supersonic jet noise. AIAA Pap 1976;76–491.

[209] Parthasarathy SP. Evaluation of jet noise source by cross-correlation of far field microphone signals. AIAA J 1974;12:583–90.

[210] Billingsley J, Kinns R. The acoustic telescope. J Sound Vib 1976;48:485–510.

[211] Fisher MJ, Harper-Bourne M, Glegg SAL. Jet engine noise source localization: the polar correlation technique. J Sound Vib 1977;51:23–54.

[212] Maestrello L. On the relationship between acoustic energy density flux near the jet axis and far-field acoustic intensity. NASA Tech Note 1973; NASA TN D-7269.

[213] Maestrello L, Pao SP. New evidence of the mechanisms of noise generation and radiation of a subsonic jet. J Acoust Soc Am 1975;57:959–60.

[214] Maestrello L. Two point correlations of sound pressure in the far field of a jet experiment. NASA Tech Memo 1976; NASA TM X-72835.

[215] Maestrello L, Fung Y-T. Quasiperiodic structure of a turbulent jet. J Sound Vib 1979;64:107–22.

[216] Schaffer M. Direct measurements of the correlation between axial in-jet velocity fluctuations and far field noise near the axis of a cold jet. J Sound Vib 1979;64:73–83.

[217] Moon LF, Zelanzy SW. Experimental and analytical study of jet noise modeling. AIAA J 1975;13:387–93.

[218] Grosche FR, Jones JH, Wilhold GA. Measurements of the distribution of sound source intensities in turbulent jets. AIAA Pap 1973;73–989.

[219] Glegg SAL. The accuracy of source distributions measured by using polar correlation. J Sound Vib 1982;80:31–40.

[220] Bogey C, Bailly C. Computation of a high Reynolds number jet and its radiated noise using large eddy simulation based on explicit filtering. Comput Fluids 2006;35:1344–58.

[221] Colonius T, Lele SK, Moin P. Sound generation in a mixing layer. J Fluid Mech 1997;330:375–409.

[222] Jaing X, Avital EJ, Luo KH. Direct computation and aeroacoustic modeling of a subsonic axisymmetric jet. J Sound Vib 2004;270:525–38.

[223] Gohil TB, Saha AK, Muralidhar K. Numerical study of instability mechanisms in a circular jet at low Reynolds numbers. Comput Fluids 2012;64:1–18.

[224] Middleton D. A note on the acoustic output from round and interfering jets. J Sound Vib 1971;18:417–21.

[225] Jordan P, Colonius T. Wave packets and turbulent jet noise. Ann Rev Fluid Mech 2013;45:173–95.

[226] Elvhammer H, Moss H. Silenced compressed air blowing. Proc Inter-noise 1977;77:B220–8.

[227] Saiyed NH, Mikkelsen KL, Bridges JE. Acoustics and thrust of separate-flow exhaust nozzles with mixing devices for high bypass-ratio engines. NASA/TM-2000-209948, June 2000.

[228] Bechara W, Lafon P, Bailly C, Candel SM. Application of a k-e turbulence model to the prediction of noise for simple and coaxial free jets. J Acoust Soc Am 1995;97: 3518–31.

[229] Ffowcs Williams JE, Gordon CG. Noise of highly turbulent jets at low exhaust speeds. AIAA J 1965;3:791–2.

[230] Ffowcs Williams JE. Jet noise at very low and very high speed. *Proc. AFOSR-UTIAS Symp. Aerodyn. Noise*. Toronto: 1968.

[231] Ffowcs Williams JE, Howe MS. The generation of sound by density inhomogeneities in low Mach number nozzle flows. J Fluid Mech 1975;70:605–22.

[232] Whitfield OJ, Howe MS. The generation of sound by two-phase nozzle flows and its relevance to excess noise of jet engines. J Fluid Mech 1976;75:553–76.

[233] Plett EG, Summerfield M. Jet engine exhaust noise due to rough combustion and nonsteady aerodynamic sources. J Acoust Soc Am 1974;56:516–22.

[234] Crighton DG, Ffowcs Williams JE. Sound generation by turbulent two-phase flow. J Fluid Mech 1969;36:585–603.

[235] Freund JB. Noise sources in a low-Reynolds-number turbulent jet at Mach 0.9. J Fluid Mech 2001;438:277–305.

[236] G.M. Lilley, On the noise from jets, noise mechanisms, AGARD-CP-13, pp 13.1–13.12, 1974

[237] Ffowcs Williams JE, Kempton AJ. The noise from large-scale structure of a jet. J Fluid Mech 1978;84:673–94.

[238] Goldstein ME. An exact form of Lilley's equation with a velocity quadrupole/temperature dipole source term. J Fluid Mech 2001;443:231–6.

[239] Ffowcs Williams JE. The noise from turbulence convected at high speed. Philos Trans R Soc London Ser A 1963;255:469–503.

[240] Tam CKW, Morris PJ. The radiation of sound by the instability waves of a compressible plane turbulent shear layer. J Fluid Mech 1980;vol. 98(part 2):349–81.

[241] Ffowcs Williams, J.E., Some thoughts on the effect of aircraft motion and eddy convection on the noise from air jets. Univ. Southampton Aero. Astr. Rep. no. 155.

[242] Gudmundsson K, Colonius T. Instability wave models for the near field fluctuations of turbulent jets. J Fluid Mech 2011;vol. 689:97–128.

[243] Pletcher RH, Tannehill JC, Anderson DA. Computational fluid dynamics and heat transfer. 3ed Boca Raton, FL: CRC Press; 2013.

[244] G.M. Lilley, On the noise from jets, noise mechanisms, AGARD-CP-13, pp 13.1–13.12, 1974

[245] Tam CKW, Morris ANDPJ. The radiation of sound by the instability waves of a compressible plane turbulent shear layer. J Fluid Mech 1980;98(part 2):349–81.

[246] Cohen J, Wygnanski IW. The evolution of instabilities in the axisymmetric jet. Part 1. The linear growth of disturbances near the nozzle. J Fluid Mech 1987;176:191–219.

[247] Cohen J, Wygnanski I. The evolution of instabilities in the axisymmetric jet. Part 2. The flow resulting from the interaction between two waves. J Fluid Mech 1987;176:221–35.

[248] Han G, Tumin A, Wygnanski I. Laminar-turbulent transition in Poiseuille pipe flow subjected to periodic perturbation emanating from the wall Part 2. Late stage of transition, J Fluid Mech 2000;419:1–27.

[249] Eliahou S, Tumin A, Wygnanski I. Laminar-turbulent transition in Poiseuille pipe flow subjected to periodic perturbation emanating from the wall. J Fluid Mech 1998;361:333–49.

[250] Oberleithner K, Paschereit CO, Wygnanski I. On the impact of swirl on the growth of coherent structures. J Fluid Mech 2014;741:156–99.

[251] H. Hubbard, Ed., Aeroacoustics of flight vehicles: theory and practice, NASA Reference Publication 1258, Technical report 90-3052, 1991.

[252] Camussi R, editor. Noise in turbulent shear flows: fundamentals and applications. Springer; 2013.

[253] Glegg S. Location of jet noise sources using an acoustic mirror [MS thesis]. Southampton, UK: University of Southampton; 1975.

[254] Goldstein ME. A generalized acoustic analogy. J Fluid Mech 2003;488:315–33.

[255] Ffowcs Williams, J.E., Some thoughts on the effect of aircraft motion and eddy convection on the noise from air jets. Univ. Southampton Aero. Astr. Rep. no. 155, 1960

[256] Leib SJ, Goldstein ME. Hybrid source model for predicting high-speed jet noise. AIAA J 2011;49:1324–35.

[257] Khavaran, A. Bridges, J. Georgiadis, N. Prediction of turbulence-generated noise in unheated jets, Part 1: JeNo technical manual (Version 1.0), NASA TM -2005-213827, 2005.

[258] Khavaran, A., Bridges, J. & Freund, J.B. 2002 A parametric study of fine-scale turbulence mixing noise. NASA/TM 2002-211696, 2002.

[259] Brown, C. and Bridges, J. Small hot jet acoustic rig validation, NASA/TM—2006-214234, April 2006.

[260] Goldstein ME, Leib SJ. The aeroacoustics of slowly diverging supersonic jets. J Fluid Mech 2008;600:291–337.

[261] Goldstein ME. Hybrid reynolds-averaged Navier–Stokes/large eddy simulation approach for predicting jet noise. AIAA J 2006;44:3136–42.

[262] Fisher MJ, Lush PA, Harper bourne M. jet noise. J Sound Vib 1973;28(3):563–85.

[263] Tanna HK, Dean PD, Fishier MJ. The influence of temperature on shock-free supersonic jet noise. J Sound Vib 1975;39(4):429–60.

[264] Viswanatha K. Scaling laws and a method for identifying components of jet noise. AIAA J 2006;44:2274–85.

[265] Viswanatha K. Improved method for prediction of noise from single jets. AIAA J 2007;45:151–61.

[266] P.-T. Lew, G.A. Blaisdelly, and A.S. Lyrintzisz, Investigation of noise sources in turbulent hot jets using large eddy simulation data. Paper AIAA 2006-16, 45th AIAA aerospace sciences meeting and exhibit, January 8-11, 2007, Reno, NV, USA.

Chapter 4

Dipole Sound From Cylinders

The sound radiated by flow across a circular cylinder, known also as the Aeolian tone, is one of the most fundamental of flow-acoustic phenomena. In spite of the apparent simplicity of some of their aspects, the aeroacoustic processes taking place in the Aeolian tone are one-dimensional examples of those which occur in almost all cases of flow-induced sound and vibration. Notably these are the generation of flow disturbances, the Reynolds number dependence of the related body forces, the spatial correlations of the forces, and the elements of dipole sound generation from nonvibrating bodies. This chapter will therefore examine the Aeolian tone in detail and then it will extend the specific results for the purpose of describing and scaling aerodynamic sound from acoustically similar air flow machinery. We will deal with the sounds emitted by cylinders held rigidly in the flow or allowed to vibrate slightly in relatively small displacements (with respect to radius of the cylinder).

The broad class of large-amplitude fluid–structure interactions (such as "galloping") are reserved for other references, e.g., Blevins [1]. Owing to the importance of vortex shedding fluid mechanics to both research and engineering, the number of publications on this subject is vast making a detailed review here out of the question. Since the focus of this chapter is on sound production review here this will be limited to a body of literature that provides insight to the fluid dynamic factors that control sound with as many references as possible.

4.1 INTRODUCTION: HISTORY AND GENERAL DESCRIPTION OF VORTEX FLOW, LIFT FLUCTUATION, AND SOUND

Possibly one of the most investigated topics of subsonic flow-induced noise generation is the flow over cylinders. Historically Strouhal [2] and shortly later Rayleigh [3] examined the frequencies of tones that emanated when air flowed past stretched wires. Strouhal noted that the pitch of the whirring sounds, which were generated by air passage transversely to the cylinder axis, was proportional to the velocity U_∞ divided by cylinder diameter d. In his experiments a wire was stretched in a frame, and the frame was rotated about an axis perpendicular to its length. Rayleigh's experiments [3,4] were conducted with a wire stretched across a chimney. He found that the frequency of the tone f_s would diminish as the kinematic viscosity of the air ν

Mechanics of Flow-Induced Sound and Vibration, Volume 1.
DOI: http://dx.doi.org/10.1016/B978-0-12-809273-6.00004-X

decreased by heating. This result and that of Strouhal [2] lead to the postulation that a dimensionless frequency depended on Reynolds number dU_∞/ν as

$$f_s\, d/U_\infty = F(dU_\infty/\nu) \tag{4.1}$$

where $F(dU_\infty/\nu)$ is an increasing function of its argument. Rayleigh postulated that the generation of the tone was in some way related to a vortex sheet instability and that the cylinder need not vibrate for the tone to be generated. Rayleigh observed that the passage of air caused vibration of the cylinder in a direction perpendicular to the wind direction.

A formal relationship between the frequency of sound radiation and the process of vortex shedding was not made until 1908, when Benard [5] associated the pitch of the note with the formation of the vortex street. He observed photographically the formation of concentrated regions of vorticity in the wake of a circular cylinder. The vortex spacing was seen to be regular. In 1912, von Karman and Rubach [6] (see also Lamb [7]) determined the conditions for the stable existence of parallel rows of vortices of alternate sign. They found that the distance between rows and the distance between vortices in a row must be in the ratio 0.281 in order for the vortex street to undergo stable translation. This simple analytical treatment of von Karman is, even today, fundamental to much of the theoretical analysis of vortex-induced forces. Contemporary computational fluid dynamics model approaches can now describe the generation, translation, and final decay of the vortex streets behind bluff bodies fully. The analytical model of von Karman permits the calculation of the translation velocity of the vortex street as well as steady drag induced on the shedding cylinder by the vortices [6–8].

In the years following von Karman's classical analysis, investigations of vortex shedding were generally experimental and devoted largely to establishing the frequencies of vortex generation and drag coefficients over various ranges of Reynolds number.

Investigations that quantified the sound levels generated by the Aeolian tone in terms of flow variables began with that of Stowell and Deming [9] in 1936. They developed empirical correlations between the sound radiated and the length, diameter, and speeds of the rods. They also determined the directivity of the sound as being radiated normal to the direction of flow. Later investigations [10,11] further determined empirically the dependence of the radiated sound on the size of the cylinder and the speed of flow over the cylinder.

Systematic analytical treatments of the noise problem awaited the attention by Phillips [12] and (apparently nearly simultaneously) Etkin et al. [13]. The analysis given by Phillips demonstrated that the sound intensity depended on the axial length scale of the force fluctuations that are generated by the vortex shedding. This was the first formal treatment of the stochastic nature of the shedding process. Phillips's result showed clearly how the physical variables, which control the intensity of the Aeolian tone, could be derived within the framework of the theory of aerodynamic noise.

In this chapter we shall examine the noise from flow over the surfaces of acoustically compact, rigid bodies from vortex–surface interactions. The flow-induced noise from circular cylinders that generate periodic wakes will be examined in depth. Formulations will be derived that estimate oscillatory lift coefficients on the wake-forming bodies in terms of the strength and geometries of the shed vortex streets. Measured parameters, including shedding frequency, lift coefficients, and axial correlation length scales, will be reviewed. Influences on these parameters of upstream turbulence and cylinder vibration will be considered from an experimental basis. The acoustic problem will be formulated as a straightforward application of the result of Section 2.4.2, Illustration I of Curle's Equation: Radiation From a Concentrated Hydrodynamic Force and extended to emphasize the importance of the flow variables in controlling the intensity of sound. Measured noise levels will then be examined in connection with the theoretical formulations for both translating and rotating cylindrical bodies. Other forms of vortex–body interaction due to eddies incident on leading edges and to turbulent flow off the trailing edges of airfoils will be examined in Chapter 11, Noncavitating Lifting Sections. Fluid–body interactions leading to vortex-induced lift enhancement will also be examined in Chapter 11, Noncavitating Lifting Sections although some aerodynamic aspects of cylinder motion will be touched on in Section 4.3.4 of this chapter.

4.2 MECHANICS OF VORTEX FORMATION BEHIND CIRCULAR CYLINDERS

4.2.1 A General Description of the Wake Structure and Vortex Generation

In the case of viscous flow over circular cylinders, there exists a sequence of flow domains that are defined in terms of R_d, where $R_d = U_\infty d/\nu$ is the Reynolds number based on the diameter of the cylinder d and the inflow velocity U_∞. It is the understanding of the physics of these domains that leads one to fluid dynamic means for controlling the flow-induced sound from cylinders. Fig. 4.1 shows a series of photographs that were taken by Homann [14]. These pictures of oil films show representative vortex streets whose patterns are sketched in Fig. 4.2. For $R_d = 32$ the flow is stable; the oil film is a single streak downstream of a pair of bound vortices. We note that although the flow is laminar, Stokes flow (ideal viscous flow) ceases to exist at R_d above unity. Even when the cylinder is vibrated, as if by plucking, as Kovasznay [15] points out, the flow pattern remains stable. At and above Reynolds numbers of 40 Kovasznay reports that sinusoidal disturbances begin to propagate and grow downstream of the cylinder. At $R_d = 65$ these disturbances change in character, forming vortices at some distance behind. Increases in R_d cause the vortices to be formed closer to the cylinder as these free shear

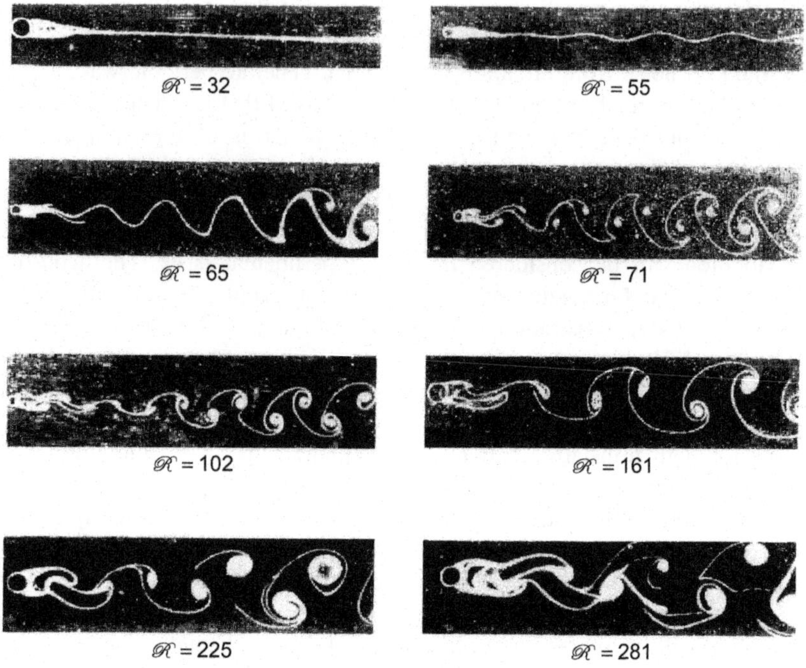

FIGURE 4.1 Homann's [14] photographs of flow around a circular cylinder.

layers between the cylinder boundary layers and the nearest vortex become turbulent. The street of regularly spaced vortices exists with laminar cores over the range of Reynolds numbers from 65 to approximately 400. At higher Reynolds numbers the regular vortex street persists; however, Roshko [16] showed that the cores of the vortices become turbulent. The process of laminar to turbulent flow transition in the shear layers leaving the cylinder surface and the vortex cores was examined in detail by Bloor [17], who has shown that some turbulent motion can actually be observed at Reynolds numbers above 200. The range of Reynolds numbers above which vortices with turbulent cores are shed periodically extends to approximately 2×10^5. At the upper extreme of this range, irregularity eventually culminates in the complete disintegration of the vortex street. For Reynolds numbers greater than 5×10^5 or 10^6 the wake is no longer fully periodic. The separation points on the cylinder occur further downstream, as shown in Fig. 4.2, because the boundary layers on the surface of the cylinder are actually turbulent. Turbulent boundary layers maintain greater attachment over more adverse pressure gradient conditions than do laminar boundary layers. In spite of the general irregularity of the shed wake, however, there are weakly periodic disturbances that persist at slightly higher dimensionless frequencies than those observed at lower Reynolds numbers. Also, due to the occurrence of attached boundary layer flow over a greater portion of the periphery of the cylinder, the drag coefficient is reduced.

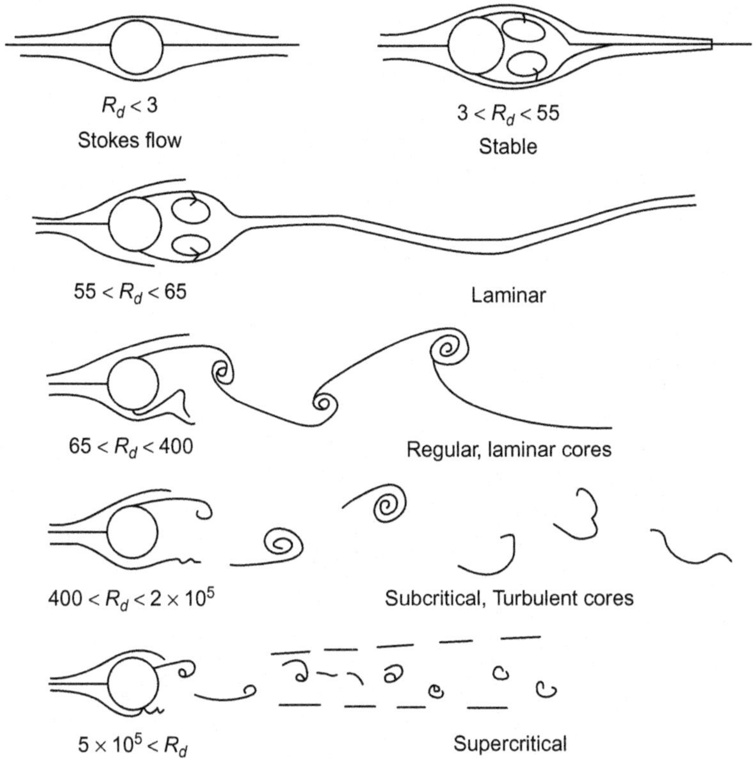

FIGURE 4.2 Diagrams of flow patterns around circular cylinders in various Reynolds number regimes.

4.2.2 Analysis of Vortex Production

The fluid dynamics of the vortex shedding is complex and without a complete and rigorous mathematical description. A detailed review of work done on vortex formation behind cylinders has been provided by Williamson [135]. An example of an early experimental and theoretical examination of vortex generation at low values of R_d is that of Sato and Kuriki [18], who have calculated the characteristic wave speeds and frequencies of the disturbances generated. Wake disturbances develop in the same manner described in Chapter 3, Shear Layer Instabilities, Flow Tones, and Jet Noise for the generation of disturbances in shear layers. The study of Sato and Kuriki supports this statement. Their experimental work was conducted on a shear layer (photographed using a visualization technique) that resembled the one shown in Fig. 4.2 for $55 < R_d < 65$. Actually, their wake was generated by laminar flow downstream of a very thin flat plate. Analyses of the type performed by Sato and Kuriki (see also Refs. [19,20]) consists of a solution of the eigenfunctions of the homogeneous Orr–Sommerfeld equation (Eq. (3.5)) for a

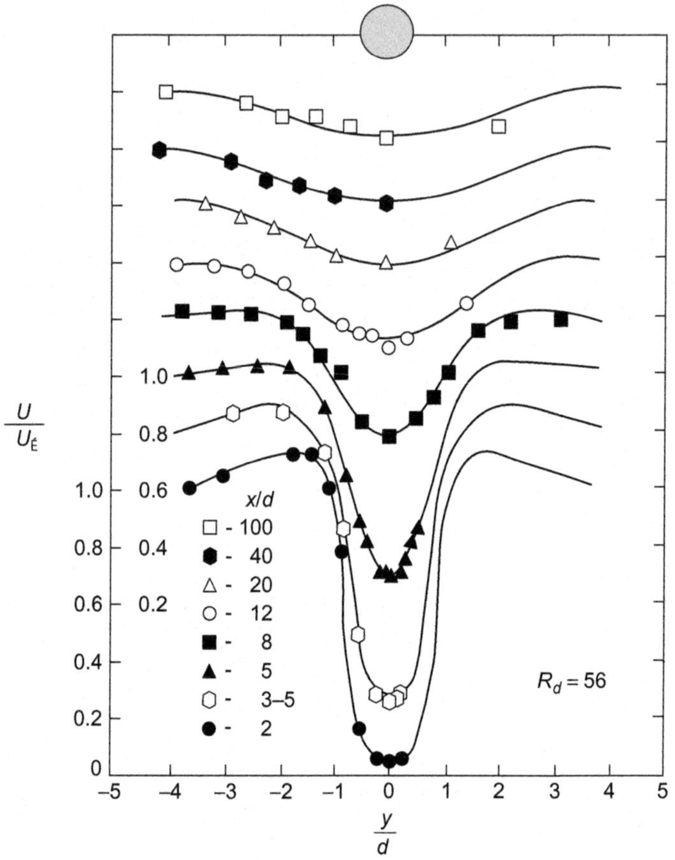

FIGURE 4.3 Kovasznay's [15] measurements of mean velocities in the wake of a cylinder $R_d = 56$.

prescribed velocity profile. This analysis shows that such wakes are naturally unstable for Reynolds numbers greater than 10.

Velocity profiles of the flow at various distances behind cylinders, measured by Kovasznay [15], are shown in Figs. 4.3 and 4.4 for $R_d = 56$. The root-mean-square fluctuating velocities show local maxima; the cross-stream distance between these maxima contracts slightly near $x/d = 5$, and the disturbances attain an absolute maximum at $x/d \simeq 7$, as shown in Fig. 4.3. These results illustrate the spatial instability of such disturbances. Comparison with the measurements of Ballou [21] (Fig. 4.5) demonstrates the intensification of this process at higher Reynolds number. It has been shown that the dependence on Reynolds number of the formation of wake vortices is strongly related to the transition to turbulence of the flow in the near wake of the cylinder (or at very high Reynolds numbers on the cylinder itself). This transition phenomenon is superimposed on the development of the primary vortex structure and has been shown to modify it. This leads to a flow behavior more complex than discussed above,

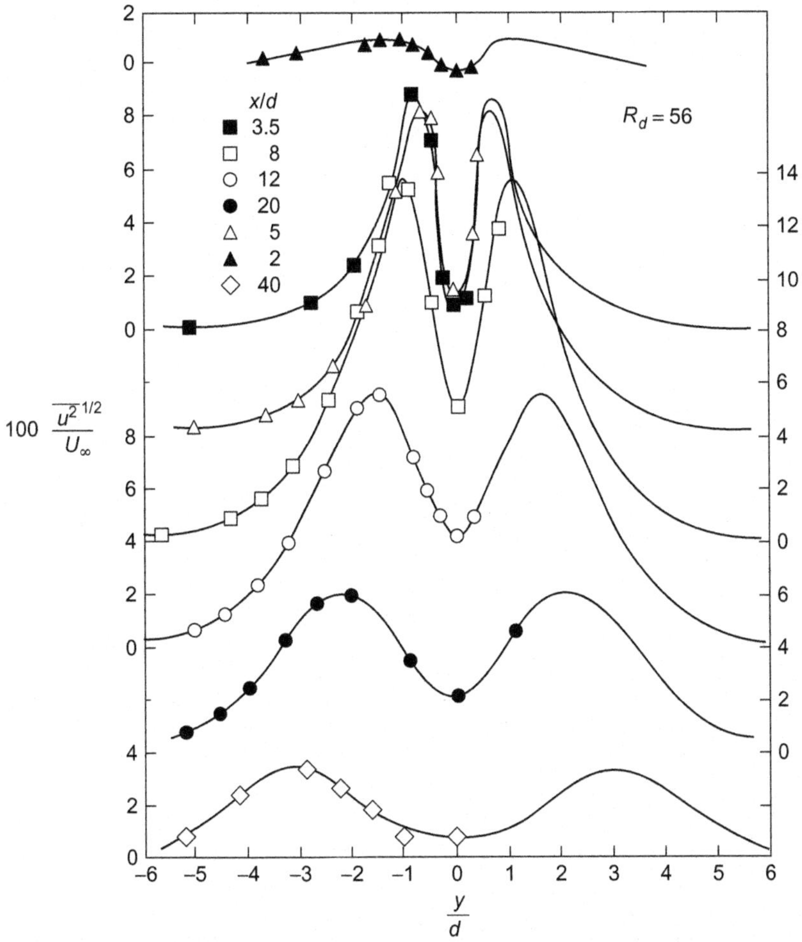

FIGURE 4.4 Kovasznay's [15] measurements of fluctuating velocity in the wake of a circular cylinder at $R_d = 56$.

which has not yet been fully analyzed either experimentally or theoretically. Roshko [16], Bloor [17], and Dale and Holler [22] have provided relatively detailed experimental descriptions of this phenomenon. Earlier, Schiller and Linke [23] had disclosed the importance of near-wake turbulence in determining the behavior of drag coefficients of cylinders with Reynolds number. The flow near a circular cylinder is now a common test case for large-scale numerical computation pertinent to the prediction of flow-induced loads on bluff bodies. Some of the earlier examples of these are the direct numerical simulations of Ma et al. [24] and the large eddy simulations of Franke et al. [25].

At Reynolds numbers less than 3×10^5 the laminar boundary layer on the cylinder separates at an angle of approximately $\pm 80°$ from the forward stagnation point. The free shear layer that is formed in the near wake is

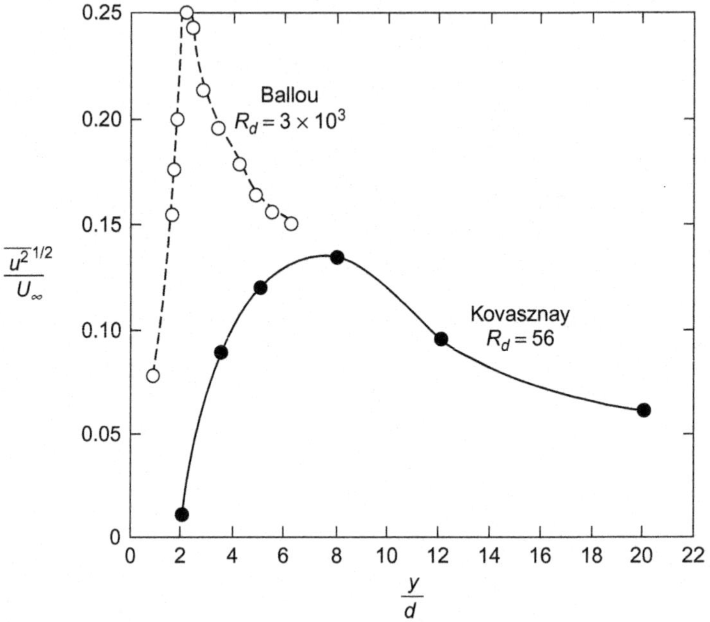

FIGURE 4.5 Downstream growth and decay of local maxima of root-mean-square fluctuating velocity in the wakes of rigid circular cylinders. *Data sources from Kovasznay LSG. Hot-wire investigation of the wake behind circular cylinders at low Reynolds numbers. Proc R Soc London Ser A 1949;198:174–90 and Ballou CL. Investigation of the wake behind a cylinder at coincidence of a natural frequency of vibration of the cylinder and the vortex shedding frequency. Acoust Vib Lab Rep 76028–2. Cambridge, MA: MIT.*

unstable with spatially and temporally growing wavelike disturbances that depend on the fluid velocity, the thickness of the shear layer, and the kinematic viscosity of the fluid. The breakdown of these transition waves to turbulence occurs prior to the formation of the primary vortex pattern at Reynolds numbers greater than 1.3×10^3 but it occurs in the cores of the vortices [16,17] after their formation at lower Reynolds numbers greater than 150–300. The frequencies of the transition waves f_t for $R_d > 1.3 \times 10^3$ have been shown by Bloor [17], Dale and Holler [22], and Roshko [26], to be a large multiple of the fundamental frequency of vortex shedding f_s such that

$$\frac{f_t}{f_s} = A \frac{f_s d}{U_\infty} \left(\frac{U_\infty d}{\nu} \right)^{1/2} \tag{4.2}$$

where A has an experimental value of approximately 2×10^{-2} to 3×10^{-2}. This behavior of the transition wave frequency f_t suggests that $f_t \delta_e / U_\infty$ is nearly a constant, where δ_e is the laminar boundary layer thickness at the point where the flow separates from the cylinder. The breakdown of these disturbances occurs closer to the cylinder as Reynolds number increases. The

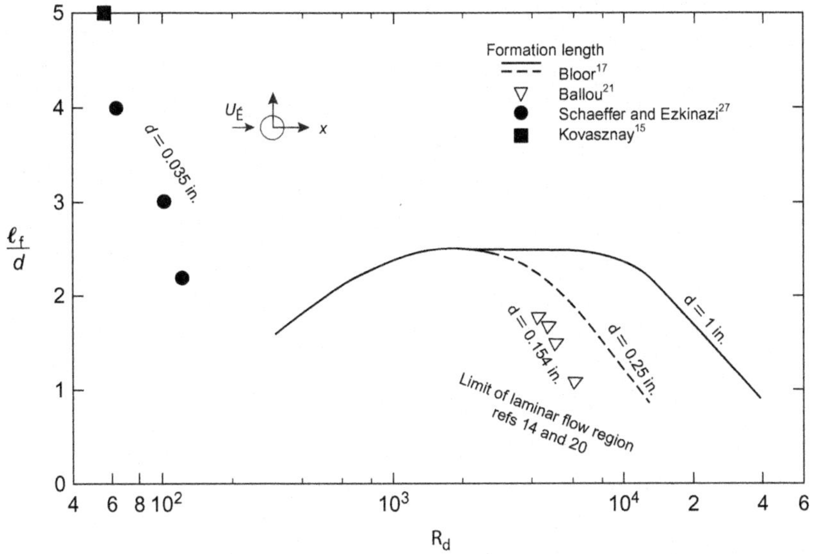

FIGURE 4.6 Vortex formation lengths behind circular cylinders as a function of Reynolds number.

limit of the region of laminar flow extends less than only 1.5 diameters downstream, as shown by experimental results of Bloor [17] and Schiller and Linke [23]. These experimental results are summarized by a single curve in Fig. 4.6. For Reynolds numbers less than 1.3×10^3 the transition waves are not observed before the formation of vortices.

Also shown in Fig. 4.6 are the distances behind the cylinder where vortices are formed. The available range of experimental data is sufficient to give a fairly complete Reynolds number behavior of the formation length. The length of the formation region can be defined in various ways. The region extends downstream of the shedding cylinder to a point where the width of the wake is a minimum, as discussed by Schaefer and Eskinazi [27]. For R_d greater than 200–400, this zone is also a region described by Bloor [17] within which irregular disturbances with characteristic frequencies which are much less than the shedding frequency exist. The end of the formation region is marked by a clearly distinguishable reduction in the amplitude of these disturbances. Ballou's [21] values of formation length were obtained by determining the location of the maximum values of fluctuating velocity in the wake as shown in Fig. 4.3. Measurements have been obtained on a range of cylinders; although the results are shown in dimensionless form, there seems to be a consistent dependence on the diameter of the cylinder. It is notable that an analogous dependence on diameter of the limit on the extent of laminar flow was not observed by Bloor, although the connection between this limit and the formation length is fairly apparent. Both distances decrease with an increase in Reynolds number. At Reynolds numbers greater than 100, the formation length is on the order of 2.5 cylinder diameters or less; as the Reynolds number decreases below 100 the distance for formation increases. This increase in formation length at low Reynolds numbers is also suggested by the photographs in Fig. 4.1.

This discussion of the near-wake dynamics and the previous one on vortex development precedes our description of the other elements of the fluid dynamics of the cylinder wake because it is fundamental to the behavior of the vortex wakes. Although it is not well explained by rigorous theory and extensive empirical characterization, there is ample evidence that this secondary transition (1) contributes to the generation of three-dimensional disturbances in the primary vortex street and (2) is the vehicle through which transverse cylinder motion and upstream turbulence (see Gerrard [28]) can influence vortex formation (see also Dale and Holler [22]). The three-dimensional nature of the vortex street, the formation length, and the vortex strength are the dominant wake characteristics that determine the unsteady lift and the axial correlation length of lift and, ultimately, the radiation of sound, as will be discussed subsequently.

4.3 MEASURED FLOW-INDUCED FORCES AND THEIR FREQUENCIES

4.3.1 Mean Drag and Vortex-Shedding Frequencies

The time-averaged drag on a circular cylinder is strongly dependent on the Reynolds number because as the characteristics of the formation of vortices change so, too, will the rate of fluid momentum that is transferred to the wake. In this and the following sections we shall review the details of wake formation and forcing to provide the quantities necessary to understand production of sound. This is a broad topic; the reader is referred to the work of Norberg [29] for further detail and references. As vortices are formed, time-dependent forces are generated on the cylinder that balance the rate of change of momentum associated with the circulation of each vortex. The force system and its relationship to a vortex pattern are diagrammed in Fig. 4.7. The illustrations of Figs. 4.1 and 4.2 show a regular vortex system exists; we have truncated the street in Fig. 4.7 to only three vortices. For the time being we shall restrict our attention to the now-classical measured characteristics of average drag force F_x designated in the figure. A mean-drag coefficient $\overline{C}_{\mathrm{D}}$ is defined as

$$\overline{C}_{\mathrm{D}} = \frac{F_x}{\frac{1}{2}\rho U_\infty^2\, dL} \tag{4.3}$$

where L is the length of the cylinder and d is its diameter. Fig. 4.8 shows the dependence on $\overline{C}_{\mathrm{D}}$ on R_d as published by Roshko [30]. The characteristic vortex-shedding regimes of Fig. 4.2 are also indicated in Fig. 4.8. The curve

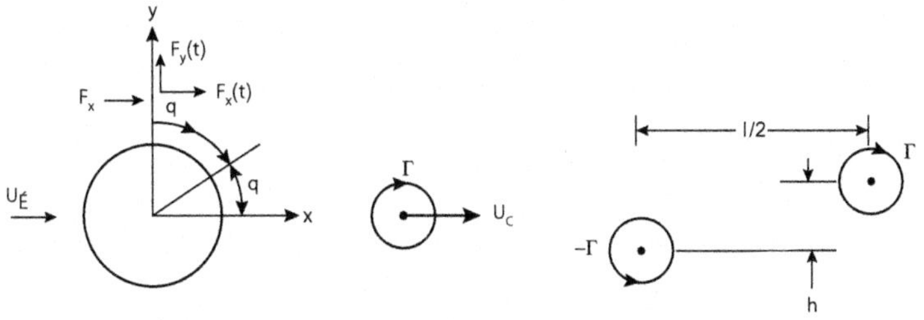

FIGURE 4.7 The vortex street and force system on a shedding cylinder. *Reprinted from Gerrard JH. A disturbance-sensitive Reynolds number range of the flow past a circular cylinder. J Fluid Mech 1965:22:187–96 by permission of Cambridge University Press.*

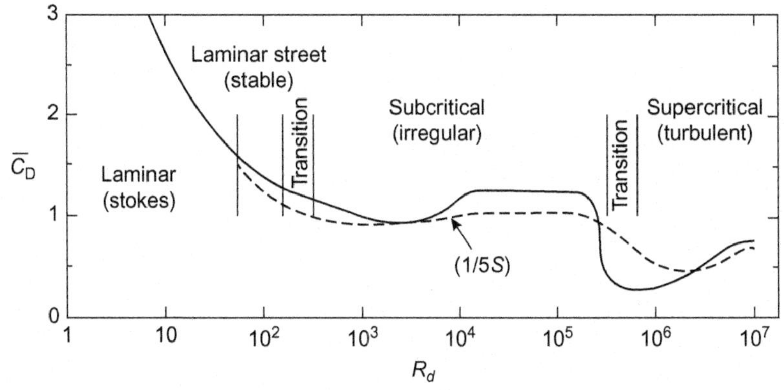

FIGURE 4.8 Drag coefficient of circular cylinder. *Reprinted from Roshko A. Experiments on the flow past a circular cylinder at very high Reynolds number. J Fluid Mech 1961;10:345–56 by permission of Cambridge University Press.*

in Fig. 4.8 is constructed from a number of experimental results; Schlichting [31] and Chen [32–34] cite many of these, beyond those included in the Chapter references. For R_d less than 20, the drag coefficient is largely controlled by viscous friction. The two transition regions that are indicated pertain to the development of turbulence in the wake. In the range of Reynolds numbers 150–300 turbulent vortex cores exist; in the range from 2×10^5 to 5×10^5 the boundary layers on the upper and lower arcs of the periphery of the cylinder become turbulent. These turbulent boundary layers separate at an angle of approximately 120° along the periphery from the forward stagnation point. The influence of this viscous effect is seen in the pressure distributions shown in Fig. 4.9. The local pressure $P(\theta)$ on the surface of the cylinder is presented as a coefficient

$$C_p = \frac{P(\theta) - P_\infty}{\frac{1}{2}\rho_0 U_\infty^2} \tag{4.4}$$

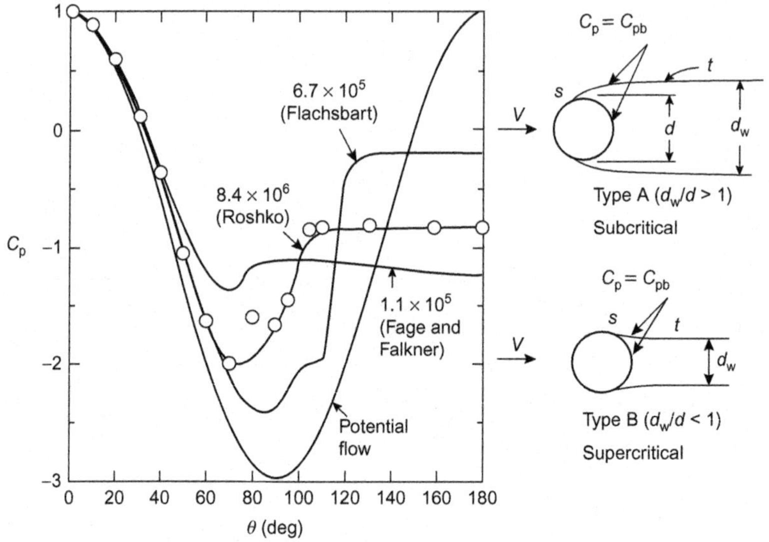

FIGURE 4.9 Pressure distributions on cylinders in the subcritical-to-supercritical flow ranges. *Reprinted from Roshko A. Experiments on the flow past a circular cylinder at very high Reynolds number. J Fluid Mech 1961;10:345–56 by permission of Cambridge University Press. Data sources: Flachsbart O, 1929. From an article by H. Muttray. "Die Experimentellen Tatasachen des Widerstandes Ohne Auftrieb," Handbuch der Experimentalphysik Hydro und Aero Dynamik vol. 4, Part 2, pp. 232–336 [35]; Leipzig, 1932 and Fage A, Falkner VM. The flow around a circular cylinder. R & M No. 369. Aeronaut. Research Council, London; 1931 [36].*

In the subcritical range $R_d < 2 \times 10^5$, the pressure coefficients reflect the separation occurring over a large sector of the cylinder than in the supercritical range. In the latter range the pressure coefficient in the separated zone, called the base pressure coefficient C_{pb}, increases and the wake contracts. This results in the observed reduction in $\overline{C}_D$. In the supercritical region the width of the wake d_w reduces in size to less than a cylinder diameter, but as R_d continues to increase the streamlines expand again. This expansion is reflected in Roshko's pressure distribution for $R_d = 8.6 \times 10^6$ and causes an increase in the drag coefficients.

In the range of $10^4 < R_d < 3 \times 10^5$ the drag coefficient is constant, probably owing to the nearly invariant wake characteristics in this range. It is also the upper extreme end of the range over which laminar-to-turbulence transition has occurred in the formation zone of the wake. For R_d greater than 8×10^3, Bloor [17] has shown that this transition occurs abruptly in the shear layer within a distance less than one-half cylinder diameter from the separation point (see Fig. 4.6). Before the formation of a vortex these transition waves fully break down to turbulence. This abrupt breakdown in the shear layer is in contrast to that which occurs in the Reynolds number range

1.3×10^3 to 8×10^3 for which the breakdown occurs more gradually and is preceded by the clearly distinguished transition waves. Thus we see that the behavior of the drag coefficient is paralleled by the occurrence of the secondary disturbances. The drag is therefore governed by the mechanics of wake vortex development. As we shall soon see, it is also related to the frequency of vortex shedding. Roughness on the surface of the cylinder has been observed [37] to decrease the base pressure coefficient and increase the steady drag at Reynolds numbers near 3×10^5 when the boundary layer on the cylinder is turbulent. The boundary layer separation points move forward when the cylinder is roughened.

The vortex-shedding frequency can be measured by placing a hot wire anemometer probe into the cylinder wake and measuring the frequency of the velocity disturbances that are sensed. The probe must be on one side or the other of the axis of the wake. The fluctuations in velocity are associated with the individual vortices, as shown in Fig. 4.7. The vortices move with a velocity U_c and vortices with the same circulation direction have a streamwise separation distance l. Two rows of vortices exist, each with vortices of a common circulation direction. Thus a fixed probe on one side of the wake will sense an undulating velocity of frequency

$$f_s = U_c/l$$

because as vortices of alternating sign pass by the probe, the direction of the velocity disturbance will alternate. If the probe were to be placed on the axis of the wake, it would sense alternating velocities of a frequency $2f_s$. The frequency of vortex passage, or of vortex production, is typically expressed in the dimensionless form

$$S = \frac{f_s d}{U_\infty} = \frac{U_c}{U_\infty}\frac{d}{l} \tag{4.5}$$

that was introduced in Eq. (4.1). Another method of measurement consists of determining the frequency of fluctuations of the oscillating lift on the shedding cylinder. This force component is $F_y(t)$ in Fig. 4.7.

Strouhal numbers in the range of R_d extending from 10^1 to 10^4 are shown in Fig. 4.10, which is reproduced from Roshko's original report. It shows Kovasznay's as well as Roshko's [16] measurements on a variety of cylinder diameters. Background turbulence levels in the wind tunnels reported by the two investigators were between 0.03% and 0.18%. It is interesting to recall now Rayleigh's [3,4] postulation (Eq. (4.1)) that S is an increasing function of R_d and to see that it applies in Fig. 4.10 from $R_d = 40$ (Kovasznay's [15] lower limit of vortex formation) until $R_d = 10^3$. Except for the Reynolds number range 140–300 the Strouhal number depends smoothly on R_d. In this short range (Roshko's transition range) Roshko [16] reports irregularities and bursting of the fluctuating velocities in the wake. At lower R_d, velocity fluctuations are sinusoidal, as well as above this range. However, although

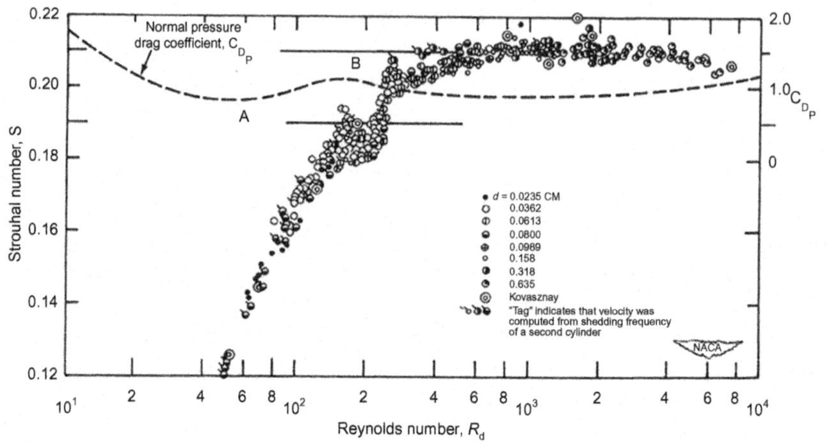

FIGURE 4.10 Strouhal numbers for vortex shedding behind circular cylinders at low Reynolds numbers (see Roshko [16]).

the vortices are shed uniformly and coherently along the entire axis of the cylinder up to $R_d = 150$, as shown visually by Hama [38], three-dimensionality begins to develop in the wake at R_d between 150 and 300. At R_d greater than 300 Hama's photographs show that a spatial periodicity occurs along the axes of the filaments of shed vortices. As the vortices are convected downstream, the vortex street becomes turbulent. The range of R_d between 4×10^2 and 1.3×10^3 has also been shown by Bloor to be that in which laminar-to-turbulent transition occurs just prior to vortex formation. It is also one in which transition waves of a separate and higher frequency are not detectable. At Reynolds numbers above 1.3×10^3, corresponding to the region of observed transition waves, the Strouhal number is nearly constant.

Measured values of Strouhal number at higher values of Reynolds number are shown in Fig. 4.11. These values have been obtained by measuring frequencies of fluctuations in lift or of wake velocity. The collection of points reflects the degree of repeatability in the values of S throughout the Reynolds number range up to the "transition" range shown in the figure. This and other names used in the text that are given to the various regions of vortex shedding are due to Roshko. However, other names have been given by other authors. To avoid confusion, names will be used here sparingly and always with reference to a figure diagram. For Reynolds numbers greater than 5×10^5 the irregularity in the cylinder wake is reflected in the wider range of reported values of Strouhal number that have been reported. The particularly large values of S that have been reported by Delany and Sorensen [39] and Bearman [40] relative to those of Jones [41] and of Roshko [42] appear to be out of line with other values. The reason for this is not understood; however, all investigators of this range of Reynolds number

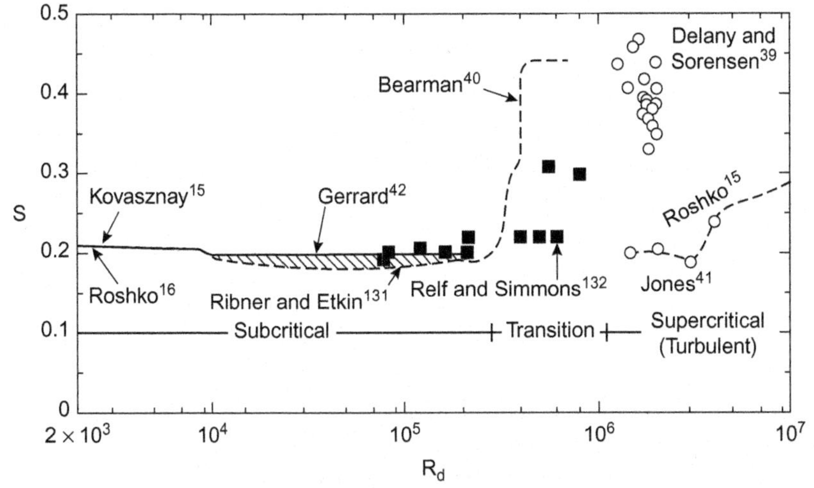

FIGURE 4.11 Strouhal numbers at high Reynolds numbers as function of Reynolds number.

have reported difficulty in establishing a predominant frequency. Jones [41], for example, quotes a range of Strouhal frequencies for the lift fluctuations of from 0.025 to 0.21 at any given Reynolds number in the range of from 10^6 to 5×10^6. At higher Reynolds numbers, in the range 5×10^6 to 2×10^7, he reports that the lift fluctuations are more periodic and with a predominant Strouhal frequency of 0.3. Bearman [40] used flow visualization to disclose the existence of two laminar separation bubbles, indicating the formation of two separate vortices above and below the stagnation point, as illustrated in Fig. 4.2 for the supercritical range.

The existence of higher Strouhal numbers at high Reynolds number has been linked with the contraction of the wake that is illustrated in Fig. 4.9. The streams are assumed to be parallel downstream of location t; separation occurs at location s. Roshko [43] has postulated that the frequency of vortex shedding is dependent on the (theoretically calculated) separation of the free stream lines of the shear layers d_w and the velocity of the free stream lines U_s. Now,

$$U_s = U_\infty \sqrt{1 - C_{pb}} \tag{4.6}$$

where C_{pb} is the base pressure coefficient. Thus Roshko's [43] definition of Strouhal number is

$$S^* = f_s d_w / U_s$$

This number has the value $S^* \simeq 0.16$ for circular cylinders, 90° wedges, and flat plates set normal to the flow. It is nearly constant over the range of $U_s d_w / \nu$ from 10^4 to 4.4×10^4. The point is that the increase in Strouhal number for R_d in the supercritical region of Fig. 4.11, as reported by Roshko, is in direct proportion to the product

$$S = S^* \frac{d}{d_w} \sqrt{1 - C_{pb}}$$

The variations in observed values of S are accompanied by corresponding variations in the mean-drag coefficient. High values of S are associated with relatively low values in $\overline{C}_D$. To see this we use von Karman's [6] formula for the drag on a bluff body shedding a vortex street (since $h/l = 0.281$)

$$\overline{C}_D = \frac{h}{d}\left[5.65\frac{U_\infty - U_c}{U_\infty} - 2.25\left(\frac{U_\infty - U_c}{U_\infty}\right)^2\right]$$

where h is the vortex row spacing shown in Fig. 4.7. For the case that the vortex velocity U_c is a constant percentage of U_∞ over the Reynolds number range, the drag coefficient is proportional to h/d. Now, further assuming that the vortex street spacing is proportional to the shear layer spacing d_w and that S^* is constant we find the result

$$\overline{C}_D = kS^{-1}$$

where k is constant over the range of Reynolds number. Arbitrarily selecting $k = \frac{1}{5}$, we use Strouhal numbers in Figs. 4.10 and 4.11 to construct the dotted line shown in Fig. 4.8. This comparison, first shown by Roshko [43], supports the speculated correspondence, and it further shows the intimacy between cylinder drag and the details that govern the vortex generation.

The interdependence of the mean-drag coefficient and the geometry of the vortex street has been further clarified by Bearman [44]. He used Kronauer's [45] stability criterion, which states that the vortex spacing ratio h/l is determined by a requirement that the drag induced by the wake is a minimum. It will be discussed in Section 4.4 (Eq. (4.20)) that the drag coefficient of the vortex street for an arbitrary spacing ratio is

$$\overline{C}_D = \frac{4\,l}{\pi\,d}\left(\frac{U_v}{U_\infty}\right)^2\left[\coth^2\frac{\pi\,h}{l} + \left(\frac{U_\infty}{U_v} - 2\right)\frac{\pi\,h}{l}\coth\frac{\pi\,h}{l}\right]$$

Kronauer's criterion states that the spacing ratio, for a given vortex speed relative to the free stream $U_v = U_\infty - U_c$ is determined by

$$\left[\frac{\partial \overline{C}_D}{\partial(h/l)}\right]_{U_v=\text{const}} = 0$$

Bearman [44] has shown that this condition predicts spacing ratios that depend on the vortex velocity, as shown in Fig. 4.12. Using measured values of U_v/U_∞ and of the streamwise vortex spacing l for a variety of vortex wakes Bearman was thus able to determine the transverse spacing h. These wakes were varied and controlled by bleeding fluid and by inserting splitter

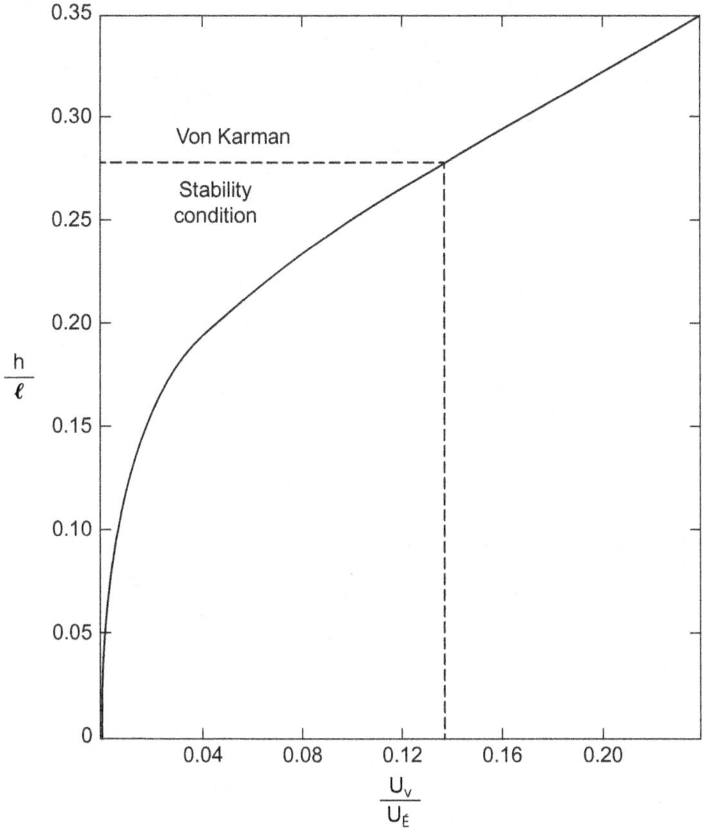

FIGURE 4.12 Vortex spacing ratio calculated by Bearman [44] as a function of vortex velocity for the Kronauer minimum-drag requirement.

plates at the base of the cylinder; see also the end of Section 4.4. Then he defined a Strouhal number, analogous to Roshko's [43], which is

$$S^{**} = \frac{f_s h}{U_s}$$

where U_s is given by Eq. (4.6). For values of U_s/U_∞ between 1.1 and 1.47 (which represents a range of base pressure coefficients of from −0.21 to −1.1) this Strouhal number is a constant equal to 0.18; see also Chapter 11, Noncavitating Lifting Sections. The significance of this number is that it applies to a wide variety of bluff body cross-section shapes: circular elliptical, ogival, prismatic, and flat plate. Also, both Roshko's [43] and Bearman's [44] definition of Strouhal number are entirely compatible, and the assertion that *h/d* is a constant (which is nearly unity) is apparently borne out by available experiments.

4.3.2 Oscillatory Lift and Drag Circular Cylinders

Our consideration of the acoustic radiation from vortex-shedding cylinders will require a knowledge of the oscillatory forces, the drag $F_x(t)$, and the lift $F_y(t)$ on the cylinder. As each vortex is shed from the cylinder it generates a lift force that changes direction as vortices are produced in each shear layer. This force may be written

$$F_y(t) = \tilde{F}_y e^{-i\omega_s t}$$

where $\omega_s = 2\pi f_s$ is the shedding frequency in radians per second. The drag fluctuation is generated as a result of oscillations in the base pressure coefficient. Since the base pressure responds to disturbances that are spatially symmetric about the wake axis and since these disturbances are generated twice as each vortex pair is produced, the frequency of the drag fluctuations will be $2f_s$, and

$$F_x(t) = \tilde{F}_x e^{-i2\omega_s t}$$

Although the vortex shedding is temporally periodic, the phase of shed vortices and therefore of the induced forces is a random variable along the axis of the cylinder for Reynolds numbers R_d greater than 150. As Hama's [38] photographs have shown, the vortices are strictly two dimensional only for $R_d < 140$. At greater values of R_d three-dimensionality occurs (see Section 4.3.3). Thus our consideration of the fluctuating forces is best founded on a statistical basis. The time-dependent pressure distribution on the cylinder we therefore write in a general form as

$$p(\theta, z, t) = [p_L(\omega_s)\cos\theta\, e^{-i\omega_s t} + p_b(2\omega_s)\sin\theta\, e^{-i2\omega_s t}]e^{i[\phi(z,\, t)]} \tag{4.7}$$

where $p_L(\omega_s)$ and $p_b(2\omega_s)$ are pressure amplitudes constant along the axis and $\phi(z, t)$ is the phase along the axis. Here θ is measured from the y axis as shown in Fig. 4.7. The expression shows that the circumferential variation of the pressure fluctuations that yield lift is maximum at $\theta = 0°$ and minimum at $\theta = 90°$. The fluctuating base pressure is given by p_b since the pressure component at $2\omega_s$ has a maximum amplitude there. The phase function $e^{i[\phi(z,t)]}$ accounts for the axial variation in vortex structure that occurs for $R_d > 150$. We assume that both the fluctuating lift and drag have the same axial correlation characteristics. Although the phase function can be a deterministic function of location (as it is at low Reynolds numbers and on some vibratory cylinders when shedding locks onto the vibration), it has been observed to be a stochastic variable at most Reynolds numbers on rigid cylinders.

The total force on the cylinder is

$$F(t) = (F_x(t),\ F_y(t)) = \int_0^{2\pi}\int_0^L p(\theta, z, t)\mathbf{n}(\theta)\frac{d}{2}\,d\theta\, dz \tag{4.8}$$

where $\mathbf{n}(\theta) = (\sin\theta, \cos\theta)$ is the normal to the surface of the cylinder and L is its length. Substitution of Eq. (4.7) into Eq. (4.8) gives an expression for the dimensionless forces:

$$\frac{F_y(t)}{\frac{1}{2}\rho_0 U_\infty^2\, dL} = \frac{1}{2}\left(\frac{p_L(\omega_s)}{\frac{1}{2}\rho_0 U_\infty^2}\right)\frac{1}{L}\int_0^L e^{i\phi(z,\,t)}\,dz\, e^{-i\omega_s t} \tag{4.9a}$$

and

$$\frac{F_x(t)}{\frac{1}{2}\rho_0 U_\infty^2\, dL} = \frac{1}{2}\left(\frac{p_b(2\omega_s)}{\frac{1}{2}\rho_0 U_\infty^2}\right)\frac{1}{L}\int_0^L e^{i\phi(z,\,t)}\,dz\, e^{-2i\omega_s t} \tag{4.9b}$$

The time-mean-square oscillatory lift coefficient is defined as

$$\overline{C_L^2} = \left(\frac{1}{2}\rho_0 U_\infty^2\, dL\right)^{-2}\overline{F_y^2(t)} \tag{4.10a}$$

and the drag coefficient is

$$\overline{C_D^2} = \left(\frac{1}{2}\rho_0 U_\infty^2\, dL\right)^{-2}\overline{F_x^2(t)} \tag{4.10b}$$

where the overbars denote a time average. Looking ahead to an interpretation of experimental data, we note that if the unsteady forces are temporally harmonic, the mean square is just $\frac{1}{2}\tilde{F}^2$. This situation is expected at Reynolds numbers less than 3×10^5, where the shedding frequency is well defined. At higher values of R_d, where the vortex shedding covers a more distributed frequency range, the mean square must be taken in its strictest meaning. This discussion is pertinent because various investigators have measured the fluctuating lift by determining either peak values or temporal mean-square values. Later the measurements will be compared.

Another difficulty in interpreting lift coefficient measurements lies in the unknown function $\phi(z, t)$. Thus if we let l_z describe a typical axial length scale over which $\phi(z, t)$ is constant, it is important that the shedding cylinder on which the lift is measured have a length less than l_z. Of course if the lift is deduced from a pressure distribution, as in the case of Gerrard [42], the lift coefficient is closely related to a pressure coefficient, $p(\frac{1}{2}\rho_0 U_\infty^2)^{-1}$, as is seen by comparing Eqs. (4.9) and (4.10).

Values of the root-mean-square oscillatory lift coefficients measured by various investigators are summarized by the points and solid lines in Fig. 4.13. (Representative values, which will be used later, are shown as dotted lines.) Of the group, Gerrard [42], McGregor [54], Koopman [46], Schmidt [47], and Schwabe [48] determined fluctuating lift by the integration of a measured fluctuating pressure distribution. Their measurements confirm our representation for the circumferential variation of pressure used in

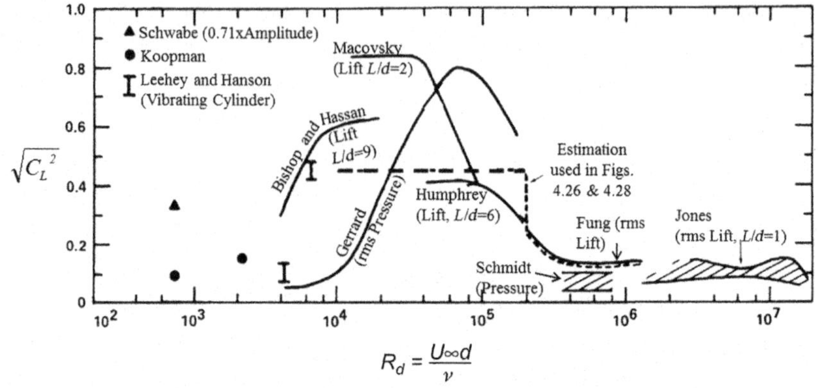

FIGURE 4.13 A summary of measured values of root-mean-square oscillatory lift coefficients on circular cylinders as a function of Reynolds number. *Data sources from Jones GW. Unsteady lift forces generated by vortex shedding about large stationary, and oscillating cylinder at high Reynolds number. ASME Symp Unsteady Flow, Pap. 68-FE-36; 1968; Koopman GH. Wind induced vibrations and their associated sound fields. PhD thesis, Catholic University: Washington, DC; 1969; Schmidt LV. Measurement of fluctuating air loads on a circular cylinder. J Aircr 1965;2:49–55; Schwabe M. Über Druckermittlung in der nichtstationären ebenen Strömung. Ing Arch 1935;6:34–50; Humphrey JS. On a circular cylinder in a steady wind at transition Reynolds numbers. J Fluid Mech 1960;9:603–12; Macovsky MS. Vortex-induced vibration studies. DTMB Rep. No. 1190. David Taylor Naval Ship R & D Center: Washington, DC; 1958; Bishop RED, Hassan AY. The lift and drag forces on a circular cylinder in a flowing fluid. Proc R Soc London Ser A 1964;277:32–50; Leehey P. Hanson CE. Aeolian tones associated with resonant vibration. J Sound Vib 1971;13:465–83; Gerrard JH. The calculation of the fluctuating lift on a circular cylinder and its application to the determination of Aeolian tone intensity. AGARD Rep. AGARD-R-463; 1963.*

Eq. (4.7). McGregor obtained the value of 0.42 for the lift coefficient in the range of R_d from 4×10^4 to 1.2×10^5, and this agrees well with Humphrey's value. The values shown here attributed to Humphrey [49] and Macovsky [50] have been deduced from their measured values of lift amplitude by assuming periodic lift fluctuations. Their published oscillographs (time histories) of lift fluctuations show that this is a reasonably accurate interpretation of the data. In Humphrey's case there was some irregular modulation, of long time scale, of an otherwise periodic lift signal; thus his quoted values of average amplitude were used to estimate a root-mean-square lift. Since $R_d = 2 \times 10^5$ marks the upper limit of observed periodicity, this approximation could not be applied to Humphrey's data in the range 2×10^5 to 6×10^5. Maximum values of lift coefficient decreased in this range of Reynolds number from 0.8 to 0.35, roughly matching the peak values obtained by Fung. Root-mean-square values reported by Fung [55] and Jones [41] appear to agree closely. Schmidt's [47] measurements were obtained in the same facility as Fung's, but he noted that the lift coefficients were reduced if the cylinder was well polished. Bishop and Hassan [51] have provided direct measurements of root-mean-square lift, and the coefficients attributed to Macovsky were determined from his quoted maximum lift levels. The single point attributed to Schwabe is the lowest value of Reynolds number for which oscillatory lift data are available: It was obtained by integrating a pressure distribution. Measurements of Schewe [56] were deduced from lift fluctuations on a 10 to 1 length-to-diameter ratio polished cylinder in a variable pressure wind tunnel. These measurements throughout $2 \times 10^4 < R_d < 7 \times 10^6$ were generally in agreement with those shown in Fig. 4.13. Finally, the measurements of Leehey and Hanson [52] were obtained by an indirect method. These measurements were accomplished during an experiment involving wind-induced cylinder vibration and Aeolian tones. They first determined the level of wind-induced vibration at a particular vortex-shedding frequency. They then excited the cylinder electromagnetically in still air at the same frequency that it was wind excited. This was accomplished by passing an alternating current through the steel wire around which had been placed a system of permanent magnets. By measuring the current through the wire and magnetic field strength they determined the force generated on the wire, which was proportional to the cylinder vibration level that they measured at the same time. From this they deduced the aerodynamic lift on the wire at any given point in the experiment. This method had also been previously used by Powell [57]. The vertical bars denote the upper and lower limits of lift coefficient that were reported. Unfortunately a certain level of cylinder vibration accompanied the measurement, and this is known to influence the lift (see Section 4.3.4).

In spite of the wide range of reported lift coefficients shown in Fig. 4.13 there appears to be a general trend toward maximum values in a range of Reynolds numbers near 4×10^4. At either extreme of this region the coefficients appear to be somewhat smaller in value. The three most plausible reasons for disagreement among investigators are as follows:

1. Environmental influences, such as upstream turbulence and cylinder motion can modify the lift (see Section 4.3.4).
2. Spatial averaging of local pressure along the axis of the cylinder can reduce the apparent lift as suggested by Eq. (4.9).
3. These measurements, being dynamical in nature, are subject to a degree of experimental inaccuracy. Scatter in individually reported data is, for example, ±12% in the case of Bishop and Hassan [51], 30% for the pressures reported by Gerrard [42], and 25% for Macovsky's [50] measurements. Errors are the limits of the spread in observed values expressed as percentages of the centroid of the population of data. The range of values shown in Fig. 4.13 is not too far out of line with the scatter.

Measured fluctuating drag coefficients are less numerous than lift coefficients. Fig. 4.14 shows measurements of Fung [55], Schmidt [47], van Nunen [58], McGregor [54], and Gerrard [42] that are substantially in agreement. There appears to be maximum at Reynolds numbers near 4×10^4. The drag fluctuations are approximately 1/10 of the lift coefficients.

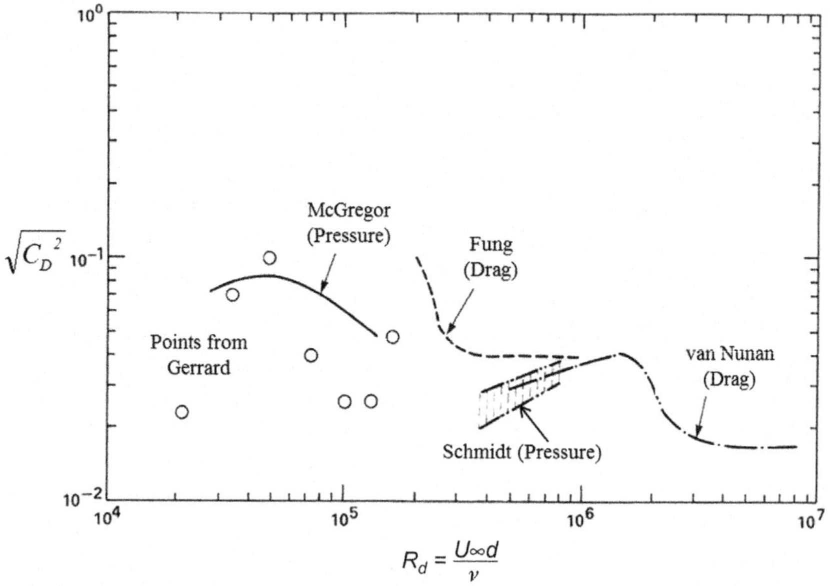

FIGURE 4.14 A summary of fluctuating drag coefficients on circular cylinders as a function of Reynolds number. *Data is from Schmidt LV. Measurement of fluctuating air loads on a circular cylinder. J Aircr 1965;2:49–55; Gerrard JH. The calculation of the fluctuating lift on a circular cylinder and its application to the determination of Aeolian tone intensity. AGARD Rep. AGARD-R-463; 1963; McGregor DM. An experimental investigation of the oscillating pressures on a circular cylinder in a fluid stream. UTIA Tech. Note No. 14; 1957; Fung YC. Fluctuating lift and drag acting on a cylinder in a flow at supercritical Reynolds numbers. J Aerosp Sci 1960;27:801–14; van Nunen JWG. Steady and unsteady pressure and force measurements on a circular cylinder in a cross flow at high Reynolds numbers. IUTAM Symp Flow-Induced Struct Vib Karlsruhe Pap. H5; 1972.*

This section has so far been devoted to an overview of the relatively classical measurements of flow-induced mean and time-dependent forces on circular cylinders. Many of these are likely done as preliminaries to later simulations done for other cross sections to the understand flow and forces. Notable among these are the large eddy simulations of Ma et al. [24] and Franke et al. [25] which agree well with the mean and turbulent wakes measured by Kovasznay [15] and the recent ones of Ong and Wallace [59]. Numerical simulations specifically devoted to calculating rms lift coefficients, e.g., of Cox et al. [60], Squires et al. [61], and Brewer [62], are well-supported by the data discussed in this section as well as the more recent results of Schewe [63], who examined a wide range of Reynolds number and provides spectral distributions of the forces, Szepessy et al. [64] who examined effects of aspect ratio on unsteady lift, Braza et al. [65] who also gave some early numerical results, and Nishimura [66] in the region of high Reynolds number.

4.3.3 Representations of Axial Phase Uniformity: Correlation Lengths

Even though the vortex-induced fluctuating pressures may be locally of the same amplitude along the cylinder axis, as noted above the phase of the pressure may vary stochastically as $\phi(z, t)$ in Eq. (4.7). The associated axial correlation lengths of the unsteady forces on the cylinder that provide the dipole strength are therefore of prime interest to acoustics and vibration. Various attempts have been made to quantify the axial phase variations and many of these measurements have been made through the use of flow visualization. Phillips [12], Macovsky [50], and Hama [38] examined the patterns of dye injected into the wake of cylinders, and Macovsky [50] also observed the three-dimensional patterns of wool tufts attached to cylinders. Roshko [16] estimated the length scales of the vortices by observing the behaviors of Lissajous patterns between two hot wire anemometer probes in a wake as the probes were moved away from one another along the axis of the cylinder. Gerrard [53] does not cite the details of any measurements, but he states that the correlation length is on the order of three diameters for R_d less than 2×10^5. Correlation measurements, in the strict sense, have been made between axially separated sensors by Prendergast [67] (pressure fluctuations), ElBaroudi [68] (velocity fluctuations near the separation point), Ballou [21] (velocity fluctuations in the wake), Leehey and Hanson [52] (velocity fluctuations in the wake), and Schmidt [47] (local lift fluctuations).

The behavior of the two-point correlation as a function only of axial separation distance $z_2 - z_1$ is a statistical reflection of the random nature of the function $\phi(z, t)$ that is employed in Eq. (4.7). Considering that the pressure is measured at two locations that are separated axially, the space–time correlation of the pressures is written

$$\begin{aligned}\hat{R}_{pp}(\theta, z_2 - z_1, \tau) &= \overline{p(\theta, z_1, t)p(\theta, z_2, t+\tau)}\\ &= \frac{1}{T}\int_0^T p(\theta, z_1, t)p(\theta, z_2, t+\tau)\, dt\end{aligned}$$

This is just the correlation that can be expressed in an approximate form in terms of Eq. (4.7) as

$$\overline{p(\theta, z_1, t)p(\theta, z_2, t)} = \frac{1}{2}\left[p_L(\omega)\cos^2\theta + p_b^2(2\omega_s)\sin^2\theta\right] \overline{\exp i[\phi_1(z_1, t) - \phi_2(z_2, t+\tau)]} \tag{4.11}$$

The averaging time T must be longer than both the characteristic shedding period $2\pi/\omega_s$ and the period of oscillation of the phase function $\phi(z, t)$. Also, for Eq. (4.11) to be valid the period of variation of the phase function must also be longer than $2\pi/\omega_s$. The overbars on the phase functions denote the replacement of the time average of the exponential by an exponential of the average phase difference. In effect, if $\omega_s >> \partial\phi/\partial t$ this assumption could be considered to be an equivalent to the use of a conditional average over time larger than $2\pi/\omega_s$, yet less than $2\pi(\partial\phi/\partial t)^{-1}$. The average of $\exp i(\phi_2 - \phi_1)$ is essentially a normalized correlation coefficient

$$R_{pp}(z_2 - z_1, \tau) = \frac{\overline{p(\theta, z_1, t)p(\theta, z_2, t+\tau)}}{[\overline{p^2(\theta, z_1, t)p^2(\theta, z_2, t)}]^{1/2}} \quad (4.12)$$

where $\overline{p^2(\theta, z, t)}$ is the mean square pressure at any location θ, z along the axis of the cylinder and

$$R_{pp}(0, 0) = 1 \quad (4.13)$$

From Eq. (4.7) we have

$$\overline{p^2(\theta, z, t)} = \frac{1}{2}[p_L^2(\omega_s)\sin^2\theta + p_b(2\omega_s)\cos^2\theta] \quad (4.14)$$

The normalized space–time correlation exposes any mean convection of the vortices along the cylinder such that letting $z_2 - z_1 = \xi$

$$R_{pp}(\xi, \tau) = R_{pp}(\xi - v_c\tau) \quad (4.15)$$

where v_c is the apparent convection velocity. Actually the vortex filaments have been observed to "peel off" the cylinder, propagating from one end of the cylinder to the other so that once formed the axis of the vortex is not quite parallel to the axis of the cylinder. The correlation function is generally dominated by the ξ dependence unless the cylinder is yawed, or unless some other asymmetry occurs in the flow. Thus we may make the approximation

$$R_{pp}(\xi, \tau) \approx R_{pp}(\xi)$$

and, using the familiar analytical form

$$R_{pp}(\xi) = e^{-\alpha|\xi|} \quad (4.16)$$

we can also obtain a good functional fit to much of the experimental data that has been collected on unyawed *rigid* cylinders. We shall define the correlation length as (Eq. (3.80))

$$2\Lambda_3 = \int_{-\infty}^{\infty} R_{pp}(\xi)\, d\xi \quad (4.17)$$

so that

$$2\Lambda_3 = \frac{2}{\alpha} \quad (4.18)$$

Thus, with negligible axial flow we take $k \ll \alpha$, and $\Lambda_3 = 1/\alpha$.

The above analysis is pertinent to the axial correlations of wake velocity as well as to lift correlations. Fig. 4.15 shows examples of correlation functions at various Reynolds numbers from 3.3×10^3 to 7.5×10^5. Note that at the lower end of the range of R_d the function is unlike the representation

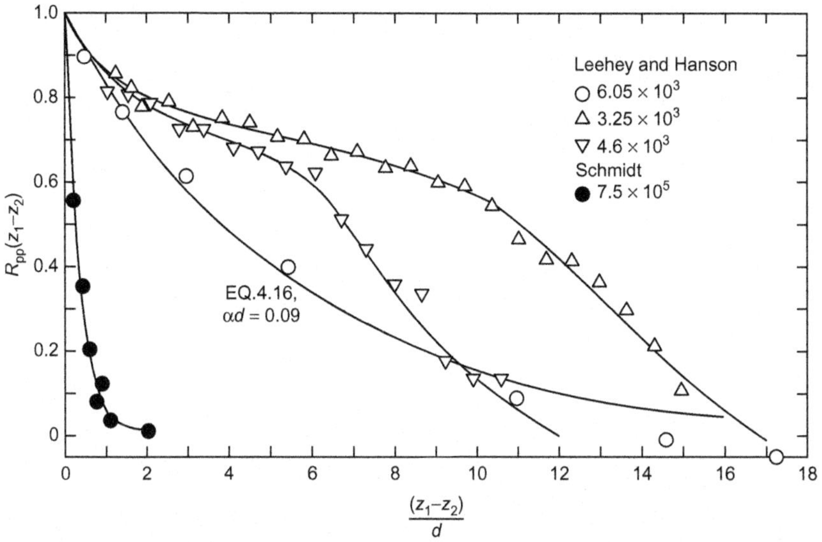

FIGURE 4.15 Axial covariance functions for fluctuating lift on circular cylinders at 4 values of Reynolds number. *Data sources from Schmidt LV. Measurement of fluctuating air loads on a circular cylinder. J Aircr 1965;2:49–55; Leehey R, Hanson CE. Aeolian tones associated with resonant vibration. J Sound Vib 1971;13:465–83.*

given in Eq. (4.16), but at the upper end the equation matches the measured functions. Fig. 4.16 shows values of $2\Lambda_3/d$ that have been reported by the various investigators. Perhaps the most striking aspect of Fig. 4.16 is that it is difficult to draw a general conclusion about the correlation length. For R_d less than 140 investigators are unanimous in reporting large correlation lengths. Hama's photographs show that vortices are correlated along the entire length of the cylinder when $R_d = 117$, i.e., $\Lambda_3/d \simeq 50$. With the exception of the correlation lengths measured by Leehey and Hanson [52], most investigators cite $\Lambda_3 \simeq d$ to $6d$ for $R_d < 2 \times 10^5$. It is tempting to explain the larger Λ_3 of Leehey and Hanson as being caused by cylinder vibration; however, they suggest that the uniform decrease of Λ_3 with increasing R_d is evidence against this explanation.

4.3.4 Other Influences on Vortex Shedding

As suggested in the Section 4.3.3, measured values of lift fluctuations can be influenced by cylinder motion, surface roughness, and upstream turbulence. There are few systematic investigations that quantify these effects. In one, Gerrard [28] has made an investigation of the effects of upstream turbulence in the range of R_d from 800 to 4×10^4. He measured the intensity of the velocity fluctuations in the cylinder boundary layer just downstream of

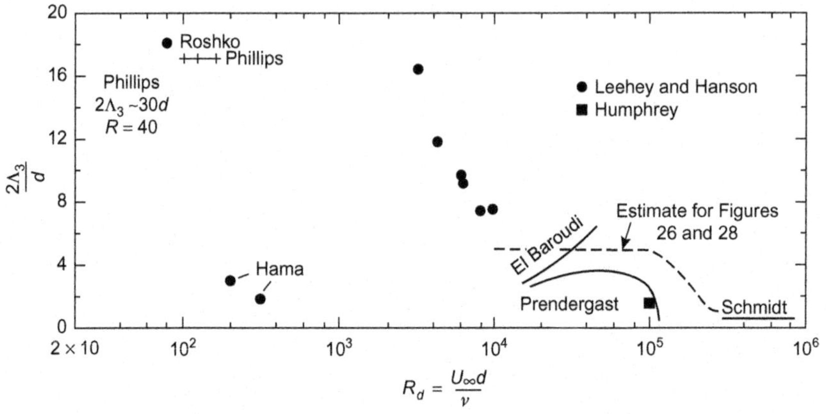

FIGURE 4.16 A summary of measured values of lift correlation lengths on circular cylinders. *Data sources from Phillips OM. The intensity of Aeolian tones. J Fluid Mech 1956;1:607–24; Roshko A. On the development of turbulent wakes from vortex streets. Natl Advis Comm Aeronaut Tech Note No. 2913; 1953; Hama FR. Three-dimensional vortex pattern behind a circular cylinder. J Aeronaut Sci 1957;24:156–8; Schmidt LV. Measurement of fluctuating air loads on a circular cylinder. J Aircr 1965;2:49–55; Humphrey JS. On a circular cylinder in a steady wind at transition Reynolds numbers. J Fluid Mech 1960;9:603–12; Leehey R, Hanson CE. Aeolian tones associated with resonant vibration. J Sound Vib 1971;13:465–83; Prendergast V. Measurement of two-point correlations of the surface pressure on a circular cylinder. UTIAS Tech. Note No. 23; 1958; ElBaroudi MY. Measurement of two-point correlations of velocity near a circular cylinder shedding a Karman vortex street. UTIAS Tech Note No. 31; 1960.*

separation, at $x = 0$ (see Fig. 4.8). This velocity increases with R_d in a similar fashion as the fluctuating lift coefficient. Fig. 4.17 shows this behavior for two levels of upstream turbulence, 0.02% and 1% of the free stream velocity. There is a consistent increase of as much as a factor of 4 in the fluctuating velocity u_s for Reynolds numbers greater than 10^3, and this increase is caused by the increase in inflow turbulence. The similar variation of u_s and C_L with R_d suggested to Gerrard that these two flow variables were related and that environmental influences on one variable reflected similar influences on the other. Furthermore, Gerrard [53] has argued that high free stream turbulence slightly increases both the vortex strength and the vortex-shedding frequency, see also Section 4.2.2. He has also speculated that the variation in formation length for cylinders of different diameters measured by Bloor [17] shown in Fig. 4.6, is attributable to a higher level of incident turbulence for the smaller diameter cylinder. This may not be entirely correct, because the measurements by Ballou [21], also shown in Fig. 4.6, were obtained with a free stream turbulence level of 0.04% of the mean velocity [21,52]. This level of turbulence is comparable to or less than that encountered in Bloor's measurements. Gerrard [69] contends that an increase in lift of a factor of 4, suggested by the increase in u_s, cannot be explained by the

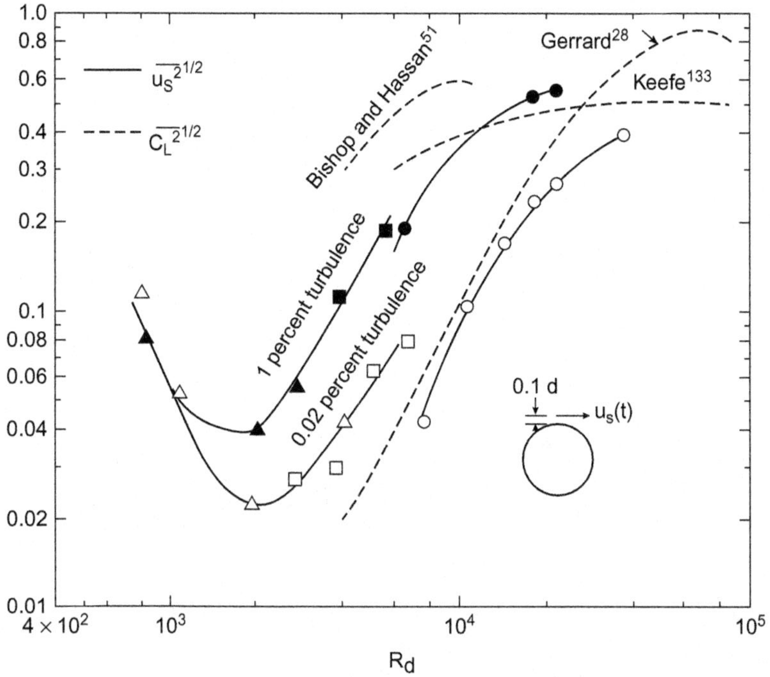

FIGURE 4.17 Comparison of unsteady lift coefficient with upstream turbulence intensity as a function of Reynolds number. *Reprinted from Gerrard JH. A disturbance-sensitive Reynolds number range of the flow past a circular cylinder. J Fluid Mech 1965;22:187–96 by permission of Cambridge University Press.*

possible reduction in formation length and increase in vortex strength that he has speculated as due to incident turbulence. It would appear that the question of the cause of increased lift by upstream disturbances can only be fully answered by an experimental investigation in which all the pertinent dependent variables, lift, formation length, vortex strength (see also Bloor and Gerrard [70]), and shedding frequency are measured as functions of both Reynolds number and upstream turbulence. A formal connection between u_s and fluctuating lift should also be established.

In addition to effects of incident free stream turbulence, Gerrard [28] has determined that the shedding velocity u_s can be increased by acoustic excitation. At a Reynolds number of 6.9×10^3, Gerrard found that sound with a root-mean-square velocity of 0.01% of the free stream velocity could increase u_s by a factor of 2.5. The effect is frequency dependent, and this increase in u_s was observed at an excitation frequency equal to the transition wave frequency of secondary disturbances [17]. This observation thus raises the question of whether acoustic excitation at frequencies given by Eq. (4.2) also increases the oscillatory lift coefficient.

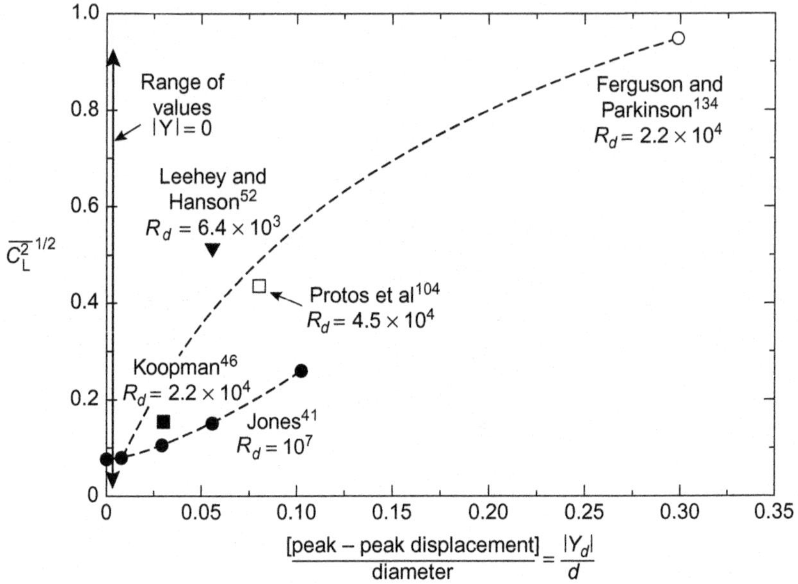

FIGURE 4.18 Dependence of root-mean-square lift coefficient on cylinder displacement.

Motion of the shedding cylinder, transverse to the direction of flow, has been observed by Jones [41] to increase the root-mean-square lift at high Reynolds numbers. Fig. 4.18 shows one set of his results. The effect is undoubtedly dependent on the Reynolds number. Not only the lift amplitude but also its phase relative to the cylinder displacement is important. This important aspect will be discussed as a nonlinear interaction in Section 5.7.3, Sound From Flow-Induced Vibration of a Circular Cylinder. Some other isolated measurements of fluctuating lift coefficients have been published with observed values of cylinder displacement. Those data are also shown in Fig. 4.18. An increase of the lift coefficient with the peak-to-peak displacement of the cylinder appears to be uniform. It is also more pronounced in range of R_d between 2×10^3 and 4.5×10^4. Note that Koopmann [46] measured fluctuating pressure and that Leehey and Hanson [52] determined an increase fluctuating lift without a corresponding increase in Λ_3.

Although we will discuss self-excited oscillation of the cylinder in Section 5.7.2, here we will discuss some of the fluid mechanical behavior caused by the vibration of the cylinder. An experimental relationship between the rate of vorticity generation and transverse displacement has been shown by Griffin and Ramberg [71]. At a Reynolds number of 144 and a forced vibration frequency equal to the vortex-shedding frequency, they determined that the rate of vortex generation increased by a factor of 1.65 over the rate for no motion when the amplitude of motion was one-half the cylinder diameter. This implies that the vortex strength is increased by this

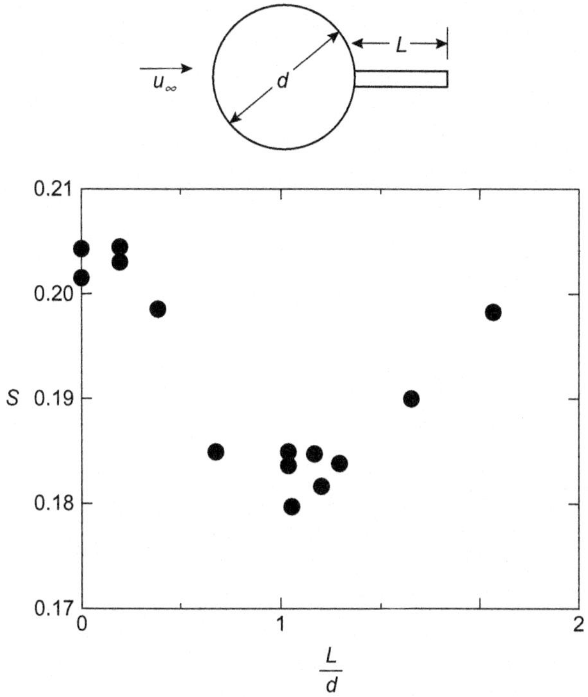

FIGURE 4.19 Variation of Strouhal number S with length of splitter plate [69]; $d = 1$ in., $R_d = 2 \times 10^4$.

type of motion. This transverse motion also organizes the vortex shedding along the axis of the cylinder, shown visually by Koopmann [72] at $R_d = 200$, as well as changing the vortex spacing in the vortex street [71]. These aspects will be further discussed in Section 5.7.3, Self-Excited Vibration. The effect of cylinder motion goes far beyond influencing the magnitude of oscillating lift. The entrainment of the wake disturbances by the motion of the cylinder brings about both a phase and an amplitude relationship between the cylinder motion and the wake-induced lift.

Another influence on vortex shedding can be effected by splitter plates, studied both experimentally, e.g., [69,73–75] and numerically to examine effects of splitter plate motion [76]. Fig. 4.19, due to Gerrard [69], shows a diagram of a cylinder with a splitter plate and the measured variation of Strouhal number as the length of splitter plate was changed. The splitter plate interferes with the cross-wake interaction of the shear layers, reducing unsteady lift and shedding frequency. The reduction in Strouhal number for a plate length equal to the cylinder diameter implies that a spreading in the wake width occurs and that the drag coefficient increases; see Section 4.3.1. The results are essentially in agreement with those of Apelt et al. [73,74].

Since a point in the wake one diameter downstream from the cylinder coincides with a formation length of $1.5d$, as shown in Fig. 4.6, the splitter plate interference is conjectured to have caused a downstream shift in formation. This downstream shift may also cause a reduction in the strengths of shed vortices. In earlier studies, Roshko [43] had found similar effects of splitter plates on Strouhal numbers in the same range of Reynolds number, and in Ref. [30] he found that a splitter plate, with $l/d = 2.65$, annihilated vortex shedding at R_d greater 3×10^6. The influence of splitter plates on vortex shedding from airfoils will be discussed in Chapter 11, Noncavitating Lifting Sections.

4.4 ESTIMATIONS OF WAKE-INDUCED FORCES IN TWO-DIMENSIONAL FLOW

In spite of the development of powerful numerical capabilities that can provide details of flow and forces on circular (e.g., Refs. [24,25,60–62]) and noncircular (e.g., Refs. [76–78]) cross-section cylinders a very simple, yet powerful, representation of the forces on long circular cylinders has been in use for many years. In this, the vortex street is considered a two-dimensional array of line vortices that trail behind the shedding body as illustrated in Fig. 4.7. The modeling of the two-dimensional vortex street wake in this manner was originally proposed by von Karman and Rubach [6], and in its original form, the von Karman vortex street, shown in Fig. 4.7, consisted of two parallel and infinite rows of vortices that are separated a distance h apart. In each row, the vortices of like sign are a distance l apart, and those in the upper and lower rows have opposite sign with the vortices alternating in position. In analytically modeling the two-dimensional wake-induced forces, the von Karman vortex street is truncated as a pair of semi-infinite streets of point vortices behind the shedding body as shown in Fig. 4.7. The analysis of the infinite vortex street is classical, with treatments in such books, e.g., of Milne-Thompson [79] and Lamb [7]. Wille [80] provides an excellent review of some of the physical aspects of the vortex street stability and configuration.

Complex variable theory is used by Sallet [81,82] to analyze the two-dimensional problem of steady drag and unsteady lift as an extension of von Karman's [6] analysis. Ruedy [83] and Chen [32–34] had previously presented similar analyses.

The stability analysis of the row of vortices as considered by von Karman and Rubach [6] and quoted in many texts (see, e.g., Milne-Thompson [79] and Lamb [7]) determines that there is a fixed relationship between h and l, which is

$$h = 0.281l$$

In a later development, however, Birkhoff [84] points out that as the vortex street moves downstream the momentum of the system, which is proportional to the moment of vorticity, Γh, must remain constant. This requires that h increases as the circulation, Γ, of each vortex diminishes under the action of viscosity in the far wake. This gives rise to the spreading of the wake that can be observed in Homann's photographs (Fig. 4.1) and that has been measured; see, e.g., Frimberger [85] and Schaefer and Eskinazi [27]. Furthermore, the pair spacing l tends to remain constant so that von Karman's relationship is only approximate. Also Bearman's [44] results and Fig. 4.12 show that the spacing ratio is dependent on the vortex velocity. Notwithstanding this argument, it is still often analytically attractive to utilize von Karman's constant because it represents a constant value that roughly agrees with measurements.

The dimensionless frequency of vortex shedding is written as [7,79]

$$S = \frac{f_s d}{U_\infty} = \frac{d}{l}\left(1 - \frac{|U_v|}{U_\infty}\right)$$

where U_v is the velocity of the vortex street relative to the free stream and toward the body,

$$|U_v| = \frac{\Gamma}{2l}\tanh\left(\frac{h\pi}{l}\right)$$

and the velocity relative to the body is $U_\infty - U_v = U_c$. The associated amplitudes of coefficients of unsteady lift and steady drag are [81,82]

$$C_L = \frac{\Gamma}{2U_\infty d}\left[1 - \frac{3|U_v|}{U_\infty}\right] \tag{4.19a}$$

but by replacing U_v and Γ this expression changes to

$$C_L = \left(1 - S\frac{l}{d}\right)\left(3S\frac{l}{d} - 2\right)\frac{l}{d}\coth\left(\frac{\pi h}{l}\right) \tag{4.19b}$$

and

$$\overline{C}_D = \frac{4l}{\pi d}\left(\frac{U_v}{U_\infty}\right)^2\left[\coth\frac{\pi h}{l} + \frac{\pi h}{l}\left(\frac{U_\infty}{|U_v|}\right)\right]\coth\frac{\pi h}{l} \tag{4.20}$$

with a mean-square value $\overline{C_L^2} = C_L^2/2$. The parameter l/d has been deduced from measurements of $(U_\infty - |U_v|)$ by a number of authors. Eq. (4.20) yields the relationships used in Section 4.3.1; von Karman's formula for $\overline{C}_D$ is obtained by letting $h/l = 0.283$, or $\coth \pi h/l = \sqrt{2}$. One of the most classical experimental surveys of wake structural parameters is that of Fage and Johansen [86,87]. Fig. 4.20A shows the experimentally determined values of l/d with R_d as summarized by Chen [32]. These values of l/d, as well as

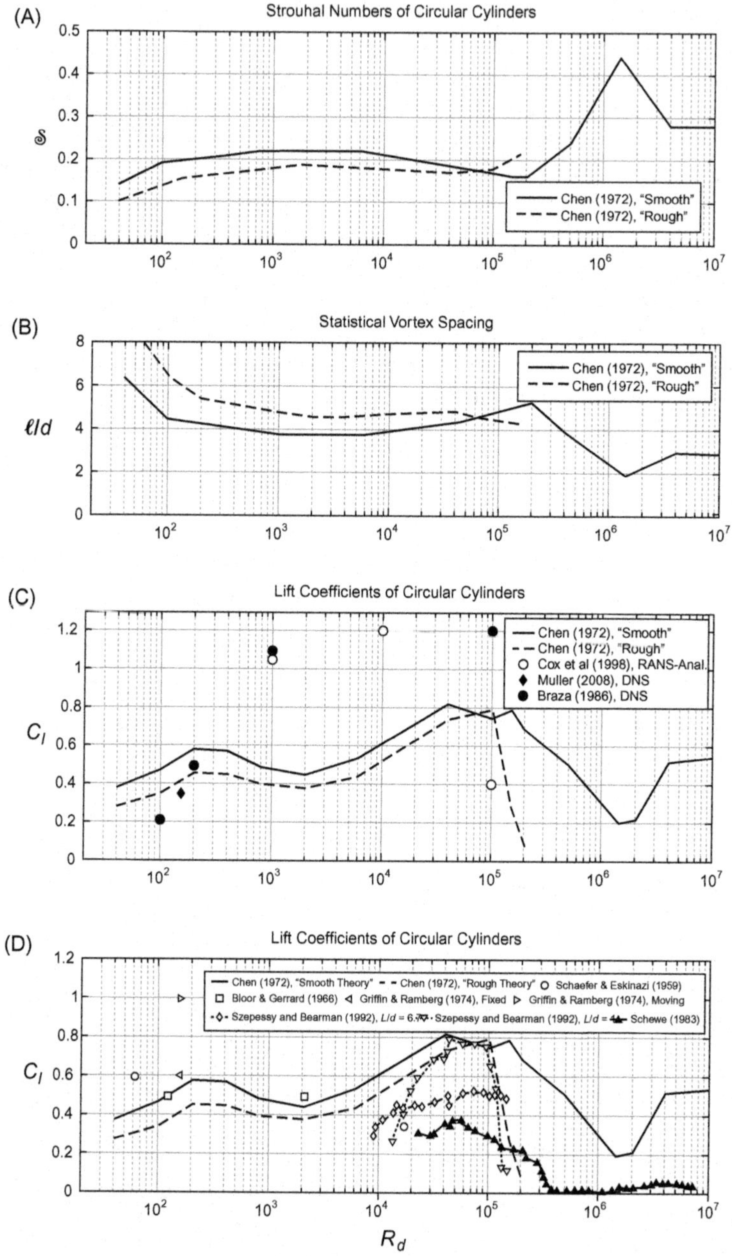

FIGURE 4.20 Calculated values of (A) Strouhal number, (B) streamwise vortex spacing, *l/d*, (C) root-mean-square oscillating lift coefficients, and (D) further measurements of root-mean-square lift coefficient beyond those in Fig. 4.13. All are shown as functions of Reynolds number. Refer to the studies of Chen [33], Cox et al. [60], Braza [65], Muller [88], Schaefer and Eskinazi [27], Griffin and Ramberg [71], Szepessy and Bearman [64], Schewe [63].

TABLE 4.1 Vortex Strengths, Velocities, and C_L Indicated by Eq. (4.19a)

R_d	$\Gamma/U_\infty d$	$\frac{\lvert U_v \rvert}{U_\alpha}$	C_L	Reference
60	2.44	0.1	0.85	[27]
120	2	0.1	0.7	[27]
144	2.5	0.1 est	0.87	[71]
2000	1.7	0.14	0.5	[70]
16,000	1.46	0.18	0.34	[70]
144	4.2	0.1 est	1.5	[71] (moving cylinder)

est means estimated

values of S in Figs. 4.10 and 4.11, may be used to estimate the root-mean-square lift coefficient as $C_L(\sqrt{2})^{-1}$ using Eq. (4.19a,b) and assuming time-harmonic forces. Fig. 4.20B and Table 4.1 show the results. The calculated root-mean-square lift coefficients are of the correct order of magnitude, but they do not precisely agree with the high values of $\overline{C_L^2}^{1/2}$ that have been reported by Gerrard [42] and Macovsky [50]. In assessing the validity of Eq. (4.19a,b), it is well to note that wake properties l/d, $\Gamma/U_\infty d$, and $|U_v|U_\infty$ vary with distance downstream of the cylinder. Furthermore, Eq. (4.19b) is sensitive to small uncertainties in l/d, which is numerically near 0.9. Values from Eq. (4.19a) are sensitive to values of $\Gamma/U_\infty d$ that must be derived from velocity measurements of the wake using appropriate modeling of vortices of finite core radius. Thus this quantity is not known to great precision. Finally, measured wake parameters and lift coefficients have not been determined together in an experiment.

The effect of transverse cylinder motion is shown in Fig. 4.20 at $R_d = 144$. Griffin and Ramberg [71] measured greater vortex strengths with a peak-to-peak cylinder displacement of $0.3d$ than with no motion. This increase in circulation, shown in Table 4.1, is responsible for the calculated increase in the lift coefficient although it should be emphasized that the data in Fig. 4.20 was obtained on fixed cylinders, with the one exception for a moving cylinder attributed to Griffin and Ramberg [71]. However, free stream turbulence also appears to increase vortex-induced lift as examined in a systematic experimental study by So et al. [89]. The effects of free stream turbulence on vortex-induced vibration (with vibration amplitudes on the order of $0.003d$ at R_d=) were observed to increase cylinder motion with increased free stream turbulence. Over the velocity range investigated, the increased vibration was deduced to be partly due to increased lift force due to turbulence-modified vortex shedding.

There has been another attempt at calculating the frequency and magnitude of fluctuating lift as reported by Gerrard [90]. This is a direct calculation of the potential field that results from the dynamical behavior of parallel shear layers. Each shear layer is modeled as a sheet of elemental vortices that are free to move under interaction with each other and with the mean flow past the cylinder. The motions of the shear layers generate a resultant set of large-sized vortices or concentrations of vorticity in the wake of the cylinder. The geometry of the vortex street and the dynamical characteristics of the calculated lift coefficients are in reasonably good agreement with measurement. A similar model of the vortex street generation had been used earlier by Abernathy and Kronauer [91]. Their calculations disclosed that the shear layer exhibited a "modelike" behavior in which concentrations of vorticity would occur in groups of 6, 4, and 2. The individual circulation of each cloud, however, increased dramatically as the number of clouds decreased to 2 per wavelength. They calculated a vortex spacing ratio $h/l = 0.28$ for this case. They also point out that the number of vortex clouds, or concentrations, times the spacing ratio h/l is roughly constant.

4.5 FORMULATION OF THE ACOUSTIC PROBLEM FOR COMPACT SURFACES

4.5.1 The General Equations

For the purposes of this chapter, we consider an acoustically compact surface to be one that has both an axial correlation length of vortices and a diameter much smaller than the observation distance from the body and an acoustic wavelength. The discussion also applies to other compact surfaces of noncircular cross section that have acoustically small thicknesses and chords. It will soon be apparent that this important class of surfaces has the special property that the radiated sound power bears a simple relationship to the statistics of alternating forces exerted by the wakes on the bodies. This is why an exhaustive treatment of the oscillating loadings has been given in the previous sections.

We start with the integral relationship for the radiated pressure that is due to Curle [92] and that was derived in Chapter 2, Theory of Sound and its Generation by Flow. There we found the linear acoustic radiated pressure disturbance, $p_a(\overline{x}, t)$ as Eq. (2.71), rewritten here for the case of compact source volumes (see Eq. (2.58))

$$\begin{aligned} p_a(\overline{x}, t) \simeq & \frac{1}{4\pi c_0^2} \frac{x_i x_j}{r^3} \iiint_v \frac{\partial^2}{\partial t^2} T_{ij}\left(\overline{y}, t - \frac{r}{c_0}\right) d^3\overline{y} \\ & + \frac{-1}{4\pi} \frac{1}{r} \iint_s \left[\frac{\partial}{\partial t}(\rho u_n)\right] dS(\overline{y}) \\ & - \frac{1}{4\pi c_0} \frac{x_i}{r^2} \iint_s n_i \frac{\partial}{\partial t}\left[\rho u^2 + p\right] dS(\overline{y}). \end{aligned} \tag{4.21}$$

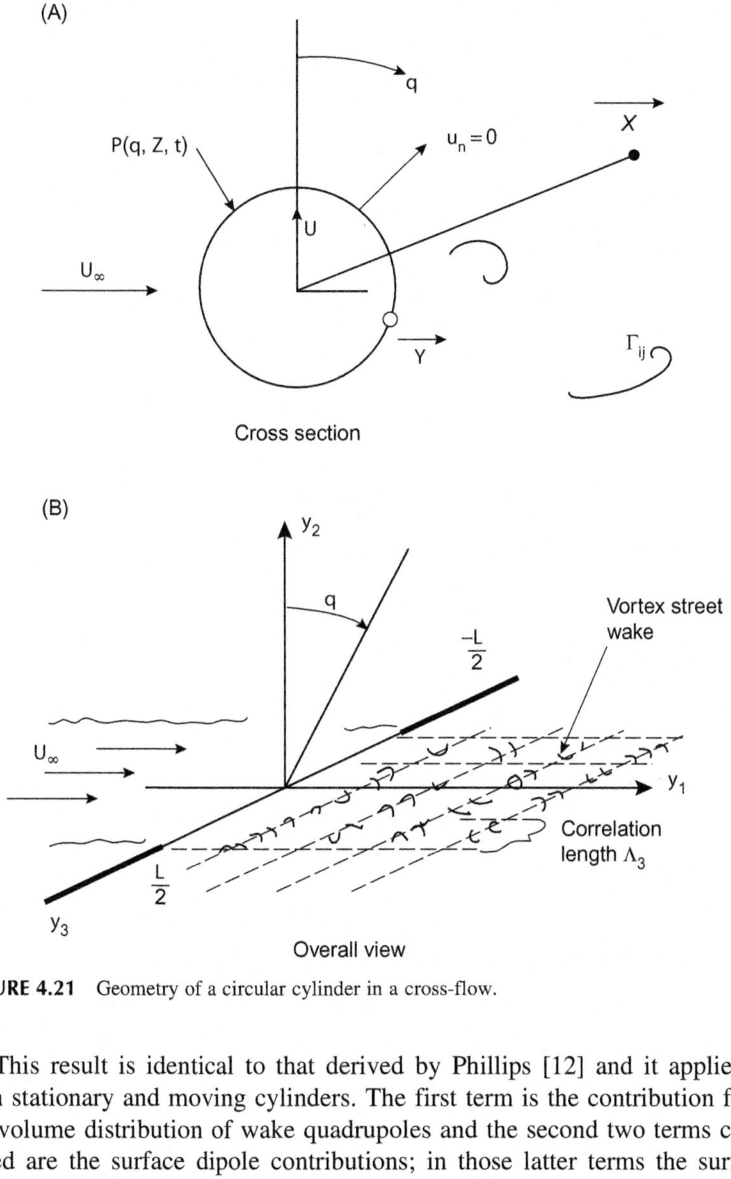

FIGURE 4.21 Geometry of a circular cylinder in a cross-flow.

This result is identical to that derived by Phillips [12] and it applies to both stationary and moving cylinders. The first term is the contribution from the volume distribution of wake quadrupoles and the second two terms combined are the surface dipole contributions; in those latter terms the surface velocity u vanishes in the case of a rigid cylinder.

For purposes of illustration the circular cylinder is examined, but generalization of the final result to other cross-section geometries is simple. In the evaluation of Eq. (4.21) the various terms will be referred to the parameters illustrated in Fig. 4.21. The forces on the cylinder (and on the fluid) are expressed by Eq. (4.7). Any motion induced in the cylinder as a result of this pressure distribution will cause a velocity $U(z)$ that is also perpendicular to the stream. Similarly, fluctuating drag forces could induce streamwise motions. The acceleration associated with this velocity causes a directly proportional fluid reaction pressure, which also has the directivity cos θ. Thus both of these integrands have the same circumferential directivity, which is identically zero in the direction of flow and which has its maximum absolute value in the cross-stream direction. In general the phases of these contributions must be considered in an evaluation of the net sound pressure.

The relative importance of the quadrupole and dipole contributions in Eq. (4.21) in the case of the rigid (stationary) cylinder was determined in Section 2.5.5. As shown at the end of that section, the dipole contribution is of order

$$p_{\mathrm{d}} \sim \rho_0 U_\infty^2 M(d/r)$$

while the quadrupole contribution is of order

$$p_{\mathrm{q}} \sim \rho_0 U_\infty^2 M^2(d/r)$$

where M is the stream Mach number. All velocity fluctuations are assumed to scale on the free stream velocity U_∞ and the forces to scale on $\rho_0 U_\infty^2 \Lambda_3^2$. Source volumes and correlation lengths are assumed to scale on S. Accordingly the ratio of these contributions is

$$\frac{p_{\mathrm{q}}}{p_{\mathrm{d}}} \sim M \tag{4.22}$$

Thus for the low mean-Mach-number flows that are the topic of this chapter, the quadrupole radiation is not a significant contributor to the total flow-induced radiation. In the remainder of our discussions the quadrupole term will be ignored.

4.5.2 Sound From a Rigid Cylinder in a Cross-Flow

A closed-form expression for the sound radiation from the rigid cylinder in a cross-flow will now be derived. By neglecting cylinder vibration, we find the dipole sound of Eq. (4.21) takes on the simple form of Eq. (2.76)

$$p_{\mathrm{a}}(\mathbf{x}, t) = -\frac{1}{4\pi c_0}\frac{x_i}{r^2}\int_{-L/2}^{L/2} \frac{\partial}{\partial t}\left(f_i\left(z, t+\frac{r}{c_0}\right)\right) dz \tag{4.23}$$

The cylinder is fixed relative to the acoustic field. The radiated sound pressure in any direction is thus seen to be proportional to the time rate of change of the total force on the cylinder in that direction. Thus there is sound radiated normal to the flow direction by the lift fluctuations as well as in the flow direction by the drag fluctuations, although these forces are only one-tenth the lift forces. Note that the radiation is also independent of the shape of the body, except that geometry influences the forces. Expressions for the forces, which permit a consideration of the axial nonuniformity of phase, are derived from Eqs. (4.7) and (4.8). The forces per unit length are of the form implied by Eq. (4.9), i.e.,

$$f_i(z,\ t) = \tilde{f}_i(\omega)e^{i[\phi(z,\ t)-\omega t]} \tag{4.24}$$

where $\phi(z,\ t)$ is the axial phase function and $\omega/2\pi$ is the frequency. The amplitude of the force per unit length at a location on the cylinder is $\tilde{f}_i(\omega)$, which for a given value of U_∞ is, in general, a function of frequency. When the vortex shedding is periodic, the lift and drag components are concentrated at the frequencies ω_s and $2\omega_s$, respectively.

The mean-square radiated sound pressure $\overline{p^2_{\text{rad}}}(x)$ is found from Eq. (4.23) by

$$\overline{p_a^2}(\mathbf{x}) = \frac{1}{16\pi^2}\frac{\omega^2}{c_0^2}\frac{x_i x_i}{r^4}\int_{-L/2}^{L/2} dz_1 \int_{-L/2}^{L/2} dz_2 \times \frac{1}{T}\int_0^T \left[f_i\left(z_1,\, t+\frac{r}{c_0}\right) f_i^*\left(z_2,\, t+\frac{r}{c_0}\right)\right] dt \tag{4.25}$$

where ω is ω_s or $2\omega_s$ depending on whether periodic lift or drag fluctuations are considered. Now, incorporating Eq. (4.24) into Eq. (4.25), we find the time-averaged far-field mean-square pressure

$$\overline{p_a^2}(\mathbf{x}) = \frac{\omega^2}{16\pi^2 c_0}\frac{x_i x_i}{r^4}\left[\frac{1}{2}|\tilde{f}_i(\omega)|^2\right]\int_{-L/2}^{L/2}\int_{-L/2}^{L/2} R_{\text{pp}}(z_1 - z_2)\, dz_1\, dz_2 \tag{4.26}$$

where we have assumed that the characteristics of the correlation function are those which were developed in Eqs. (4.11)–(4.16). This equation also holds as long as the observation point is not so close to the axis of the cylinder that the projection of the acoustic wavelength onto the axis is much larger than the correlation length $2\Lambda_3$. The integral in Eq. (4.26) can be evaluated in terms of the correlation length as

$$\begin{aligned}\int_{-L/2}^{L/2}\int_{-L/2}^{L/2} R_{\text{pp}}(z_1 - z_2)\, dz_1\, dz_2 &= \int_{-L}^{L} dr \int_{-L/2+r}^{L/2} dz_2\, R_{\text{pp}}(r)\\ &= 2[L\Lambda_3 - \gamma_c\Lambda_3]\end{aligned}$$

where Λ_3 is the correlation length as defined by Eq. (4.17) and

$$\gamma_c = \frac{1}{\Lambda_3}\int_0^\infty rR_{\text{pp}}(r)\, dr$$

is the centroid of the correlation function. The centroid of the correlation function is on the order of $2\Lambda_3/3$ as indicated in Ref. [52]. Therefore under these simplifications Eq. (4.26) reduces to

$$\begin{aligned}\overline{p_a^2}(\mathbf{x}) &= \frac{\omega^2}{16\pi^2 c_0^2}\frac{x_i x_i}{r^4}\left\{\overline{C_L^2}\left(\frac{1}{2}\rho_0 U_\infty^2\right)^2 2d^2\Lambda_3(L-\gamma_c)\right\}\\ \overline{p_a^2}(\mathbf{x}) &= \frac{\rho_0^2}{16c_0^2}\frac{\cos^2\theta}{r^2}\overline{C_L^2}\, U_\infty^6 S^2 2\Lambda_3(L-\gamma_c)\end{aligned} \tag{4.27a}$$

for the mean-square sound pressure radiated into the field directly above the cylinder by sinusoidal lift fluctuations; for sound radiated by drag fluctuations we replace $\overline{C_L^2}$ by $\overline{C_D^2}$, ω_s by $2\omega_s$, and $\cos\theta$ by $\sin\theta$. An analogous two-dimensional result is derived in the appendix. Using the directivity factors in Table 2.1 and Eq. (2.15) the mean-square sound pressure averaged over all angles θ at a distance $r \gg L$ is

$$\overline{p_a^2} = \frac{1}{12}\left(\frac{L}{r}\right)^2 q_\infty^2 M^2 S^2 \overline{C_L^2}\left(\frac{2\Lambda_3}{L}\right) \tag{4.27b}$$

or, using the definitions of Section 1.5.1, Sound Levels the sound pressure level is

$$L_s = -11 + L_q + L_M + 10\log\left\{S^2 C_L^2 \frac{2\Lambda_3}{d}\right\} + 10\log\frac{dL}{r^2}$$

and the sound power level is

$$L_\Pi = -2.8 + L_P + 1.5L_M + 10\log\left\{S^2 \overline{C_L^2} \frac{2\Lambda_3}{d}\right\}$$

where L_P is a mechanical power factor

$$L_P = 10\log\left(\frac{1}{2}\rho_0 U_\infty^3 dL\right)/P_0$$

and

$$L_M = 20\log M$$

with $\mathbb{P}_0 = 10^{-12}$ W and L_q is given in Fig. 1.12.

In the range of Reynolds number for which the vortex shedding is irregular and weakly periodic, the frequency spectrum of the forces $[\Phi_{FF}(\omega)]$ in the i direction replaces the bracketed term in Eq. (4.27a).

That is,

$$[\Phi_{FF}(\omega)]_i = \left[\frac{1}{2}\rho_0\, U_\infty^2\right]^2 2d^2\Lambda_3(L-\gamma_c)\overline{C_{L,D}^2}\phi_{FF_i}(\omega) \tag{4.28}$$

where $\phi_{FFi}(\omega)$ may be either narrow or broad over a frequency range centered on $\omega_s = 2\pi S U_\infty/d$ and $i = $ L or D depending on whether interest is in lift or drag-related sound. This function $\phi_{FFi}(\omega)$ has the dimensions of time, it is defined over $-\infty < \omega < \infty$, and its integral over all frequencies is normalized to unity, i.e.,

$$\int_{-\infty}^{\infty} \phi_{FF_i}(\omega)\, d\omega = 1 \tag{4.29a}$$

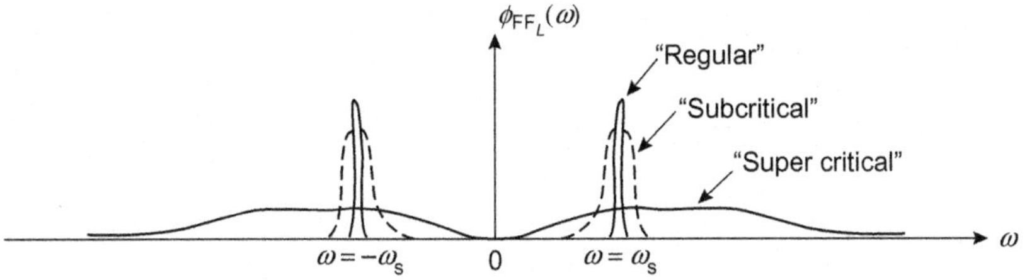

FIGURE 4.22 Schematic of spectrum forms of $\varphi_{FF_L}(\omega)$ to be expected on rigid cylinders in the various vortex-shedding zones shown in Fig. 4.2.

This relationship can be deduced by comparing Eqs. (4.28) and (4.10). In the regular vortex-shedding zone the lift spectrum will be nearly tonal as illustrated by the sketch in Fig. 4.22, while as Reynolds number increases into the supercritical range, say $3\times10^5<R_d<3\times10^6$ as indicated by Schewe [56], the spectrum broadens. Since the spectrum is normalized to unity, the level at $\omega=\omega_s$ decreases as the bandwidth increases. The spectrum is shown to extend over positive and negative frequencies as discussed in Section 1.4.4.1, Descriptions of Linear Bandpass Filters. In the case of purely harmonic shedding, as described in Eq. (4.7), the lift function is given by

$$\varphi_{FF_L}(\omega)=\frac{1}{2}\{\delta(\omega-\omega_s)+\delta(\omega+\omega_s)\} \tag{4.29b}$$

and similarly for the fluctuating drag function with ω_s replaced by $2\omega_s$.

Utilization of these relationships in Eq. (4.26) gives the sound emitted from the unsteady lift in the form

$$\begin{aligned}\overline{p_a^2}(\mathbf{x})=\frac{\cos^2\theta}{16\pi^2r^2}\left(\frac{U_\infty}{c_0}\right)^2\left(\frac{1}{2}\rho_0U_\infty^2\right)^2\left[\int_{-\infty}^{\infty}\left(\frac{\omega d}{U_\infty}\right)^2\phi_{FF_L}(\omega)\,d\omega\right]\\ \times\overline{C_L^2}x\left[\int_{-L}^{L}(L-r)R_{pp}(r)\,dr\right]\end{aligned} \tag{4.30}$$

These variations, Eqs. (4.27) and (4.30), in the expressions for the radiated sound intensity can be used to estimate sound levels from the known properties of the flow-induced forces over rather extensive ranges of the Reynolds number.

4.5.3 Review of Measured Acoustic Intensities

Measurements of sound levels from circular cylinders in a cross-flow have been conducted by Holle [93], Gerrard [94], Phillips [12], Leehey and Hanson [52], Koopmann [46], Etkin et al. [13], and Guedel [95]. Computational efforts to calculate the sound include direct integration of the Navier Stokes equations by Muller [88] and Inoue et al. [96]. We examine the experimental results to facilitate comparisons with theory of dipole sound and follow Phillips's example and rewrite Eq. (4.30) in the form

$$\overline{p_a^2}\frac{c_0^2}{\rho_0^2} = U_\infty^6\left[\frac{S^2 L\, d}{r^2}\right]\frac{\cos^2\theta}{16}\overline{C_F^2}\frac{2\Lambda_3}{d}\left(1-\frac{\gamma_c}{L}\right) \tag{4.31}$$

which clearly exposes the sound pressure level as a function of the lift coefficient and the axial correlation length. Fig. 4.23 shows the measurements of Phillips [12], Holle [93], and Gerrard [94], as presented by Phillips in the form of a linear function of $U_\infty\left[\frac{S^2 L\, d}{r^2}\right]^{1/6}$. The measurements were conducted at $\theta = 0°$ and, collectively, over a wide range of Reynolds numbers. The slopes of the lines in Fig. 4.23 are given by

$$\text{Slope} = \left[\frac{1}{16}\overline{C_L^2}\frac{2\Lambda_3}{d}\left(1-\frac{\gamma_c}{L}\right)\right]^{1/6}$$

In the Reynolds number range between 10^3 and 10^4 a typical value of the lift coefficient is 0.3 (Fig. 4.13) and of the correlation length is $10d$ (Fig. 4.16). This gives a slope 0.6, which agrees favorably with the slope of the line through the data of Holle [93] and Gerrard [94].

This correspondence between the measured sound pressure and the calculations that utilize the parameters of Figs. 4.12 and 4.15 is approximate. Inconsistencies in the measurements are minimized by taking the sixth root of the pertinent parameters $\overline{C_L^2}$ and Λ_3/d. A precise verification of Curle's [92] equation and of the integrated result for the Aeolian tone, Eqs. (4.27) and (4.31), awaited the nearly simultaneous attentions of Leehey and Hanson [52] and of Koopman [46]. In both cases simultaneous measurements of the fluctuating lift coefficient and of the correlation length provided a critical comparison of experiment and theory. Table 4.2 shows the ratios of calculated-to-measured sound intensities for the investigations. In all cases the flow-induced peak-to-peak cylinder displacement $2a$ was not zero so that the measured values of $\overline{C_L^2}$ may have been influenced by this effect. Recall that it was the contention of Leehey and Hanson [52] that cylinder motion did not influence the correlation length in their experiment. The agreement shown in this table, however, clearly shows that calculations of sound levels may be made quite reliably, given accurate measurements of the cylinder forcing parameters.

Although Eq. (4.31) shows a U_∞^6 increase for a given cylinder diameter d borne out by the collection of experimental results in Fig. 4.23, in some instances, say, for an elastic cylinder over a considerable speed range, this dependence on speed may not be observed. An example is shown in Fig. 4.24, where the measured sound level increases with a substantially greater speed dependence than U_∞^6. The change in the dependence shown in the figure occurs at a speed for which the vortex-shedding frequency nearly equals the resonance frequency of the cylinder. This effect reflects the

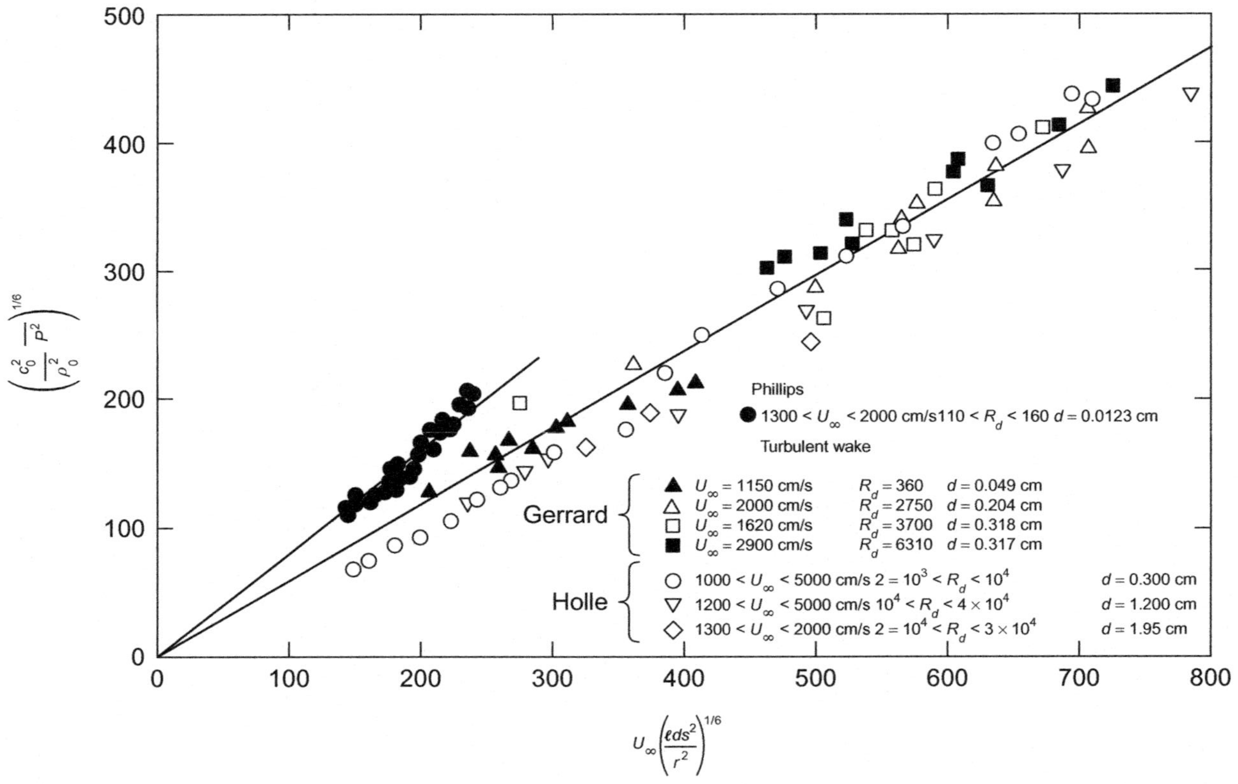

FIGURE 4.23 Far-field sound pressure levels for Aeolian tones shown as a function of wind speed to demonstrate the sixth-power law for rigid cylinders. *Data sources from Phillips OM. The intensity of Aeolian tones. J Fluid Mech 1956;1:607–24; Gerrard JH. Measurements of the sound from circular cylinders in an air stream. Proc Phys Soc London B 1955;68:453–61; Holle W. Frequenz- und Schallstarkemessungen an Hiebtonen. Akust Z 1938;3:321–31. Reprinted from Phillips (1956) by permission of Cambridge University Press.*

TABLE 4.2 Comparison of Measured and Calculated Sound Intensities

R_d	$\frac{2a_a}{d}$	$\overline{C_L^2}^{1/2}$	$\frac{2\Lambda_3}{d}$	$10\log\frac{(p_a^2)_{calc}}{(p_a^2)_{meas}}$	Reference
21,000	0.03	0.15	4(= L/d)	3	[46]
4000	*	0.04	15	1	[52]
4090	*	0.03	13	0	[52]
4140	*	0.08	12.5	1	[52]
6050	*	0.42	9.7	2	[52]
6260	*	0.43	9.2	2	[52]
6450	0.056	0.51	8.5	3	[52]

[a]An asterisk indicates that cylinder displacements were not measured.

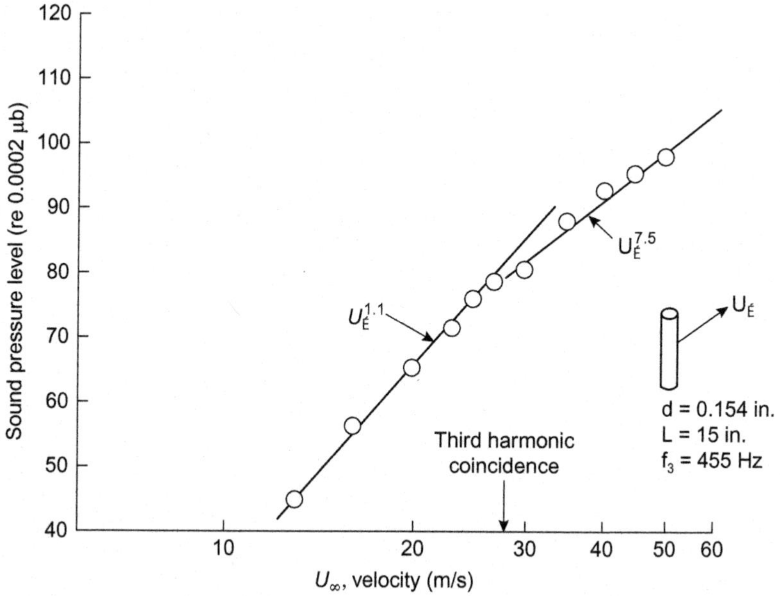

FIGURE 4.24 Sound pressure levels from a taught cylinder in a cross-flow from Leehey and Hanson [52]. These levels show the importance of vibration in causing departure from the sixth-power law.

change in the oscillatory lift coefficient as the speed changes and vibratory fluid–structure interaction alters the vortex shedding.

In spite of this apparent complication in a simplified interpretation of the relationship, Eqs. (4.26) and (4.27) are powerful for making simple acoustic estimates. Eq. (4.26) states that the radiated sound intensity from any acoustically compact rigid body is directly proportional to the time derivative of the mean-square fluctuating force so that, from Eq. (2.74),

$$\overline{p_a^2}(\mathbf{x}) = \frac{1}{16\pi^2 c_0^2} \frac{\cos^2\theta}{r^2} \overline{\left(\frac{\partial F(t)}{\partial t}\right)^2} \tag{4.32}$$

where θ is measured from the direction of the fluctuating force vector. Thus, in a practical situation in which the radiating surface is small, the acoustic intensity could be estimated by calculating the magnitude of the mean-square fluctuating force.

The sound radiation from a circular cylinder whose axis is set at a yaw angle to the flow direction has been measured experimentally by Guedel [95]. We let ϕ be the angle between the flow direction and the normal to the axis of the cylinder so that the velocity component normal to the cylinder is $U_\infty \cos\phi$. Assuming that the root-mean-square periodic forces are proportional to $[U_\infty \cos\phi]^2$ and that the vortex-shedding frequency is proportional to $U_\infty \cos\phi$, Eq. (4.31) shows that

$$\overline{p_a^2} \sim U_\infty^6 (\cos\phi)^6$$

Guedel's experimental results substantially support this result, yielding an exponent that varies from 5.34 to 6.

Measurements of shedding frequencies by Smith et al. [97] and Ramberg [98] have verified that $f_s d = SU_\infty \cos\phi$ increases slightly with the yaw angle ϕ; these observations have been corroborated by Williamson [135]. The increase is approximately 15% with $\phi = 30°$ and 25% with $\phi = 60°$. The implication of this can be seen referring to Eq. (4.19a). Given that the Strouhal number based on $U_\infty \cos\phi$ is constant, an increase in C_D based on $U_\infty \cos\phi$ would imply an increase in $\Gamma/(U_\infty d \cos\phi)$ and therefore an increase in the fluctuating lift coefficient. Guedel's results, however, indicate that the product $\overline{C_L^2}\Lambda_3$ decreases with $\cos\phi$ since the dependence of sound intensity is less than $(\cos\phi)^6$. This discrepancy can be interpreted as indicating a loss in axial coherence as ϕ is increased. Indications of a decrease in Λ_3 with increased yaw angle have been shown by Smith et al. [97]. The results of Ramberg [98] show that the wake thickness at the end of the formation zone is independent of yaw ($\phi < 50°$) and that the formation length is only weakly influenced by yaw. The base pressure coefficient, defined with the velocity $U_\infty \cos\theta$ is generally less than the coefficient at zero yaw angle by a factor as large as 1.2 at $\phi \sim 50°$. For yaw angles greater than 50°, Ramberg finds for cylinders without end plates that the character of shedding changes from one in which the vortices are nearly parallel to the cylinder, to one in which they peel off from the tip and are aligned with the flow direction. Vortex-induced vibrations of yawed cylinders due both to oscillating lift fluctuations and oscillating drag fluctuations have been studied experimentally by King [99], and numerically by Zhao et al. [100] using direct numerical simulation. Zhao et al. also examine a number of numerical issues in the simulation of $R_d = 1000$ flow, presenting lift coefficients, flow structures and comparisons with experimental results. The subject of yawed lifting surfaces will be taken up in Chapter 11, Noncavitating Lifting Sections.

We revise Eq. (4.31) to account for the effects of yaw by introducing the yaw angle

$$\overline{p_a^2}\frac{c_0^2}{\rho_0^2} = [U_\infty \cos\theta]^6 \left[\frac{S^2 L\, d}{r^2}\right] \frac{\cos^2\theta}{16}\, \overline{C_F^2}\, \frac{2\Lambda_3}{d}\left(1 - \frac{\gamma_c}{L}\right) \tag{4.33}$$

for yaw angles less than 45° as indicated by Ramberg's measurements [98] and Zhao's simulation [100].

To summarize the effects of real flow on vortex-shedding sound, departures from this simple power laws on, S, Λ_3,γ_c, $\sqrt{\overline{C_L^2}}$, as indicated by this equation, and as illustrated in Fig. 4.23, will accompany changes due to

1. Reynolds number and surface finish on the cylinder which will alter vortex formation;
2. vibration motion of the cylinder;
3. flow and pressure interaction with other cylinders and surfaces;
4. induction along the axis of the cylinder due to yaw, finite length, presence of end plates.

The cross-section geometry also plays a role as we shall discuss in Section 4.7.1. It is well to note that while in recent years little has been added to the knowledge of mechanism related to the circular cross-section geometry, strides in computational methods have enabled rather accurate simulations of two-dimensional and three-dimensional bluff-body flow and forces as encountered in practical engineering applications.

4.6 RADIATION FROM ROTATING RODS

We turn now to an examination of the acoustic intensity from cylinders rotating transversely to their axes. Fig. 4.25 shows the spin axis to be perpendicular to the axis of the cylinder with the angular velocity of spin denoted by Ω and the diameter is taken as constant along the length. The local tangential mean velocity is

$$U(z) = \Omega\, z$$

and at the tip of the cylinder

$$U(Z_t) = \Omega Z_t = U_t$$

where z is the radial coordinate measured from the spin axis. We assume that if there is a mean advance velocity parallel to the spin axis it is uniform and negligible compared to the tangential velocity at $z = Z_t$, where Z_t

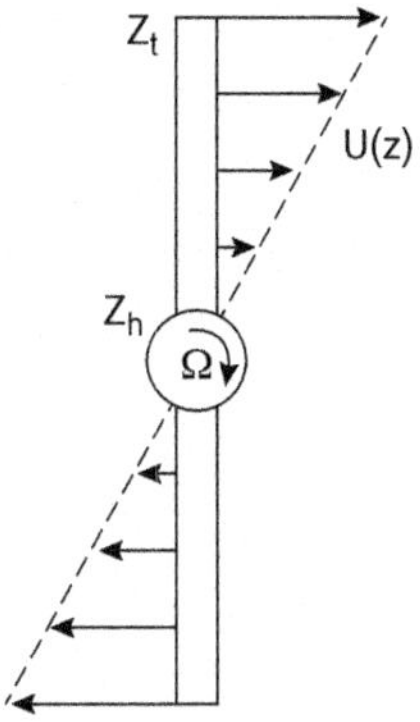

FIGURE 4.25 Geometry of a circular cylinder rotating about an axis perpendicular to its length.

is the tip coordinate, or the radius of spin. The rod extends from a hub of radius Z_h. We shall ignore until Chapter 12, Noise From Rotating Machinery the effects of Doppler shift and the temporal variation in range r due to the rotation of the acoustic sources. Both of these factors may cause a spreading of the frequency spectrum of the far-field radiated sound or the occurrence of side bands at frequencies both above and below the fundamental ω_s.

The analysis consists of writing Eq. (4.23) in a slightly more general form to account for the fact that the fundamental frequency of shedding and the magnitudes fluctuating forces increase outward along the cylinder. The linear increase in shedding frequency with distance from the hub is given by

$$\omega_s d = 2\pi S U(z) \quad \text{for} \quad Z_h < z < Z_t \tag{4.34}$$

we are essentially assuming that the vortices are shed from the cylinder continuously along its length, although experiments by Maul and Young [101] suggest that this assumption is not strictly correct. The vortices are shed in a stepwise fashion along the length with lengths of individual patterns on the order of $4d$ at $R_d = 2.85 \times 10^4$. Also, since the shedding frequency varies along the rod, we must maintain the general interpretation of the oscillating force spectrum that is given in Eq. (4.28). Now, combining Eqs. (4.25) (with ω moved inside the integral), (4.28), and (4.34) we obtain the total far-field acoustic intensity as

$$\overline{p_a^2}(\boldsymbol{x}) = \frac{1}{16\pi^2 c_0^2} \frac{\cos^2 \beta}{r^2} \int_{-\infty}^{\infty} d\omega \int_{Z_h}^{Z_t} dz_1 \int_{Z_h}^{Z_t} dZ_2 \left[\tfrac{1}{2}\rho_0 U^2(z)\right]^2 d^2 \times \overline{C_L^2} \omega^2 R_{pp}(z_1 - z_2) \phi_{pp}(\omega) \tag{4.35}$$

where β is the polar angle measured from the axis of rotation. The spectral density $\phi_{pp}(\omega) = \phi_{FF_L}(\omega)$ is the surface pressure spectrum and is strongly peaked near $\omega = \omega_s$, as described in connection with Eq. (4.28) and it can be expressed in terms of a universal function with a dimensionless frequency as its argument. Thus

$$\phi_{pp}(\omega) = \frac{d}{U}\phi_{pp}\left(\frac{\omega d}{U}\right) \tag{4.36}$$

and the dimensionless pressure spectral density satisfies the boundedness condition

$$\phi_{pp}(\omega\, d/U) < \phi_{pp}(2\pi\, S)$$

In order to substitute Eq. (4.36) into Eq. (4.35) conveniently we must further assume that the correlation length is small relative to the length of the cylinder $Z_t - Z_h$ and that the shedding frequency is constant over the correlation length. Under this further assumption we now obtain the frequency spectral density of acoustic intensity in the far field as

$$\Phi_{p_{\rm rad}}(\boldsymbol{x},\ \omega) = \frac{\cos^2\beta}{16\pi^2 c_0^2 r^2}\omega^2 \int_{Z_h}^{Z_t}\left[\frac{1}{2}\rho_0 U(z)^2\right]^2 d^2 2\Lambda_3 \frac{d}{U(z)}\phi_{pp}\left(\frac{\omega d}{U(z)}\right) dz \tag{4.37}$$

where β is the polar angle off the axis of rotation. The total acoustic intensity is given by an integral over frequency,

$$\overline{p_a^2(\boldsymbol{x})} = \int_{-\infty}^{\infty} \Phi_{p_{\rm rad}}(x,\ \omega)\, d\omega$$

In deriving Eq. (4.37) it was assumed for simplification that the centroid of the correlation is negligible in comparison with the length of the cylinder. If also Λ_3 is independent of z, we can rewrite the intensity spectrum in terms of the rotational tip speed U_T

$$\begin{aligned}\Phi_{p_{\rm rad}}(\boldsymbol{x},\ \omega) &= \frac{\cos^2\beta\, d^2}{16\pi^2 r^2}\left(\frac{U_T}{c_0}\right)^2\left(\frac{\omega\, d}{U_T}\right)^2 \frac{2\Lambda_3}{d}\frac{2L}{d}\overline{C_L^2} \\ &\times \left[\frac{1}{2}\rho_0 U_T^2\right]^2 \frac{d}{U_T}\int_{Z_h/L}^{Z_t/L}\left[\frac{U(z)}{U_T}\right]^4 \frac{U_T}{U(z)}\phi_{pp}\left(\frac{\omega d}{U(z)}\right)\frac{dz}{L}\end{aligned} \tag{4.38}$$

where $L = (Z_t - Z_h)$ is one-half of the total length of the rotating cylinder, Fig. 4.25. Eq. (4.38) clearly shows how the acoustic intensity depends on both the Mach number and the dynamic pressure based on tip speed as well as on a dimensionless frequency based on tip speed. The integral signifies the summing of the local oscillating pressures along the cylinder; the magnitude of the mean square pressure increases with $U^4(z)$ while the time scale of the vortex shedding decreases as $[U(z)]^{-1}$. The integral is a net oscillating pressure spectrum that has been made dimensionless on the cylinder diameter and the rotational tip speed. This net spectral density has the form

$$\Gamma\left(\frac{\omega d}{U_{\rm T}}\right)=\int_0^1\left[\frac{z}{L}\right]^3\phi_{\rm pp}\left[\left(\frac{\omega d}{U_{\rm T}}\right)\left(\frac{L}{z}\right)\right]d\left[\frac{z}{L}\right] \tag{4.39}$$

if $Z_{\rm h} \ll Z_{\rm t}$ so that since $L = Z_{\rm t}$ then $U(Z_{\rm h})$ is effectively 0 and $U(L) = U_{\rm T}$.

The spectral density of the radiated sound pressure from the rotating rod is distributed over frequency even if the vortex shedding at a point on the radius of the rod is locally a pure sinusoid. This can be seen more clearly by considering a simple analytical example of a locally periodic spectrum given previously in Eq. (4.28). The pressure fluctuations at a radial point z on the rod occur at discrete frequencies given by

$$\frac{\omega_{\rm s} d}{U(z)} = \pm 2\pi\, S$$

The spectral density of radiated sound for the rotating system is obtained by substitution into Eq. (4.39) as

$$\begin{aligned}\frac{\Gamma(\omega d/U_{\rm T})}{\overline{C_{\rm L}^2}} &= \frac{1}{2}\left|\frac{\omega d}{2\pi U_{\rm T} S}\right|^4\frac{1}{2\pi S} \quad \text{for} \quad \left|\frac{\omega d}{2\pi U_{\rm T} S}\right| \le 1 \\ &= 0 \quad \text{for} \quad \left|\frac{\omega d}{2\pi U_{\rm T} S}\right| > 1\end{aligned} \tag{4.40}$$

so that the acoustic intensity spectrum is

$$\begin{aligned}\Phi_{p_{\rm rad}}(\boldsymbol{x},\omega) &= \frac{\cos^2\beta\, d^2}{16\pi^2 r^2}(M)^2(2\pi\, S)^2\frac{2\Lambda_3}{d}\frac{2L}{d}q^2\overline{C_{\rm L}^2} \\ &\times \frac{d}{U_{\rm T}}\left(\frac{\omega d/U_{\rm T}}{2\pi}\right)^2\begin{Bmatrix}\dfrac{1}{2}\dfrac{1}{2\pi\, S}\left|\dfrac{\omega d/U_{\rm T}}{2\pi}\right|^4 & \text{for} \quad \dfrac{\omega d}{U_{\rm T}} < 2\pi \\ 0 & \text{for} \quad \dfrac{\omega d}{U_{\rm T}} > 2\pi\end{Bmatrix}\end{aligned} \tag{4.41}$$

and the total intensity of the sound is determined by integration to be

$$\overline{p_{\rm a}^2(\boldsymbol{x})} = \frac{\cos^2\beta}{28}\left(\frac{d}{r}\right)^2(M)^2(S)^2\frac{2\Lambda_3}{d}\frac{2L}{d}q^2\overline{C_{\rm L}^2} \tag{4.42}$$

where $M_{\rm T} = U_{\rm T}/c_0$ and $q = \frac{1}{2}\rho_0 U_{\rm T}^2$. These relationships will be compared to measured sound levels.

The validation of these relationships is provided by experiments conducted by Yudin [10], Stowell and Deming [9], Scheiman et al. [102], and results of the author [20]. In the 1930–50 time span, most quantitative measurements of radiated dipole sound from cylinders were obtained on rotating cylinders. This was the only practical means of achieving high Reynolds number flow without high background acoustic levels. Stowell and Deming's results, for example, demonstrate that the radiated intensity behaves as

$$\overline{p_a^2(\boldsymbol{x})} \propto \cos^2 \beta \, U^6 L / r^2$$

Examples of measured auto-spectral densities of the radiated sound intensity are shown in Fig. 4.26. Reynolds numbers R_T based on the diameter of the cylinder and the tip speed range from 2×10^4 to 4.9×10^5. The measurements were obtained on rods of length $2L$ that were spun about their centers either outdoors or in an anechoic room. That cylinder vibration did not influence the measurements is indicated by the absence of resonance effects in the measured spectral densities. Scheiman et al. [102] and this case found the sound levels to be insensitive to a superimposed axial mean velocity with magnitudes at least as large as $U_T/10$. The measured spectral densities are shown to be dimensionless functions of q_T and M_T, which are more peaked at the lower values of R_T. Comparison to Eq. (4.41), with $\overline{C_L^2}^{1/2} = 0.45$, $S = 0.2$, and $2\Lambda_3 = 5d$, is shown in Fig. 4.26A. Theoretically, if the vortex shedding is temporally sinusoidal at all points along the entire radius of rotation of the cylinder, a spectral peak will occur at $\omega d/U_T = 2\pi S = 1.26$ with sound absent at higher frequencies. The measured spectral density at $R_T = 2 \times 10^4$ increases as $(\omega d/U_T)^6$, and then rapidly decreases for $\omega d/U_T > 1.0$ in rough agreement with the simple theory. The disagreement for $\omega d/U_T \simeq 1$ is probably due to a spectral broadening of the local vortex shedding such as has been experimentally observed by Maul and Young [101] for a cylinder shear flow that varies uniformly along the axis. As Reynolds number increases, the spectral densities of the sound intensities broaden in a systematic fashion until at $R_T = 4.9 \times 10^5$ the spectrum has a greatly suppressed peak.

Various directivity measurements are shown in Fig. 4.27; they generally support the cos β dependence. The discrepancy at $\beta = 30°$ is unexplained particularly in light of the observations of Gerrard [94] and Guedel [95] on a stationary rigid cylinder that confirm the basic cosine directivity.

Total radiated intensities for a variety of experimental cases are shown as a function of tip speed Mach number in Fig. 4.28. The measured intensities are in close agreement with values that were calculated using Eq. (4.42) and the measured parameters in Figs. 4.11, 4.13, and 4.16 at the appropriate values of R_T. Estimated values of the measured $\overline{C_L^2}^{1/2}$ and $2\Lambda_3$ are shown in the respective figures. An average value of a given parameter was selected for a range of Reynolds number between $\frac{1}{2}R_T$ and R_T for each estimate. Although the precise values of the shedding parameters are perhaps somewhat questionable, the trends of computed and measured intensities shown in Fig. 4.28 demonstrate an important point. The speed dependence of the radiated sound is U_T^6 for a limited range of M_T that is bounded by $R_T < 4$ to

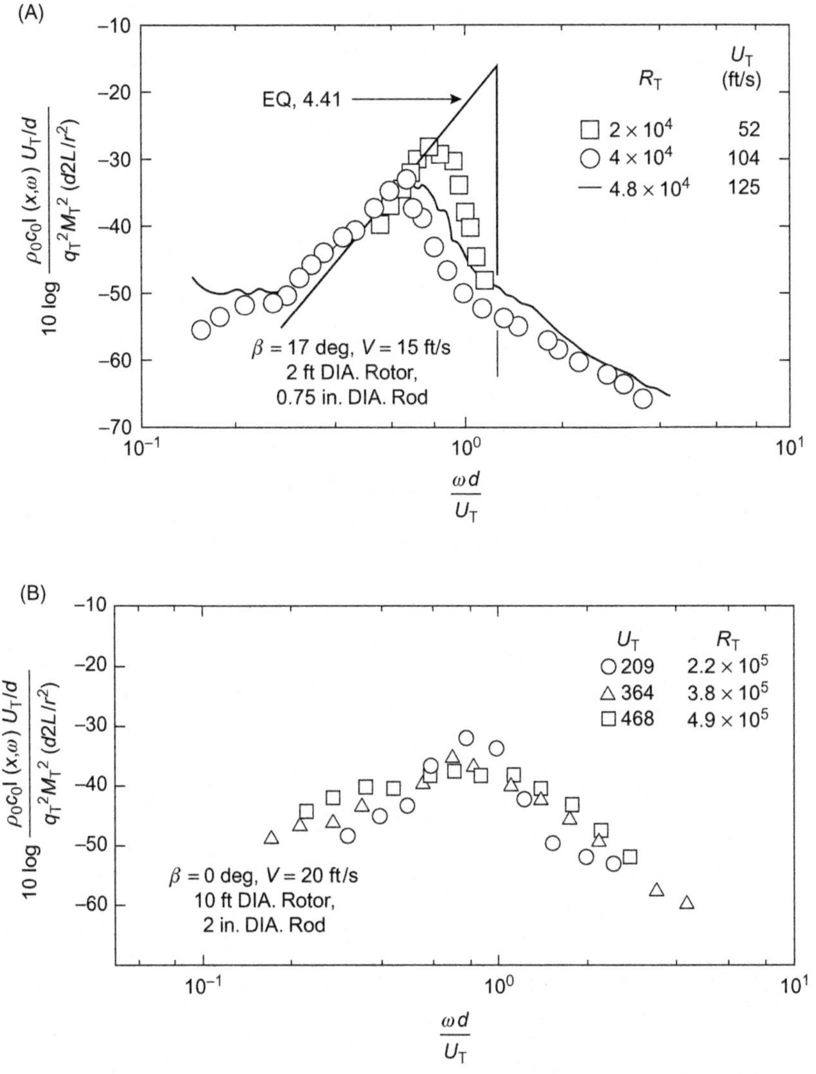

FIGURE 4.26 (A) Spectral density of noise from a rotating rod (see Blake [20], $L/d = 36$). (B) Spectral density of noise from a rotating rod (see Scheiman et al. [102], $L/d = 30$).

10×10^4. For higher values of Mach (and Reynolds) numbers the speed dependence falls off, becoming more like U^4_T. This change in speed dependence is matched by the calculated intensities. The calculations show that the reduction in both the fluctuating life coefficient and the axial correlation length at high Reynolds numbers can account for this observed change in speed dependence; see the trends shown in Figs. 4.13, and 4.16. This result

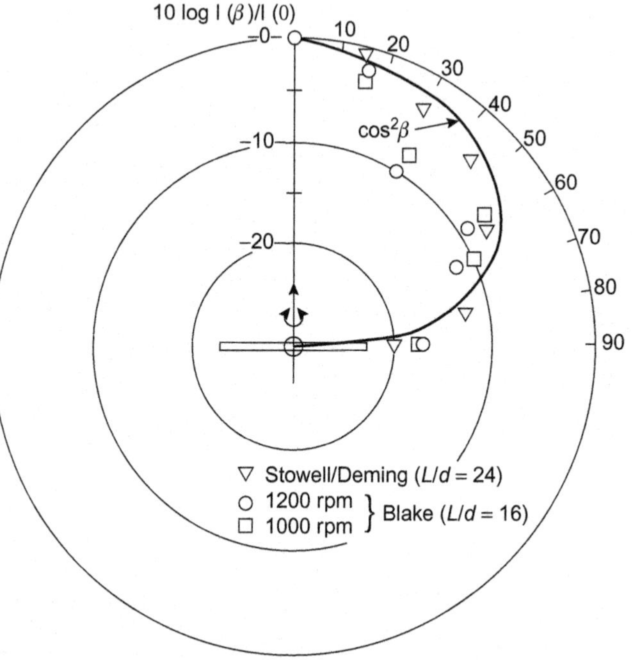

FIGURE 4.27 Noise from rotating rods: directivity patterns, $\beta = 0$ coincides with the axis of rotation. *Data sources from Stowell EZ, Deming AF. Vortex noise from rotating cylindrical rods. J Acoust Soc Am 1936;7:190–8 and Blake WK. Aero-hydroacoustics for ships, 2 vols. NSRDC Report 84–010; 1984.*

amplifies the statement made in the last section regarding Gerrard's [94] observed U^5 speed dependence.

In certain practical situations, the speed dependence of a particular sound source does not always uniquely describe the physical nature of the source. As seen in these instances the nature of the noise mechanism is identical in each case. The observed, U^4, U^5, U^6 speed dependences are the result of the shedding process itself being strongly Reynolds number dependent over a large range as delineated at the end of the preceding section.

4.7 OTHER TOPICS IN VORTEX-INDUCED NOISE

The problems of acoustic radiation from and the forced vibration of cylinders of noncircular cross section are rather specialized to specific instances. Also, extensions of the acoustics and vibration problem to multiple tube banks have been given by some investigators. In those cases for which data are nonexistent, the fundamental formulations of the previous sections can be used to

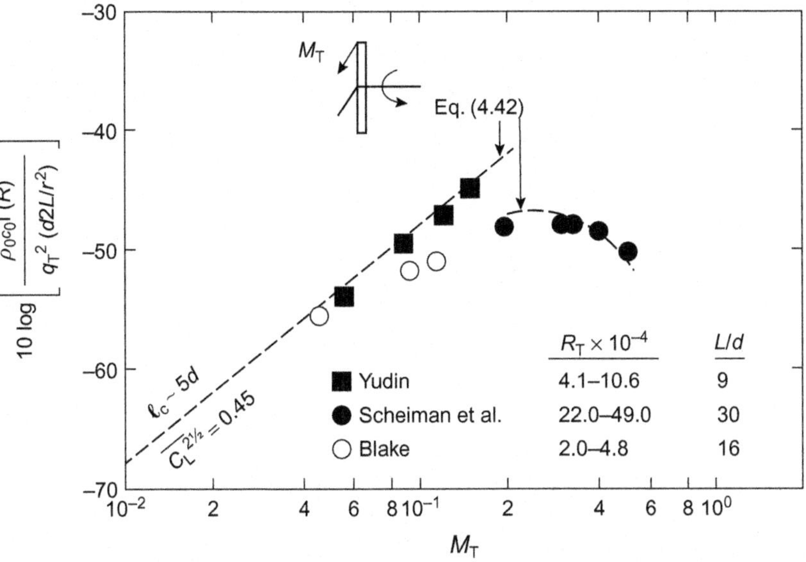

FIGURE 4.28 Variation of sound intensity level with tip Mach number. *Data sources from Yudin EY. On the vortex sound from rotating rods. Zh Tekhn Fiziki 1944;14(9):561; translated as NACA TM 1136; 1947; Scheiman J, Helton DA, Shivers JP. Acoustical measurements of the vortex noise for a rotating blade operating with and without its shed wake blown downstream. NASA Tech. Note TN D-6364; 1971; Blake WK. Aero-hydroacoustics for ships, 2 vols. NSRDC Report 84–010; 1984.*

provide estimates of force coefficients. When vortex structures are known, the estimated and measured magnitudes and correlation lengths of lift can be used to predict acoustic intensities for these, more complex, instances.

4.7.1 Cylinders With Noncircular Cross Sections

There have been a number of measurement programs in recent years that have yielded data useful for acoustic estimations. The measurements often have had application to the prediction of wind-induced fluctuating forces on architectural structures. Generally, the measurements are limited in scope so that wide ranges of Reynolds number or larger numbers of unsteady parameters have not been reported. Table 4.3 summarizes a number of useful results on some important shapes. More recently King and Pfitzenmeier [108] provided a systematic series of measurements on cross sections of varying shapes and varying values of span. Those results are generally in line with those of Table 4.3. In all cases, the oscillatory lift coefficient was determined using direct force measurement, and with the obvious exception of shape (d) in Table 4.3, sharp corners existed on the shapes. These results

TABLE 4.3 Lift Coefficients on Cylinders

Shape	R_d	$\overline{C_L^2}^{1/2}$	$2\Lambda_3/d$	S	Source
a 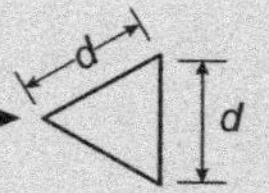	$1 \times 10^4 - 3 \times 10^4$	0.054	–	–	Integrated pressure [103]
b 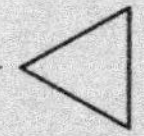	4.5×10^4	0.054	–	0.23	Measured lift [104] ($L/d = 6$)
c 	4.5×10^4	0.52	–	0.14	Measured lift [104] ($L/d = 6$)
d 	4.5×10^4	0.28	–	0.21	Measured lift [104] ($L/d = 6$)
e 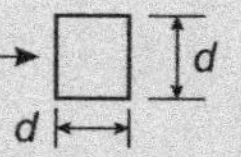	$3 \times 10^4 - 11 \times 10^4$		2.5		Integrated pressure [105]
f 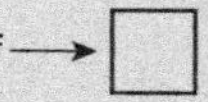	1×10^4	0.7 − 1.3	3–6	0.12	Integrated pressure [106]
g 	$3.3 - 13 \times 10^4$	1.0	–	0.125	Measured lift [107] ($L/d = 4.65$)
h 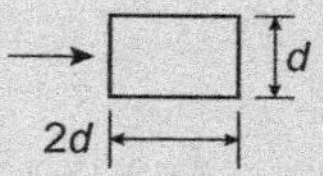	$3.3 - 13 \times 10^4$	0.35	0	0.083	Measured lift [107] ($L/d = 4.65$)
i 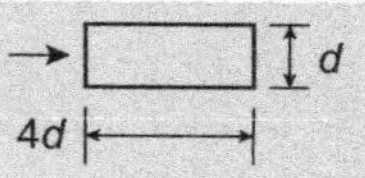	$3.3 - 13 \times 10^4$	0.05	–	0.118	Measured lift [107] ($L/d = 7.8$)

have been made dimensionless on the thickness d and the total span of lift measurement L. In all cases, the highest reported value of the lift coefficient occurred when the flat side of the sharp-edged cylinder faced the wind. According to Vickery [106] the magnitude and the correlation length of the lift are reduced by free stream turbulence; both values are shown in the table. The periodicity of the lift is also reduced. Bearman and Trueman [109] show separation to occur at the forward sharp edge of the square cross section, which causes a strong wake vortex field that is strongly correlated along the span. In case (h) the separated shear layers still do not reattach [107] before the trailing edge, but in case (i) they reattach before clearing the downstream trailing edge. In this case the lift coefficient is small perhaps, due to the formation of a weaker, more irregular, wake than in the other cases.

The coefficients of average drag and the vortex-shedding frequencies have been obtained on a variety of shapes by Delany and Sorensen [39]. The largest values of drag were observed on shapes such as (c), (e), (f), and (g) in the table with sharp corners. Drag coefficients of approximately $C_D = 2$ were measured in the nominal Reynolds number range 10^4-10^6. Rounding the corners, to the extent that the radius of curvature was 0.25–0.3 of the length d, resulted in values of C_D and S that were similar to those observed on cylinders of circular cross section. In the absence of measurements of fluctuating lift coefficients in these cases of Delany and Sorensen [39] we can only surmise that oscillating lift on the noncircular cylinders reduces and becomes comparable to those measured on circular cylinders as the edges are somewhat rounded. Rockwell [110] has found that oscillating pressures on square cylinders depend on the angle of incidence to the flat facing side α. The pressures increase somewhat as α reaches 4–6°, but drops to less than 0.1 of the value at $\alpha = 0$ when $\alpha > 10°$.

Strouhal numbers and vortex spacings have been measured in the wakes of elliptic cylinders of Modi and Dikshit [111]. Elliptical cross sections of various eccentricities e were used: $e = [1 - (b^2/a)^2]^{1/2}$, where a and b are the major and minor axes, respectively; $e = 0$ for circular cylinders; and $e = 1$ for a flat plate. The cylinders were oriented with their major axes aligned with the direction of flow. Strouhal numbers defined as $R = fh/U_\infty$, where h is the projected height in the direction normal to the flow, ranged from 0.20 to 0.22 for angles of attack 0° through 90° and eccentricities $e = 0.8$, 0.6, and 0.44. The Reynolds number $U_\infty a/\nu$ was 68,000. Unsteady lifts, or vortex strengths, were not measured, but the steady drag coefficient based on h was reported to increase with angle of attack. At zero angle of attack, C_D was on the order of 0.67–0.8 for the values of e cited above. Longitudinal vortex spacing in the far wake was found to be approximately $5h$. Finally, the vortex convection velocity, $(U_\infty - U_v)/U_\infty$, was found to be approximately 0.9 for $e = 0.92$ and 0.95 for $e = 0.44$. With these parameters, the equations in Section 5.5, Essential Features of Structural Radiation can be used to make order-of-magnitude estimates for the oscillating lift coefficient.

4.7.2 Unsteadiness in Tube Bundles

This topic is of importance in diagnosing vibration and noise in some heat exchanger applications. However, the fluidic interactions of parallel-oriented vortex-shedding bodies in close proximity are of sufficient general importance that we shall briefly discuss some aspects of the sources of unsteadiness in tube banks. Other, very much related, vortex-shedding phenomena are known to control the edge tone and the jet tone; these are discussed in Chapter 3, Shear Layer Instabilities, Flow Tones, and Jet Noise. Also of significance, but of less direct relevance to acoustics are flows with large-amplitude aero/hydroelastic effects. Review of this subject can be found in Ref. [112].

Flow across two parallel cylinders that are displaced a distance G perpendicular to the flow direction causes of unsteadiness with a range of frequencies. Fig. 4.29 shows the frequencies observed by Spivack [113] in the range of R_d from 10^4 to 10^5. As the gap thickness between the cylinder surfaces increases from zero the Strouhal number, based on the single cylinder diameter, increases from approximately 0.1. This is because the vortex shedding, which occurs on the outer surfaces of the cylinders, when the gap is closed is determined by twice the cylinder diameter. As the gap opening increases, the velocity of flow through the gap increases. For $G/d > 0.5$, disturbances of two frequencies were sensed throughout the region of flow surrounding the cylinders. These higher-frequency disturbances were not restricted to

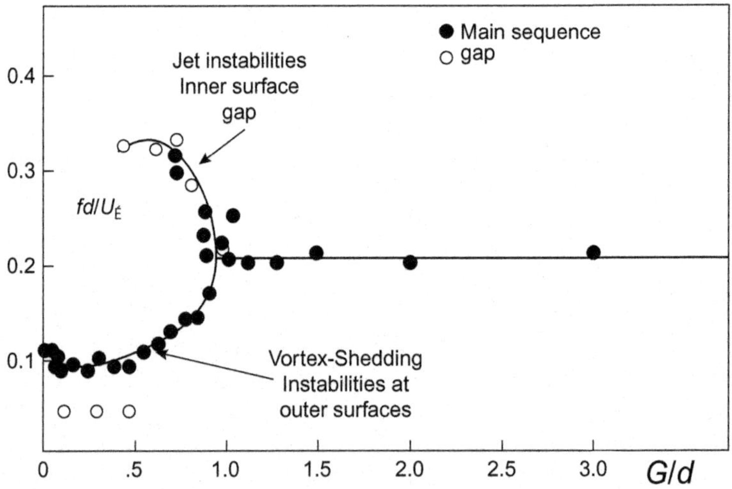

FIGURE 4.29 Strouhal number as a function of separation of two cylinders (for air speeds from 20 of Aeronautics and Astronautics; reprinted with to 140 ft/s with $\frac{11}{8}$-in. diameter cylinders). *From Spivack HM. Vortex frequency and flow pattern on the wake of two parallel cylinders at varied spacing normal to an air stream. J Aeronaut Sci 1946;13:289–301; © American Institute permission.*

locations immediately behind and on the center line of each cylinder, but rather they existed at locations both above and below the cylinder pair. The source of the higher-frequency disturbances is believed by Spivack to be related to jet instabilities associated with gap flow, analogous to those discussed at the end of Section 3.4.2, Dimensionless Frequencies of Jet Tones. Unfortunately his measurements of mean velocity were not conducted near enough to the gap to disclose the existence of a jet-like mean-velocity profile that would have generated these disturbances.

As the gap width was increased to a critical value $G/d = 1$ the disturbances attained a single frequency because the vortex shedding occurred independently on each cylinder. Measurements of unsteady forces on this geometry are not available; however, it is likely that force fluctuations will occur at both frequencies because both of the disturbances were easily detected throughout the flow field.

When the tubes are grouped in bundles, the tube interactions are more complex and dependent on whether the cylinders in adjacent rows are staggered or in line. The work of Chen [32–34] has provided unsteady lift coefficients and Strouhal numbers useful for the purpose of making predictions. Fig. 4.30 shows the general arrangements of in-line and staggered tube groupings. The pertinent parameters are the dimensionless transverse, T/d, and longitudinal, L/d, tube spacings. For either arrangement, the wakes of forward cylinders impinge on those downstream influencing the shedding from those cylinders. Measured spectral densities of velocity fluctuations in the tube banks have disclosed disturbances of increasing degrees of periodicity as the tube spacings increase [34]. At small tube spacings, such that the clearance magnitudes are less than the diameters of the cylinders, the fluid velocity disturbances can be a continuous spectrum [114]. Fig. 4.31 shows the Strouhal numbers, defined as $S = f_s d/U_\infty$, as obtained by Chen [115,136] from measurements of the frequencies of flow-induced vibration in tube bundles with both the in-line and staggered arrangements. The increase in the Strouhal number with a reduction in L/d for a constant value of T/d near 2 can be explained by postulating [34] that the length scale that is pertinent

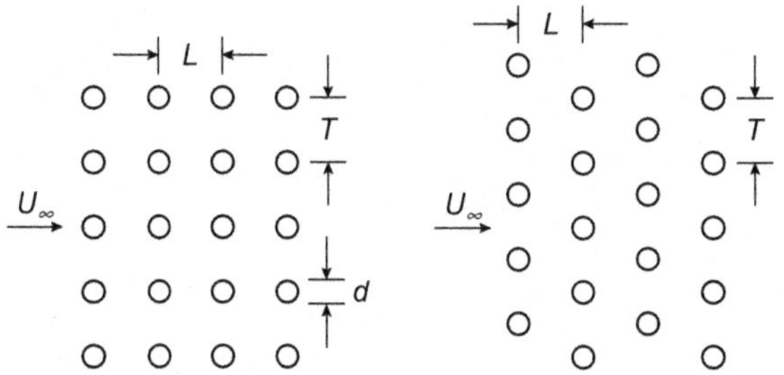

FIGURE 4.30 Arrangements of tube bundles.

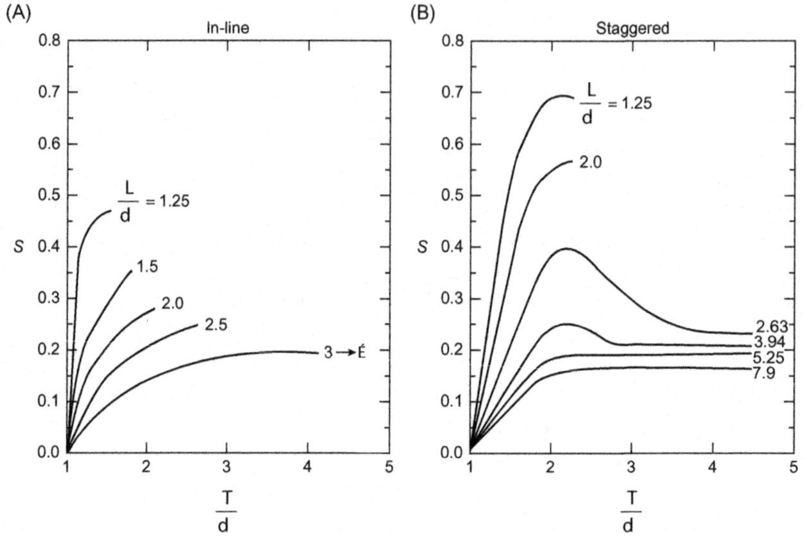

FIGURE 4.31 Strouhal numbers for tube bundles as functions of tube spacing parameter. *From Chen YN. Flow-induced vibration and noise in tube-bank heat exchangers due to von Karman streets. J Eng Ind 1968;90:134–46.*

to the fluid disturbances is proportional to L rather than d. This is because L determines the intertube gap distance. Thus as L/d decreases, a Strouhal number f_sL/U_∞, where U_∞ is the mean velocity into the tube bank, will be roughly a function of T/d.

The amplitudes of unsteady forces on tubes have been deduced from the measured values of vortex-induced tube vibration by Chen [34]. His results, which are summarized in Fig. 4.32, show higher amplitudes of lift coefficients in staggered tube alignments. For L/d greater than 3.0 (a gap distance $G/d \geq 2$) the lift coefficients increase with T/d in a regular fashion. Chen's values attained 0.5–0.9 for R_d between 10^3 and 4×10^4 depending on the geometry of the bundle. In this Reynolds number range, as we have already shown in Sections 5.3 and 5.4, General Features of Structures Driven by Randomly Distributed Pressure Fields; Modal Shape Functions for Simple Structures and Figs. 4.13 and 4.17, the lift fluctuations on single cylinders are particularly sensitive to environmental disturbances. In his method of measuring the lift fluctuations, Chen [115] necessarily recorded levels of transverse vibration whose peak-to-peak displacements were on the order of $0.5d$. Motion of this magnitude is certainly capable of causing an augmentation of fluctuating lift as indicated by Fig. 4.18. Also, as Leehey and Hanson [52] have shown, lift augmentation can occur without substantial increases in correlation length and without the attendant "lock-in" that often occurs in cases of cylinder motion. This "lock-in" is the condition, discussed

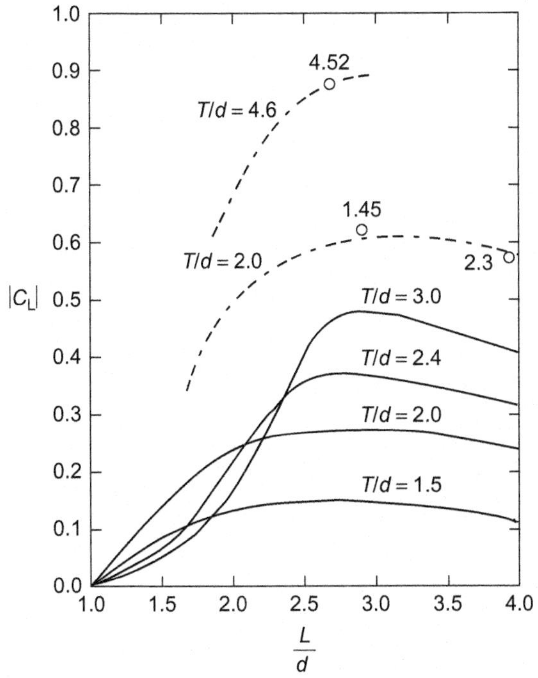

FIGURE 4.32 Amplitude of the fluctuating lift coefficient on cylinders in tube bundles shown as a function of L/d with T/d as parameter: (———), in-line tube spacing; (– – – – –), staggered tube spacing. *From Chen YN. Fluctuating lift forces of the Karman vortex streets on single circular cylinders and in tube bundles, Part 3—Lift forces in tube bundles. J Eng Ind 1972;94:623—8. © American Society of Mechanical Engineers.*

in Chapter 5, Fundamentals of Flow-Induced Vibration and Noise for which the frequency of vortex shedding nearly coincides with the resonance frequency of the cylinder and, over a small speed range, becomes independent of speed. Thus it is plausible that cylinder vibration influenced Chen's measurements and that the results in Fig. 4.32 are upper bounds.

There is little experimental information on correlation lengths in tube bundles. However, Chen [115] reports complete correlation along the entire cylinder axis, but he does not state what the tube length was. Finally, we note that since the tube banks are contained within enclosures, self-excited aeroacoustic resonances are possible. Thus, as Chen [115] suggests, vortex shedding may couple with acoustic cross modes of the chamber at flow rates exceeding an empirically determined threshold.

In another source, forces on tubes both in the flow direction (drag) and perpendicular to the flow cause a whirling motion of the tube in its fundamental mode when those forces attain the appropriate relative phase and magnitude with respect to each other and to the motion of the tube. A condition of self-excitation has been derived by Blevins [1,116,117], which states that the critical velocity through the tube bundle is given by

$$\frac{U_{\text{crit}}}{f_r d} = \frac{2(2\pi)^{1/2}}{(C_x K_y)^{1/4}} \left(\frac{\pi m \eta}{\rho_0 d^2} \right)^{1/2}$$

where m is the mass per unit length of the tube, U is the average velocity in the tube gaps, f_r is the mechanical resonance frequency of the tubes with loss factor η. The denominator of the first term contains dimensionless force coefficients C_x and K_y for forces aligned with and normal to the flow direction, respectively. This term, called the whirling parameter, is a function of transverse spacing T/d as shown in Fig. 4.33. When the velocity exceeds U_{crit}, the self-excitation occurs. The source of excitation, as postulated by Blevins [1,116] has to do with motion-dependent transverse and streamwise forces induced by changes in gap clearance as the tubes vibrate. Modulation of the gap clearance alters the throughflow and therefore the lift and drag of the tubes. Although the theory has some proponents, it is apparently not without question [116,117]. See also Refs. [118,119].

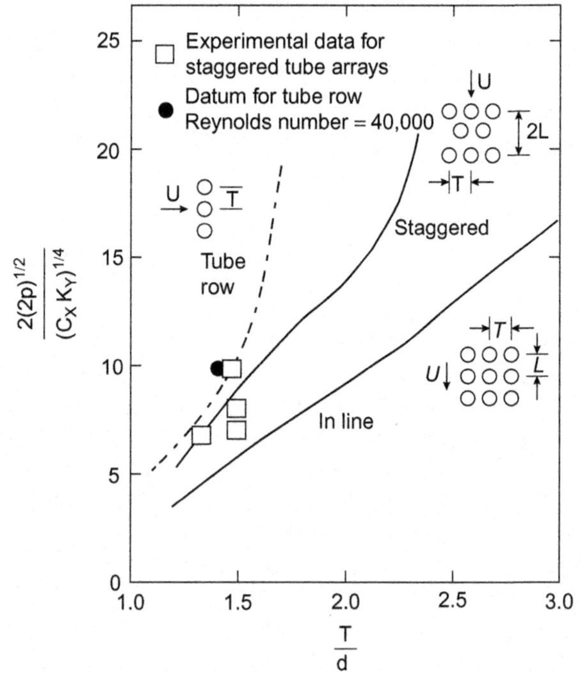

FIGURE 4.33 Whirling parameter for tube arrays. *From Blevins RD. Fluid elastic whirling of tube rows and tube arrays. J Fluids Eng 1977;99:457–61. © American Society of Mechanical Engineers.*

4.7.3 Methods of Reducing Vortex-Induced Forces

We shall now consider methods of reducing the flow-induced sound from cylinders; many of these are now well-known and will be recognized by the reader. These reductions are achieved by reducing the vortex forces and as well the axial correlation length of those forces. This has been accomplished by Scruton [120] and Weaver [121], who installed a helical strake consisting of a smaller-diameter cylinder wrapped around the larger cylinder in a helical pattern. Sharp-edged strakes are more effective than those with cylindrical cross section. Strakes of height $0.1d$ and pitch $5 - 10d$ have been found to be effective in completely reducing vibrations.

Axial taper of the cylinder is also effective since it allows for a variation of vortex-shedding frequencies. Since the correlation length of vortices is on the order of $5-8d$, the taper should be sufficient to change the vortex-shedding frequency by at least 30% in distances of $5d$. This is equivalent to a 6% taper.

Splitter plates, discussed in Section 4.3.4, are well known to modify the shedding frequency and forces. Experimentally-determined forces are contained in Ref. [75]; large eddy simulations can be found in Refs. [62,76]. The efficacy of splitter plates is based on the fact that formation lengths extend farther downstream of the base of the cylinder than naturally-occurring, as was shown in Fig. 4.6. The splitter plate of length on the order of 3 cylinder diameters will interfere with wake formation, delaying it and interfering with the development of circulation into the wake. Also, shear layer reattachment may occur on opposite sides of the splitter plates perhaps prohibiting the formation of the vortex street altogether. Splitter plates will be further discussed in Chapter 11, Noncavitating Lifting Sections.

Since free stream turbulence apparently increases alternating lift forces (see Section 4.4 and Fig. 4.17) in the Reynolds number range 10^3-10^5, reductions of upstream turbulence may reduce alternating forces. Accordingly, it is expected that slightly rough cylinders and tubes will generate larger lift coefficients than smooth ones [33,66]. Schmidt [47] has observed reductions by nearly $\overline{C_L^2}$ with smoothing of the surface of the cylinder. His data, shown in Fig. 4.12, were obtained with polished surfaces. The effect could be more pronounced at lower Reynolds numbers. Alternatively, we would expect large-size roughness to cause decorrelation of the vortex structure. Prismatic forms located at the separation position could be expected to generate trailing vortices, that decorrelate the vortex street. Unfortunately, such shapes would also cause cavitation in liquid applications. Finally, we note that Vickery [106] has found that increases in upstream turbulence result in decreases in lift coefficients on prismatic forms.

A particularly useful method of reducing vortex shedding is to install fairings to reduce "bluntness." These, then, are classed as aerofoil (or hydrofoil) shapes, which will be discussed in Chapter 11 *Noncavitating Lifting Surfaces.*

4.7.4 Sounds From Ducted Elements

Dipole sound is often generated by flow in pipes and ducts from flow spoilers and pressure reducers such as throttles, silencers, diffusers, flow straighteners, and (at the exit) gratings. Fig. 4.34 shows illustrations of the types of duct elements that are common. The disturbing bodies are generally not aerodynamically shaped so that the considerations of this chapter apply. We shall consider some of the engineering aspects of such sources as an example of the principles of this chapter and assume for the present that interest is in noise at frequencies low enough that the geometric extent of the flow disturbance in the duct is smaller than an acoustic wavelength. It will be further assumed that interest is confined to the sound power radiated from the open end of a duct to free field. The analysis from Section 2.8, Sources in Ducts and Pipes applies, and all that is needed is a measure of the force spectrum in the appropriate i direction $[\Phi_{FF}(\omega)]_i$; see, e.g., Eq. (4.28). The predominant frequency of this force will be of order

$$f_s \sim \frac{SU_0}{d} \tag{4.43}$$

where U_0 is the local velocity past the body and d is the lateral dimension; see also Table 4.3 for values of d of simple prismatic forms. Now, according to Eq. (4.28), the force spectrum will then behave parametrically as

$$\Phi_{FF}(\omega) \sim \frac{1}{4}\rho_0^2 U_0^4 d^2 \Lambda_3\, L_3\, \phi_{FF}(\omega) \tag{4.44}$$

In the case of plane wave propagation in the duct with subsequent radiation into a free space from an open end, only the fluctuating drag on the throttling element is important in estimating the sound power from the equations of Section 2.8.2, Radiation From Multipoles in an Infinitely Long Pipe. If the spoiler is a circular cylinder, drag coefficients of Fig. 4.14 may be used in ducts with fairly clean (nonturbulent) flow. Heller and Widnall [122]

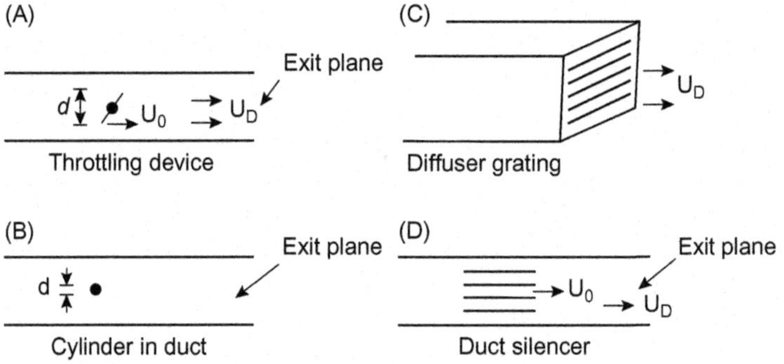

FIGURE 4.34 Illustrations of typically encountered ducted throttling and diffusing elements.

published some of the earliest work on this subject and indicated for a single spoiler spanning a turbulent pipe flow that $\overline{C_D^2}^{1/2} \simeq 0.0023$ on various types of blunt cylinders of both circular and prismatic cross section. The spectrum $\Phi_{FF}(\omega)$ was found to be a broad arch-shaped curve centered on $S = 0.1-0.2$ and with -3 dB down points at $S \simeq 0.05$ and $S \simeq 0.7$.

For obstructions that are *not* aerodynamically shaped a relation was derived for obstructions in pipes that states that the sound power radiated from the open end depends on square of the pressure drop across the constriction. Essentially, it can be assumed that the root-mean-square fluctuating drag per unit length is proportional to the steady drag per unit length across the constriction. Couched in the context of a grid of N cylinders across the duct in the exit plane assumed to be non interacting, this means that

$$\Phi_{FF_D}(\omega) \sim N\overline{C_D^2}\left(\frac{1}{2}\rho_0 U_0^2\right)^2 d^2 L_3 \Lambda_3\, \phi_{FF}(\omega) \tag{4.45a}$$

$$\sim \overline{C_D^2}\left(\frac{1}{2}\rho_0 U_0^2\right)^2 A_g d^2\, \phi_{FF}(\omega) \tag{4.45b}$$

or approximately

$$\sim (\Delta P)^2 A_D (A_g/A_D) d^2 \phi_{FF}(\omega) \tag{4.45c}$$

where ΔP is the static pressure drop across obstruction in the duct, A_g is the frontal area of the constricting body or grating, d is an appropriate dimension of a grid element, and $\Lambda_3 \sim d$. Now, assuming that Eq. (4.45c) holds we find using any of the relationships of Section 2.8.2, Radiation From Multipoles in an Infinitely Long Pipe that the radiated sound power in bandwidth $\Delta\omega$ is of the form

$$\mathbb{P}(\omega, \Delta\omega) \sim \frac{1}{\rho_0 c_0^3}\left(\frac{SU_D}{\mathrm{d}}\right)^2 (\Delta P)^2 A_g d^2 \phi_{FF}(\omega)\, \Delta\omega$$

Rearranging to emphasize the pressure coefficient

$$\mathbb{P}(\omega, \Delta\omega) \sim \frac{\rho_0 S^2}{c_0^3} U_D^6 \xi^2 A_g \phi_{FF}(\omega)\, \Delta\omega \tag{4.46}$$

where $\xi = \Delta P(\frac{1}{2}\rho_0 U_D^2)^{-1}$. This parametric form applies to the general problem of sound radiated from the open end of a duct into a free field whether or not the sound in the duct is plane. This is because of the functional equivalence of Eqs. (2.166), (2.174), and (2.175). Therefore putting Eq. (4.46) in terms of the sound power level in a one-third-octave frequency band, $L_N(f)$ where

$$L_N(f) = 10 \log[\mathbb{P}(\omega, \Delta\omega)/\mathbb{P}_0] \tag{4.47}$$

and $\mathbb{P}_0$ is conventionally 10^{-12} W for aeroacoustic applications; Eq. (4.46) indicates that the quantity

$$L_N(f) - 60\log(U_D/U_0) - 10\log(A_g/A_0) - 20\log\xi = S(fd/U_D) \quad (4.48)$$

where $U_0 = 10$ m/s and $A_0 = 1$ m^2 are constant reference quantities, should be universal for a particular geometry of diffuser-duct configuration. An expression for the overall sound power level that is consistent with the above is

$$L_N = 60\log(U_D/U_0) + 10\log(A_g/A_0) + 20\log(\xi/\xi_0) + 10 \quad (4.49)$$

Indeed, this is the case also with duct diffusers and gratings as measurements by Hubert [123] show. Fig. 4.35 shows that two samples of grids and grating give similarly shaped curves of $S(fd/U_D)$. For each grating indicated, the $S(fd/U_D)$ has a spread of only ±2 dB for a range of velocities from 5 to 30 m/s. It applies to radiation into a free field by a grating placed across the duct exit.

The above analysis is the framework of other examples of empirical correlations of noise from diffusers and gratings such as given in Beranek [124]; however, a modification of Eq. (4.48), namely,

$$L_N(f) - 60\log(U_D/U_0) - 10\log(A_D/A_0) + 30\log(\xi/\xi_0) = S(fd/U_D), \quad (4.50)$$

seems to fit a wider range of geometries than those indicated in Fig. 4.35. Eq. (4.49) would accordingly change only by the factor $30\log(\xi/\xi_0)$; for most practical duct diffusers $\xi \sim 3-7$.

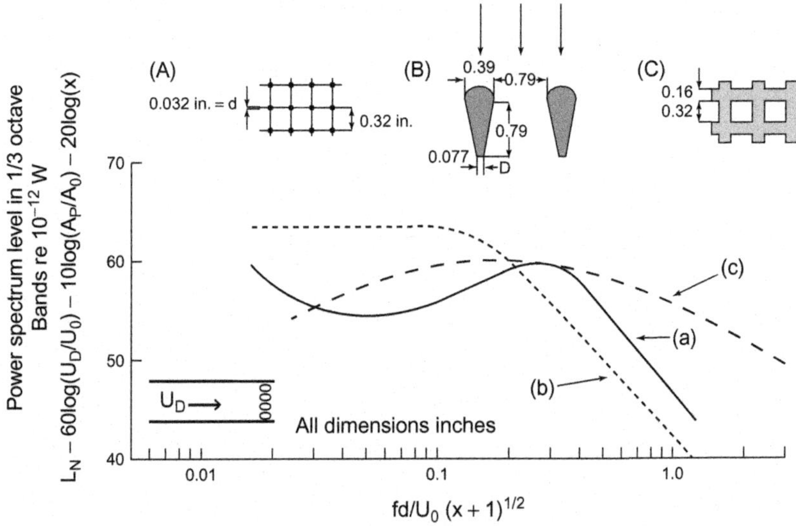

FIGURE 4.35 Radiated sound power spectrum levels from three geometries of simple gratings. *From Hubert M. Strömungsgerausche. VDI-Bildungswerk. Düsseldorf, Contribution BW 406 in training book "Technical measures to control machinery noise," p. 1–21; 1969.*

A similar type relationship as Eq. (4.48) has been found to apply to duct silencers. Ver [125] has found that

$$L_N(f) - 55\log(U_D/U_0) - 10\log(A_D/A_0) + 45\log(P/100) \simeq S(f) \quad (4.51a)$$

where $(P/100)$ is the percentage open area of the silencer and

$$S(f) \simeq 38 \quad (4.51b)$$

in the frequency range of 125–8000 Hz. Eq. (4.51a) including the ratio of open area to cross-section area accounts for the constriction of the silencer. Accordingly it compensates in rough fashion for the ratio U_0/U_D being greater than unity. The radiated sound power is independent of the axial length of a silencer so that is believed that the noise is produced only in the vicinity of the trailing edges. A more contemporary use of these methods had been published giving an analogous functional form to the ones above by Gurein et al. [126].

To examine the throttling noises of flow spoilers, say for A of Fig. 4.34, Gordon [127,128] uses Eq. (4.45b) to give the dimensionless power spectrum function ($\Lambda_3 \sim D$)

$$\frac{\mathbb{P}(f,\Delta f)\rho_0^2 c_0^3}{(P_0 - P_e)^3 D^2} = \phi\left(\frac{fd}{\sqrt{2(P_0 - P_e)/\rho_0}}\right) \quad (4.52)$$

where P_0 is the total head upstream of the constriction,

$$P_0 = P_D + \frac{1}{2}\rho_0 U_1^2$$

He has assumed that Bernoulli's equation holds; i.e., $\frac{1}{2}\rho_0 U_0^2 = P_D + \frac{1}{2}U_D^2 - P_e$, where P_e is the static pressure in the exit plane and P_D and U_D are the static pressure and the velocity upstream of the throttle. The spectrum function $\phi(fd/\sqrt{2(P_0 - P_e)/\rho_0})$ is another arch-shaped curve with a maximum of 10^{-5} to 10^{-4} at $fd/\sqrt{2(P_0 - P_e)/\rho_0} \simeq 0.4$ and with -3 dB down points roughly at $fd/\sqrt{2(P_0 - P_e)/\rho_0} = 0.15$ and 1.0.

Eq. (4.52) can be rearranged to the form of Eq. (4.48)

$$L_N - 60\log(U_D/U_0) - 10\log(A_p/A_0) - 30\log(\xi + 1) = S(fd/U_D\sqrt{\xi + 1}) \quad (4.53)$$

where $S(fd/U_D\sqrt{\xi + 1})$ is given in Fig. 4.36 and shows a close resemblance to the spectrum shown in Fig. 4.35. The spread of spectrum values arises from a variety of arrangements and speeds without any systematic variation. Eq. (4.53) reduces to Eq. (4.50) when $\xi > 1$. Comparison of Eq. (4.51a,b) with Figs. 4.35 and 4.36 allows an assessment of the relative dipole strengths of the various duct elements. The self-noise of silencers is seen to be generally dominated by the noises of diffusers and throttling devices.

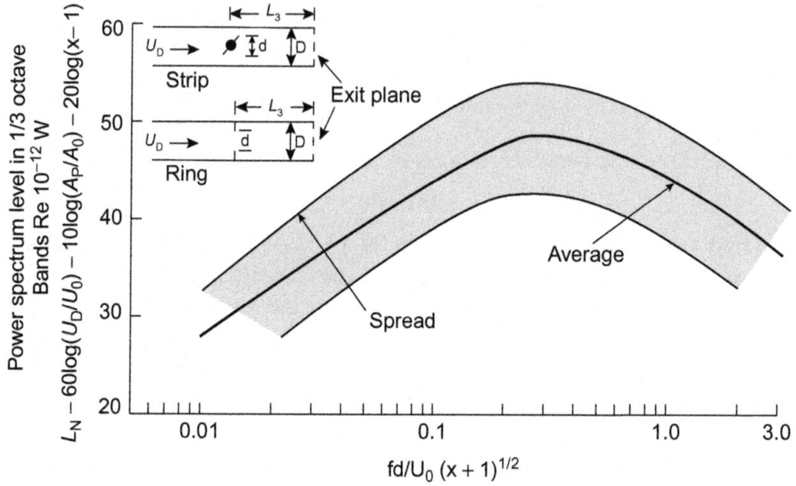

FIGURE 4.36 Radiated sound pressure levels from unflanged pipes with ring and strip throttles. Strip throttles: plate angles, 0°, 22.5°, 90°; strip lengths, 0.26D, 0.53D; distance from plane $L_3 = 3.2D$. Ring throttle: $d = 0.8D$, $L_3 = 3.2D$. *Adapted from Gordon CG. Spoiler-generated flow noise. II. Results. J Acoust Soc Am 1969;45:214–23.*

4.8 APPENDIX: THE SOUND FIELD OF A TWO-DIMENSIONAL DIPOLE

Occasionally use is made of two-dimensional acoustic modeling. Physically, these models apply to axially (or laterally) uniform sources whose extent L_3 greatly exceeds the range r from the axis of the source to the acoustic observation point (see Fig. 2.1). Such problems may apply to the acoustic far field ($r \gg \lambda_0$), but to the geometric near field ($r < L_3$) of some acoustic radiators. Two-dimensional modeling will be used in Chapter 10, Sound Radiation From Pipe and Duct Systems for pipe-generated noise and in Chapter 11, Noncavitating Lifting Sections for the acoustic field of a line vortex convected past the edge of an infinite half plane. It is instructive therefore to derive an expression for the two-dimensional field of the Aeolian tone and compare it to Eq. (4.27).

Assuming harmonic ($e^{-i\omega t}$) time dependence, the Helmholtz equation (Section 2.6.1: General Implications) describes the acoustic field. The associated free-space Green function for the unit line source generating a field that is uniform in azimuth (i.e. n = 0) and satisfying Eq. (2.106) is of the form

$$g(r, \omega) = AH_0^{(1)}(k_0 r)$$

where the Hankel function $H_0^{(1)}(k_0 r)$ is a solution of the homogeneous wave equation for an axially symmetric source

$$\frac{1}{r}\frac{\partial}{\partial r}\left(r\frac{\partial H_0^{(1)}(k_0 r)}{\partial r}\right) + k_0^2 H_0^{(1)}(k_0 r) = 0$$

and A is a numerical coefficient. The Hankel function has the useful limiting forms [129]:

$$\lim_{k_0 r \to 0} H_0^{(1)}(k_0 r) \to 1 + \frac{2i}{\pi}\ln(k_0 r)$$
$$\lim_{k_0 r \to \infty} H_0^{(1)}(k_0 r) \to \frac{2}{\pi k_0\, r} e^{i(k_0\, r - \pi/4)} \tag{4.54}$$

Solutions of Eq. (2.115) consist of the radiation field of the unit source for which $g(r, \omega)$ is normalized as in the case of the spherical source. For a spherical source the normalization is

$$\iint_s \nabla_r g(r, \omega)\, d^2 s = 1$$

The analogous normalization in the case of the unit line source reduces to

$$\int_{2\pi} \nabla_r g(r, \omega) r\, d\theta = 1$$

Thus in the limit $k_0 r \to 0$

$$\partial g(r, \omega)/\partial r = A2i/\pi r$$

so that

$$A \cdot 2i/\pi r \cdot r2\pi = 1$$

and

$$A = -i/4$$

and the two-dimensional free-space Green function is therefore (see also Ref. [130])

$$g(r, \omega) = -(i/4) H_0^{(1)}(k_0 r) \tag{4.55}$$

The two-dimensional analog of Eqs. (2.118) and (4.21) in which the wake quadrupoles are neglected ($\sigma(\mathbf{y}, \omega) = 0$) and the cylinder is rigid ($\partial p/\partial n = 0$) is

$$p_a(\mathbf{x}, \omega) = -\iint p(\mathbf{y}, \omega)\frac{\partial}{\partial y_r}[g(r, \omega)]\, n(\theta) a d\theta dz \tag{4.56}$$

where $\mathbf{y}$ lies on the cylinder (a, θ_0) and $\mathbf{x}$ on the cylinder in the field (r, θ). Since $r \gg a$ the Hankel function may be replaced by its large argument limit to give the far-field approximation to $g(r, \omega)$,

$$g(\mathbf{r}, \omega) \approx \frac{-i}{4}\left(\frac{2}{\pi k_0 r}\right)^{1/2} e^{i(k_0 R - \pi/4)} e^{-ik_0 a \cos(\theta-\theta_0)} \tag{4.57}$$

where we made use of the far-field approximation introduced following Eq. (2.25), and where $r = |\mathbf{x}|$ is the field point radius. Note also that

$$e^{-ik_0 a \cos(\theta-\theta_0)} = e^{-i(k_0 y_1 \cos\theta + k_0 y_2 \sin\theta)} = e^{-i\mathbf{k}_0 \cdot \mathbf{y}}$$

where $\mathbf{k}_0 = (k_0 \cos\theta, k_0 \sin\theta)$, and $\mathbf{y} = (y_1, y_2)$ so that $g(r, \omega)$ is now really in the separable form given by Eq. (2.120). Since $P(\mathbf{y}, \omega) = P(\omega_s) \cos\theta$ and since $|\mathbf{y}| = a$, and $k_0 a \ll 1$, Eq. (4.56) reduces to (letting the 1,2 directions coincide with the cross-stream and streamwise directions, respectively)

$$P_a(\mathbf{x}, \omega) = \frac{-i}{2}\sqrt{\frac{k_0}{2\pi R}}\left(\frac{dF_2(\omega_s)}{dy_3}\right)\cos\theta\, e^{i(k_0 R + \pi/4)} \tag{4.58}$$

where $dF_2(\omega_s)/dy_3 = P(\omega_s) \cdot a$ is the force per unit length (see Eq. (4.9a)). Eq. (4.58) differs functionally from its three-dimensional counterpart, which may be written directly from Eqs. (2.74) or (4.23) as

$$P_a(\mathbf{x}, \omega) = \frac{-i\, k_0}{4\pi\, r} F_2(\omega_s) \cos\theta\, e^{ik_0 r} \tag{4.59}$$

where $F_2(\omega_s)$ is the net force on the cylinder. This relationship applies for an axially constant force for which $r \gg L_3$. The three-dimensional result includes an additional $(k_0/r)^{1/2}$ dependence, which results in an additional speed dependence. This can be seen by writing the mean-square sound pressure in the two-dimensional case parametrically as

$$\overline{p_a^2} \propto \frac{\rho_0^2 U_\infty^5 d}{R c_0} \cos^2\theta$$

while in the three-dimensional case for a cylinder of length $L_3 \ll r$

$$\overline{p_a^2} \propto \frac{\rho_0^2 U_\infty^6 L_3^2}{r^2 c_0^2} \cos^2\theta$$

where it is assumed that ω_s and $dF_2/dy_3 \simeq F_2/L_3$ scale on ρ_0, U_∞, and d as discussed in Section 4.5. Since the sound power from a cylindrically symmetric source radiates through a surface $2\pi R L_3$ and that from a spherically symmetric source region radiates through a surface $4\pi r^2$, both of these relationships satisfy the requirements for acoustic energy conservation.

REFERENCES

[1] Blevins RD. Flow-induced vibration. 2nd ed. Malabar, Fla: Krieger Pub. Co; 2001.

[2] Strouhal V. Uber eine besondere ort der Tonne regung. Ann Phys Chem 1878;5(10): 216–51.

[3] Rayleigh L. Acoustical observations. Philos Mag 1879;7(42):149–62.

[4] Rayleigh, L. The theory of sound, vol. 2. Dover, NY; 1945.
[5] Benard H. Formation de centres de giration à l'arrière d'un obstacle en movement. C R Hebd Seances Acad Sci 1908;147:839–42.
[6] von Karman T, Rubach H. Uber den Mechqnismus des Flussigkeits und Luftwiderstandes. Phys Z 1912;13(2):49–59.
[7] Lamb, H. Hydrodynamics. Dover, NY; 1945.
[8] Goldstein, S. Modern developments in fluid dynamics, vol. 2. Dover, NY; 1965.
[9] Stowell EZ, Deming AF. Vortex noise from rotating cylindrical rods. J Acoust Soc Am 1936;7:190–8.
[10] Yudin EY. On the vortex sound from rotating rods. Zh Tekhn Fiziki 1944;14(9):561; translated as NACA TM 1136 (1947).
[11] Blokhintsev, D.I. Acoustics of a nonhomogeneous moving medium. NACA TIM 1399 (1956). Translation of "Akustica Neodnorodnoi Dvizhushchcisya Sredy." Ogiz, Gosudarstvennoe Izdatel'stvo, Tekhniko-Teoreticheskoi Literatury, Moskva, 1946, Leningrad.
[12] Phillips OM. The intensity of Aeolian tones. J Fluid Mech 1956;1:607–24.
[13] Etkin ET, Korbacher GK, Keefe RT. Acoustic radiation from a stationary cylinder in a fluid stream (Aeolian tones). J Acoust Soc Am 1957;29:30–6.
[14] Homann F. Einfluss grosser Zähigkeit bei Strömung um Zylinder. Forsch Ingenieurwes 1936;7:1–10.
[15] Kovasznay LSG. Hot-wire investigation of the wake behind circular cylinders at low Reynolds numbers. Proc R Soc London, Ser A 1949;198:174–90.
[16] Roshko A. On the development of turbulent wakes from vortex streets. Natl Advis Comm Aeronaut Tech Note No. 2913; 1953.
[17] Bloor SM. The transition to turbulence in the wake of a circular cylinder. J Fluid Mech 1964;19:290–304.
[18] Sato H, Kuriki K. The mechanism of transition in the wake of a thin flat plate placed parallel to a uniform flow. J Fluid Mech 1961;11:321–52.
[19] Mattingly GE, Criminale WO. The stability of an incompressible two-dimensional wake. J Fluid Mech 1972;51:233–72.
[20] Blake, W.K. Aero-hydroacoustics for ships, 2 vols. Naval Ship Research and Development Center, Washington, D.C., NSRDC Report 84–010; 1984.
[21] Ballou, C.L. Investigation of the wake behind a cylinder at coincidence of a natural frequency of vibration of the cylinder and the vortex shedding frequency. Acoust Vib Lab Rep 76028–2. Cambridge, MA: MIT.1967.
[22] Dale JR, Holler RA. Secondary vortex generation in the near wake of circular cylinders. J Hydronaut 1969;4:10–15.
[23] Schiller L, Linke W. Pressure and frictional resistance of a cylinder at Reynolds numbers 5,000 to 40,000. NACA Tech Memo No. 715 1933.
[24] Ma X, Karamanos G-S, Karniadakis GE. Dynamics and low-dimensionality of a turbulent near wake. J Fluid Mech 2000;410:29–65.
[25] Franke J, Frank W. Large eddy simulation of the flow past a circular cylinder at $Re_D = 3900$. J Wind Eng Ind Aero 2002;90:1191–206.
[26] Roshko A. Transition in incompressible near-wakes. Phys Fluids 1967;10:SI81–3.
[27] Schaefer JW, Eskinazi S. An analysis of the vortex street generated in a viscous fluid. J Fluid Mech 1959;6:241–60.
[28] Gerrard JH. A disturbance-sensitive Reynolds number range of the flow past a circular cylinder. J Fluid Mech 1965;22:187–96.

[29] Norberg C. Fluctuating Lift on a circular cylinder: review and new measurements'. J Fluids Struct 2003;17:57–96.
[30] Roshko A. Experiments on the flow past a circular cylinder at very high Reynolds number. J Fluid Mech 1961;10:345–56.
[31] Schlichting H. Boundary layer theory. New York: McGraw-Hill; 1979.
[32] Chen YN. Fluctuating lift forces of the Karman vortex streets on single circular cylinders and in tube bundles, Part 1–The vortex street geometry of the single circular cylinder. J Eng Ind 1972;94:603–12.
[33] Chen YN. Fluctuating lift forces of the Karman vortex streets on single circular cylinders and in tube bundles, Part II, Lift forces of single cylinders. J Eng Ind 1972;613–22.
[34] Chen YN. Fluctuating lift forces of the Karman vortex streets on single circular cylinders and in tube bundles, Part 3–Lift forces in tube bundles. J Eng Ind 1972;94:623–8.
[35] Flachsbart O. From an article by H. Muttray. "Die Experimentellen Tatasachen des Widerstandes Ohne Auftrieb,". Handbuch der Experimentalphysik Hydro und Aero Dynamik 1929;vol. 4(Part 2):232–336, Leipzig, 1932.
[36] Fage A, Falkner VM. The flow around a circular cylinder. R & M No. 369. London: Aeronaut. Research Council; 1931.
[37] Guven O, Patel VC, Farell C. A model for high-Reynolds number flow past rough-walled circular cylinders. J Fluids Eng 1977;99:486–94.
[38] Hama FR. Three-dimensional vortex pattern behind a circular cylinder. J Aeronaut Sci 1957;24:156–8.
[39] Delany, N.K., Sorensen, N.E. Low-speed drag of cylinders of various shapes. Natl Advis Comm Aeronaut Tech Note No. 3038; 1953.
[40] Bearman PW. Vortex shedding from a circular cylinder in the critical Reynolds number regime. J Fluid Mech 1969;37:577–85.
[41] Jones GW. Unsteady lift forces generated by vortex shedding about large stationary, and oscillating cylinder at high Reynolds number. ASME Symp Unsteady Flow 1968; Pap. 68-FE-36.
[42] Gerrard JH. An experimental investigation of the oscillating lift and drag of a circular cylinder shedding turbulent vortices. J Fluid Mech 1961;2:244–56.
[43] Roshko, A. On the drag and shedding frequency of two-dimensional bluff bodies. Natl Advis Comm Aeronaut Tech Note No. 3169; 1954.
[44] Bearman PW. On vortex street wakes. J Fluid Mech 1967;28:625–41.
[45] Kronauer, R.E. Predicting eddy frequency in separated wakes. IUTAM Symp Cone Vortex Motions Fluids. Ann Arbor, MI; 1964.
[46] Koopman GH. Wind induced vibrations and their associated sound fields. PhD thesis. Washington, DC: Catholic University; 1969.
[47] Schmidt LV. Measurement of fluctuating air loads on a circular cylinder. J Aircr 1965;2:49–55.
[48] Schwabe M. Über Druckermittlung in der nichtstationären ebenen Strömung. Ing Arch 1935;6:34–50.
[49] Humphrey JS. On a circular cylinder in a steady wind at transition Reynolds numbers. J Fluid Mech 1960;9:603–12.
[50] Macovsky MS. Vortex-induced vibration studies. DTMB Rep. No. 1190. Washington, DC: David Taylor Naval Ship R & D Center; 1958.
[51] Bishop RED, Hassan AY. The lift and drag forces on a circular cylinder in a flowing fluid. Proc R Soc London Ser A 1964;277:32–50.

[52] Leehey R, Hanson CE. Aeolian tones associated with resonant vibration. J Sound Vib 1971;13:465–83.

[53] Gerrard, J.H. The calculation of the fluctuating lift on a circular cylinder and its application to the determination of Aeolian tone intensity. AGARD Rep. AGARD-R-463; 1963.

[54] McGregor, D.M. An experimental investigation of the oscillating pressures on a circular cylinder in a fluid stream. UTIA Tech. Note No. 14; 1957.

[55] Fung YC. Fluctuating lift and drag acting on a cylinder in a flow at supercritical Reynolds numbers. J Aerosp Sci 1960;27:801–14.

[56] Schewe G. On the force fluctuations acting on a circular cylinder in cross flow from sub-critical up to transcritical Reynolds numbers. J Fluid Mech 1983;133:265–85.

[57] Powell A. On the edge tone. J Acoust Soc Am 1962;34:163–6.

[58] van Nunen, J.W.G. Steady and unsteady pressure and force measurements on a circular cylinder in a cross flow at high Reynolds numbers. IUTAM Symp Flow-Induced Struct Vib Karlsruhe Pap. H5; 1972.

[59] Ong L, Wallace J. The velocity field of the turbulent very near wake of a circular cylinder. Exp Fluids 1996;20:441–53.

[60] Cox J, Brentner KS, Rumsey CL. Computation of vortex shedding and radiated sound for a circular cylinder: subcritical to trans-critical Reynolds numbers. Theor Comp Fluid Dyn 1998;12:233–53.

[61] Squires KD, Krishman V, Forsythe JR. Prediction of the flow over a circular cylinder at high Reynolds number using detached eddy simulation. J Wind Eng Aerodynamics 2008;96:1528–36.

[62] Breuer M. A challenging test case for large eddy simulation: high Reynolds number circular cylinder flow. Int J Heat Fluid Flow 2000;21:648–54.

[63] Schewe G. On the force fluctuations acting on a circular cylinder in cross flow from sub-critical up to transcritical Reynolds numbers. J Fluid Mech 1983;133:265–85.

[64] Szepessy S, Bearman PW. Aspect ratio and end plate effects on vortex shedding from a circular cylinder. J Fluid Mech 1992;234:191–217.

[65] Braza M, Chassaing J, Minh HHa. Numerical study and physical analysis of the pressure and velocity fields in the near wake of a circular cylinder. J Fluid Mech 1986;165:79–130.

[66] Nishimura H, Taniike Y. Aerodynamic characteristics of fluctuating forces on a circular cylinder. J Wind Eng Ind Aerodynamics 2001;89:713–23.

[67] Prendergast, V. Measurement of two-point correlations of the surface pressure on a circular cylinder. UT1AS Tech. Note No. 23; 1958.

[68] ElBaroudi, M.Y. Measurement of two-point correlations of velocity near a circular cylinder shedding a Karman vortex street. UT1AS Tech Note No. 31; 1960.

[69] Gerrard JH. The mechanics of the formation region of vortices behind bluff bodies. J Fluid Mech 1966;25:401–13.

[70] Bloor MS, Gerrard JH. Measurements on turbulent vortices in a cylinder wake. Proc R Soc London Ser A 1966;294:319–42.

[71] Griffin OM, Ramberg S. The vortex-street wakes of vibrating cylinders. J Fluid Mech 1974;66:553–76.

[72] Koopman GH. The vortex wakes of vibrating cylinders at low Reynolds numbers. J Fluid Mech 1967;28:501–12.

[73] Apelt CJ, West GS, Szewczyk AA. The effects of wake splitter on the flow past a circular cylinder on the range $104 < R < 5 \times 104$. J Fluid Mech 1973;61:187–98.

[74] Apelt CJ, West GS. The effects of wake splitter plates on bluff body flow in the range 104 < R < 5 × 104, Part 2. J Fluid Mech 1975;71:145–61.

[75] Qiu Y, Sun Y, Wu Y, Tamura Y. Effects of splitter plates and Reynolds number on the aerodynamic loads acting on a circular cylinder. J Wind Eng Ind Aerodynamics 2014;127:40–50.

[76] Sudhakar Y, Vengadesan S. Vortex shedding of a circular cylinder with an oscillating splitter plate. Comput Fluids 2012;53:40–52.

[77] Vakil A, Green SI. Drag and lift coefficients of inclined circular cylinders at moderate Reynolds numbers. Comput Fluids 2009;38:1771–81.

[78] Spalart PR, Shur ML, Strelets M,K, Travin AK. Initial-noise-predictions-for-rudimentary-landing-gear. J Sound Vib 2011;330:4180–95.

[79] Milne-Thompson LM. Theoretical hydrodynamics. 4 ed. New York: Macmillan; 1960.

[80] Wille R. Karman vortex streets. Adv Appl Mech 1960;6:273–87.

[81] Sallet DW. The lift force due to von Karman's vortex wake. J Hydronaut 1973;7:161–5.

[82] Sallet, D.W. On the prediction of flutter forces. IUTAM Symp Flow-Induced Struct Vib Karlsruhe Pap. B-3; 1972.

[83] Ruedy R. Vibration of power lines in a steady wind. Can J Res 1935;13:82–98.

[84] Birkhoff G. Formation of vortex streets. J Appl Phys 1953;24:98–103.

[85] Frimberger R. Experimentelle Unterschungen an Karmanschen Wirbelstrassen. Z Flugwiss 1957;5:355–9.

[86] Fage A, Johansen FC. On the flow of air behind an inclined flat plate of infinite span. Proc R Soc London Ser A 1927;116:170–97.

[87] Fage A, Johansen FC. The structure of vortex streets. R&M No. 1143. London: Aeronaut. Research Council; 1928.

[88] Muller B. High order numerical simulation of Aeolian tones. Comput Fluids 2008;37:450–62.

[89] So RMC, Wang XQ, Xie W-C, Zhu J. Free-stream turbulence effects on vortex-induced vibration and flow-induced force on an elastic cylinder. J Fluids Struct 2008;24:481–95.

[90] Gerrard JH. Numerical computation of the magnitude and frequency of the lift on a circular cylinder. Proc R Soc London Ser A 1967;118(261):137–62.

[91] Abernathy FH, Kronauer RE. The formation of vortex streets. J Fluid Mech 1962;13:1–20.

[92] Curle N. The influence of solid boundaries upon aerodynamic sound. Proc R Soc London Ser A 1955;231:505–14.

[93] Holle W. Frequenz- und Schallstarkemessungen an Hiebtonen. Akust Z 1938;3:321–31.

[94] Gerrard JH. Measurements of the sound from circular cylinders in an air stream. Proc Phys Soc London B 1955;68:453–61.

[95] Guedel, G.A. Aeolian tones produced by flexible cables in a flow stream. U.S. Marine Lab. Rep. 116/67; 1967.

[96] Inoue O, Hatakeyama N. Sound generation by a two-dimensional circular cylinder in a uniform flow. J Fluid Mech 2002;471:285–314.

[97] Smith, R.A., Moon, W.T., Kao, T.W. Experiments on the flow about a yawed cylinder. Catholic Univ Am, Inst Ocean Sci Eng Rep, 70–7; 1970; also J Basic Eng Dec., 771–776; 1972.

[98] Ramberg SE. The influence of yaw angle upon the vortex wakes of stationary and vibrating cylinders. PhD thesis. Washington, DC: Catholic University of America; 1978.

[99] King R. Vortex excitation of yawed circular cylinders. J Fluids Eng 1977;99:495–502.

[100] Zhao M, Cheng L, Zhou T. Numerical simulation of three-dimensional flow past a yawed circular cylinder. J Fluids Struct 2009;25:831–47.

[101] Maul DJ, Young RA. Vortex shedding from bluff bodies in a shear flow. J Fluid Mech 1973;60:401–9.

[102] Scheiman, J., Helton, D.A., Shivers, J.P. Acoustical measurements of the vortex noise for a rotating blade operating with and without its shed wake blown downstream. NASA Tech. Note TN D-6364; 1971.

[103] Twigge-Molecey, C.F.M., Baines, W.D. Measurements of unsteady pressure distributions due to vortex-induced vibration of a cylinder of triangular section. IUTAM Symp Flow-Induced Struct Vib Karlsruhe. Pap. El; 1972.

[104] Protos. A, Goldschmidt VW, Toebs. GH. Hydroelastic forces on bluff cylinders. ASME J Basic Eng Pap. 68-FE-12; 1968.

[105] Wilkinson, R.H., Chaplin, J.R., Shaw, T.L. On the correlation of dynamic pressures on the surface of a prismatic bluff body. IUTAM Symp Flow-Induced Struct. Vib Karlsruhe Pap. E-4; 1972.

[106] Vickery BJ. Fluctuating lift and drag on a long cylinder of square cross section in a smooth and turbulent stream. J Fluid Mech 1966;23:481.

[107] Nakamura Y, Mizota T. Unsteady lifts and wakes of oscillating rectangular prisms. J Eng Mech Div Am Soc Civ Eng 1975;101(No. EM6):855–71.

[108] King WF, Pfitzenmeier E. An experimental study of sound generated by flows around cylinders of different cross section. J Sound Vib 2009;vol. 328:318–37.

[109] Bearman PW, Trueman DM. An investigation of the flow around rectangular cylinders. Aeronaut Q 1972;23:229–37.

[110] Rockwell DO. Organized fluctuations due to flow past a square cross section cylinder. J Fluids Eng 1977;99:511–16.

[111] Modi VJ, Dikshit AK. Near wakes of elliptic cylinders in subcritical flow. AIAA J 1975;13:490–6.

[112] Blevins RD. Review of sound by vortex shedding from cylinders. J Sound Vib 1984;92:455–70.

[113] Spivack HM. Vortex frequency and flow pattern on the wake of two parallel cylinders at varied spacing normal to an air stream. J Aeronaut Sci 1946;13:289–301.

[114] Owens PR. Buffeting excitation of boiler tube vibrations. J Mech Eng Sci (London) 1965;7:431–8.

[115] Chen YN. Flow-induced vibration and noise in tube-bank heat exchangers due to von Karman streets. J Eng Ind 1968;90:134–46.

[116] Blevins RD. Fluid elastic whirling of a tube row. J Pressure Vessel Technol 1974;96:263–7.

[117] Blevins RD. Fluid elastic whirling of tube rows and tube arrays. J Fluids Eng 1977;99:457–61.

[118] Savkar SD. A note on the phase relationships involved in the whirling instability in tube arrays. J Fluids Eng 1977;99:727–31.

[119] Savkar SD. A brief review of flow induced vibrations of tube arrays in cross-flow. J Fluids Eng 1977;99:517–19.

[120] Scruton, C. On the wind-excited oscillations of stacks, towers, and masts. Int Conf Wind Eff Build Struct, Natl Phys Lab, Middlesex, Engl. Pap. 16; 1963.

[121] Weaver W. Wind-induced vibrations in antenna members. J Eng Mech Div Am Soc Civ Eng 1961;87:141–65.

[122] Heller HH, Widnall SE. Sound radiation from rigid flow spoilers correlated with fluctuating forces. J Acoust Soc Am 1970;47:924–36.

[123] Hubert, M. Strömungsgerausche. *VDI*-Bildungswerk. Düsseldorf, Contribution BW 406 in training book "Technical measures to control machinery noise," p. 1–21; 1969.

[124] Beranek L. Noise and vibration control. New York: McGraw-Hill; 1971.

[125] Ver, I.L. Prediction scheme for the self-generated noise of silencers. Proc Inter-Noise '12. Washington, DC; 1972. p. 294–8.

[126] Guerin S, Thomy E, Wright SE. Aeroacoustics of automotive vents. J Sound Vib 2005;285:859–75.

[127] Gordon CG. Spoiler-generated flow noise. I. The experiment. J Acoust Soc Am 1968;43:1041–8.

[128] Gordon CG. Spoiler-generated flow noise. II. Results. J Acoust Soc Am 1969;45:214–23.

[129] Abramowitz, M., Stegun, I.A. Handbook of mathematical functions. National Bureau of Standards Applied Mathematics Series. No. 55. Washington, DC; 1965.

[130] Morse PM, Ingard KU. Theoretical acoustics. New York: McGraw-Hill; 1968.

[131] Ribner, H.S., Etkin, B. Noise research in Canada. Proc Int Congr Aeronaut Sci, 1st. Madrid; 1959.

[132] Relf EF, Simmons LFG. The frequency of eddies generated by the motion of circular cylinders through a fluid. R & M No. 917. London: Aeronaut. Research Council; 1924.

[133] Keefe, R.T. An investigation of the fluctuating forces acting on a stationary circular cylinder in a subsonic stream and of the associated sound field. UTIA Rep. 76. University of Toronto; 1961.

[134] Ferguson N, Parkinson GV. Surface and wake flow phenomena of the vortex-excited oscillation of a circular cylinder. J Eng Ind 1967;89:831–8.

[135] Williamson CHK. Vortex dynamics in the cylinder wake. Ann Rev Fluid Dyn 1996;28:477–539.

[136] Chen SS. Dynamics of heat exchanger tube banks. J Fluids Eng 1977;99:462–8.

Chapter 5

Fundamentals of Flow-Induced Vibration and Noise

Flow-induced sound often involves more than direct dipole sound of the type described in Chapter 4, Dipole Sound From Cylinders. It is common for flow-induced forces on bodies and structures to generate vibration, and this vibration in turn induces additional sound and may even modify the unsteady fluid forces. This chapter, then, will develop the theory of flow-induced random vibrations of structures using as examples the rectangular panel and the taut string. The elements of noise and vibration control will follow directly as will many of the essential features of fluid loading of vibrating structures. The subject of the Aeolian tone will be revisited as an example of the application of the results of this chapter to problems of flow-induced noise that involve vibration.

5.1 INTRODUCTION

Flow-induced noise from an interaction of a structure and its bounding fluid is related to both the forces of interaction and the vibration of the surface. We have already seen the elementary form of this interaction in Chapter 4, Dipole Sound From Cylinders: the Aeolian tone intensity is proportional to the mean-square force exerted on the fluid by the cylinder. In more complicated fluid–structure interactions the structure is most certainly excited to motion, and this motion causes additional sound. In general it is possible for these two contributions to be equal in magnitude, in which case they may constructively or destructively interfere, thus causing a modification of the total sound power radiated. Often, however, in many practical situations one of these contributions will dominate. The determination of which component is dominant may be ascertained, of course, by separate evaluations of the respective sound powers. A second commonly encountered situation is that of boundary layer induced noise. If the surface is flat and rigid and if the boundary layer is also homogeneous in the plane of the surface, then the direct radiation from the boundary layer is, by Powell's reflection principle (Chapter 2: Theory of Sound and its Generation by Flow), quadrupole. If the bounding surface is allowed to vibrate, but not so much as to alter the

Mechanics of Flow-Induced Sound and Vibration, Volume 1.
DOI: http://dx.doi.org/10.1016/B978-0-12-809273-6.00005-1

motions of the boundary layer, additional noise will be generated by the vibration of the surface. In practical occurrences this noise will generally overwhelm the direct quadrupole noise from the boundary layer itself. In the case of flow in cylindrical shells such as ducts and pipes, the internal sources will drive the shell walls to produce sound on the outside (Chapter 10: Sound Radiation From Pipe and Duct Systems). In yet another example (Chapter 11: Noncavitating Lifting Sections), when a lifting surface encounters a gust, the reaction force between the body and the fluid radiates sound. However, given a finite structural impedance, the lifting surface will vibrate and radiate additional sound.

In this chapter general relationships that describe flow-induced vibration and structural radiation will be derived to express a response variable, say, surface vibration velocity, to a driving variable, say, fluid surface pressure. We define a *blocked surface pressure* as that generated by the fluid on the surface, but with the surface rigid. We shall assume in this chapter that whatever surface motion occurs does not influence the fluid dynamics of flow that may bound the surface. This assumption is essential to keep the analysis straightforward, and it is physically valid for most fluid–structure interactions except those involving vortex shedding and hydrofoil singing, which will be discussed in Chapter 11, Noncavitating Lifting Sections and Chapter 12, Noise From Rotating Machinery. Furthermore, we shall assume that the response of the structure is linear; that is, the driving force and response velocity are linearly proportional. The structure, unless it is highly damped (we shall define "highly" in due course), will be assumed to respond in its "normal" modes of vibration. Therefore, given a known description of the blocked fluid characteristics on the surface, we ask the questions: What surface motion will result, and what acoustic radiation will this motion induce?

Since the subject of fluid–structure interaction is complex and many-faceted, it is therefore expedient at this point to cite several well-known texts (see Junger and Feit [1], Skelton and James [2], Fahy and Guardonio [3], as well as some older texts, e.g., Lin [4], Crandall [5,6], Cremer et al. [7], and Skudrzyk [8]). The analysis that will be presented in this chapter is to be considered somewhat introductory, and only the results that are necessary to understand the principles in these references will be developed. The problems of structural acoustics can be approached in varying levels of approximation. The response of each resonant structural mode is formally written as a blocked force coefficient divided by the impedance of the mode and the total response of the structure is found by a summation of the responses of the individual modes.

To summarize some of the earlier work on which most statistical formulations are based, Lin [4] extensively examined the statistical formulations for both stationary and nonstationary random driving fields as well as linear and nonlinear vibrating systems; references [5,6] are collections of contributions by experts and include discussions of vibration caused by jet noise and turbulent boundary layer pressures. Cremer et al. [7] emphasizes the statistical description of structural vibration (including the use of damping treatments) for both local and distributed force excitation. Skudrzyk [8] emphasizes more the single-mode character of vibrating systems. Once the vibration velocity is known, the acoustic radiation can be determined either for each mode [9], or for the over all response using an average radiation coefficient [1,2,3,10,11]. Basically, the far-field acoustic pressure is determined by the response vibration velocity normal to the surface times a radiation impedance as implied by Eq. (2.33). Junger and Feit [1] present the general methods of calculating the radiated fields for known vibration distributions on plates and shells. Although these are deterministic analytical problems, the methods can be used to derive the radiation impedance for use in statistical analyses. Maidanik [10,11] provides radiation impedances for individual modes of essentially flat rectangular panel members, but he also shows how to determine average impedances which are averaged over a multiplicity of modes. Lyon and DeJong [11] and Lyon and Maidanik [12] show how to avoid considering the responses of individual modes by using energy balances. Here the kinetic energy of response is assumed to be shared equally by all modes of vibration in the structure and the power dissipated by the damping in excited modes is equal to input power from the fluid excitation. Since the dissipated power is proportional to the time and space-averaged mean-square vibration velocity, the average vibration level of complex structures can be estimated once the input power is known. Radiated power is proportional to the product of the mean-square vibration velocity and the average radiation impedance of the modes in analogy to Eq. (2.34) derived for the oscillating sphere. This approach, known as statistical energy analyses (or SEA, see Lyon and DeJong [11]), is successful for highly complex structures for which many modes are excited by the flow.

Analysis of individual modes of structures is necessary when considering narrow enough frequency bands that only a few modes are concerned. Now, with the development of user-friendly finite element routines as well as powerful engineering analysis software packages, contemporary analysis of fluid–structure interaction has a starting point that is, perhaps, more fundamentally deterministic. The response of structures to random flow forcing can now be examined on a modal basis as well as in broad frequency bands. In this chapter we shall develop the general relationships for describing the response of single and multimodal structures and use the case of the simply supported rectangular plate as an example of a specific result. The chapter will end with example analyses of the forced vibration sound of flat plate structures and of flow-induced vibration and sound field of a circular cylinder as an example of a total examination of a one-dimensional fluid-loaded flow-driven structure.

5.2 RESPONSE OF SINGLE-DEGREE-OF-FREEDOM SYSTEMS TO TEMPORALLY RANDOM EXCITATION

The steady-state response of complex structures to random excitation may be expressed in mathematical forms that are identical to those of the linear single-degree-of-freedom oscillator. Therefore the fundamental properties of this elementary system will be reviewed in some detail; further discussion can be found in standard texts on vibration, e.g., Newland [13] and den Hartog [14].

Consider the motion of a spring-mass system that is illustrated in Fig. 5.1. The mass M is excited to a displacement $x(t)$ by a force $f(t)$. The force on the mass by the spring is opposite $f(t)$ and equals $k_{sp}x(t)$, where k_{sp} is the spring constant. The damping is assumed to be linear and viscous so that the damping force of the "dash pot" is $C_d dx(t)/dt$. The force balance on the mass, which leads to its acceleration $d^2x(t)/dt^2$, is

$$M\frac{d^2x(t)}{dt^2} = f(t) - k_{sp}x(t) - C_d\frac{dx(t)}{dt}$$

thus

$$M\ddot{x}(t) + C_d\dot{x}(t) + k_{sp}x(t) = f(t) \tag{5.1}$$

where the overdots denote time derivatives. This system represents the most simple dynamic oscillator which will be used below as a building block for the description of more complex structural systems.

Now, consider the force to be steady state and random in time so that we introduce the generalized Fourier transform

$$x(t) = \int_{-\infty}^{\infty} X(\omega)e^{-i\omega t}d\omega \tag{5.2}$$

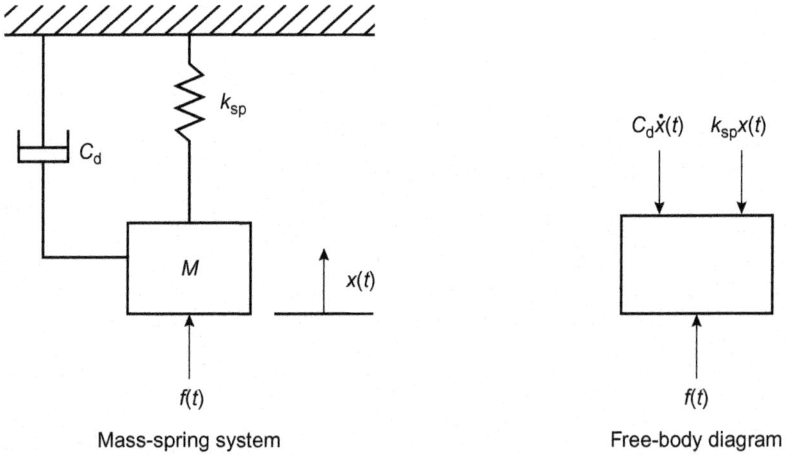

FIGURE 5.1 Mass-spring system and free-body diagram.

If we were interested in transient motion, we would necessarily have to introduce initial conditions and then use the Laplace transform. Here we assume that the motion has lasted a number of cycles so that transients have diminished.

Substitution of Eq. (5.2) into Eq. (5.1) leads to

$$[-M\omega^2 - iC_d\omega + k_{sp}]X(\omega) = F(\omega) \tag{5.3}$$

or

$$-\frac{i\omega X(\omega)}{F(\omega)} = \frac{i\omega}{M(\omega_0^2 - \omega^2 - i\eta\omega_0\omega)} \tag{5.4}$$

where

$$\omega_0 = \sqrt{k_{sp}/M} \tag{5.5}$$

is the resonance frequency. The loss factor η is defined as

$$\eta\omega_0 = C_d/M \tag{5.6}$$

The Fourier transform of the velocity of the mass is $-i\omega X(\omega)$, and the ratio of the velocity to the force is the admittance of the spring-mass system. When the frequency of the force coincides with the resonance frequency, the velocity is largest.

The spectral densities of the force and velocity are simply related. Using the definitions of the generalized Fourier transform and the auto-spectral density developed in Eqs. (2.100), (2.125), and following the remarks after Eq. (2.143), the relationship between the force and velocity spectra is found

$$\frac{\Phi_{vv}(\omega)}{\Phi_{FF}(\omega)} = \frac{\omega^2}{M^2[(\omega_0^2 - \omega^2)^2 + \eta^2\omega_0^2\omega^2]} \tag{5.7}$$

or

$$\Phi_{vv}(\omega) = |Y(\omega)|^2\Phi_{FF}(\omega)$$

where $\Phi_{vv}(\omega) = \omega^2\Phi_{xx}(\omega)$ and $|Y(\omega)|^2$ may be called the admittance of the oscillator. The mean-square velocity is

$$\overline{V^2} = \int_{-\infty}^{\infty} \Phi_{vv}(\omega)d\omega \tag{5.8}$$

The admittance function is defined over the entire positive and negative frequency domain. The maximum response occurs when $\omega = \pm\omega_0$. For frequencies slightly above and below ω_0, i.e., for $\pm\omega = \omega_0 \pm \omega_0\eta/2$, $\Phi_{vv}(\omega) = \frac{1}{2}\Phi_{vv}(\omega_0) = 1/2;\ M^2\eta^2$ as illustrated in Fig. 5.2 for the positive frequency domain. The behavior for negative frequencies is just the mirror of that shown for positive frequencies; this is illustrated with less detail in Fig. 5.3.

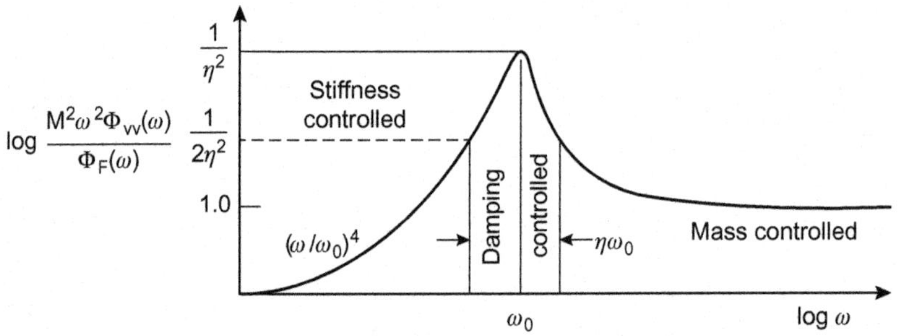

FIGURE 5.2 The magnitude squared of the admittance of a simple harmonic oscillator consisting of a mass, linear spring, and viscous damping.

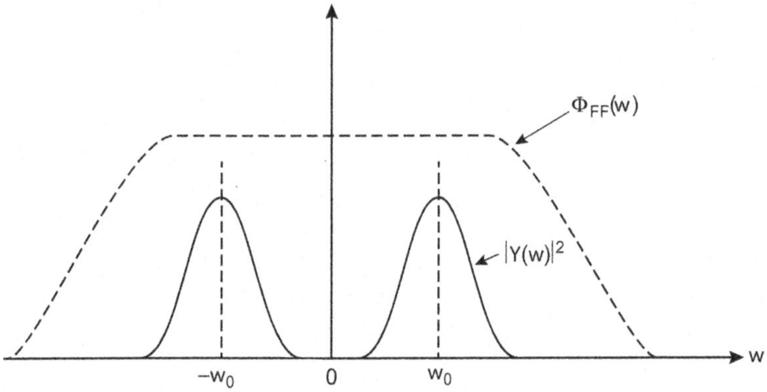

FIGURE 5.3 Illustration of the frequency spectra of the driving force and admittance of a simple harmonic oscillator.

For cases in which the force spectrum is broad compared to the resonance band, Fig. 5.2 represents the spectrum of the acceleration of the mass made dimensionless on the mass M and the exciting force spectrum. At low frequencies, $\omega \ll \omega_0$, the acceleration response is proportional to ω^4 since $\Phi_{aa}(\omega)$, the acceleration spectrum, is equal to $\omega^2\Phi_{vv}(\omega) = \omega^4\Phi_{xx}(\omega)$. Further, since $\Phi_{xx}(\omega) = k^{-2}\Phi_{FF}(\omega)$ when $M\Phi_{aa}(\omega) \ll k^2\Phi_{xx}(\omega)$ (i.e., $M\ddot{x} \ll kx$), the acceleration response is stiffness-controlled. On the other hand, when $M\Phi_{aa}(\omega) > > k^2\Phi_{xx}(\omega)$, (i.e., $\omega \gg \omega_0$), the response is determined by the inertia of the mass. In the vicinity of the resonance, the response is sensitive to the damping in the system as represented by the loss factor η. The quality factor of the system Q is

$$Q = 1/\eta \tag{5.9}$$

and is a measure of the ratio of resonant vibration to the mass-controlled vibration of the system; while the fraction of critical damping is defined

$$C_d/C_0 = \frac{1}{2}\eta \tag{5.10}$$

where $C_0 = 2\sqrt{Mk_{sp}} = 2M\omega_0$ is the critical damping coefficient. The critical damping is that required to just prevent oscillatory motion. For light damping, $\eta \ll 1$ (typically, η lies between 10^{-3} and 10^{-2} unless special lossy treatments are applied to the structure), the acceleration response to a broadband force excitation is dominated by the peaks of breadth $\eta\omega_0$ at $\omega = \pm\omega_0$. The mean-square response to a broadband excitation force is found by integrating over all frequencies as in Eq. (5.8) and as illustrated in Figs. 5.2 and 5.3; thus

$$\overline{V^2} = 2\frac{\omega_0^2 \Phi_F(\omega_0)}{M^2\eta^2\omega_0^4}\frac{\pi}{2}\eta\omega_0$$

$$\overline{V^2} = \frac{\pi\Phi_F(\omega_0)}{M^2\eta\omega_0} \tag{5.11}$$

The first factor of two arises from the double peaks at $\omega = \pm\omega_0$, and the effective bandwidth of the resonance is $(\pi/2)\eta\omega_0$; thus we require that the bandwidth of the force $(\delta\omega)_F$ satisfy the condition

$$(\delta\omega)_F \gg \frac{1}{2}\pi\eta\omega_0 \tag{5.12}$$

It is instructive to note that the time-averaged power dissipated $\mathbb{P}_D$ in the system in a time $T \gg 1/\omega_0$ is just equal to the power applied to the mass P_{in}; i.e.,

$$\mathbb{P}_D = \frac{1}{T}\int_{-T/2}^{T/2} f(t)\dot{x}(t)dt = \overline{f(t)V(t)} \tag{5.13}$$

or

$$\mathbb{P}_D = \mathbb{P}_{in} \tag{5.14}$$

For the single mode of vibration that we have here, the dissipated power

$$\mathbb{P}_D = C_d\overline{V^2}$$

which is rewritten in terms of the mass and loss factor

$$\mathbb{P}_D = M\eta\omega_0\overline{V^2} \tag{5.15}$$

These relationships can be seen by multiplying Eq. (5.1) by $\dot{x}(t)$ and integrating. For simple harmonic motion, the displacement velocity $\dot{x}(t)$ may be described by the continuous waveform

$$\dot{x}(t) = V_0 \cos \omega_0 t \tag{5.16}$$

so that the time average of $x(t)$ and its quadrature vanishes and we have $\overline{x(t)\dot{x}(t)} = \overline{\ddot{x}(t)\dot{x}(t)} = 0$. Then we have from Eq. (5.1)

$$\frac{1}{2}C_d V_0^2 = C_d\overline{V^2} = \frac{1}{T}\int_0^T F(t)\dot{x}(t)dt$$

and we use Eq. (5.6) to eliminate the damping coefficient C_d. Combining Eqs. (5.14) and (5.15) we have a simple expression for the mean-square velocity in terms of the input power

$$\overline{V^2} = \frac{\mathbb{P}_{\text{in}}}{M\eta\omega_0} \tag{5.17}$$

Now we see that if we have an estimate of either the average input power or the spectrum of the oscillating force we can find the mean-square response. However, although Eqs. (5.11) and (5.17) differ in the analytical starting point of their respective derivations, they are both general. Any advantage of one over the other is provided by the particular problem and the simplicity of providing $\mathbb{P}_{\text{in}}$ or $\Phi_{FF}(\omega)$. The method of statistical energy analyses is a most useful tool for estimating mean-square response using as a basis the input power and equations such as Eq. (5.17); the development of this method is offered by Lyon [11,15] and the developmental work of Lyon and Maidanik [12], Smith and Lyon [16], Fahy [17], and Maidanik [18]. Eq. (5.17) is just the simplest of many more general relationships.

We shall now consider the simple vibrating plate excited by a spatially and temporally stochastic forcing field to see the extent to which these notions apply. This example will serve to illustrate both general limitations and extensions of these methods of modal analysis as they may apply to more complex structures.

5.3 GENERAL FEATURES OF STRUCTURES DRIVEN BY RANDOMLY DISTRIBUTED PRESSURE FIELDS

5.3.1 Modal Velocities and Excitation Functions

We will now develop a mathematical description of the forced response of structures in terms of the linear simple harmonic oscillator using a technique known as normal mode analysis. The method of normal mode analysis used in this section has been developed and used in a number of references [19–31]; a text on the method is that of Lin [4]. A motion response variable for the vibrating surface is expanded in its normal modes; each mode describes a condition of resonance. This variable may be the transverse displacement, the acceleration, or the strain. The vibration of the surface is therefore regarded as a summation over the contributions of the complete set of these modes. The equation of motion for the flexural displacement $\xi = \xi(\mathbf{y}_{13}, t)$ which is normal to the static position of the surface is

$$m_s\ddot{\xi} + C_d\dot{\xi} + L(\xi) = -p(\mathbf{y}_{13}, t) \tag{5.18}$$

The convention sign we use is positive up, but we let p be directed downward, therefore the minus sign. m_s is the mass per unit area of the surface,

C_d is an *ad hoc* viscous damping coefficient, $\mathbf{y} = (y_1, y_3)$, and $p(\mathbf{y}, t)$ is the fluctuating load per unit area, which is assumed to be distributed over the surface of the structure. The coordinate $\mathbf{y}_{13}$ lies in the plane of the structure. $L(\boldsymbol{\xi})$ is a linear differential operator specific to the type of structure [1,32,33] and which is determined by the conditions of elastic deformation. For example,

$$L(\xi) = T_e \nabla^2 \xi \text{ (membrane of uniform tension per unit length, } T_e)$$

$$= D_s \nabla^4 \xi = D_s \left[\frac{\partial^4 \xi}{\partial y_1^4} + \frac{\partial^4 \xi}{\partial^2 y_1 \partial^2 y_3} + \frac{\partial^4 \xi}{\partial y_3^4} \right] \text{ (Timoshenko–Mindlin plate of uniform stiffness, } D_s)$$

$$= T_e \frac{\partial^2 \xi}{\partial y_1^2} \text{ (string of uniform tension per unit length, } T_e)$$

$$= D_{sb} \frac{\partial^4 \xi}{\partial y_1^4} \text{ (Bernoulli–Euler beam of uniform stiffness, } D_{sb})$$

The plate bending stiffness is

$$D_s = \frac{Eh^3}{12(1 - \mu_p^2)} \tag{5.19}$$

where E is Young's modulus, h is the thickness of the plate or beam, μ_p is Poisson's ratio; and κ is the radius of gyration

$$\kappa = h/\sqrt{12} \tag{5.20}$$

For the one-dimensional beam, the bending stiffness does not include the Poisson ratio so,

$$D_{sb} = Eh^3/12.$$

The Laplacian used above is

$$\nabla^2 = \frac{\partial^2}{\partial y_1^2} + \frac{\partial^2}{\partial y_3^2}$$

and the biharmonic operator is

$$\nabla^4 = \frac{\partial^4}{\partial y_1^4} + 2\frac{\partial^4}{\partial y_1^2 \partial y_3^2} + \frac{\partial^4}{\partial y_3^4}$$

Solutions to Eq. (5.18) may be obtained for steady-state vibration most simply for our purposes by taking the generalized Fourier transform and defining the displacement

$$\xi(y_{13}, t) = \int_{-\infty}^{\infty} \xi(y_{13}, \omega) e^{-i\omega t} d\omega$$

where we have also dropped the $\mathbf{y}_{13}$ notation and let $\mathbf{y}$ lie in the plane of the plate until otherwise noted. Substitution into Eq. (5.18) gives

$$-\omega^2 m_s \xi(\mathbf{y},\omega) - i\omega C_d \xi(\mathbf{y},\omega) + L(\xi(\mathbf{y},\omega)) = -P(\mathbf{y},\omega) \tag{5.21}$$

where $P(\mathbf{y}, \omega)$ is the generalized Fourier frequency transform of the driving pressure. In the case of negligible damping ($\omega C_d \ll \omega^2 m_s$) free vibration (i.e., $P(\mathbf{y}, \omega) = 0$) may be described by

$$L(\xi(\mathbf{y},\omega)) - \omega^2 m_s \xi(\mathbf{y},\omega) = 0$$

The four operators defined below Eq. (5.18) fall into two general classes according to their order, i.e., according to whether $L(\xi)$ contains a second- or fourth-order spatial differential operator. For membranes, this equation is second order and becomes

$$\nabla^2 \xi(\mathbf{y},\omega) - \omega^2 (m_s/T_e)\xi(\mathbf{y},\omega) = 0$$

and we have defined a characteristic wave number, k_T, in terms of the mass per unit area, m_s, and the tension per unit length, T_e, as

$$k_T^2 = \omega^2 \left(\frac{m_s}{T_e}\right) \tag{5.22}$$

Roots of k_T^2 are all real for a real frequency so that solutions to the second-order equation will be formed as combinations of the harmonic functions. For example, for the taut string these harmonics are functions of one dimension, say where y is aligned with the axis, and are $\sin k_T y$ and $\cos k_T y$. An alternative harmonic solution pair are the exponential forms $\exp(ik_T y)$ and $\exp(-ik_T y)$. Which of these functions is included in the solution is determined by the boundary condition of the string.

Analogously with the case of the one-dimensional beam we have the fourth-order operator which gives for the 2 dimensional vector position, $\mathbf{y}$,

$$\nabla^4 \xi(\mathbf{y},\omega) = k_b^4 \xi(\mathbf{y},\omega)$$

where we have also defined a characteristic wave number

$$k_b^4 = \omega^2 (m_s/D_{sb}) \tag{5.23}$$

for flexural waves. In this case roots k_b consist of $\pm k_b$ and $\pm ik_b$ so that the class of functions which may be used to describe the beam flexure will either include combinations of

$$e^{ik_b y_i}, e^{-ik_b y_i}, e^{k_b y_i}, \text{and } e^{-k_b y_i}$$

or, as above, the class will include a linear superposition of products of harmonic and hyperbolic functions $\sin k_b y_i$, $\sinh k_b y_i$, $\cos k_b y_i$, and $\cosh k_b y_i$,. Here y_i represents either y_1 or y_3 in the plane of the late as depicted in Fig. 5.4, and $\mathbf{y} = (y_1, y_3)$. The two-dimensional cases of plates

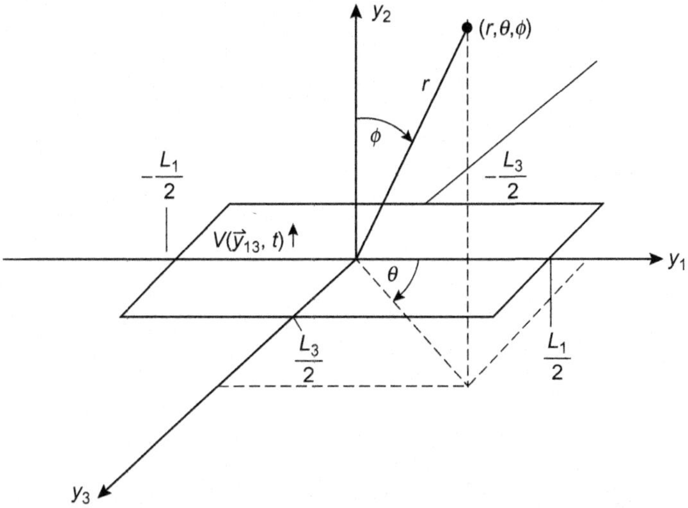

FIGURE 5.4 Coordinates of the rectangular plate and the contiguous medium.

and membranes involve more complicated functions of two space variables and it is not in general possible to construct eigenfunctions for the two-dimensional cases by building on one-dimensional functions, see Leissa [38]. It is however possible to express these functions in closed form for the simply supported rectangular and circular plate.

Free vibration displacement patterns of flat plates may be written as a superposition of two-dimensional normal modes

$$\xi(\mathbf{y}, \omega) = \sum_{m,n=1}^{\infty} W_{mn}(\omega)\Psi_{mn}(\mathbf{y}) \tag{5.24a}$$

where the subscripts m, n are used to simply order the progression of modes and $\Psi_{mn}(\mathbf{y})$ is composed of the functions listed above in the case of rectangular plates. For two-dimensional structures, the double index m, n uses the single index m for orders along y_1 and n along y_3. For a one-dimensional system, a string or a beam, the single-order n would be retained. Each of the $\Psi_{mn}(\mathbf{y})$ satisfy the respective equation, e.g., for beams

$$L\left[\Psi_n(\mathbf{y})\right] = k_n^4\Psi_n(\mathbf{y})$$

and for thin plates of uniform thickness and properties [1,32]

$$L[\Psi_{mn}(\mathbf{y})] = \nabla^4\Psi_{mn}(\mathbf{y}) = k_{mn}^4\Psi_{mn}(\mathbf{y}) \tag{5.24b}$$

for each order m, n (see Section 5.3.2)

$$k_{mn} = k_\mathrm{p} = \sqrt[4]{\omega_{mn}^2 m_s/D_s} \tag{5.25}$$

and ω_{mn} is the resonance frequency of the mode of order mn. A similar relationship applies for frequencies of one-dimensional modes of order n. In all cases, the specific values of k_{mn} and forms of $\Psi(\mathbf{y})$ are determined by the geometries and boundary conditions of the structure; they are called eigenvalues and eigenfunctions, respectively. The modes $\Psi(\mathbf{y})$ are orthogonal modes, i.e., for a rectangular plate

$$\iint_{A_\mathrm{p}} \psi_{mn}(\mathbf{y})\psi_{op}(\mathbf{y})d^2\mathbf{y} = N_\mathrm{p}\delta_{mnop}$$

where $\delta_{mnop}=1$ when $m=n$ and $o=p$, and $\delta_{mnop}=0$ when $m\neq n$ and $o\neq p$ if the modes are uncoupled. N_p is a numerical value that is associated with the plate geometry and boundary conditions and is proportional to the area of the plate A_p. Only in simple cases is it independent of mode order. A general proof of this condition of orthogonality even for one-dimensional structures is beyond this text (see Timoshenko [34]). However, the value of N_p may be associated with the areal integration of the eigenfunction and its operator, e.g., for a plate,

$$N_\mathrm{p} = \frac{1}{k_{mn}^4}\iint_{A_\mathrm{p}} \psi_{mn}(y)\nabla^4\psi_{mn}(y)d^2y$$

A particular example for a simply supported rectangular plate will be derived in Section 5.3.2. In this text the eigenfunctions will be normalized simply according to

$$\iint_{A_\mathrm{p}} \psi_{mn}(\mathbf{y})\psi_{op}(\mathbf{y})d^2\mathbf{y} = \delta_{mnop}A_\mathrm{p} \tag{5.26}$$

and it will be understood that the numerical coefficients $(N_\mathrm{p})_{mnop}/A_\mathrm{p}$ will be incorporated in the definition of $\psi_{mn}(\mathbf{y})$. Finally, it should be recognized that modes may become physically coupled by spatially nonuniform damping, fluid loading, and areal distributions of mass and stiffness. For structures of varying thickness or stiffness or for curved plates, the fourth-order operators given above also do not apply in general. However, curved plates may be approximated as vibrating flat plates as long as the condition $k_\mathrm{p}a \gg \sqrt[4]{12}\sqrt{a/h}$ holds where a is the radius of curvature and h is the thickness of the plate. This high-frequency condition may be deduced by the leading terms of the curved plate equation as presented in Chapter 10, Sound Radiation From Pipe and Duct Systems, and by Leissa [38], Junger and Feit [1] or Timoshenko and Woinowsky-Krieger [32].

We can use the above statements to reduce the problem of flow-excited structural vibration of a lightly-damped panel into the framework of the simple harmonic oscillator. Substitution of Eq. (5.24a) into Eq. (5.21), multiplication of the resulting equation by $\psi_{mn}(\mathbf{y})$ followed by integration over A_p yields in the case of forced vibration of plates

$$\begin{aligned}\left[k_{mn}^4 D_\mathrm{s} - i\eta_\mathrm{s}\omega\omega_{mn}m_\mathrm{s} - m_\mathrm{s}\omega^2\right]W_{mn}(\omega) &= Z_{mn}(\omega)[(-i\omega)W_{mn}(\omega)] \\ &= \frac{1}{A_\mathrm{p}}\iint_{A_\mathrm{p}} P(\mathbf{y},\omega)\psi_{mn}(\mathbf{y})d^2\mathbf{y} \\ &= P_{mn}(\omega)\end{aligned} \tag{5.27}$$

where the modal impedance function has been written as

$$Z_{mn}(\omega) = \frac{k_{mn}^4 D_\mathrm{s} - i\eta_\mathrm{s}\omega\omega_{mn}m_\mathrm{s} - m_\mathrm{s}\omega^2}{-i\omega} \tag{5.27a}$$

and where C_d has been replaced by an *ad hoc* loss factor that we define as

$$\eta_\mathrm{s} = \frac{C_\mathrm{d}}{m_\mathrm{s}\omega_{mn}}$$

A similar equation would be obtained for the membrane, but with $k_{mn}^2 T_\mathrm{e}$ replacing $k_{mn}^4 D_\mathrm{s}$.

Now, Eq. (5.27) is in the form of Eq. (5.3) where a_{mn} replaces the modal displacement $X(\omega)$ and $-P_{mn}(\omega)$ replaces the applied force coefficient $F(\omega)$ to the oscillator. In the case of the plate, the load per unit area is applied normal to the plane of the structure; it may consist of a superposition of locally applied forces, random (in space and time) pressure fields, and fluid back reaction to the motion ξ. The fluid back reaction included in $p(\mathbf{y}, t)$ can be viscous (damping), inertial (added fluid mass), and acoustic radiation (appearing as a damping because it represents energy radiated from the plate). In more complicated situations in which multiple structures enclose a fluid, the fluid reaction on each structure is dependent on the motions of adjacent structures in which case the whole fluid–structure system becomes coupled; see, e.g., Strawderman [35], Obermeier [36], Arnold [37], and especially White and Powell [25] for general treatments. Only the single structure and the unbounded acoustic medium will be considered in this chapter. The nature of the fluid back reaction on the structure will be discussed for plates in Section 5.6 and for flow-induced vibration of cylinders in Section 5.7. Thus, the response of a lightly damped uniform structure may be thought of as comprised of a set of oscillators, all of which respond simultaneously and independently if the modes are uncoupled.

The special case of the infinite plate provides wave-mechanical interpretations of the motion parameters that were introduced above and associated limiting behavior of large finite plates. The assumed displacement is of the exponential form

$$\xi = \xi_0 e^{i(\mathbf{k}\cdot\mathbf{y} - \omega t)}$$

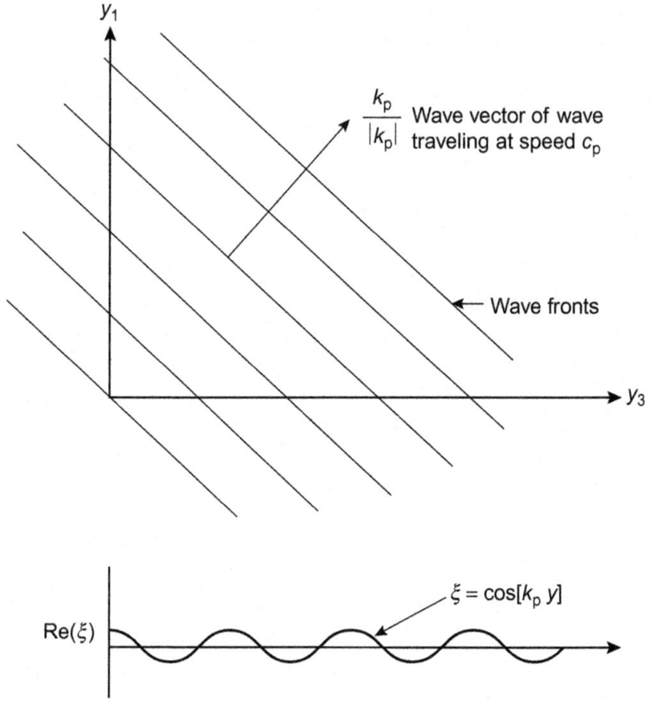

FIGURE 5.5 Diagram of the amplitude of straight-crested bending waves on a flat plate. The function that is illustrated is the real part of ξ.

which represents waves traveling on the plate in the direction **k**/|**k**| as illustrated in Fig. 5.5. By substituting this function into Eq. (5.18) we find

$$\xi_0(-m_s\omega^2 - iC_d\omega + D_s k^4)e^{i(k\cdot y-\omega t)} = 0$$

for the free traveling waves that exist very far from the source of wave excitation. For the case of light damping we can neglect the term $iC_d\omega$; leaving the criterion that the wave number of the waves is related to the frequency by Eq. (5.25), i.e., $|\mathbf{k}| = |\mathbf{k}_p| = k_p$. The phase speed of the bending waves is defined as

$$c_p = \frac{\omega}{k_p} = \sqrt{\frac{\omega\kappa c_L}{\sqrt{1-\mu_p^2}}} \qquad (5.28)$$

where $c_L = \sqrt{E/\rho_p}$ is the bar wave speed in the material, which gives waves of the form

$$\xi = \xi_0 \exp\left\{ik_p \frac{\boldsymbol{k}_p}{|\boldsymbol{k}_p|}\cdot y - c_p t\right\}$$

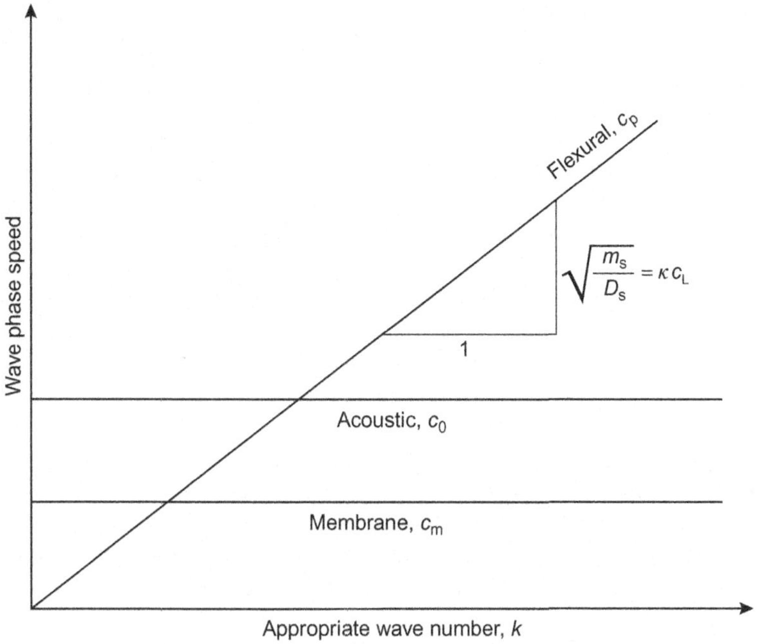

FIGURE 5.6 Diagram of phase speeds as a function of wave number for bending waves on plates, c_p, membranes, c_m, and acoustic (dilatational) waves, c_0.

where $|\mathbf{k}_p| = k_p$. The term in parentheses defines the phase fronts of the traveling waves.

The magnitude of the wave vector $|\mathbf{k}_p|$ is related to the wavelength by

$$k_p = 2\pi/\lambda_b$$

and Eq. (5.28) shows that the shorter bending waves travel faster than the longer bending waves. Fig. 5.6 illustrates the behavior of c_p with k_p. For membranes and strings, the second-order operator $L(\xi)$ gives rise to a k_p^2 or a k_T^2 rather than a k_p^4 in the unnumbered equation before Eq. (5.23). Thus for membranes the phase speed is

$$c_m = \sqrt{T_e/m_s} \tag{5.29}$$

which is independent of wave number. Flexural waves in membranes are similar to dilatational acoustic waves in this respect.

Another feature to be noted from Eqs. (5.28) and (5.29) is that when the structure is stiffened by increasing D_s or stretched to increase T_e, the bending wave speeds are increased. We shall see in Section 5.5 that the wave number dependence of flexural waves of plates (these are called dispersive waves) complicates the acoustic radiation phenomenon because as frequency is increased the bending waves could travel faster than the speed of sound in the fluid.

The eigenfunctions of the motion of finite panels are constructed as a linear superposition of free bending waves. Depending on the shape of the panel and the manner in which its boundaries are constrained, the superimposed waves will either reinforce or interfere. For some types of boundaries, waves will be generated which will decay in amplitude as they move from the boundaries. Such waves are called "evanescent," and such boundaries are said to generate "near fields." The hyperbolic functions introduced below Eq. (5.23) describe such waves. In any case only at preferred frequencies will there be a large group of waves reinforced with a discrete value of $k_p = k_{mn}$. At some other frequencies these waves will interfere. The frequencies of reinforcement are ω_{mn} and they are related to the values of k_{mn} by Eq. (5.25). The functions $\Psi_{mn}(\mathbf{y})$ describe the spatial variation of amplitude for the reinforcing wave systems, and they are also called mode shapes of the vibrating plate.

The modeling used in this chapter is particularly valid at high enough frequencies that the boundaries of the structure are separated by more than a vibration wavelength. The Fourier inversions used then represent a wave analogy in which the response is just the continuum of all possible contributions launched from the drive points and the reflecting boundaries.

The derivations for each mode of the finite plate proceeds along lines similar to that in Section 5.2. It is convenient, however, to deal with the velocity of the mode. Analogously to the derivation of Eqs. (5.4) and (5.7) we introduce the flexural velocity analog to Eq. (5.21)

$$v(\mathbf{y}, t) = \dot{\xi}(\mathbf{y}, t)$$

and

$$v(\mathbf{y}, t) = \int_{-\infty}^{\infty} V(\mathbf{y}, \omega) e^{-i\omega t} d\omega \tag{5.30}$$

so that

$$V(\mathbf{y}, \omega) = -i\omega\xi(\mathbf{y}, \omega)$$

and in accordance with Eq. (5.24a)

$$V(\mathbf{y}, \omega) = \sum_{mn} V_{mn}(\omega)\Psi_{mn}(\mathbf{y}) \tag{5.31}$$

The auto-spectral density of the velocity can be described in terms of the auto-spectral density of surface pressure and impedance of the surface following the methods leading to Eq. (5.7). From Eq. (5.31) the velocity spectral density is just a summation over the mn modes. We will be assuming that the modes are uncoupled in order to have, in accordance with Eq. (2.121),

$$\begin{aligned}\Phi_{\mathrm{vv}}(\omega) &= \lim_{T\to\infty}\sum_{mn} 2\pi\frac{|V_{mn}(\omega)|^2}{T} \\ &= \sum_{mn}\Phi_{mn}(\omega)\end{aligned} \tag{5.32}$$

For the mnth mode of a vibrating structure Eq. (5.27) gives

$$V_{mn}(\omega) = \frac{P_{mn}(\omega)}{Z_{mn}(\omega)} \tag{5.33}$$

The spectral density of the mode mn of the panel velocity is found by invoking Eq.(1.53) or (2.130) to both sides

$$\Phi_{mn}(\omega) = \frac{\Phi_{p_{mn}}(\omega)}{\left|Z_{mn}(\omega)\right|^2} \tag{5.34}$$

This relationship is general, but in the case of fluid loading that is not negligible, m_{s} should be increased by the added fluid mass and η_{s} should be increased by the radiation loss factor η_{rad} so that it is replaced by a total loss factor

$$\eta_{\mathrm{T}} = \eta_{\mathrm{s}} + \eta_{\mathrm{rad}} \tag{5.35}$$

as described and illustrated in Section 5.6.

The function $\Phi_{p_{mn}}(\omega)$ is the auto-spectral density of the modal pressure $P_{mn}(\omega)$ that was introduced in Eq. (5.27). Physically, it expresses the degree with which the excitation field couples to the mnth mode of the structure. As shown by the integral that defines $P_{mn}(\omega)$, this coupling is spatial; that is, it depends on the spatial phase of the driving force relative to the spatial phase of the vibration mode. A simple example for a one-dimensional mode would be for $P(y_1,\ \omega) = P_0 \sin(2k_n y_1)$ and $\Psi_n(y_1) = \sin k_n y_1$; then $P_n(\omega)$ would be identically zero. The modal pressure for this mode would be greatest for when $P(y_1,\ \omega) = P_0 \sin k_n y_1$ because the spatial phases would coincide. For random response estimates it is necessary to express the auto-spectrum of modal pressure in terms of the statistics of the surface pressure induced on the plate by the flow or other form of excitation. Information about this pressure field may be in the form of space−time correlations. Thus it is well to derive alternative representations of $\Phi_{p_{mn}}(\omega)$. We may define an autocorrelation function

$$\hat{R}_{p_{mn}p_{mn}}(\tau) = \langle P_{mn}(t+\tau)P_{mn}(t)\rangle$$

where we define

$$P_{mn}(t) = \frac{1}{A_{\mathrm{p}}}\int_{A_{\mathrm{p}}} p(\mathbf{y},t)\psi_{mn}(\mathbf{y})d^2\mathbf{y} \tag{5.36a}$$

or in the frequency domain

$$P_{mn}(\omega) = \frac{1}{A_\mathrm{p}} \int_{A_\mathrm{p}} p(\mathbf{y}, \omega)\psi_{mn}(\mathbf{y})d^2\mathbf{y} \tag{5.36b}$$

is a stochastic variable and the inverse transform of $P_{mn}(\omega)$. This correlation function is expressed in terms of the space−time correlation of surface pressure between two locations on the surface, $\mathbf{y}_1$ and $\mathbf{y}_2$; in the case of statistical homogeneity

$$\hat{R}_{p_{mn}p_{mn}}(\tau) = \frac{1}{A_\mathrm{p}^2} \iint_{A_{pn}} \cdot \iint_{A_{pn}} \hat{R}_{\mathrm{pp}}(\mathbf{y}_1 - \mathbf{y}_2, \tau)\Psi_{mn}(\mathbf{y}_1)\Psi_{mn}(\mathbf{y}_2)d^2\mathbf{y}_2 d^2\mathbf{y}_1 \tag{5.37}$$

We will specialize this to the case of point drives in Section 5.6.2. For now, the quantity of interest from the point of view of modal response is the Fourier transform of $\hat{R}_{p_n p_n}(\tau)$, which will be called the modal pressure spectrum, i.e.

$$\Phi_{p_{mn}}(\omega) = \frac{1}{2\pi} \int_{-\infty}^{\infty} e^{i\omega\tau} \hat{R}_{p_{mn}p_{mn}}(\tau)d\tau$$

A relationship for $\Phi_{p_n}(\omega)$ will now be derived in terms of the mode shape and the surface pressures correlation function. The wave number frequency spectrum of the surface pressure is related to its correlation function by

$$\hat{R}_{\mathrm{pp}}(\mathbf{r}, \tau) = \iint \int_{-\infty}^{\infty} e^{i(\mathbf{k}\cdot\mathbf{r} - \omega\tau)} \Phi_{\mathrm{pp}}(k, \omega)d^2\mathbf{k}d\omega \tag{5.38}$$

in the case of the correlation being spatially homogeneous in the sense of Eqs. (2.139) or (2.123). Modal shape functions are also defined as Fourier spatial transforms of the mode shapes, i.e.,

$$S_n(\mathbf{k}) = \iint_{A_\mathrm{p}} e^{-i\mathbf{k}\cdot\mathbf{y}} \Psi_n(\mathbf{y})d^2\mathbf{y} \tag{5.39}$$

Substitution of Eqs. (5.38) and (5.39) into Eq. (5.37) and combination with Eq. (5.37) gives

$$\Phi_{p_{mn}}(\omega) = \frac{1}{A_\mathrm{p}^2} \iint_{A_p} \cdot \iint_{A_p} \Phi_{\mathrm{pp}}(\mathbf{y}_1, \mathbf{y}_2, \omega)\Psi_{mn}(\mathbf{y}_1)\Psi_{mn}(\mathbf{y}_2)d^2\mathbf{y}_2 d^2\mathbf{y}_1 \tag{5.40a}$$

For either spatially homogeneous or inhomogeneous forcing and invoking statistical homogeneity we have the auto-spectrum of the modal pressure analogously to Eq. (2.144) as

$$\Phi_{p_{mn}}(\omega) = \frac{1}{A_\mathrm{p}^2} \int\int_{-\infty}^{\infty} \Phi_{\mathrm{pp}}(\mathbf{k}, \omega)|S_{mn}(\mathbf{k})|^2 d^2\mathbf{k} \tag{5.40b}$$

$$\int\int_{-\infty}^{\infty} |S_n(\mathbf{k})|^2 d^2\mathbf{k} = (2\pi)^2 A_\mathrm{p}$$

Eqs. (5.37), (5.40a), and (5.40b) represent generalizations of Parseval's theorem. Eqs. (5.34), (5.38), (5.40a), and (5.40b) provide cardinal results of this section, and these results may be further used to derive general input functions to the modal equations as shown below.

We now approach calculation of modal response using a hybrid of statistical modeling of the forcing pressure and the deterministic result of a finite element analysis of the structure. Combining Eqs. (5.31)–(5.33) we obtain a result for the vibration velocity at location $\boldsymbol{x}$,

$$V(\boldsymbol{x},\omega) = \sum_{mn} \frac{P_{mn}(\omega)}{Z_{mn}(\omega)} \Psi_{mn}(\boldsymbol{x}) \tag{5.40c}$$

This modal summation, though formal for the specifics considered here, is equivalent to a result for $V(\boldsymbol{x},\omega)$ that might be obtained with a finite element model using a unit point force. We note that a point force applied normal to the plate at a location $\boldsymbol{x}_0$ can be expressed as the fluctuating load appearing in Eq. (5.21) by

$$P(\boldsymbol{x},\omega) = F(\omega)\cdot\delta(\boldsymbol{x}-\boldsymbol{x}_0) \tag{5.40d}$$

so that in Eq. (5.27)

$$P_{mn}(\omega) = \frac{F(\omega)}{A_\mathrm{p}}\cdot\Psi_{mn}(\boldsymbol{x}_0) \tag{5.40e}$$

and

$$V(\boldsymbol{x},\boldsymbol{x}_0,\omega) = \frac{F(\omega)}{A_\mathrm{p}}\cdot\sum_{mn}\frac{\Psi_{mn}(\boldsymbol{x})\Psi_{mn}(\boldsymbol{x}_0)}{Z_{mn}(\omega)} \tag{5.40f}$$

The velocity response to a unit point force to define the "frequency response function"between forcing at $\boldsymbol{x}_0$ and response at $\boldsymbol{x}$,

$$\frac{V(\boldsymbol{x},\boldsymbol{x}_o,\omega)}{F(\omega)} = Y(\boldsymbol{x},\boldsymbol{x}_o,\omega) = \frac{1}{A_\mathrm{p}}\sum_{mn}\frac{\Psi_{mn}(\boldsymbol{x})\Psi_{mn}(\boldsymbol{x}_o)}{Z_{mn}(\omega)} \tag{5.40g}$$

which can be obtained in a matrix form with finite element models to give a series of responses each due to a distribution of drive points over a discretized surface, i.e., to form $Y(\boldsymbol{x},\boldsymbol{x}_o,\omega)$ and a similar function between pairs of points $\boldsymbol{y}$ and $\boldsymbol{y}_0$. In those cases admitting an eigenmode expansion, the modal admittance is

$$Y_{mn}(\boldsymbol{x},\boldsymbol{x}_0,\omega) = \frac{1}{A_\mathrm{p}}\frac{\Psi_{mn}(\boldsymbol{x})\Psi_{mn}(\boldsymbol{x}_0)}{Z_{mn}(\omega)} \tag{5.40h}$$

A cross-spectral density of the response velocities at positions $\boldsymbol{x}$ and $\boldsymbol{y}$ on the plate due to distributed forcing can be seen by using Eqs. (5.40f) and (5.40g) to form a correlation function analogous to Eq. (5.37)

$$\begin{aligned}\Phi_{vv}(\boldsymbol{x},\boldsymbol{y},\omega) &= \langle V^{*}(\boldsymbol{x},\omega)V(\boldsymbol{y},\omega)\rangle \\ &= \sum_{m,n}\iint_{A_p}\iint_{A_p}\Phi_{pp_{mn}}(\boldsymbol{x}_0,\boldsymbol{y}_0,\omega)\cdot Y_{mn}(\boldsymbol{y},\boldsymbol{y}_0,\omega)Y_{mn}^{*}(\boldsymbol{x},\boldsymbol{x}_0,\omega)d^2\boldsymbol{x}_0 d^2\boldsymbol{y}_0\end{aligned} \tag{5.40i}$$

The use of digital solutions requires a discretized form of Eq. (5.40j) which is a summation over all (m,n) computed eigenmodes,

$$\begin{aligned}\Phi_{vv}(\boldsymbol{x}_i,\boldsymbol{y}_j,\omega) = \sum_{m,n}\sum_{u}\sum_{v}\Phi_{pp_{mn}}(\boldsymbol{x}_{ou},\boldsymbol{y}_{ov},\omega)\cdot Y_{mn}(\boldsymbol{y}_j,\boldsymbol{y}_{ov},\omega)\\ Y_{mn}^{*}(\boldsymbol{x}_{oi},\boldsymbol{x}_{ou},\omega)(\Delta\boldsymbol{x}_o)_u(\Delta\boldsymbol{y}_0)_v\end{aligned} \tag{5.40j}$$

where $(\Delta\boldsymbol{x}_0)_u$ and $(\Delta\boldsymbol{y}_o)_v$ are the areas associated with the index pair of the discretization, and $\boldsymbol{x}_i$ and $\boldsymbol{y}_i$ are the vector coordinate locations at index positions i and j in a discretized response space. In recent years with the developments in large-scale computation, relationships such as Eq. (5.40j) are now commonly used in fully analytical [43,44] and hybridized finite element (and the related energy finite element) method analysis to calculate sound from flow-driven plates, shells, and pipes. See, e.g., Refs. [45–50] which describe the use and technique of using computer models for frequency response functions and statistically-based flow forcing functions. The statistical elements of the calculation are also used, especially for broad bandwidth applications and high frequencies on which we will focus the following section.

5.3.2 Response Estimates for Structures of Many Modes

As long as the mode shape function $\Psi_{mn}(y)$ and the statistical properties of the excitation may be described with sufficient accuracy, the velocity spectrum of the mode can be calculated using Eq. (5.40d). To do this, mean-square velocity of mode mn can be determined from Eqs. (5.8) and (5.11)

$$\overline{V_{mn}^2} = \frac{\pi A_p^2 \Phi_{p_{mn}}(\omega_{mn})}{M^2\eta_s\omega_{mn}} \tag{5.41}$$

where $M = m_s A_p$. The velocity is also related to the time-averaged input power to the mode, $(\mathbb{P}_{in})_n$, by Eq. (5.17), rewritten here to apply to a single mode:

$$\overline{V_{mn}^2} = \frac{(\mathbb{P}_{in})_{mn}}{M\eta_s\omega_{mn}} \tag{5.42}$$

By equating Eqs. (5.41) and (5.42) we obtain an explicit relationship for the time-averaged input power into the mode in terms of the statistics of the exciting pressure field; i.e.,

$$(\mathbb{P}_{\text{in}})_{mn} = \frac{\pi A_{\text{p}}^2 \Phi_{p_n}(\omega_{mn})}{M} \tag{5.43}$$

The function $A^2{}_p\Phi_{p_{mn}}(\omega)$ is the spectral density of the mean-square modal force applied to the nth mode. In comparing Eqs. (5.41), (5.42), and (5.43) we see that the average kinetic energy of mode $M\overline{V_{mn}^2}$ is determined by the resonant motion in the frequency band $\eta_s\omega_{mn}$ and that the power into the mode depends on the spectrum level of the modal force $A_p^2\Phi p_{mn}(\omega)$ and the mass of the structure. We will apply these concepts to the statistical representations of structural response of complex structures.

In some practical circumstances the response characteristic of interest involves large groups of modes of a structure rather than the response of a single mode. This interest would come from a need for a broadband rather than a narrowband description or attention directed to high enough frequencies that modal averages must be used. Recall from Eq. (5.25) that the structure (either one- or two-dimensional, membrane or plate) will resonate at specific frequencies ω_{mn} that are uniquely determined by a discrete set of one- or two-dimensional wave numbers k_{mn}. The characteristic wave numbers of a structure depend on its geometry and the nature of its supports. Although a description of the motion as a superposition of modes in the manner above is possible for many structures, exact closed-form expressions for eigenfunctions and eigenvalues are known only for rectangular plates with simple supports, free and simply supported circular plates, and rectangular and circular membranes. Solutions for other geometries and boundary conditions must be obtained numerically, or approximated with analytical functions which apply only over limited frequency ranges. In this section, then, continuing with descriptions of systems with uncoupled (i.e., orthogonal) modes we shall determine the eigenfunctions for the simply supported rectangular plate and use this particular result to illustrate the response characteristics of more complex multimodal structures throughout this chapter.

The geometry under consideration is shown in Fig. 5.4 and it consists of a rectangular plate element mounted in a plane infinite rigid surface which coincides with the $y_2 = 0$ plane. There is a fluid medium above the plate ($y_2 > 0$) which is assumed to be compressible with field points designated by (r, θ, ϕ), although for now, the presence of fluid above the plate will not figure in the determination of its modes. This point will be saved for Section 5.6 where we will develop the fluid loading impedance in terms of these mode shapes. The excitation response of the plate and the subsequent acoustic radiation to the region $y_2 > 0$ will be dependent on the eigenfunctions $\psi_{mn}(\mathbf{y})$. The plate is simply supported such that at the edges the plate velocity normal to the (y_1, y_3) plane is zero and no moment constraint or in-plane resistance is applied. Thus the boundary conditions in $V(\mathbf{y}, \omega)$ of Eq. (5.30) along $y_1 = \pm L_1/2$ or $y_3 = \pm L_3/2$ are

$$V\left(y_1 = \pm\frac{L_1}{2}, y_3, \omega\right) = 0 \quad \text{and} \quad \frac{\partial^2 V}{\partial y_1^2}\left(y_1 = \pm\frac{L_1}{2}, y_3, \omega\right) = 0$$

and

$$V\left(y_1, y_3 = \pm\frac{L_3}{2}, \omega\right) = 0 \quad \text{and} \quad \frac{\partial^2 V}{\partial y_3^2}\left(y_1, y_3 = \pm\frac{L_3}{2}, \omega\right) = 0$$

The appropriate solutions to Eqs. (5.21) and (5.24a) are such that the boundary conditions apply to the eigenfunctions $\psi_{mn}(\mathbf{y})$. These solutions will all be subject to an arbitrary constant coefficient, e.g., one such solution is

$$\psi_{mn}(\mathbf{y}) = A \sin \alpha_1 y_1 \sin \alpha_3 y_3$$

and the boundary conditions give

$$\alpha_1 = k_1 = m\pi/L_1 \quad \text{and} \quad \alpha_3 = k_3 = n\pi/L_3$$

The coefficient A is determined by the normalization condition (5.26) which gives

$$A = 2$$

This particular solution applies only to modes which are odd in both y_1 and y_3. The other solutions involve combinations of sines and cosines, thus for the coordinate system originating at the center of the panel the entire set may be written

$$\psi_{mn}(\mathbf{y}) = 2\sin(k_m(y_1 + L_1/2))\sin(k_n(y_3 + L_3/2)), \quad |y_1| \le L_1/2 \quad |y_3| \le L_3/2 \tag{5.44a}$$

and example sketches of node lines are shown in Fig. 5.7B. The wave number parameters are

$$k_m L_1 = \pm m\pi, \quad m = 1, 2, \ldots \tag{5.44b}$$

and

$$k_n L_3 = \pm n\pi, \quad n = 1, 2, \ldots \tag{5.44c}$$

In this case n and m denumerate the number of half-waves across each dimension as shown. Eq. (5.44) is also a solution to the second-order analog for the rectangular membrane.

The eigenfunction given above, Eq. (5.44a), is a particular simple example of a more general class of solutions for rectangular plates [38] with other boundary conditions. Other closed form exact and approximate solutions

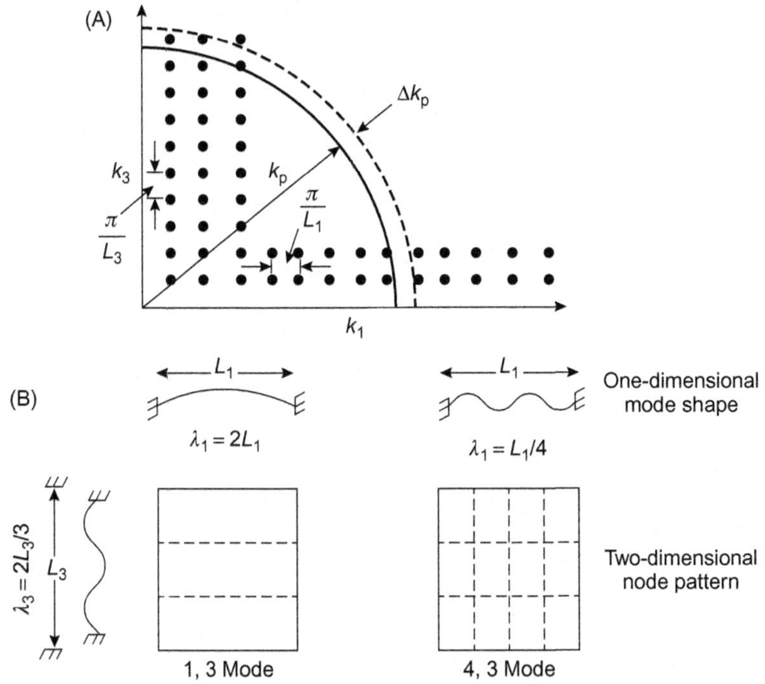

FIGURE 5.7 Illustration of mode orders of structural vibration of rectangular panels which are simply supported.

may be written down for circular plates with simple support or free boundaries and for special cases of other plate elements. These solutions are characterized by their separable nature as a product of two functions, i.e.,

$$\psi_{mn}(\mathbf{y}) = \psi_m(y_1)\psi_n(y_3) \tag{5.45}$$

When each of $\psi_m(y_1)$ and ψ_n (y_3) are harmonic functions, then

$$\nabla^2\psi_{mn}(\mathbf{y}) = (k_m^2 + k_n^2)\psi_{mn}(\mathbf{y})$$

for membranes, or

$$\nabla^4\psi_{mn}(\mathbf{y}) = (k_m^2 + k_n^2)^2\psi_{mn}(\mathbf{y})$$

for plates. Thus the wave numbers of the modes are the resultants of individual eigenvectors, i.e.,

$$\begin{aligned} k_{mn}^2 &= k_m^2 + k_n^2 \\ &= \left(\frac{m\pi}{L_1}\right)^2 + \left(\frac{n\pi}{L_2}\right)^2 \end{aligned}$$

in our example. In the two diagrams shown in Fig. 5.7, the 1,3 mode has $m = 1$ half-wave in the one-direction and $n = 3$ half-waves in the three-direction. For the m,n mode, there will be a resonance frequency ω_{mn} that is given by the transpose of Eq. (5.25).

The total response of a panel of many modes will be a summation over all modes mn of the individual modal responses as given by Eq. (5.32). As shown later in this section this is done in the context of a frequency band $\Delta\omega$ in which there are multiple resonant modes. This is also illustrated in the lattice shown in Fig. 5.7 by an annular element of radius k_p corresponding to the frequency of interest (Eq. (5.25)) with a differential wave number Δk_p corresponding to $\Delta\omega$. Ways of expressing the number of resonant modes in this annular region will now be developed. As $\Delta\omega$ increases or decreases, the number of modes in the lattice will change accordingly. The number of modes included in the quarter circle of radius k_p (i.e., in the range of $0 < k < k_p$) is [4,12]

$$N = \frac{(\pi/4)k_p^2}{(\pi/L_1)(\pi/L_3)} = \frac{k_p^2 L_1 L_3}{4\pi}$$

because the area of the quarter circle is $\pi k_p^2/4$ and the area defined by each of the interstices in the lattice is $(\pi/L_1)(\pi/L_3)$. Resonance conditions exist whenever $k_p = k_{mn}$. The number of modes per unit increase in wave number is therefore

$$n(k_p) = \frac{dN}{dk_p} = \frac{k_p L_1 L_3}{2\pi} \tag{5.46}$$

and, accordingly, the number of modes per unit increase in frequency is

$$n(\omega) = \frac{dN}{dk_p}\frac{dk_p}{d\omega} = \frac{n(k_p)}{c_g}$$

This function is called the frequency "mode density" and it is, specifically, for a flat plate

$$n(\omega) = \frac{L_1 L_3}{4\pi}\left(\frac{m_s}{D_s}\right)^{1/2}$$

where c_g is the group velocity of the waves,

$$n(\omega) = \frac{n(k_p)}{c_g} = \frac{1}{2\pi\sqrt{\omega\kappa c_L}}$$

Since the longitudinal wave speed in a plate is

$$c_l = \sqrt{\frac{E}{\rho_p(1-\mu_p^2)}} = \sqrt{\frac{D_s}{m_s\kappa^2}} \tag{5.47}$$

we can write the mode density as

$$n(\omega) = \frac{A_p}{4\pi\kappa c_l} \tag{5.48a}$$

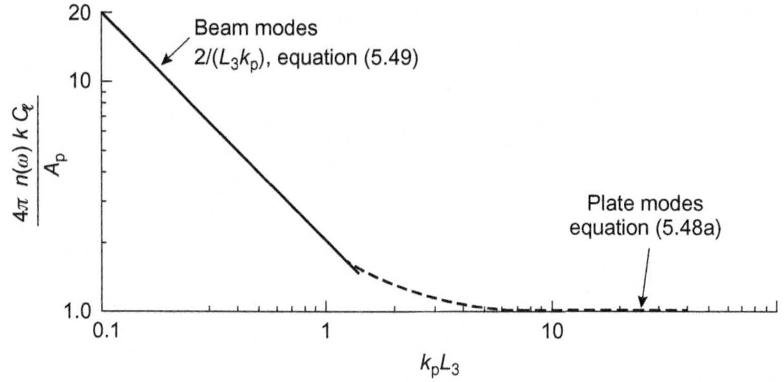

FIGURE 5.8 Mode density for rectangular plate for which $k_pL_1 > \pi$ and $L_3 < L_1$. Note regions of plate-like and beam-like vibration.

where $A_p = L_1L_3$ is the area of the plate. The mode density is a crucial factor used in describing the multimode response of vibrating structures using statistical methods, see Lyon and DeJong [11] and in following chapters we will use it in building predictive models of broadband forced vibration and sound.

Eq. (5.48a) shows that the mode density for a plate is independent of frequency. Also, the plate area has been introduced instead of the length and width because Eq. (5.48a) applies generally for the higher modes of all single plates. For membranes still,

$$n(k_p) = \frac{k_pA_p}{2\pi}$$

but since

$$k_p = \omega\sqrt{m_s/T_e} = \omega/c_m$$

the frequency mode density increases with frequency as

$$n(\omega) = \frac{A_p\omega}{2\pi c_m} \tag{5.48b}$$

In like manner, the wave number mode density for a beam of length L is

$$n(k_p) = L/\pi$$

and introducing the bar wave speed c_L

$$n(\omega) = \frac{n(k_p)}{c_g} = \frac{1}{2\pi\sqrt{\omega\kappa c_L}} \tag{5.49}$$

which shows that the frequency mode density decreases as frequency increases. For a long, narrow structure one would expect $n(\omega)$ to follow Eq. (5.49) for frequencies above the fundamental resonance frequency and below the first lateral mode for which $k_pL_3 < \pi$. For both $k_pL_3 > \pi$ and $k_pL_1 > \pi$, Eq. (5.49) applies. Fig. 5.8 illustrates this dependence of the mode density on the width, k_pL_3, of a rectangular panel for which $L_1 > L_3$ for all modes $k_pL_1 > \pi$.

Relationships (5.48a) and (5.49) are independent of the nature of the constraint applied to the panel or beam (clamped, free, etc.) for modes that have more than a half-wave between boundaries. Thus we need not be concerned with the details of constraint at this point, but only with the one- or two-dimensionality of the structure and whether the bending waves are nondispersive with c_g constant, or dispersive with c_g dependent on wave number.

Thus we are now in a position to illustrate the methods of estimating the flow-induced response of structures observed in a physical measurement using a finite bandwidth filter. We shall invoke the filter band $\Delta\omega_f$, centered on ω_f for which and $0 \leq \omega f = 2\pi ff \leq \infty$ to be consistent with the definitions introduced in Section 1.4.4.1. The resonance frequency ω_{mn} still represents the resonance frequency and in our presentation of response we will average over all responses of modes resonant in the bandwidth $\Delta\omega f$.

$$n(\omega_f)\Delta\omega_f > 1$$

As indicated by Eqs. (5.31) and (5.32) the flexural velocity spectral density on a two-dimensional structure is a summation over all modes of the modal velocity spectra $\Phi_{mn}(\omega)$. It has been assumed that the modes are uncoupled. As illustrated in Figs. 5.2 and 5.3 and shown in Eq. (5.34), the $\Phi_{mn}(\omega)$ for each mode is sharply peaked about all frequencies for which $\omega = \pm\omega_{mn}$. Thus where we consider the mean-square velocity in the broad frequency band of a filter, $\Delta\omega_f$, centered on ω_f, $\Delta\omega_f$ must be large enough that $\Delta\omega_f \gg (\pi/2)\eta_T\omega_{mn}$ and $n(\omega_f)\Delta\omega_f > 1$. The total mean-square velocity in the $\Delta\omega_f$ is just the summation over all ω_{mn} lying in $\omega_f \pm \Delta\omega_f$. The frequency interval between modes must be small enough that the set of resonances resembles a continuum. Using Eq. (5.32)

$$\begin{aligned}\overline{V^2}(\omega_f, \Delta\omega) &= 2\int_{\omega_f-\Delta\omega_f/2}^{\omega_f+\Delta\omega_f/2} \Phi_{vv}(\omega)d\omega \\ &= 2\sum_{mn}\int_{\omega_f-\Delta\omega_f/2}^{\omega_f+\Delta\omega_f/2} \Phi_{mn}(\omega)d\omega \\ &= \sum_{\substack{\text{all modes} \\ \text{in } \Delta\omega}} \overline{V_{mn}^2}\end{aligned}$$

where ω_f is the center frequency of the filter band, the factor of 2 folds the negative frequency axis on to the positive axis to form the one-sided spectrum, and $\overline{V_{mn}^2}$ is given by Eq. (5.41). We already specified that the modal force spectrum, $\Phi_{p_{mn}}(\omega)$ must be roughly constant over the resonance frequency band $(\pi/2)\eta_T\omega_{mn}$, but now if we also require that it is roughly constant over the entire filter bandwidth of interest, $\Delta\omega_f$, then we can approximate the summation by an integration over all ω_{mn} lying in $\Delta\omega_f$ since $n(\omega_f) \gg 1$; thus

$$\overline{V^2}(\omega_f, \Delta\omega_f) \simeq 2\int_{\omega_f-\Delta\omega_f/2}^{\omega_f+\Delta\omega_f/2} V_{mn}^2(\omega)n(\omega)d\omega \\ \simeq \overline{\overline{V_{mn}^2}}(\omega_f)n(\omega_f)\Delta\omega_f \tag{5.50}$$

where $\overline{\overline{V_{mn}^2}}(\omega_f)$ is again given by Eq. (5.41), but it now represents the typical mean-square modal velocity generalized as a continuous function of frequency and represents a mean square response of all modes that are resonant in the band $\Delta\omega_f$. If the structure is a plate of area A_p, then Eqs. (5.8), (5.11), (5.41), (5.48a), (5.49), and (5.50) combine to give the mean-square (time- and area-averaged) panel velocity as

$$\overline{V^2}(\omega_f, \Delta\omega_f) \simeq \frac{\pi\overline{\Phi_{p_{mn}}(\omega_f)}}{m_s^2\eta_T}\frac{A_p}{4\pi\kappa c_l}\frac{\Delta\omega_f}{\omega_f} \tag{5.51}$$

where c_l is given by Eq. (5.47).

"Modal average"

The function $\overline{\Phi_{p_{mn}}(\omega)}$ represents the average of the modal pressure spectrum over all m,n modes that are resonant in the frequency band $\Delta\omega_f$; see Section 5.5.5. This spectrum is found as

$$\overline{\Phi_{p_{mn}}(\omega)} = \frac{1}{N}\sum_{m,n}\Phi_{p_{mn}}(\omega)$$

where

$$(N = n(\omega)\Delta\omega_f)$$

An example of such a calculation is given in Chapter 9 for the case of structures driven by a turbulent boundary layer.

The time-averaged *power in* $\Delta\omega_f$, $\mathbb{P}_{in}(\omega_f)$, that is supplied to the plate by the flow is

$$\mathbb{P}_{in}(\omega_f) = 2\int_{\omega_f-\Delta\omega_f/2}^{\omega_f+\Delta\omega_f/2}\left[\iint_{A_p}\langle p^*(\boldsymbol{x},\omega)V(\boldsymbol{x},\omega)\rangle d^2\boldsymbol{x}\right]d\omega \tag{5.52a}$$

$$\mathbb{P}_{in}(\omega_f) = \sum_{\substack{\text{modes}\\ \text{in } \Delta\omega}} (\mathbb{P}_{in})_n \\ = \int_{\omega_f-\Delta\omega_f/2}^{\omega_f+\Delta\omega_f/2}\mathbb{P}_{in}(\omega_f)n(\omega_f)d\omega_f$$

Using Eq. (5.41) we find the power spectral density *per unit area* to be $\pi_{\rm in}(\omega_{\rm f}) = \mathbb{P}(\omega_{\rm f})/(2A_{\rm p}\Delta\omega_{\rm f})$ and it is defined ONLY over the positive frequency axis for which $\omega_{\rm f} = 2\pi {\rm f}_{\rm f}$ and recalling $0 < \omega_{\rm f} < \infty$,

$$\pi_{\rm in}(\omega_f) = \frac{\pi\overline{\Phi_{p_{mn}}(\omega_f)}}{m_{\rm s}} n(\omega_f) \tag{5.52b}$$

$$= \frac{\pi\overline{\Phi_{p_{mn}}(\omega_f)}}{m_{\rm s}} \frac{A_{\rm p}}{4\pi\kappa c_{\rm l}} \tag{5.52c}$$

This can be rewritten

$$\pi_{\rm in}(\omega_f) = \frac{\overline{\Phi_{p_{mn}}(\omega_f)}A_{\rm p}}{R_\infty} \tag{5.52d}$$

where

$$R_\infty = 8m_{\rm s}\kappa c_{\rm l} \tag{5.52e}$$

is the point input impedance of an infinite plate equal to the average (over all modes) resistance of an infinite plate [7,11,17].

The total input power per unit area given is by

$$\pi_{\rm in} = \int_0^\infty \pi_{\rm in}(\omega_f)d\omega_f$$

and $\Phi_{p_{mn}}(\omega)$ is given by Eq. (5.40).

In Section 5.6 we shall illustrate that the high-frequency multi-mode point impedance (or frequency response function) also approaches that of an infinite plate even in narrow frequency bands. The relationships just derived, (5.50), (5.51), and (5.52b), are subject to rigid restrictions as discussed fully by Lyon and DeJong [11]. It must be established before these are valid that $\overline{V_{mn}^2}(\omega)$ is indeed nearly independent of the mode order. This, in turn, requires that the modal damping is roughly constant for all modes in the frequency band. If one mode is very lightly damped compared to others, its response will overwhelm the vibration of others in the band, and it must therefore be considered separately. A less obvious but often more important restriction to the use of these approximations is that the auto-spectrum of the modal excitation pressure, $\Phi_{p_{mn}}(\omega)$ of Eq. (5.40), is the same for all modes in the band. Under this broad set of requirements the input power, Eq. (5.52d), is related to the mode-ensemble mean-square velocity $\overline{V^2}(\omega, \Delta\omega)$, Eq. (5.50), by a relationship analogous to Eq. (5.42),

$$\overline{V^2}(\omega, \Delta\omega) = \frac{\mathbb{P}_{\rm in}(\omega)}{Mn_{\rm T}\omega} \tag{5.53}$$

This mean-square velocity is dependent on the boundary conditions of the plate by influencing the details of the shape function $S_{mn}(\boldsymbol{k})$ appearing in the integral for input pressure, $P_{\rm in}$, as we shall see in the next section.

The restrictions that allow straightforward modal averaging may not be followed by a number of situations. If, for example, the wave number spectrum of the excitation pressure may have elevated values for only a restricted range of wave numbers; then only certain modes in the frequency band will be selectively excited. This situation can arise when boundary layers excite structures. This modal selectivity is also likely to arise when vortex-induced pressures excite the trailing edges of hydrofoils. Another restriction can occur when acoustic radiation from the structure is estimated. As we shall see in Section 5.5, certain classes of modes more effectively radiate sound than others. In these cases modal averaging must be undertaken selectively according to a strategy that accounts for clusters of radiating and nonradiating modes.

5.4 MODAL SHAPE FUNCTIONS FOR SIMPLE STRUCTURES

The input power and the mean-square flow-induced velocity depend on the spatial matching of the exciting pressure and the mode shape (or frequency response function). This matching depends on shape of the modal shape function $S_{mn}(\mathbf{k})$, which appears in Eq. (5.40) and is given by Eq. (5.39). It is worthwhile to explore this interaction for which $S_{mn}(\mathbf{k})$ behaves as a linear filter and may be determined in principle by a Fourier wave number transformation of the $\psi_{mn}(\boldsymbol{y})$ or, more generally, $Y(\boldsymbol{y},\boldsymbol{x},\omega)$. To illustrate the method, we refer again to the simply supported rectangular plate for which the mode shape is given by Eq. (5.44). The related modal shape function is, by Eq. (5.39) [16,28,31]

$$S_{mn}(\boldsymbol{k}) = 2A_{\mathrm{p}}\frac{\left[e^{ik_1L_1/2}-(-1)^m e^{-ik_1L_1/2}\right]}{(k_mL_1)(1-(k_1/k_m)^2)}\cdot\frac{\left[e^{ik_3L_3/2}-(-1)^n e^{-ik_3L_3/2}\right]}{(k_nL_3)(1-(k_3/k_n)^2)} \tag{5.54a}$$

This function is defined over the positive and negative values of wave number and positive and negative values of n and m. Being consistent with the separable functions, Eqs. (5.44) and (5.45), separate into the two functions

$$S_{mn}(\boldsymbol{k}) = S_m(k_1)S_n(k_3) \tag{5.54b}$$

$S_{mn}(\mathbf{k})$ is peaked about $k_1 = k_m$ and $k_3 = k_n$, which defines its main acceptance region. We consider the separate functions $S_m(k_1)$ for the odd-order mode ($m = 1, 3, 5, \ldots$), which can be written

$$\frac{2S_m(k_1)}{L_1} = \frac{4\sqrt{2}\sin(k_1L_1/2)}{(k_mL_1)(1-(k_1/k_m)^2)}$$

and, similarly for the odd order function where cosine replaces the sine. Fig. 5.9 shows $2S_m(k_1)/L_1$ as a function of k_1/k_m. By virtue of the

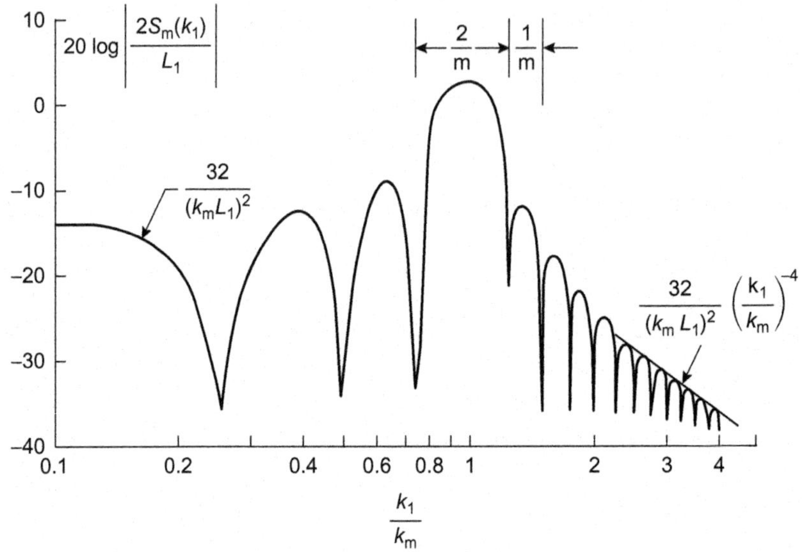

FIGURE 5.9 Modal shape function for a one-dimensional odd-order mode shape for a simply supported panel shown here with asymptotic functions identified for $k_mL_1 = m\pi = 7\pi$; simply supported panel.

boundary condition of vanishing displacement at $y_1 = \pm L_1/2$, $k_mL_1 = m\pi$. The maximum occurs when $k_1 = k_m$ so

$$20\log\left|\frac{2S_m(k_1)}{L_1}\right| \tag{5.55}$$

$$\left|\frac{2S_m(k_m)}{L_1}\right|^2 = 2 \tag{5.56}$$

and the wave number bandwidth of this main acceptance region is

$$\Delta k_1/k_m = 2/m \quad \text{or} \quad \Delta k_1 = 2\pi/L_1 \tag{5.57}$$

In the limit of $k_1 \ll k_m$

$$\left|\frac{2S_m(k_1)}{L_1}\right|^2 \sim \frac{32}{(k_mL_1)^2}\sin^2\frac{k_1L_1}{2} \tag{5.58}$$

and the cosine replaces the sine for odd-order modes. When $k_l >> k_m$ L'Hopital's rule is used to find

$$\left|\frac{2S_m(k_1)}{L_1}\right|^2 \sim \frac{32}{(k_mL_1)^2}\left(\frac{k_m}{k_1}\right)^4\sin^2\frac{k_1L_1}{2} \tag{5.59}$$

for both even and odd-order modes. The only difference between the even- and odd-order modes lies in the appearance of the sine or cosine function, and accordingly in the behavior as k_1 approaches zero; $S_m(k_1)$ and $S_n(k_3)$ for even-order modes vanish in the limits of zero wave number. Otherwise the approximate relationships (5.56)–(5.59) apply, see ahead to Fig. 5.10. Below we will summarize shape functions for various mode classes.

For a clamped beam, Aupperle and Lambert [39] and Martin [40] have approximated the shape function for a clamped–clamped beam of length L:

$$\left|\frac{2S_m(k_1)}{L_1}\right|^2 \simeq \frac{32}{1-(k_mL_1)^{-1}}\left[\frac{(k_mL_1)^2}{(k_mL_1)^2+(k_1L_1)^2}\right]^2 \times \left[\frac{\sin(k_m-k_1)L_1/2}{(k_m-k_1)L_1}+\frac{\sin(k_m+k_1)L_1/2}{(k_m+k_1)L_1}\right]^2 \tag{5.60}$$

where $k_mL_1 = [(m+1)/2]\pi$ for $m>3$. The value of $|2S_m(k_m)/L_1|^2$ is 2 just as it is for simply supported beams. However, the limiting values are

$$\left|\frac{2S_m(k_1)}{L_1}\right|^2 \sim \frac{64}{(k_mL)^2}\left(\frac{k_m}{k_1}\right)^6 \sin^2\frac{k_1L_1}{2}, \quad k_1 \gg k_m \tag{5.61}$$

and

$$\left|\frac{2S_m(k_1)}{L_1}\right|^2 \sim \frac{64}{(k_mL)^2}\sin^2\frac{k_1L_1}{2}, \quad k_1 \ll k_m \tag{5.62}$$

For the clamped plate, the shape function is just the product of two of these each for the appropriate mode order and length. The shape function squared for the clamped–clamped boundary condition at low wave numbers is thus a factor of 2 greater than for the simply supported boundary condition, but it is functionally identical. These limiting behaviors are illustrated for both even m and odd m orders in Fig. 5.10. The asymptotic functional behaviors that were derived above are illustrated in each case. Notably for consideration in the next section, the behaviors at low wave numbers are dependent on mode order; for $k_1 \geq k_m$ the responses are independent of oddness/evenness, but strongly dependent on boundary condition. The case of free edges, for which the magnitude of the function is [40,52]

$$\left|\frac{2S_m(k_1)}{L_1}\right|^2 \simeq \frac{8(k_mL_1)^2[1-(-1)^m\cos(k_1L_1)]}{(k_mL_1)^2+(k_1L_1)^2} \tag{5.63}$$

has understandably the greatest magnitude which is particularly of importance for leading and trailing edge noise, see Chapter 11, Noncavitating Lifting Sections.

The asymptotic of $S_m(k_1)$ for $k_1 \gg k_m$ can be more generally considered by noting that for $k_1/k_m > 1$ integration of Eq. (5.39) yields, e.g., [41],

$$\frac{2S_m(k_1)}{L_1} = \sum_{j=0}^{N-1}(i)^{j+1}\left(\frac{k_1L_1}{2}\right)^{-j-1}\left[e^{i(k_1L_1/2)}\psi_m^{(j)}(1) - e^{-i(k_1L_1/2)}\psi_m^{(j)}(-1)\right] + \theta\left[\left(\frac{k_1L_1}{2}\right)^{-N}\right] \tag{5.64}$$

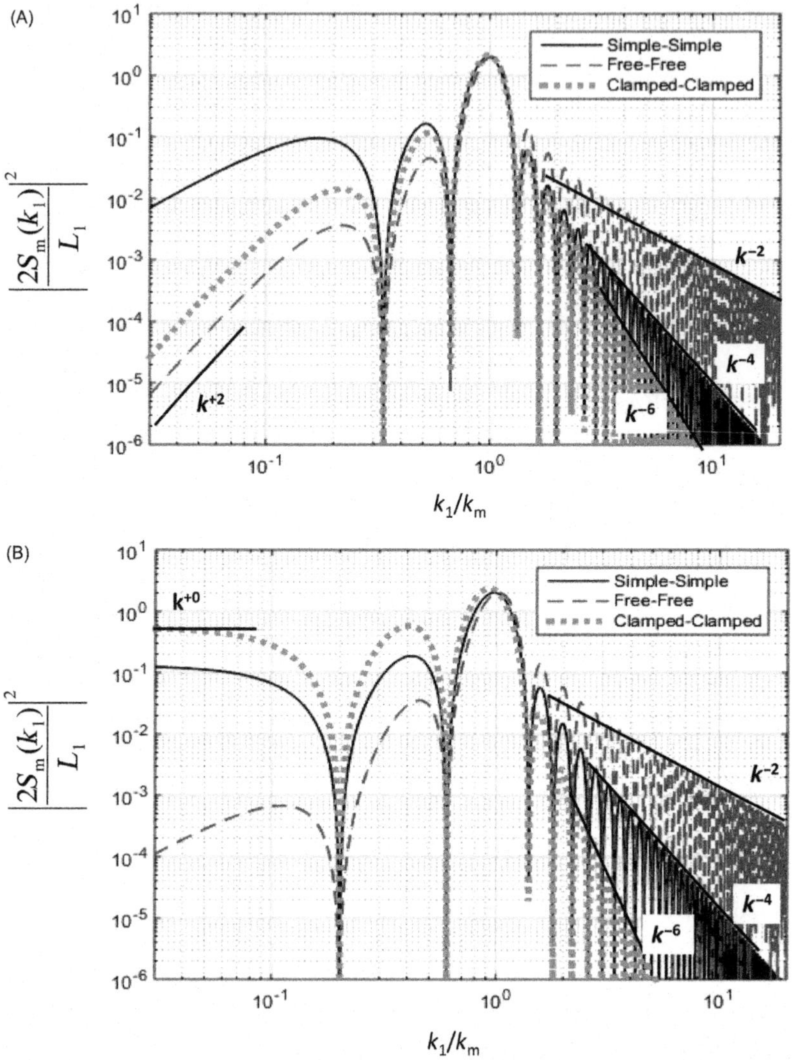

FIGURE 5.10 Modal shape functions for simple−simple, clamped−clamped, and free−free boundary conditions. Functions typical of even-order modes are shown at the top , for odd-order modes at the bottom.

where the short hand notation is used,

$$\psi_m^{(j)}(\pm 1) = \left(\frac{L_1}{2}\right)^j \frac{\partial^{(j)}\psi_m(\pm L_1/2)}{\partial y_1^{(j)}}$$

and similarly for $S_n(k_3)$. Chase [51] has derived expressions similar to Eq. (5.64) for both circular axisymmetric surfaces and rectangular surfaces. The boundary conditions determine the values of the edge derivatives, $\psi_m^{(j)}\ (\pm L_i/2)$.

The symmetry of mode order about the center of the panel determines the combination of the terms in the brackets. One needs only to consider the lowest-order nonzero derivative in evaluating Eq. (5.64). Substituting into Eq. (5.61) we find for $\psi_m(y_1)$ harmonic in $k_m y_1$ that

$$\psi_m^{(j)}(1) \propto \left(\frac{k_m L_1}{2}\right)^j, \quad k_1 \gg k_m$$

which makes

$$\left|\frac{2S_m(k_1)}{L_1}\right|^2 \sim \frac{16}{(k_m L_1)^2}\left(\frac{k_m}{k_1}\right)^{2(j+1)}, \quad k_1 \gg k_m \tag{5.65}$$

where j is the lowest-order nonvanishing derivative boundary condition. For a clamped plate $j=2$, for a simple support $j=1$, and for a free motion (or nearly free allowing some displacement) at the edge $j=0$.

Some general comments about the behavior of $S_m(k_1)S_n(k_3)$ may now be made on the strength of the above results. Thus we have the following possibilities:

i. $\psi_m\,(\pm L_i/2)=0$, zero displacement at the edges,
ii. $(\partial\psi_m/\partial y_i)(\pm L_i/2)=0$, zero slope for a clamped boundary condition requiring a nonvanishing moment,
iii. $(\partial^2\psi_m/\partial y_i^2)(\pm L_i/2)=0$, zero curvature for a simple (pinned) boundary requiring a vanishing moment.

Fig. 5.10 illustrates the example curves for the canonical boundary conditions: simple supports (i, iii), clamped boundaries (i, ii), and free edges (ii, iii). At very low wave numbers, such that $k_1 \ll k_m$, $k_3 \ll k_n$, or $k \ll k_{mn}$ the acceptance $|S_m S_n|$ of a given mode is proportional to the area of the plate and inversely proportional to the number of half-wavelengths of vibration across either dimension L_1 and L_3. At wave numbers $k_1 \approx k_m$ and $k_3 \approx k_n$, equivalently $k \approx k_p$, the plate accepts its excitation in direct proportion to its area and independently of mode order. Its response is greatest when excited by a disturbance whose wave number equals the free bending wave number. At very high wave numbers the acceptance is strongly dependent on the boundary conditions as well as the number of half-waves as measured by $k_m L$ or $k_n L_3$. The annotations in Fig. 5.9 illustrate that as the size of the plate increases $k_m L$ increases the bandwidth of the main acceptance lobe narrows and its magnitude rises above the other lobes.

A direct consequence of the normalizing equation (5.26) is that the integral of $|S_{mn}(\mathbf{k})|^2$ over all $\mathbf{k}$ is

$$\frac{1}{(2\pi)^2}\int\!\!\int_{-\infty}^{\infty} |S_{mn}(\mathbf{k})|^2 d^2\mathbf{k} = \iint |\psi_{mn}(\mathbf{y})|^2\, d^2\mathbf{y}$$

Therefore

$$\iint_{-\infty}^{\infty} |S_{mn}(\boldsymbol{k})|^2\, d^2\mathrm{k} = (2\pi)^2 A_p \tag{5.66}$$

as long as Eq. (5.26) holds regardless of either the shape of the panel or its dynamic boundary conditions. To show the derivation for Eq. (5.66), substitute Eq. (5.39) and its complex conjugate into Eq. (5.66) to obtain

$$\iint_{-\infty}^{\infty} |S_{mn}(\boldsymbol{k})|^2\, d^2\boldsymbol{k} = \iint_{-\infty}^{\infty} d^2\boldsymbol{k} \iint_{A_p} e^{-i\boldsymbol{k}\cdot\boldsymbol{y}}\psi_{mn}(\boldsymbol{y})d^2\boldsymbol{y} \iint_{A_p} e^{-i\boldsymbol{k}\cdot\boldsymbol{y}'}\psi_{mn}(\boldsymbol{y}')d^2\boldsymbol{y}'$$

Rearranging the right hand side

$$\iint_{-\infty}^{\infty} |S_{mn}(\boldsymbol{k})|^2\, d^2\boldsymbol{k} = \iint_{A_p}\iint_{A_p} \psi_{mn}(\boldsymbol{y})\psi_{mn}(\boldsymbol{y}')\left[\iint_{-\infty}^{\infty} e^{i\boldsymbol{k}\cdot(\boldsymbol{y}-\boldsymbol{y}')}d^2\boldsymbol{k}\right] d^2\boldsymbol{y}'\, d^2\boldsymbol{y}$$

Now introducing Eq. (1.64) for the two-dimensional Dirac delta function and carrying out the integrations of Eq. (1.62) and Eq. (5.26) we obtain Eq. (5.66).

In view if the prominence of a main lobe, e.g., at $k_1 = k_m$ for a rectangular panel as illustrated in Figs. 5.9 and 5.10 we can approximate the integrated filtering behavior of $S_{mn}(\boldsymbol{k})$ as

$$\left|S_{mn}(\boldsymbol{k})\right|^2 = \pi^2 A_p[\delta(k_1 + k_m) + \delta(k_1 - k_m)][\delta(k_1 + k_n) + \delta(k_1 - k_n)] \tag{5.67a}$$

More generally for an extended homogeneous plate of any planform geometry it is clear that the integrated behaviour of the admittance function approaches a delta function, so in line with Eq. (5.66), following Jones [53, Ch. 8], and noting that $\iint \cdots d^2\boldsymbol{k} \to 2\pi \int \cdots |k| dk$, we have

$$\left|S(\boldsymbol{k})\right|^2 \approx 2\pi A_p |\boldsymbol{k}|^{-1}\delta(|\boldsymbol{k}| - k_p) \tag{5.67b}$$

The integrated behavior of the integral behaves in the limit essentially as free waves on an infinite plate since the resultant resonance wave numbers, k_{mn}, are nearly equal to the prevailing plate wave number, k_p, at the frequency of interest. At very high wave numbers such that $k_p h$, where h is thickness of the plate, approaches unity, the Timoshenko-Mindlen plate formulation is not valid so that the limiting forms given above [Eqs. (5.60), (5.62), (5.64), and (5.65)] do not apply, see e.g. Junger and Feit [1] or Fahy and Gardonio [3].

5.5 ESSENTIAL FEATURES OF STRUCTURAL RADIATION

5.5.1 Acoustic Radiation From a Simply Supported Panel

The problem of determining the sound that is radiated into a quiescent fluid by a body with a known velocity normal to its surface is a deterministic one. The

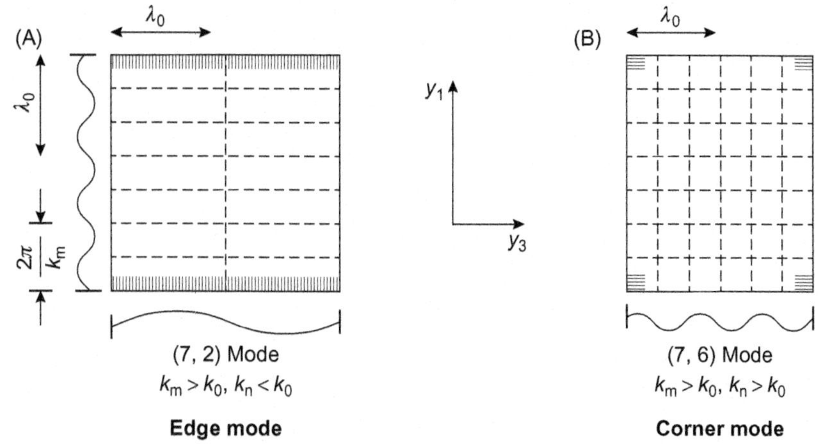

FIGURE 5.11 Illustration of "edge" and "corner" mode radiation for a simply supported rectangular panel. Shaded areas indicate uncanceled regions of volume velocity.

prescription of the velocity of the fluid at the surface is identified as a Neumann boundary value problem; the complexity of the solution of the acoustic wave equation depends on the geometry of the body as well as on the variation of the velocity over the surface. General methods of solution for flat, baffled surfaces, and spherical and cylindrical shells have been described by Junger and Feit [1], Skelton and James [2], and Fahy and Guardonio [3]. We will defer to these monographs for further detail and scope than that given here.

In order to show application of the other published accounts on structural acoustics per se, it is instructive to outline here as an example the solution of the problem of radiation from a rectangular baffled plate; we will also discuss the radiation from cylindrical structures in Chapter 10, Sound Radiation From Pipe and Duct Systems. Fig. 5.11 illustrates the geometry: a flat panel with simply supported boundaries at coordinates $\pm L_1/2$, $\pm L_3/2$ lying in the $y_2 = 0$ plane which is rigid except for the panel area. Fluid exists above the plane; the field point at which the sound pressure is to be evaluated is $y' = y'_1, y'_2, y'_3$. The velocity of the panel normal to the $y_2 = 0$ plane is $v(y, t)$ only within the region belonging to $|y_1| < L_1/2$ and $|y_3| < L_3/2$. Outside this region the velocity is identically zero; i.e., it is baffled. The velocity field on the panel can be described in terms of its temporal Fourier transform $V(\mathbf{y}, \omega)$, Eq. (5.30), and it is a superposition of the normal modes of the surface, $\psi_{mn}(\mathbf{y})$, as in Eq. (5.31). The equation for the acoustic pressure in the fluid $p_a(\mathbf{y}, t)$ is Eq. (2.7), repeated here:

$$\nabla^2 p_a - \frac{1}{c_0^2}\frac{\partial^2 p_a}{\partial t^2} = 0$$

In as much as $p_a(\mathbf{y}, t)$ is a stochastic function of time, it is convenient to use the Fourier transform of the acoustic pressure

$$P_a(y_2, \boldsymbol{k}, \omega) = \frac{1}{(2\pi)^3} \iint\int_{-\infty}^{\infty} e^{-i\boldsymbol{k}\cdot\boldsymbol{y}+i\omega t} p_a(y_1, y_2, y_3, t) dy_1\, dy_3\, dt \tag{5.68}$$

so that Eq. (2.7) becomes

$$\frac{\partial^2 P_a(y_2, \boldsymbol{k},\ \omega)}{\partial y_2^2} + \left(k_0^2 - k^2\right) P_a = 0 \tag{5.69}$$

which has a solution

$$P_a(y_2, \boldsymbol{k}, \omega) = A e^{i\sqrt{k_0^2 - k^2} y_2}, \tag{5.70}$$

where $k^2 = k^2{}_1 + k^2{}_3$ and we use the convention

$$\begin{aligned} \sqrt{-1} &= i \quad \text{for} \quad y_2 > 0 \\ \sqrt{-1} &= -i \quad \text{for} \quad y_2 < 0 \end{aligned} \tag{5.71}$$

The linearized boundary condition for the fluid (for which the convection term $(\mathbf{V}\cdot\nabla)\cdot\nabla$ in Eq. (2.4) is ignored) is

$$\frac{\partial V_2}{\partial t} = -\frac{1}{\rho_0}\frac{\partial P_a}{\partial y_2} \tag{5.72}$$

so now we can relate the field acoustic pressure to the panel velocity. To do this we write the space–time Fourier transform of the velocity $V(\mathbf{k}, \omega)$ introduced at the end of Section 2.6.2, Derivation of the Wave Equation With Vortical Sources

$$V(\boldsymbol{k},\ \omega) = \frac{1}{(2\pi)^3} \iiint_{-\infty}^{\infty} V(\mathbf{y},\ t) e^{-i(\boldsymbol{k}\cdot\boldsymbol{y}-\omega t)} dy_1\, dy_2\, dt$$

$$V(\boldsymbol{k},\ \omega) = \frac{1}{(2\pi)^3} \sum_{mn} V_{mn}(\omega) S_{mn}(\boldsymbol{k}) \tag{5.73}$$

Now, the double sum over m and n replaces the simple sum over n, and $V(\mathbf{k}, \omega)$ is subject to the requirement that $V(\mathbf{y}, t)$ vanishes on the baffle. Eq. (5.72) becomes

$$-i\omega V(\boldsymbol{k},\ \omega) = -\frac{1}{\rho_0}\frac{\partial P_a(y_2, \boldsymbol{k}, \omega)}{\partial y_2}, \quad y_2 > 0 \tag{5.74}$$

Combining Eqs. (5.70), (5.73), and (5.74), we find the Fourier transform of the pressure to be a summation of influences of all modes,

$$P_a(y_2, \boldsymbol{k}, \omega) = \sum_{mn} (2\pi)^{-2} \rho_0 c_0 \frac{V_{mn}(\omega) S_{mn}(\boldsymbol{k})}{\sqrt{1 - k^2/k_0^2}} e^{i\sqrt{k_0^2 - k^2} y_2} \tag{5.75a}$$

$$P_a(y_2, \boldsymbol{k}, \omega) = \sum_{mn} (2\pi)^{-2} Z_a(\omega, \boldsymbol{k}) V_{mn}(\omega) S_{mn}(\boldsymbol{k}) e^{i\sqrt{k_0^2 - k^2}y_2} \tag{5.75b}$$

for which we have the acoustic impedance function

$$Z_a(\omega, \boldsymbol{k}) = \frac{\rho_0 c}{\sqrt{1 - k^2/k_0^2}} \tag{5.75c}$$

The acoustic pressure $p_a(\mathbf{y}, t)$ is the inverse Fourier transform of the wave number transform. The function $P_a(\mathbf{y}, \omega)$ is

$$P_a(\mathbf{y}, \omega) = \sum_{mn} (2\pi)^{-2} \rho_0 \omega V_{mn}(\omega) \iint_{-\infty}^{\infty} \frac{S_{mn}(\boldsymbol{k}) e^{+i\left(k_1 y_1 + k_3 y_3 + \sqrt{k_0^2 - k_1^2 - k_3^2} y_2\right)}}{\sqrt{k_0^2 - k_1^2 - k_3^2}} dk_1\, dk_3 \tag{5.76}$$

The integral in Eq. (5.76) is of the form

$$g(\mathbf{y}, \omega) = \iint_{-\infty}^{\infty} f(\boldsymbol{k}) e^{i\phi(\boldsymbol{k})} dk_1\, dk_3 \tag{5.76a}$$

where $\phi(\mathbf{k})$ goes through a large number of periods as k goes through its limits. With the exception of a square-root singularity, $f(\mathbf{k})$ is more smoothly valued in this interval of k. Then if there are any values of $\mathbf{k}$, say $\overline{\mathrm{k}}$, for which $\phi(\boldsymbol{k})$ has a minimum, then $g(\mathbf{y}, \omega)$ will be controlled by the value of its integrand at those points. The method of stationary phase [41], developed by Lamb [42] and extended by Junger and Feit [1] to two dimensions, is the procedure used for evaluating the integral at these points. This procedure will be developed here since it is crucial to multiple derivations to follow in later chapters; to this end Eqs. (5.76) and (5.76a) are in a typical form.

The phase, $\phi(\boldsymbol{k})$, may be expanded in a two-dimensional Taylor's series about points $k_1 = \overline{k_1}$ and $k_3 = \overline{k_3}$

$$\phi(k_1, k_3) \approx \phi(\overline{k_1}, \overline{k_3}) + \frac{1}{2}\frac{\partial^2 \phi}{\partial k_1^2}(k_1 - \overline{k_1})^2 + \frac{1}{2}\frac{\partial^2 \phi}{\partial k_3^2}(k_3 - \overline{k_3}) + \frac{\partial^2 \phi}{\partial k_1\, \partial k_3}(k_1 - \overline{k_1})(k_3 - \overline{k_3})$$

where the "stationary phase" points are determined by the vanishing slopes

$$\left.\frac{\partial \phi}{\partial k_1}\right|_{k_1 = \overline{k_1}} = \left.\frac{\partial \phi}{\partial k_3}\right|_{k_3 = \overline{k_3}} = 0$$

These points can be found by writing the phase in spherical coordinates, Fig. 5.4,

$$\begin{aligned} y_2 &= r \cos \phi \\ y_1 &= r \sin \phi \cos \theta \\ y_3 &= r \sin \phi \sin \theta \end{aligned}$$

differentiating the phase, and equating the derivatives to zero, giving

$$\begin{aligned} \overline{k_1} &= k_0 \sin \phi \cos \theta \\ \overline{k_3} &= k_0 \sin \phi \sin \theta \end{aligned} \tag{5.77}$$

and

$$\overline{k}_1^2 + \overline{k}_3^2 = k_0^2 \sin^2 \phi$$

These wave numbers $\overline{\mathrm{k}}_1$ and $\overline{\mathrm{k}}_3$ are the trace wave numbers of the acoustic wave front projected on the panel. The equation for $g(\mathbf{y}, \omega)$ now becomes

$$\begin{aligned} g(\mathbf{y}, \omega) &= f(\overline{\mathbf{k}}) e^{i\phi(\overline{\mathbf{k}})} \iint_{-\infty}^{\infty} dk_1\, dk_3 \\ &\times \exp\left\{ i \left[\frac{1}{2} \frac{\partial^2 \phi}{\partial k_1^2} (k_1 - \overline{k}_1)^2 + \frac{1}{2} \frac{\partial^2 \phi}{\partial k_3^2} (k_3 - \overline{k}_3)^2 \right.\right. \\ &\left.\left. + \frac{\partial^2 \phi}{\partial k_1 \partial k_3} (k_1 - \overline{k}_1)(k_3 - \overline{k}_3) \right] \right\} \\ &= \frac{\pm 2\pi i}{|D(\overline{k}_1, \overline{k}_3)|^{1/2}} f(\overline{\mathbf{k}}) e^{i\phi(\overline{\mathbf{k}})} \end{aligned}$$

where the positive or negative sign is chosen to match the sign of $D(\overline{k}_1, \overline{k}_3)$. This factor is

$$D(\overline{k}_1, \overline{k}_3) = \left(\frac{\partial^2 \phi}{\partial k_1 \partial k_3} \right)^2 - \left(\frac{\partial^2 \phi}{\partial k_1^2} \right) \left(\frac{\partial^2 \phi}{\partial k_3^2} \right)$$

and evaluated at $k = \overline{k}$

$$D(\overline{k}_1, \overline{k}_3) = -(r/k_0 \cos \phi)^2 \phi(\overline{k}_1, \overline{k}_3) = ik_o R$$

The stationary phase result is valid when

a. $k_0 r \gg 1$
b. third-order and higher derivatives of ϕ must be negligible
c. $f(\mathbf{k})$ must not have poles in the interval

making the necessary substitutions the approximate integral in Eq. (5.76) is [1]

$$P_a(R, \theta, \phi, \omega) \sim \sum_{mn} -i(2\pi)^{-1} \rho_0 \omega \frac{P_{mn}(\omega) S_{mn}(\overline{k_1}, \overline{k_3})}{Z_{mn}(\omega, \overline{k}_1, \overline{k}_3)} \frac{e^{+ik_0 r}}{r} \tag{5.78a}$$

or

$$P_a(R, \theta, \phi, \omega) \sim \sum_{mn} -i(2\pi)^{-1} \rho_0 \omega V_{mn}(\omega) S_{mn}(\overline{k_1}, \overline{k_3}) \frac{e^{+ik_0 r}}{r} \tag{5.78b}$$

which is valid only when $k_0 r \gg 1$.

The sound pressure in the far field will be the linear superposition of contributing pressures from all modes which are resonant at the frequency. The directivity of the sound from each mode is associated with the coincidence of trace wave numbers $\overline{k}_1$ and $\overline{k}_3$ with mode wave numbers k_m and k_n, respectively. There will be distinct points in space (θ, ϕ) at which each mode radiates most effectively. These points occur because of the local reinforcement of sound waves at (R, θ, ϕ) radiated outward to that point from the mosaic of half-waves that compose the mode of the plate. The individual waves will be most in phase when k_m and k_n coincide with the trace wave numbers $\overline{k_1}$ and $\overline{k_3}$, respectively. We shall discuss this coincidence further in Section 5.5.3.

Eq. (5.78) is completely general, applying to the far-field pressure of a rectangular panel of any modal character. All that is required is a specification of $S_{mn}(\mathbf{k})$ for that panel. For flat panels that are not rectangular [1], the pressure is functionally similar to Eq. (5.78), but there are differences in numerical coefficients and in the detailed form of $S_{mn}(\mathbf{k})$.

Similar expressions can be derived for curved surfaces [1], see Chapter 10, Sound Radiation From Pipe and Duct Systems but the presence of the modal acceptance function $S_{mn}(\mathbf{k})$ in the integrand is common to all cases. The differences lie in the replacement of the harmonic function

$$e^{i\sqrt{k_0^2-k^2}y_2}(k_0^2-k^2)^{-1/2}$$

which is characteristic of the planar radiator, with functions that are appropriate to other coordinate systems: spherical, cylindrical, etc. This shall be illustrated in Chapter 10, Sound Radiation From Pipe and Duct Systems, for the cylindrical radiator.

5.5.2 The Fluid Impedance of a Simply Supported Panel

In our solution of Eq. (5.21) no specifications were placed on the components of the pressure $p(\mathbf{y}, t)$. It must now be recognized that in general $p(\mathbf{y}, t)$ includes a number of contributions. The first is a pressure, say, $p_{b1}(\mathbf{y}, t)$, which is the primary pressure caused by hydrodynamic flow. This will be discussed in subsequent chapters. The fluid adjacent to the panel offers reaction pressures to the panel motion. Most often these are acoustic and inertial, and they are governed by equations of the form just described. In specialized cases fluid viscous damping in still liquids or hydrodynamic damping for lifting surfaces moving in liquids (see Chapter 11: Noncavitating Lifting Sections; and Blake and Maga [54,55]) adds to the total damping of hydrofoil structures.

We shall now restrict attention only to the impedance offered by the acoustic reaction pressures on both sides of the panel and separate them from the primary driving pressure; so in Eq. (5.18) we assume flow on one side of the plate and replace the single pressure there by

$$p_{a+}(y, t) = -p_{a+}(y, t) + p_{b1}(y, t) \tag{5.79}$$

and

$$p(\mathbf{y},\ t) = p_{\underline{a}}(\mathbf{y},\ t)$$

on the opposite side where there is no flow. We are ultimately (Here, again, we must respect the sign convention: ξ is positive upwards, and $p(\mathbf{y}, t)$ is directed downwards; however, the reaction pressure will be opposite the excitation pressure, and therefore we have utilized the minus sign.) interested in the vibration, $v(\mathbf{y}, t)$, and sound, $p_a(\mathbf{y}, t)$ induced by the hydrodynamic pressure field, $p_{b1}(\mathbf{y}, t)$. We still assume that the flow is unaffected by the panel motion and that the acoustic radiation is unaffected by the presence of mean flow. Furthermore, we assume that $p_a(\mathbf{y}, t)$ acts on both sides of the plate and that

$$p_{a+}(\mathbf{y},\ t) = \lim_{y_2\to 0^+} p_a(\mathbf{y},\ t) = -\lim_{y_2\to 0^-} p_a(\mathbf{y}, t) = -p_{\underline{a}}(\mathbf{y},\ t)$$

The acoustic fields above and below the plate are thus equal in magnitude and π out of phase.

The modal acoustic pressure, following Eq. (5.27), is

$$P_{a_{mn}}(y_2, \omega) = \frac{1}{2\pi}\int_{-\infty}^{\infty}\int_{A_p} p_a(y_1, y_2, y_3, t)\Psi_{mn}(\mathbf{y}_{13})e^{+i\omega t}d^2\mathbf{y}_{13}\, dt$$

or

$$P_{a_{mn}}\left(y_2, \omega\right) = \iint_{A_p} P_a(\mathbf{y}, \omega)\Psi_m(\mathbf{y}_{13})d^2\mathbf{y}_{13}$$

where we have now made a distinction between position vectors $\mathbf{y}_{13}$ lying in the surface and $\mathbf{y} = (\mathbf{y}_{13}, y_2)$ in the field. Since

$$P_a\left(\mathbf{y}, \omega\right) = \iint_{-\infty}^{\infty} e^{i\mathbf{k}\cdot\mathbf{y}_{13}} P_a(y_2, \mathbf{k},\ \omega)\, d^2\mathbf{k}$$

substitution gives

$$P_{a_{mn}}\left(y_2, \omega\right) = \iint_{-\infty}^{\infty} P_a(y_2, \mathbf{k},\ \omega)S^*_{mn}\left(\mathbf{k}\right)d^2\mathbf{k} \tag{5.80}$$

for the modal acoustic pressure in terms of the acceptance function. Now, combining Eqs. (5.75) and (5.76) gives the required expression for the modal acoustic pressure on the excitation side ($y_2 \geq 0$)

$$P_{a_{mn}}(y_2, \omega) = (2\pi)^{-2}\rho_0 c_0 \sum_{op} V_{op}(\omega)\cdot \iint_{-\infty}^{\infty} \frac{S^*_{mn}(\mathbf{k})S_{op}(\mathbf{k})}{\sqrt{1 - k^2/k_0^2}} e^{iy_2\sqrt{k_0^2 - k^2}} d^2\mathbf{k}$$

The fluid reaction pressure $P_a(y_2 \to 0^+, \omega)$ on the upper side of the panel (with an equal, but opposite value on the bottom side) involves the integral of the combinations of $S^*_{mn}(\mathbf{k})S_{op}(\mathbf{k})$ for all indices o, p. Since the integral is not identically zero for mn different than op, it is clear that the m, n mode is influenced by the motion of the o, p mode; i.e., the modes are coupled by the reaction of the fluid. This modal coupling by the fluid in unbounded fluids has been discussed in similar terms by Davies [56,57,58]; coupling by enclosed fluids has been discussed by White and Powell [25], Obermeier [36], and Arnold [37] and inertial coupling of modes in cantilever plates by Blake and Maga [54]. If the fluid is light enough, we can ignore the coupling of modes; for flat plates the coupling by acoustic radiation appears to be much exceeded by inertial coupling at low frequencies. The equation for the modal pressure on the surface of the plate becomes

$$P_{a_{mn}}(y_2 \to 0, \omega) = \left\{ (2\pi)^{-2} \rho_0 c_0 \iint_{-\infty}^{\infty} \frac{|S_{mn}(\boldsymbol{k})|^2}{\sqrt{1 - k^2/k_0^2}} d^2\boldsymbol{k} \right\} V_{mn}(\omega) \tag{5.81}$$

The dominant contribution in the integral will come from wave numbers near $k_1 = k_m$ and $k_3 = k_n$. This is because these are the large acceptance regions of $S_{mn}(\mathbf{k})$; see Fig. 5.9. Physically, the fluid and structure can most effectively transfer energy when the length scales of motion are well matched. If both k_m and k_n are less than k_0, the integral is primarily real, meaning that the fluid reaction appears as a pure resistance because power is radiated away from the plate. However, if either k_m or k_n is larger than k_0, then $k_{mn} = k$ is larger and the radical in the denominator provides an imaginary or inertial term that decays as $\exp(-\sqrt{k^2 - k_0^2} y_2)$ with increasing $(y_2 > 0)$ distance from the plate. The modal reaction pressure on each side of the plate can therefore be written in the convenient form of an impedance with multiple interpretations

$$A_p P_{a_{mn}}(0, \omega) = A_p \frac{\rho_0 c}{\sqrt{1 - k^2/k_0^2}} V_{mn}(\omega) = A_p Z_a(\omega, k) V_{mn}(\omega) \tag{5.82a}$$

$$A_p P_{a_{mn}}(0, \omega) = A_p (r_{mn} - i\omega m_{mn}) V_{mn}(\omega) \tag{5.82b}$$

$$= A_p (\rho_0 c_0 \sigma_{mn} - i\omega m_{mn}) V_{mn}(\omega) \tag{5.82c}$$

if we ignore modal coupling in Eq. (5.81). If such coupling exists, then the r_{mn} and m_{mn} are really parts of an impedance matrix with the maximum values on the diagonal $m,n = o,p$ and lesser values at off-diagonal terms. The term r_{mn} is the radiation resistance per unit area [9] (see Fahy and Guardonio [3], given by the real part of the integral) and m_{mn} is the added mass or accession to inertia [1] per unit area (given by the imaginary part of the integral). The radiation resistance per unit area has been further reduced from $\rho_0 c_0$ according to

$$r_{mn} = \rho_0 c_0 \sigma_{mn} \tag{5.83}$$

where the dimensionless coefficient σ_{mn} is called the radiation efficiency of the mode. Recall that a similar factor had been identified in connection with Eq. (2.33) which may now be interpreted as the radiation efficiency of the heaving sphere.

5.5.3 Radiated Acoustic Power

The time-averaged acoustic power radiated to one side of the plate is defined as

$$\mathbb{P}_{\text{rad}} = \lim_{T\to\infty}\frac{1}{T}\int_{-T/2}^{T/2}\iint_A p_a(y,\, t)v(\mathbf{y}_{13},\, t)dt\, d^2\mathbf{y}_{13}$$

By substitution of the inverse transform of Eq. (5.68) for the pressure on $y_2 \to 0$ and the representations of the velocity given Eqs. (5.30), (5.31), and (5.73) we find the average radiated power as the real part of the integral

$$\mathbb{P}_{\text{rad}} = \sum_{mn}\iint_{-\infty}^{\infty} d^2\boldsymbol{k}\int_{-\infty}^{\infty} d\omega\int_{-\infty}^{\infty} d\omega' V_{mn}^{*}(\omega)S_{mn}^{*}(\boldsymbol{k})P_{a_{mn}}(y_2\to 0, \boldsymbol{k},\, \omega)$$
$$\times \lim_{T\to\infty}\frac{\sin(\omega-\omega')T/2}{(\omega-\omega')T/2}$$

or since

$$\lim_{T\to\infty}\frac{\sin(\omega-\omega')T/2}{(\omega-\omega')T/2} = \frac{2\pi}{T}\delta(\omega-\omega')$$

The total acoustic power is the real part of

$$\mathbb{P}_{\text{rad}} = \sum_{mn} 2\pi\iint_{-\infty}^{\infty} d^2\boldsymbol{k}\int_{-\infty}^{\infty} d\omega P_{a_{mn}}(y_2\to 0, \boldsymbol{k},\, \omega)\frac{V_{mn}^{*}(\omega)}{T}S_{mn}^{*}(\boldsymbol{k})$$

Using Eq. (5.75) for the wave number transform of the pressure we have a formal expression for the radiated power spectral density (two-sided),

$$\mathbb{P}_{\text{rad}}(\omega) = (2\pi)^{-2}\rho_0 c_0\sum_{mn}\sum_{op}\iint_{k<k_0} d^2\boldsymbol{k}\int_{-\infty}^{\infty} d\omega\frac{(2\pi/T)V_{mn}(\omega)V_{op}^{*}(\omega)S_{mn}(\boldsymbol{k})S_{op}^{*}(\boldsymbol{k})}{\sqrt{1-(k/k_0)^2}}$$

As before, we can ignore the cross-coupling and make use of the auto-spectral density of the velocity, Eq. (5.32), in order to simplify the relationship. Furthermore, since $\mathbb{P}_{\text{rad}}(\omega)$ is the sum of all modal contributions of the (two-sided) power spectral density $\mathbb{P}_{\text{rad}_{mn}}(\omega)$, then we have the final result that

$$\mathbb{P}_{\text{rad}_{mn}}(\omega) = \rho_0 c_0 (2\pi)^{-2}\iint_{k<k_0} |S_{mn}(\boldsymbol{k})|^2\left[1-\frac{k}{k_0}^2\right]^{-1/2}\Phi_{mn}(\omega)d^2\boldsymbol{k}$$

or

$$\mathbb{P}_{\text{rad}_{mn}}(\omega) = \rho_0 c_0 A_p \sigma_{mn}\Phi_{mn}(\omega) \tag{5.84}$$

is the radiated sound power spectral density of the mn mode, where

$$\sigma_{mn} = \frac{1}{A_p(2\pi)^2} \iint_{k<k_0} |S_{mn}(\boldsymbol{k})|^2 \left[1 - \frac{k}{k_0}^2\right]^{-1/2} d^2\boldsymbol{k} \tag{5.85}$$

is the modal radiation efficiency. This factor has already appeared in Eqs. (5.81) and (5.82). The flow-induced vibration velocity spectrum for the *mn* mode is given by Eq. (5.34).

The radiation efficiencies of baffled flat plates have been determined by Maidanik [9,10], Davies [28], and Wallace [59,60] for various ratios of length to width. Radiating modes have been classified by Maidanik [9] into surface edge, and corner modes depending on the relationships between k_0 and k_m and k_n. We have already seen the importance of these relationships in determining the directivity of sound in Eq. (5.78) as well as the oddness or evenness of mode orders in Fig. 5.10. Fig. 5.11 illustrates the edge and corner mode classifications that arise from these relationships. First we recall our discussion of multipoles as illustrated in Fig. 2.2. There it was said that for two sources that are separated a distance d the sound pressure increased as $k_0d = 2\pi d/\lambda_0$ until $k_0d \geq 1$, in which case the two sources radiated without interaction. Similarly in the case of a radiating surface, we have already said that the modal pattern represents a mosaic of alternately phased pistons that are spaced $\lambda_m/2 = \pi/k_m$. If $k_0/k_m = \lambda_m/\lambda_0 < 1$, the fluid can pass from one piston to the other before they can oscillate through one cycle because the characteristic wave speed in the fluid c_0 is faster than the wave speed of piston motion, $2\pi\omega/\lambda_m$. Therefore the pistons effectively cancel each other. Maidanik argues that this cancellation occurs everywhere between adjacent pistons except at the baffled edge of the pistons at each end of the array. Also, at very low frequencies so that, say $L_l/\lambda_0 < 1$, even-order modes will vanish from relevance since the $S_m(k_1)$ also vanishes at vanishing k_1 as shown in Fig. 5.10. If $\lambda_m/\lambda_0 > 1$, then the adjacent pistons can radiate more independently because they cannot interfere. In the illustrated case of an edge mode there is cancellation along y_1, but not along y_3. In the case of a corner mode, there is cancellation along both coordinate directions, and in the case of surface modes we have $k_0 > k_m$ and $k_0 > k_n$ so there is no cancellation. In Fig. 5.5 we see that for wave numbers above a threshold $k_p = c_0\sqrt{m_s/D_s}$ the phase speeds of waves on a given plate will be supersonic. This is because of the frequency-dependent character of the phase speed. The frequency at which the in vacuo bending wave speed equals the acoustic wave speed is called the acoustic coincidence frequency

$$f_c = \frac{c_0^2}{2\pi}\left(\frac{D_s}{m_s}\right)^{-1/2} = \frac{c_0^2}{\pi}\frac{\sqrt{3(1-\mu_p)}}{c_L h}$$

These frequencies are shown for a variety of materials in air or water media in Fig. 5.12. For membranes, on the other hand, if the tension is low enough the phase speed will be subsonic at all frequencies, thus the panels are subsonic as illustrated.

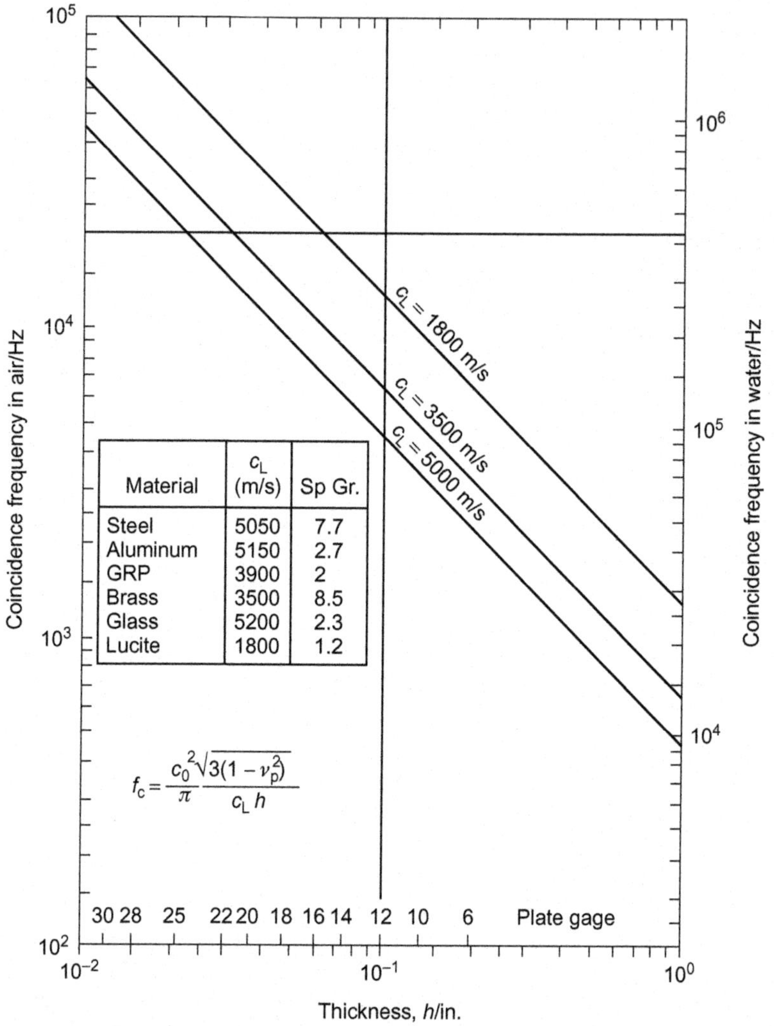

FIGURE 5.12 Acoustic coincidence frequencies for plates of various materials shown as a function of thickness or sheet gage.

The mode classifications are illustrated in the wave number plane in Fig. 5.13. The loci of edge modes lie along the k_m and k_n coordinate axes. When both k_m and k_n are less than k_0, all modes are well-radiating surface modes, and they radiate analogously to the infinite plate on which the bending wave speeds exceed the speed of sound, i.e., $k_p < k_0$ or $c_p > c_0$. The evaluation of the integral in Eq. (5.81) is controlled by location of modes in the wave number plane. Fig. 5.13 illustrates the critical regions of k_m, k_n that will be evaluated below.

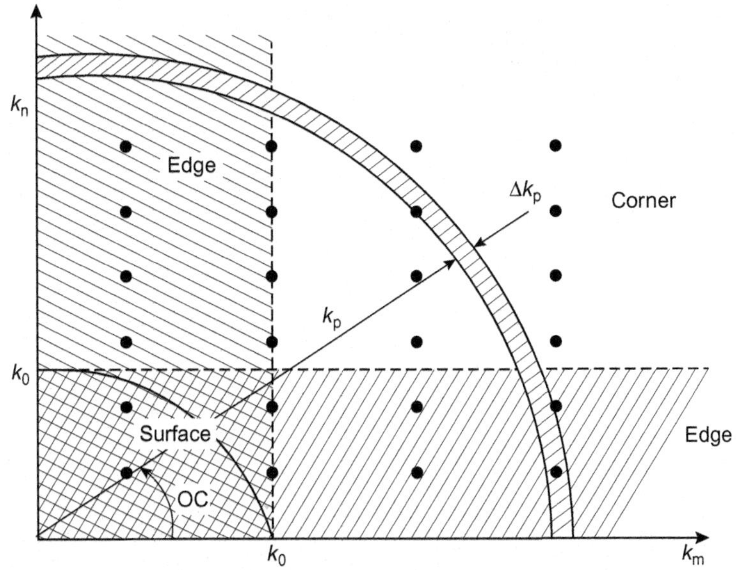

FIGURE 5.13 Illustration of the radiation classifications for rectangular plates shown in the wave number plane.

5.5.4 Radiation Efficiencies of Simple Structures

For corner modes $k_m > k_0$ and $k_n > k_0$

$$k_{mn} = \sqrt{k_m^2 + k_n^2} > k_0$$

lies outside the region of integration in Eq. (5.85) and the modal radiation efficiency is controlled by the low wave number tail of the acceptance functions (see Fig. 5.9 and Eq. (5.54)). Thus

$$\begin{aligned} \sigma_{mn} &\simeq \frac{1}{A_p(2\pi)^2} \frac{1}{k_m^2} \frac{4}{k_n^2} 2\pi k_0 \int_0^{k_0} \frac{k\,dk}{\sqrt{k_0^2 - k^2}} \\ &\simeq \frac{32\pi}{A_0(2\pi)^2} \frac{k_0^2}{k_m^2 k_n^2}, \quad \begin{cases} k_m > k_0 \\ k_n > k_0 \end{cases} \end{aligned} \tag{5.86}$$

and at very low frequencies

$$\sigma_{00} \simeq \frac{8}{\pi^5} k_0^2 A_p \quad \text{for } k_m = \pi/L_1,\ k_n = \pi/L_3, \text{ and } k_0^2 A_p \ll 1$$

For edge modes, say, $k_n < k_0$ and $k_m > k_0$, a crude approximation is

$$\sigma_{mn} \simeq \frac{2k_0}{A_p(2\pi)^3}\frac{1}{k_m^2}\int_{-\infty}^{\infty}|S_n(k_3)|dk_3\int_0^{k_0^2-k_n^2}\frac{dk_1}{\sqrt{k_0^2-k_3^2-k_1^2}}$$

$$\sigma_{mn} \simeq \frac{2\pi}{A_p k_m^2}\frac{k_0 L_3}{\pi}, \quad \begin{cases} k_n < k_0 \\ k_m > k_0 \end{cases} \tag{5.87a}$$

Nearer to coincidence a more exact relationship is

$$\sigma_{mn} \simeq \frac{2\pi}{A_p k_m^2}\frac{k_0 L_3}{\pi}\left[\frac{1+[(k_{mn}^2-k_0^2)/k_m^2]}{[(k_{mn}^2-k_0^2)/k_m^2]^{3/2}}\right] \tag{5.87b}$$

Finally for surface modes, using Eq. (5.66) since $k_0 \gg k_{mn}$

$$\sigma_{mn} \simeq 1.0, \quad k_0 \gg k_{mn} \tag{5.88}$$

These formulas have been derived as examples. More exact formulas will be found in the study of Maidanik [9,10] and Davies [28]. Fig. 5.13 shows examples of radiation efficiencies for the cases $k_{mn} > 2k_0$, $k_0 L_1 > \pi$, and $k_m > 2k_0$. One can clearly see the dependence of σ_{mn} on mode order. For clamped panels σ_{mn} should be increased by 6 dB for corner modes and by 3 dB for edge modes because of the difference between the low wave number acceptance functions (Fig. 5.14).

The modal radiation efficiencies of unbaffled plates and beams have been derived by Blake [61,62]. For an unbaffled plate for which $k_0 L_1 > \pi$, $k_0 L_3 > \pi$ and $k_n/k_0 < 1$, $k_m/k_0 > 1$ it is found [59]

$$\sigma_{mn} \simeq \frac{1}{8}\left(\frac{2\pi}{k_m^2 A_p}\frac{k_0 L_3}{\pi}\right)\left(\frac{k_0}{k_n}\right) \tag{5.89a}$$

closer to coincidence $k_0 \to k_m$

$$\sigma_{mn} \simeq \frac{\sqrt{2}+1}{16\pi}\frac{2\pi k_0 L_3}{k_m^2 A_p}\frac{k_0}{k_m}\left[\frac{-\ln(1-k_0/k_m)}{1-(k_0/k_m)^{1/2}}\right] \tag{5.89b}$$

The unbaffling introduces an additional (k_0/k_m)-dependence. Measured radiation efficiencies of an unbaffled steel panel measuring 0.6 m × 0.4 m × 0.0127 m in water are compared to theoretical expressions for baffled and unbaffled plates in Fig. 5.15. Note that when $k_m \approx k_p$ in the vicinity of coincidence there is little difference between the measured points and those calculated by either the baffled or unbaffled theory. Although a peak in σ does not occur near k_0/k_{mn}, a value greater than unity was reported there.

The radiation efficiency for an unbaffled beam for which $k_0 L_3 < 1$, $k_0 L_1 > 1$, and $k_0/k_n \ll 1$ is [62]

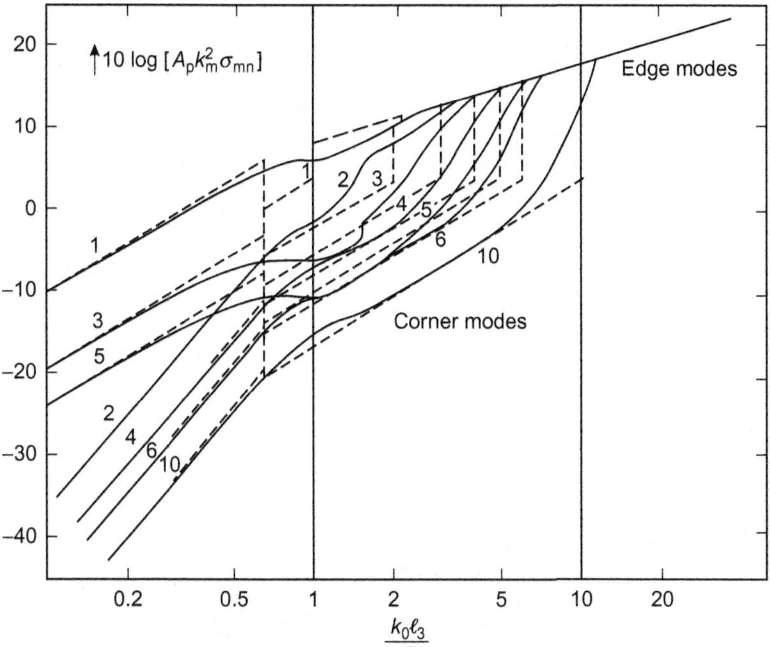

FIGURE 5.14 Values of the radiation efficiency σ_{mn} for large odd m and various n:, $m = k_1 L_1 / \pi$, $n = k_3 L_3 / \pi$, $k_m > 2k_0$, $k_p > 2k_0$. Panel is rectangular in a rigid baffle. *From Davies HC. Sound from turbulent-boundary-layer excited panels. J Acoust Soc Am 1971;49:878–89.*

$$\sigma_n \simeq \frac{\pi^2}{192}(k_0 w)^2 \frac{w}{\pi L}\left(\frac{k_0}{k_n}\right)^2 \tag{5.90a}$$

which, when compared to the baffled case [12],

$$\sigma_n \simeq \frac{w}{\pi L}(k_0/k_n)^2 \tag{5.90b}$$

shows additional $(k_0 w)^2$ dependence due to baffling. Other radiation efficiencies of cylindrical shells are given by Junger and Feit [1] and Manning and Maidanik [63]. Radiation from prolate spheroids has been calculated by Chertock [64,65].

5.5.5 Relationships for Estimating Total Acoustic Power

In Section 5.3 relationships were derived for estimating the mean-square flexural velocity averaged over the structure in terms of the input power accepted from the flow. Equivalently, the mean-square velocity of the structure could have been derived from a knowledge of the modal excitation force. Either way, the response of both simple and complex structures can be estimated. In like manner the power in a frequency

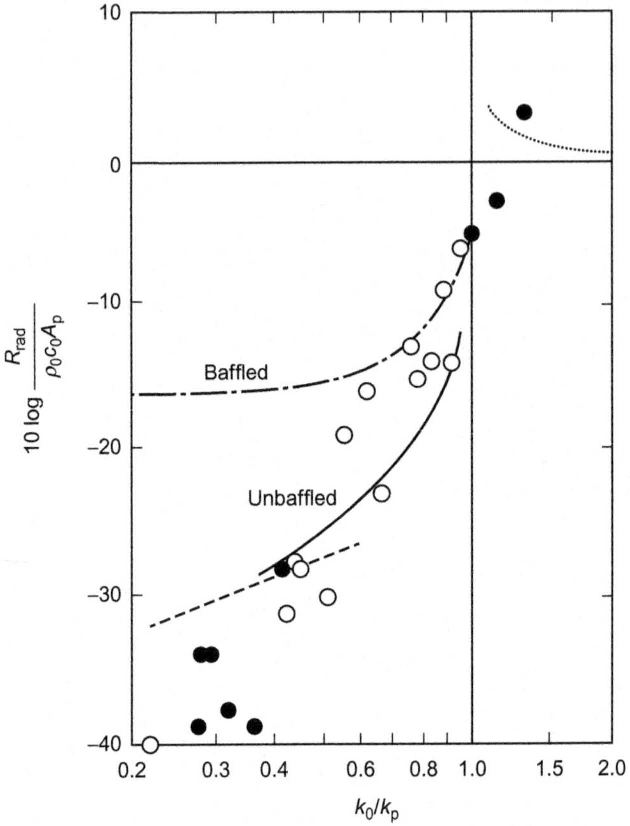

FIGURE 5.15 Measured and theoretical radiation resistances of the unbaffled 1.32-ft × 2-ft × $\frac{1}{2}$-in. steel plate in water. The points are from (○) 50-Hz analysis and (•) $\frac{1}{3}$-octave band levels. The curves are (————————), Eq. (5.89); (— —), Eq. (5.89b); (— · —), Eq. (5.87); and (· · ·), Eq. (5.92). *From Blake WK. The acoustic radiation from unbaffled strips with application to a class of radiating panels. J Sound Vib 1975;39:77–103.*

band $\Delta\omega$ radiated from the structure with a high mode density can be estimated as [9]

$$\mathbb{P}_{\text{rad}}(\omega_f, \Delta\omega_f) = 2\int_{\omega_f-\Delta\omega_f/2}^{\omega_f+\Delta\omega_f/2} \sum_{\substack{mn \\ \Delta\omega}} \rho_0 c_0 A_p \sigma_{mn} \Phi_{mn}(\omega) d\omega \tag{5.91a}$$

$$\simeq \rho_0 c_0 A_p \overline{\sigma}(\omega_f) \overline{V_{mn}^2}(\omega_f) n(\omega_f) \Delta\omega_f$$

$$\simeq \rho_0 c_0 A_p \overline{\sigma}(\omega_f) \overline{v^2} \tag{5.91b}$$

as long as there are many modes in the band, $n(\omega_f)\Delta\omega_f \gg 1$, and $\Delta\omega_f$ is larger than the band of resonance. (Note that in the above the functions Φ_{mn} and $\overline{\sigma}(\omega)$ are all symmetric for $\pm\omega$. The total power in the band is therefore twice that given by the integral.) The mean-square modal velocity $V^2{}_{mn}(\omega_f)$ is the average over modes so that

$$\overline{\upsilon^2} = \overline{V_{mn}^2}(\omega_f) n(\omega_f)\Delta\omega_f$$

is the total reverberant velocity in the bandwidth. Furthermore, it is assumed that the mean-square velocity of edge modes is the same as that of corner modes so that we can integrate the radiation efficiency over all regions in the wave number domains illustrated in Fig. 5.12. The mean-square velocity $\overline{\upsilon^2}$ is taken as the physically measurable motion that would be deduced from a set of accelerometers. It is also that which can be estimated from Eq. (5.51).

The average radiation efficiency of a simply supported rectangular panel has been determined by Maidanik [9,10] and Davies [28]. This quantity is

$$\overline{\sigma} = \frac{1}{N} 2 \int_0^{\alpha} \underset{\text{edge}}{\sigma_{mn}} \; n(k)k \, dk \, d\alpha + \int_{\alpha_e}^{\pi/2-\alpha} \underset{\text{corner}}{\sigma_{mn}} \; n(k)k \, dk \, d\alpha \tag{5.92a}$$

where $n(k)$ is the wave number mode density (Eq. (5.45)) and $N = \mathrm{n(k)}\Delta \mathrm{k}^2$ is the total number of modes, i.e., k_{mn}, included in the annular wave number region

$$(\Delta k)^2 = \frac{1}{4}\pi k_p \Delta k_p$$

where

$$\Delta k_p = \frac{1}{2}\Delta\omega(\omega\kappa c_l)^{-1/2}$$

and c_l is given by Eq. (5.47). The angle α_e is the arc through the edge mode region $\sin\alpha_e = k_0/k_p$, as can be deduced in Fig. 5.13. Using the approximate relationships of the Section 5.5.4 we find for edge and corner modes combined,

$$\overline{\sigma} \simeq \frac{32k_0}{2\pi A_p k_p^3} + \frac{2}{\pi}\left(\frac{k_0}{k_p}\right)^2 \frac{2(L_1 + L_3)}{k_p A_p} \tag{5.92b}$$

for $k_0/k_p < 1$ and either or both $k_0 L_1 > 2$, $k_0 L_3 > 2$. Maidanik [9,10] has provided this and additional formulas, e.g., above acoustic coincidence,

$$\overline{\sigma} \simeq (1-(k_p/k_0)^2)^{-1/2}, \quad k_p < k_0 \tag{5.92c}$$

at acoustic coincidence,

$$\overline{\sigma} \simeq \sqrt{2}(\sqrt{k_m L_1} + \sqrt{k_n L_3}), \quad k_p = k_0 \tag{5.92d}$$

and for acoustic corner modes

$$\overline{\sigma} \simeq \frac{16}{\pi^3}\frac{L_1 + L_3}{A_p k_0}\left(\frac{k_0}{k_p}\right)^2, \quad k_0 L_1,\ k_0 L_3 \ll 2 \tag{5.92e}$$

while Davies [28] obtains for still lower wave number corner modes

$$\overline{\sigma} \simeq \frac{32}{\pi^3}\frac{L_1 + L_3}{A_p k_0}\left(\frac{k_0}{k_p}\right)^3, \quad k_0 L_1,\ k_0 L_3 \ll 3\pi \tag{5.92f}$$

Eq. (5.92b) shows the important result that adding rib stiffeners to a panel will increase the radiation efficiency of the panel. This increase is brought about by increasing the total perimeter $2(L_1 + L_3)$ of the edges while keeping the total radiating area constant. The second term of Eq. (5.92b) is controlled by the edge modes, and it is the magnitude of this term that is increased by ribbing.

Maidanik [9] has experimentally verified these equations using the arrangement illustrated in Fig. 5.16. A reverberant vibration was generated in the ribbed aluminum test panel by a mechanical shaker; radiated sound power was measured in an acoustically reverberant chamber while the

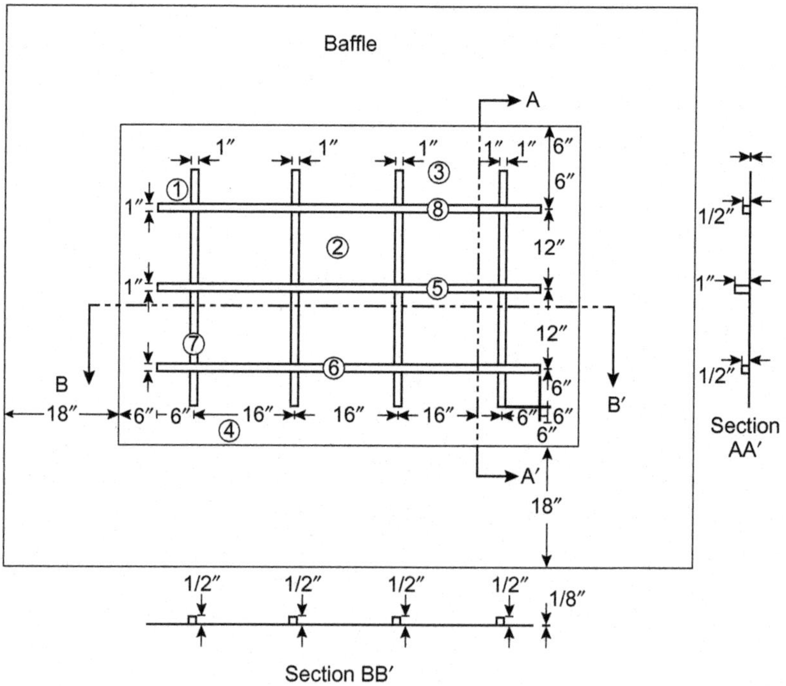

FIGURE 5.16 Diagram of aluminum test panel, steel ribs, and wooden baffle. Circled numbers indicate accelerometer positions. *From Maidanik G. Response of ribbed panels to reverberant acoustic fields. J Acoust Soc Am 1962;34:809–26.*

mean-square velocity $\overline{v^2}$ was determined on the panel. The radiation efficiency was determined using Eq. (5.91b). Fig. 5.17 shows the measured values of a with and without the baffle in place. Generally the vibration level of the ribs was from 6 to 16 dB lower than that of the panels. The radiation efficiencies as determined from the mechanically driven panel are in close agreement with the theory. Also shown are $\overline{\sigma}$ for the unbaffled panel without ribs. These values are notably even less than those of the ribbed panel without a baffle. This is because adjacent subpanels provide baffling to their neighbors. For frequencies less than 250 Hz there are a number of limitations to the experiment including the fact that rib and plate vibration levels were comparable. Other points included in Fig. 5.17 were determined by exciting the panel with a reverberant acoustic field and measuring the response. The agreement of these points with those measured by direct shaking shows that the individual panels inside the ribs were acoustically independent of each other and responded to the sound field essentially as described in Section 5.3.

Another example of a radiation efficiency measurement is provided by Manning and Maidanik [63] for a cylindrical shell. The geometry of the shell is shown in Fig. 5.18. The flanges were removable so that the effects of adding rib stiffeners could be determined. The ends were baffled with plywood boards. Measured radiation efficiencies are shown in Fig 5.19. The ring frequency f_r is

$$f_r = c_L/2\pi a$$

where a is the radius of the cylinder; f_c is the acoustic coincidence frequency of the cylinder. The modes that radiate most efficiently are those that form circumferential strips at the ends of the cylinder. These modes are such that

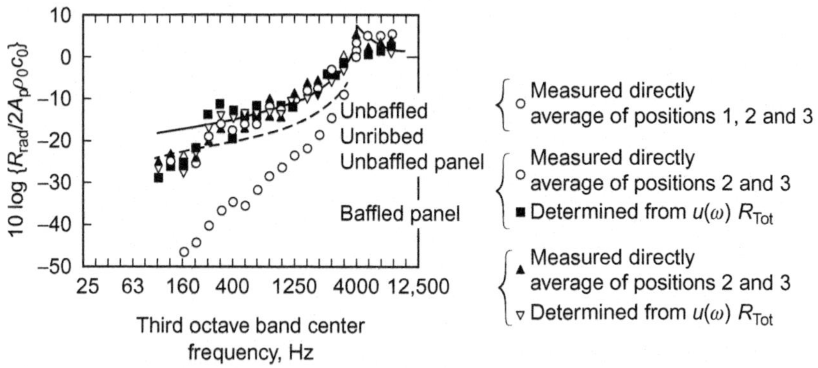

FIGURE 5.17 Normalized radiation resistance (defined by Eq. (5.83)) of test panel of Fig. 5.16. (————), Theoretical curve for the ribbed test panel from Eq. (5.92); (— — — —), theoretical curve for the unribbed baffled test panel from Eq. (5.92). *From Maidanik G. Response of ribbed panels to reverberant acoustic fields. J Acoust Soc Am 1962;34:809–26.*

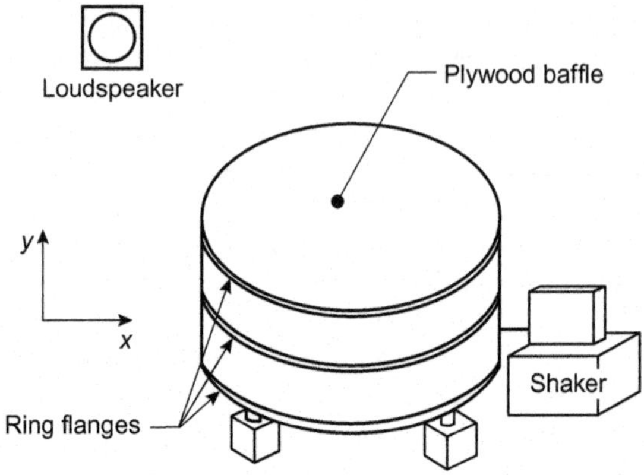

FIGURE 5.18 Sketch of test cylinder for radiation efficiencies of Fig. 5.19. *From Manning JE, Maidanik G. Radiation properties of cylindrical shells. J Acoust Soc Am 1964;36:1691–8.*

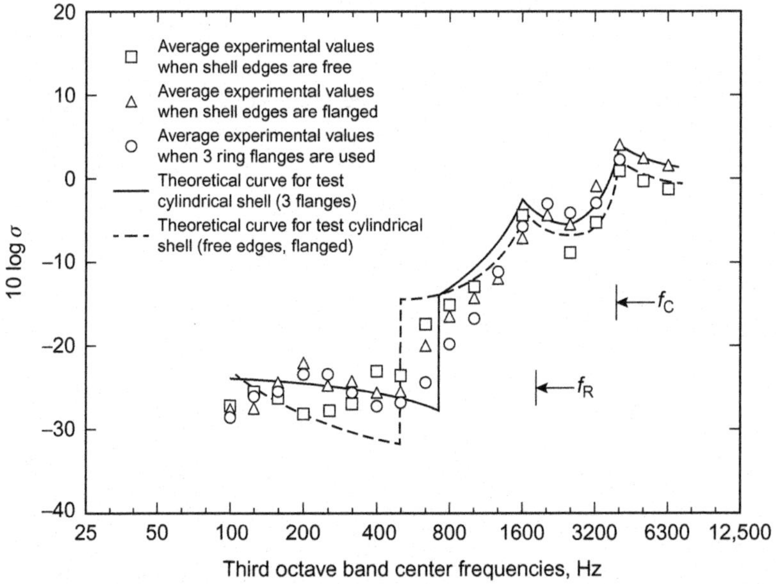

FIGURE 5.19 Average values of the measured radiation efficiency for three different boundary conditions. *From Manning JE, Maidanik G. Radiation properties of cylindrical shells. J Acoust Soc Am 1964;36:1691–8.*

there is little acoustic cancellation around the circumference. Theoretical estimates are based on applying the appropriate Eq. (5.92) for the classes of modes that were calculated for the cylinder.

The approach to be used in roughly estimating structural radiation and response is to first determine the excitation spectrum $\Phi_p(\mathbf{k}, \omega)$ that is included in Eq. (5.40). The auto-spectrum of the modal pressure $\Phi_{p_{mn}}(\omega)$ can then be approximated using the limiting functions for $|S_{mn}(k)|^2$ shown in Section 5.4. The input power (Eq. (5.52b)) or the mean-square velocity (Eq. (5.51)) can then be estimated. The radiated sound power can be estimated from Eq. (5.89) using the approximate relationships for the radiation efficiency $\overline{\sigma}$.

In succeeding chapters some estimates will be compared to measured flow-excited vibration and sound for simply defined situations. Another and perhaps more potent use of the equations is their use in scaling one known circumstance to another. Often it is desired to conduct an experiment on a prototype and extrapolate the results to another size. These relationships provide guidelines for planning and conducting those experiments. It is hoped that the examples of acoustic measurements shown exemplify the character of precision to be expected in other similar experiments of the future.

5.5.6 Added Masses of Simple Structures

The added mass per unit area on one side of the panel can be determined from Eq. (5.82c) using the integral in Eq. (5.81) for $k > k_0$. In the case of low frequencies for which $k_{mn} > k_0$ the integral gives the mass per unit area as

$$\begin{aligned} m_{mn} &= \rho_{0/kmn} \quad && k_{mn} > k_0 \\ &\sim 0, && k_{mn} < k_0 \end{aligned} \tag{5.93}$$

This function is universal, applying to both baffled [1,56] and unbaffled [61] plates. In the case of beams the added mass per unit area is [62] for $k_m > k_0$

$$m_m = \frac{1}{4}\pi\rho_0 L_3, \quad k_m L_3 < 1 \tag{5.93a}$$

and

$$m_m = \frac{1}{2}\pi\rho_0 L_3 (1 + k_m L_3)^{-1}, \quad k_m L_3 > 1$$

where L_3 is the width of the beam. Similarly, for vibrating circular cylinders [1] of radius a the added mass impedance per unit area is

$$\begin{aligned} m_m &\simeq \rho_0 a, \quad && k_m > k_0 \\ &\simeq 0, && k_m < k_0 \end{aligned} \tag{5.94}$$

5.6 SOUND FROM FORCED VIBRATION OF STRUCTURES IN HEAVY FLUIDS

5.6.1 Vibration of the Point-Driven Plate

Using the model problem of the forced vibration of a simply supported panel in a heavy fluid we will illustrate several concepts which have broad significance even with complex structures. Among other concepts, we shall illustrate that structures of many modes become effectively infinite. This occurs when the damping is great enough, or if fluid loading is great enough, that significant flexural wave damping occurs between boundaries (e.g., $\eta k_p L > 1$). In these cases, generally at higher frequencies, the response to a localized excitation is dominated by deformation at the drive point because the reverberant (resonant) motion in the structure is weakened by both damping and fluid loading. The case of simple supports is frequently used, see, e.g., Refs. [1,66–68], and has appeal in that the solutions are all analytically tractable and results are easily dissected for study. Fig 5.20 shows the coordinate system of a rectangular plate in an otherwise rigid plane that forms an acoustic baffle to the sound emitted from the plate vibration. This is an extension of Fig. 5.4 which now shows the location of the point driving force. The relationships for the modal response and acoustic radiation all apply here.

To examine the response of the panel in a fluid, we rewrite Eq. (5.21) incorporating Eqs. (5.24), (5.30), and (5.31) and including the decomposition of pressures:

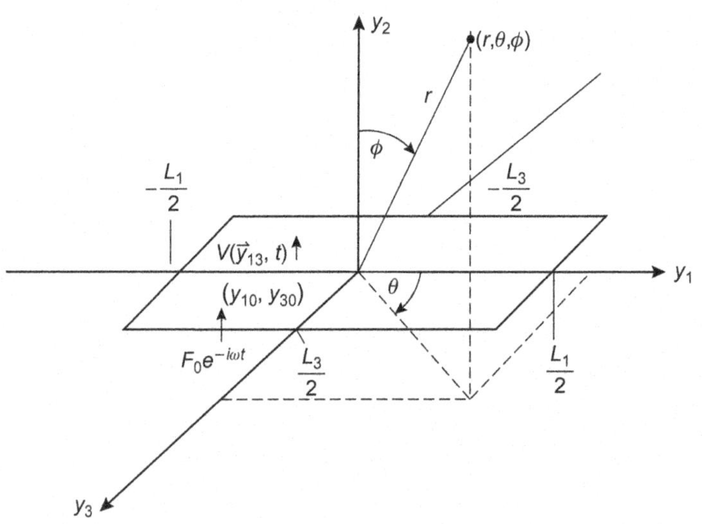

FIGURE 5.20 Coordinate system for the field of a plate of dimensions L_1 and L_3, point-driven at y_{10}, y_{30}.

$$[-m_s\omega^2 + im_s\eta_s\omega_{mn}\omega + \omega_{mn}^2 m_s]\, V_{mn}(\omega) = -i\omega[p_{b_{mn}}(\omega) - p_{a_{mn}}(\omega)] \quad (5.95)$$

where $p_{b_{mn}}(\omega)$ is the modal excitation pressure and $p_{a_{mn}}(\omega)$ is the fluid-loading pressure given terms of the inertia and fluid-loading coefficients, Eq. (5.82). Introducing these coefficients and the fluid-loaded modal impedance function (Eq. (5.75c)) symbolically, we have a modified form of Eq. (5.95) for fluid loading on the flow-excitation side

$$[Z_{mn}(\omega)]_f = \frac{m_s\omega^2(1-i\eta_s) - D(k_{mn}^2)^2}{-i\omega} + \frac{i\rho_0\omega}{\sqrt{k_{mn}^2 - k_0^2}} \quad (5.96)$$

The modal velocity response is still be given by Eq. (5.33), but using the fluid-loaded impedance instead. Under the influence of the added fluid impedance, the form of Eq. (5.95) assumes a similar form, but with additional terms,

$$[(-m_s - m_{nm})\omega^2 - i\omega(m_s\eta_s\omega_{mn} + \rho_0 c_0\sigma_{mn}) + \omega^2 m_s]\, V_{mn}(\omega) = i\omega p_{b_{mn}}(\omega) \quad (5.97)$$

where ω_{mn} is the in vacuo resonance frequency of the plate, but under the influence of fluid loading we have a new resonance condition

$$\omega_{mn_{FL}}^2 = \omega_{mn_{VAC}}^2\left(\frac{m_s}{m_s + m_{mn}}\right) \quad (5.98a)$$

with a new effective bending wave speed of

$$c_{b_{FL}} = c_{b_{VAC}}(m_s/(m_s + m_{mn}))^{1/4} \quad (5.98b)$$

and an effective, or total, loss factor, η_T which is defined as

$$\eta_T = \left[\eta_s + \frac{\rho_0 c_0 \sigma_{mn}}{m_s\omega_{mn}}\right]\left[\frac{m_s}{m_s + m_{mn}}\right] \quad (5.98c)$$
$$= \eta_s + \beta\sigma_{mn} = \eta_s + \eta_{rad}$$

The fluid loading factor β defined here has a general importance in hydroacoustics, that should be emphasized; even for finite plates the value of β determines the extent of fluid loading. The factor η_{rad} is the radiation loss factor of the structure. Thus the magnitude of the fluid-loading factor determines the level of radiation damping to a structure. In the equations of Section 5.3 η_s should be replaced by η_T if fluid loading is to be accounted for. For a two-sided fluid loading the σ_{mm}, m_{mn}, and η_{rad} in the expressions below are replaced by two times these quantities.

Simple as these relationships are, they give the basics of an important relationship governing the utility of mechanical damping. We note that the ratio of the modal input power to the modal acoustic radiated power is, by Eqs. (5.42), (5.84), and (5.96),

$$\frac{(\mathbb{P}_{\text{rad}}(\omega))_{mn}}{(\mathbb{P}_{\text{in}}(\omega))_{mn}} = \frac{\rho_0 c_0 A_{\text{p}} \sigma_{mn} \overline{V_{mn}^2} n(\omega)\Delta\omega}{m_{\text{s}} A_{\text{p}} \eta_{\text{T}} \omega_{mn} \overline{V_{mn}^2} n(\omega)\Delta\omega} = \frac{\rho_0 c_0 \sigma_{mn}}{m_{\text{s}} \omega_{mn} \eta_{\text{T}}} = \frac{(\eta_{\text{rad}})_{mn}}{(\eta_{\text{rad}} + \eta_{\text{s}})_{mn}}. \tag{5.99}$$

If the assumption of equal modal energies applies for all modes in a band and if the modal excitation force is the same for all the modes, then [11] Eq. (5.99) applies for average power levels of all modes in large frequency bands; i.e.,

$$\frac{\mathbb{P}_{\text{rad}}(\omega)\Delta\omega}{\mathbb{P}_{\text{in}}(\omega)\Delta\omega} = \frac{\overline{\eta}_{\text{rad}}}{\overline{\eta}_{\text{rad}} + \overline{\eta}_{\text{s}}} \tag{5.100}$$

where the overbars denote modal average values. Structures may be considered to be lightly radiation loaded when $\overline{\eta}_{\text{rad}} \ll \overline{\eta}_{\text{s}}$. Only in these cases do increases in structural damping result in commensurate reductions in radiated sound power. In the alternative case of $\overline{\eta}_{\text{rad}} > \overline{\eta}_{\text{s}}$ structural damping is ineffective, and all power into the structure is radiated as sound.

Fig. 5.21 shows this impedance as an admittance, for the lowest transverse wave number, $n = 1$, and normalized on the inertial impedance,

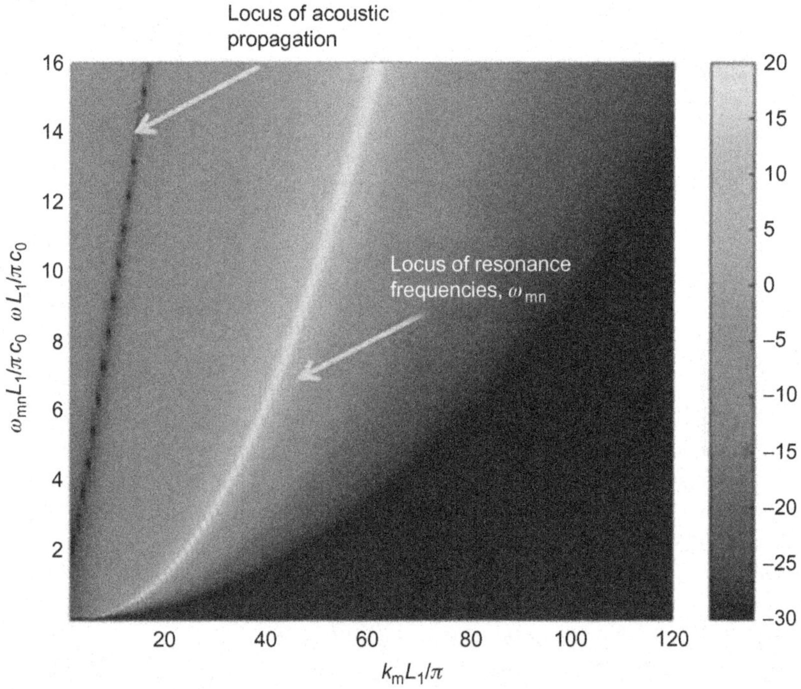

FIGURE 5.21 Normalized modal admittance of a steel plate ($L_1 = 2$ m, $L_3 = 1$ m, $h = 3.2$ mm), driven at its center by a point force of 1 N. The admittance in the color bar is presented in the form of $10^*\log[|m_s w/Z_{m1}(w)|^2)$ at $k_n L_3/\pi = 1$.

$im_s\omega/Z_{m1}(\omega)$. The frequencies are all presented normalized on L_1 and the speed of sound in the fluid, the independent variable k_m is normalized on L_1 and all variables are divided by π to represent mode order, m, along the abscissa. The bright line corresponds to the resonance frequencies for any wave number, k_m, which is

$$\omega_{mn} = (k_m^2 + k_{n=1}^2)(D/(m_s + m_{mn}))$$

and the frequency for which $\omega_a/c_0 = k_{mn}$ is given by the dark line on the left, i.e.

$$\omega_a = \sqrt{(k_m^2 + k_{n=1}^2)}c_0$$

The value of the maximum admittance (i.e., along the locus of resonance) is of order $|1/Z_{m1}(\omega)| \approx 1/\eta s m_s \omega_{m1}$. The dark line is so since these acoustic propagation wave numbers represent the values of high impedance due to inertial fluid loading, i.e., in this case $m_s\omega < \rho_0 c_0$.

Fig. 5.22 shows values of the drive point admittance at the location $x_0 = 0.1L_1$, $y_0 = 0.2L_3$ on the plate summed over all modes:

$$\frac{V(\mathbf{x}_0,\omega)}{F(\mathbf{x}_0,\omega)} = Y(\mathbf{x}_0,\omega)$$
$$= \frac{1}{A_p}\sum_{mn}\frac{|\psi_{mn}(\mathbf{x}_0)|^2}{[Z_{mn}(\omega)]_f} \tag{5.101a}$$

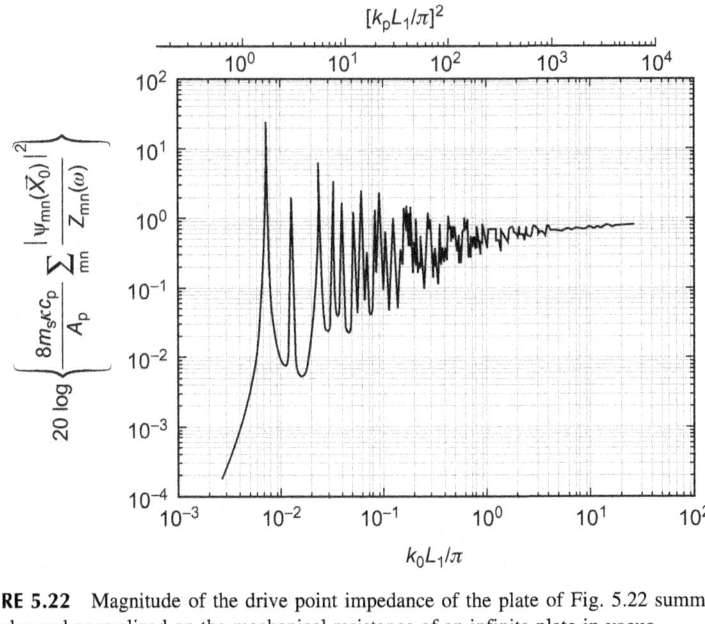

FIGURE 5.22 Magnitude of the drive point impedance of the plate of Fig. 5.22 summed over all modes and normalized on the mechanical resistance of an infinite plate in vacuo.

with a limit at high frequencies (wave numbers) of

$$\rightarrow \frac{1}{8m_s\kappa c_p} \quad \text{as } k_pL_1 \rightarrow \infty \tag{5.101b}$$

and normalized on the resistance of the infinite plate given by Eq. (5.52e). The frequency is expressed in two dimensionless forms: the lower abscissa as the acoustic wave number which corresponds to the presentation of Fig. 5.21, the upper frequency scale as the plate wave number

$$(k_pL_1/\pi)^2 = (\omega\kappa c_p)(L_1/\pi)^2$$

The figure shows that when $k_p > 10\pi/L_1$ or equivalently for this example either the bending wave length $\lambda_p < L_1/5$ or the modal mode order exceeds approximately 10, the plate admittance approaches that of the infinite plate. As the lower abscissa of Fig. 5.22 shows this is also when the acoustic wave number becomes less than twice the length of the plate. At low frequencies, the graph shows the dominance of specific modes.

5.6.2 Sound From the Locally-Driven Fluid-Loaded Plate

The stationary phase solution for the far-field sound, at $k_0r >> 1$, can be evaluated using Eq. (5.40e) by substituting the wave numbers of stationary phase, Eq. (5.77), into Eqs. (5.54a) and (5.96). Using Eqs. (5.36b), (5.45) and (5.40e) we obtain the modal excitation pressure

$$P_{mn}(\omega) = \left[\frac{F}{A_p}\right]\psi_{mn}(\vec{x}_0)$$

for the point force spectrum. The resulting modal radiated sound pressure re $p_{ref} = 1\ \mu\text{Pa}$ at 1 m. We use the summation over all m,n modes to give the total pressure.

$$\frac{P_{mn}(r,\theta,\phi,\omega)r}{p_{ref}F(\omega)} = \frac{\rho_0\omega}{2\pi}\left[\frac{1}{A_p}\Psi_m(x_{1_o})\Psi_n(x_{3_o})\right]\frac{S_{mn}(\overline{k_1},\overline{k_3})}{Z_{mn}(\omega,\overline{k_1},\overline{k_3})}$$

and (5.102)

$$\left|\frac{P_a(r,\theta,\phi,\omega)r}{p_{ref}F(\omega)}\right| \approx \left|\sum_{mn}\frac{\rho_0\omega}{2\pi}\left[\frac{1}{A_p}\Psi_m(x_{1_o})\Psi_n(x_{3_o})\right]\frac{S_{mn}(\overline{k_1},\overline{k_3})}{Z_{mn}(\omega,\overline{k_1},\overline{k_3})}\right|$$

Fig. 5.23 shows examples of both the modal sound pressure and the total of the modal sum sound pressure scaled as above for individual modes as a function of normalized frequency, $\omega L_1/2\pi c_0 = k_0L_1/2\pi$. The drive is 1 N applied to the plate, water is on one side of the plate. The values of sound pressure are all for the indicated values of lateral mode orders and shown as

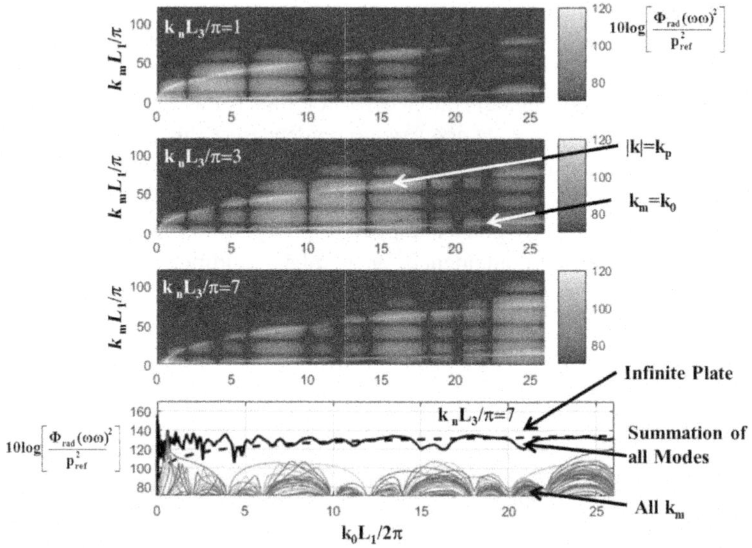

FIGURE 5.23 Modal radiated sound pressure level referred to 1 m from simple-supported rectangular plate with unbounded water on 1 side; $L_1 = 1$m, $L_3 = 0.5$m, $h = 3.2$mm with a 1 N force applied off center at ($y_{10} = 0.05L_1$, $y_{30} = 0.1L_3$). Sound is evaluated at $\theta = 45°$ and $\phi = 45°$ off the normal to the plate. The top three panels are contours of sound level vs. both longitudinal and acoustic wave number at specific orders of transverse modes. The lowest graph shows overall sound (solid line) and sound from an infinite plate of $h = 3.2$mm with water on one side (dotted line) and line plots of individual modal contributions from one transverse mode at all longitudinal wave numbers.

continuous functions of longitudinal mode wave numbers and frequency expressed as the acoustic wave number. Both of these are normalized on the length of the panel, L_1. Radiated sound pressure is evaluated at $\theta = 45^o$ and $\phi = 45^o$ off the normal to the plate, all dB re 1 μPa. The top three illustrations are modal sound pressures for m, $n = (m, 1)$, $(m, 3)$, and $(m, 7)$, respectively. The plate is driven off center in order to simultaneously excite as many modes as possible. The bottom illustration plots all modal sound pressures for $(m,7)$ modes as well as total pressure summed over all modes (solid heavy line); these are compared with the sound from a point dipole applied to an infinite plate (the dashed line, which will be discussed in some detail below).

The spectra shown in Fig. 5.23 illustrates a number of important structural acoustic mechanisms. In each of the top 3 two dimensional spectra shown in color scale are identified the straight line traces of the acoustic coincidence, $k_m = k_0$ near the k_0 coordinate. Highlights are apparent as this coincidence passes through the various resonances. The wave number pass bands of the various mode shapes (discussed previously in connection with Fig. 5.9) are shown by the scalloped patterns in the figures. Of these modes, the resonant ones are highlighted by enhanced radiated sound. The upper boundary of the (k_0, k_m) patterns for each lateral wave number, k_n, is determined by the quadratic function relating frequency to plate wave number, the $|\mathbf{k}| = \mathbf{k}_p$ line as noted. Finally, as shown in the lower line graph, the $k_n L_3/\pi = 7$ modes are apparently responsible for the sound at approximately $k_0/2\pi = 0.6$ and approximately 26 where, for this value of k_n, $k_m = k_p$ at a plate resonance and $k_m = k_0$ at acoustic coincidence for this mode.

In the case of the point-driven infinite plate without boundaries, and of course no modes, the expression that is equivalent to Eq. (5.102) effectively replaces $\Psi_m(x_{1_o})\Psi_n(x_{3_o})$ by unity and $S_{mn}(\overline{k_1}, \overline{k_3})$ by A_p in the limit of $k_0 < k_p$; i.e., well below acoustic coincidence for the plate bending,

$$\frac{P_a(r,\theta,\phi,\omega)r}{p_0 F(\omega)} \approx \sum_{mn} \frac{\rho_0\omega}{2\pi} \frac{1}{Z_{mn}(\omega,\overline{k_1},\overline{k_3})} e^{ik_0 r} \tag{5.103a}$$

or

$$\frac{P_a(r,\theta,\phi,\omega)r}{p_0 F(\omega)} \approx \frac{-i\rho_0\omega}{2\pi} \frac{1}{\left[m_s\omega - \frac{i\rho_0\omega}{k_0\cos\phi}\right]} e^{ik_0 r} \tag{5.103b}$$

This equation is identical to the classically known function [66–68]

$$p(r,\phi,t) = \frac{-ik_0F_0}{2\pi} \frac{\beta\cos\phi}{\cos\phi - i\beta} \frac{e^{i(k_0 r - \omega t)}}{r} \tag{5.104}$$

and

$$\beta = \rho_0 c_0 / m_s \omega \tag{5.105}$$

is the fluid loading factor for fluid on one side.

If the fluid was on both sides, then 2β replaces β in Eq. (5.104). Eq. (5.103) applies when the area over which the force extends is smaller than the bending wavelength, and when the frequency of excitation is below coincidence so that $k_p > k_0$. Eq. (5.103b) is the dashed line in Fig. 5.23 and forms somewhat of a mean line with the contributions from specific modes causing excursions above and below the values for the infinite plate. When a large (infinite) plate is driven at a point at low frequencies, the sound radiated is independent of the plate material and thickness (β is large) and it is double that emitted by just the force as a free dipole in the water. This is shown in Fig. 5.24 where several example cases are plotted for various plate thicknesses to illustrate the point. To this point, compare this expression to Eq. (2.75) for the dipole radiation from a point force in an unbounded fluid. When β becomes small (less than unity) the force on the plate ceases to act as an amplified free dipole and takes on a quite different character.

When we add complexity in the form of boundaries (as with a simple-supported plate in a rigid baffle) and thereby introduce modes, the picture

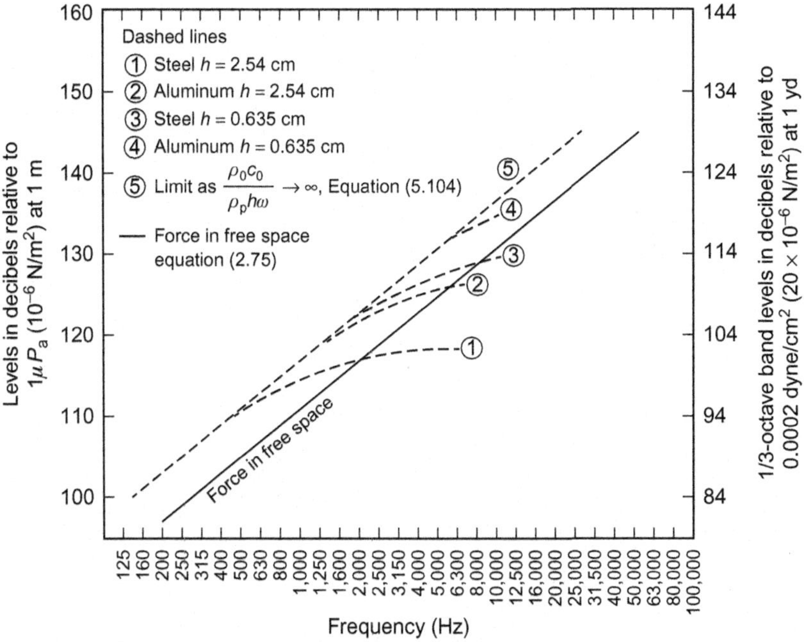

FIGURE 5.24 Radiated sound pressure level in unbounded water at 1 m on the axis of a 1 N point force in applied to infinite plates of various area densities in water. The point force is applied with fluid on *one* side of the plate.

changes by the introduction of resonance character to the spectrum of the sound and vibration as discussed in connection with Figs. 5.21 and 5.22. Fig. 5.25 further illustrates this in presenting the modal sum of all contributing modes for the same example of a simple supported plate that was discussed above. This is an example of a plate that has a high mode density in the frequency range that we are discussing. Upon summing all modes, three distinct frequency regions emerge for which the physics of sound generation changes. At low frequencies, below the fundamental mode of the plate, the sound increases as (frequency)^4 since the displacement pattern of the plate is dominated by a simple half-wave for which the plate displacement represents a net volume change on the fluid side. At frequencies between fundamental resonance and that for which $k_0 L_1/\pi \sim 1$ the plate modes dominate the sound character. For frequencies for which $k_0 L_1/\pi > 1$ the plate behaves as infinite, see also Fig. 5.22 and the fluid loading factor also begins its approach toward unity. In this upper frequency range, the radiated sound also behaves as if the plate is infinite. At these high frequencies the plate is beginning to appear acoustically large as well as noted by the convergence of the two lines.

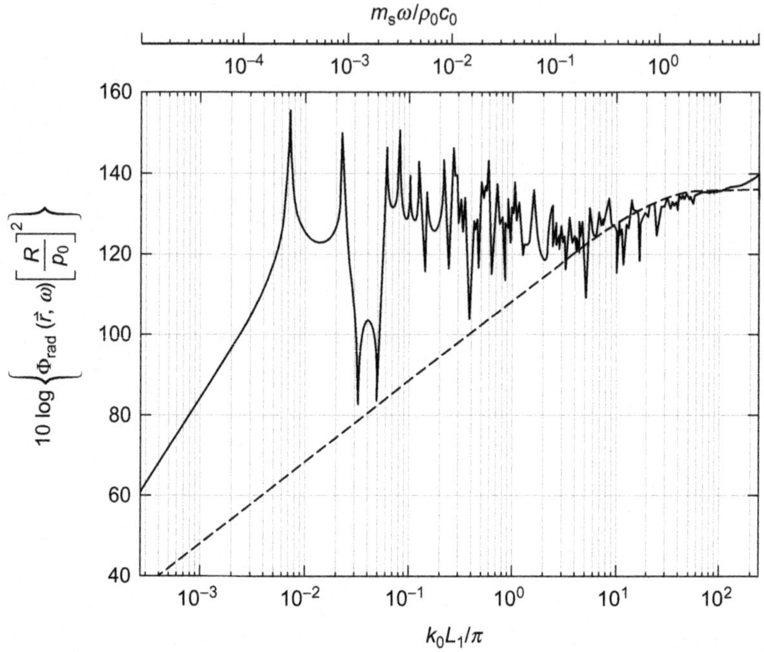

FIGURE 5.25 Radiated sound pressure at $\theta = 45°$ and $\phi = 45°$ off the normal to the plate, all dB re 1 μPa and as a function of dimensionless acoustic wave number summed over all modes. Dashed line is sound from a point force applied to an infinite plate; the solid line is the simple-supported plate with drive just off center ($x_0 = 0.05L_1$, $y_0 = 0.1L_3$) so that many modes are driven.

5.7 SOUND FROM FLOW-INDUCED VIBRATION OF A CIRCULAR CYLINDER

5.7.1 General Formulation for a One-Dimensional Structure

If a cylinder in a cross-flow is not rigidly fixed, it will undergo some degree of vibration under the influence of the unsteady lift and drag forces. This motion will generally be dominated by the lift direction unless, for some reason, the cylinder is constrained in the lift direction yet unconstrained in the drag direction.

The subject of flow-excited cylinder vibration provides a perfect example of the application of the principles of this chapter in analytically modeling flow-induced vibration and sound. Mathematically it is a much simpler problem than problems of two-dimensional structures that form the subjects of Chapter 8, Essentials of Turbulent Wall-Pressure Fluctuations, Chapter 9, Response of Arrays and Structures to Turbulent Wall Flow and Random Sound, Chapter 10, Sound Radiation From Pipe and Duct Systems, Chapter 11, Noncavitating Lifting Sections. The practical problems of vibration of cylinders and cylinder-like structures that may be solved by the methods of this section are vibrations of heat-exchanger tube bundles of structural members and cables in truss works, wind-induced vibration of buildings and architectural members, and current-induced vibration of ocean structures. It is instructional to formulate the problem from the fundamentals of Chapter 2, Theory of Sound and its Generation by Flow, in order to emphasize the physical aspects of vibration, fluid loading, and sound production.

The situation under consideration is illustrated for lift forces in Fig. 5.26. The force per unit length $f_{2h}(y_3,t)$ acts on the cylinder of mass per unit length m_s causing a velocity of the center of mass $u_2(y_3, t)$ perpendicular to the direction of flow. Since the fluid is not massless, it will induce a reaction force per unit length $f_{2r}(y_3, t)$ that opposes both the motion and the original hydrodynamic exciting force. Acoustic radiation therefore consists of three contributions: the first is from the original hydrodynamic force dipole, the second is from the vibration of the surface of the body, and the third is from the fluid reaction force, which partially cancels the originating excitation force. The net force per unit length *acting on the fluid* is

$$f_2(y_3,t) = f_{2h}(y_3,t) - f_{2r}(y_3,t) \tag{5.106}$$

while the net force per unit length acting *on the cylinder* is $-f_2(y_3, t)$. Recall the pressures introduced in Eqs. (5.79) and (5.95) for the two-dimensional problem.

We shall deal with the acoustic pressure as a function of frequency and therefore introduce the Fourier coefficient

$$F_2(y_3,\omega) = \frac{1}{2\pi}\int_{-\infty}^{\infty} f_2(y_3,t)e^{i\omega t}\,dt$$

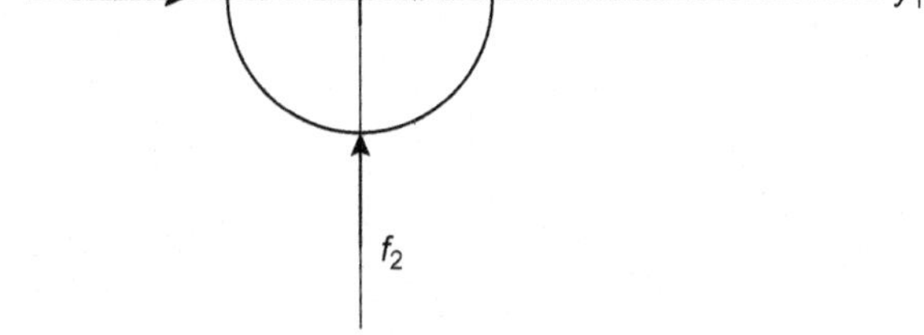

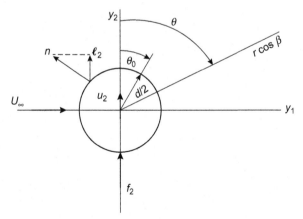

FIGURE 5.26 Cross-section coordinates for vibrating cylinder.

The Fourier coefficient of the sound pressure resulting from the surface contributions given from Eq. (2.70) in the form of Eq. (2.105) is

$$P_a(x, \omega) = -\iint_S \rho_0 l_2(-i\omega U_2(y_3, \omega)) \frac{e^{ik_0 r'}}{4\pi r'} dS(y) + \frac{\partial}{\partial x_2} \iint_S l_2 P(y) \frac{e^{ik_0 r'}}{4\pi r'} dS(y) \tag{5.107}$$

where (Eq. (4.8))

$$F_2(y_3, \omega) = \int_0^{2\pi} P(\theta_0, y_3, \omega) \cos\theta_0 \frac{d}{2} d\theta_0$$

is the lift force per unit length applied to the fluid. Since $l_2 = \cos\theta_0$ the integration around $0 < \theta_0 < 2\pi$ in the integral involving $l_2 U_2$ in Eq. (5.107) will be identically zero (no net mass influx) unless proper account is taken of the small phase variations of $k_0 r'$ in the exponential. The distance r' from a point on the cylinder and a point in the far field can be seen from the geometry shown in Fig. 5.27 to be for $r' \gg d$ and L

$$r' \simeq r - \frac{d}{2} \sin\phi \cos(\theta - \theta_0) - y_3 \cos\phi$$

(see also the arguments leading to Eq. (2.26)). Accordingly the first term of Eq. (5.104) is the radiated pressure due directly to surface vibration,

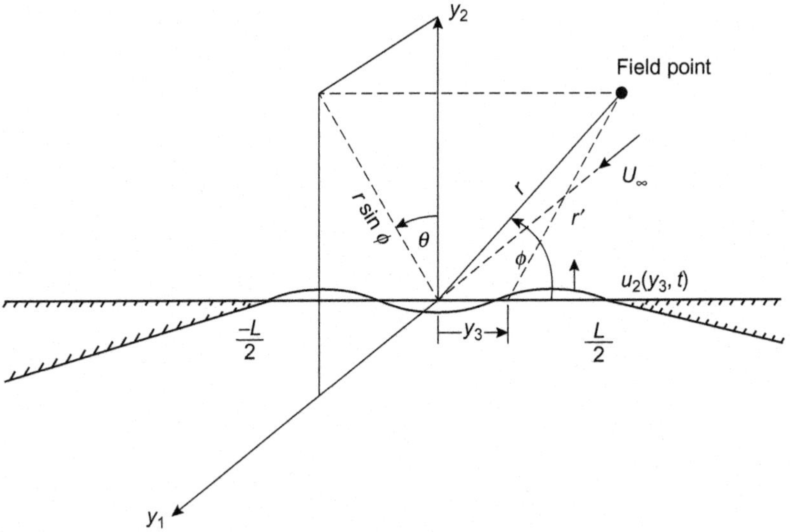

FIGURE 5.27 Coordinate system for sound radiation from vibrating cylinder.

$$P_{\mathrm{a}_u}(x,\omega) = -\int_{-L/2}^{L/2} e^{-ik_0 y_3\cos\phi}\rho_0(-i\omega)U_2(y_3,\omega)\frac{e^{ik_0 r}}{4\pi r}$$
$$\times\int_{-\pi/2}^{\pi/2} -2ik_0\frac{d}{2}\sin\phi\cos(\theta-\theta_0)\cos\theta_0\, d\theta_0\, dy_3$$

The range $0<\theta_0<2\pi$ is reduced to integration over a single range $-\pi/2\leq\theta_0\leq\pi/2$ by dividing the full circle into the two ranges of $-\pi/2<\theta_0<\pi/2$ and $\pi/2<\theta_0<3\pi/2$ for which $\cos\theta_0$ has positive and negative values, respectively, and making the necessary substitutions; the exponentials $\exp(ik_0 r')$ subtract giving a form resembling Eq. (2.26). Carrying out the integration over θ_0,

$$P_{\mathrm{a}_u}(x,\omega) = ik_0\cos\theta\sin\phi\frac{e^{ik_0 r}}{4\pi r} - i\omega\pi\frac{\rho_0 d}{2}$$
$$\times\left[\int_{-L/2}^{L/2} U_2(y_3,\omega)e^{-ik_0 y_3\cos\phi}dy_3\right] \tag{5.108}$$

The acoustic dipole contribution from the resultant forces on the cylinder can be written from Eq. (4.23)

$$P_{\mathrm{a}_h}(x,\omega) = ik_0\cos\theta\sin\phi\frac{e^{ik_0 r}}{4\pi r}\left[\int_{-L/2}^{L/2} F_2(y_3,\omega)e^{-ik_0 y_3\cos\phi}dy_3\right] \tag{5.109}$$

The terms in brackets in these equations are Fourier wave number transforms evaluated at $k_3 = k_0\cos\phi$; however, for simplification if we assume that $k_0 L\ll 1$, these terms represent only the net surface velocity and the net force acting on the fluid.

5.7.2 Expressions for the Cylinder Vibration and Net Sound Pressure ($k_0 L\ll 1$)

We now involve the modal analysis of Section 5.3 for the one-dimensional elastic structure considered here as a sum over orthogonal modes

$$U_2(y_3,\omega) = \sum_{n=0}^{\infty} U_{2n}(\omega)\Psi_n(y_3) \tag{5.110}$$

where n is the order of the mode and $\Psi_n(y_3)$ is the mode shape; $\Psi_n(y_3) = 2\cos(k_m y_3)$ as in Eq. (5.44). This shape has been illustrated in Fig. 5.27 for $n=3$. The modal response is given by Eq. (5.33) in this case rewritten

$$U_{2n}(\omega) = \frac{i\omega\int_{-L/2}^{L/2}[-F_{2h}(y_3,\omega)]\Psi_n(y_3)\,dy_3}{m_c L\omega^2[1-(\omega_n/\omega)^2 + i\eta_T(\omega_n/\omega)]} \tag{5.111}$$

where $m_c = m_a + m_s$ and m_a is the added mass of the cylinder. The integral

$$[F_h(\omega)]_n = \int_{-L/2}^{L/2} [-F_{2h}(y_3, \omega)]\Psi_n(y_3)\, dy_3 \tag{5.112}$$

is recognized as the one-dimensional analog of $P_n(\omega)$ appearing in Section 5.3 and it expresses the spatial matching of the force field on the cylinder with the axial variation of the admittance of the cylinder expressed as $\Psi_n(y_3)$. The force appearing in Eq. (5.109) is the sum of the hydrodynamic force and the reaction force to the cylinder motion that constitute the net force. At low Mach numbers and frequencies, the fluid reaction is primarily mass-like so that the reaction force on the cylinder is given by

$$F_{2r}(y_3, \omega) = i\omega m_a U_2(y_2, \omega)$$

where m_a is the added mass per unit length on the cylinder. Combining Eq. (4.8) with Eq. (5.94),

$$m_a = \int_0^{2\pi} \rho_0 (d/2)^2 \cos^2\theta \, d\theta$$
$$m_a = \rho_0 \pi (d/2)^2 \tag{5.113}$$

and the force on the fluid is $-F_{2r}(y_3,\ \omega)$. Consequently, part of the force appearing in Eq. (5.106) will include the vibration characteristics of the cylinder as per Eq. (5.112). Making all the necessary substitutions will yield an expression for the net radiated sound pressure.

Although a complete formulation for the vibration-induced acoustic field would entail a substitution of Eqs. (5.106), (5.109), and (5.109) into Eq. (5.107), instructive simplification can be introduced by taking $\Psi_n(y_3)$ as unity along the axis. This approximates the cylinder motion as a rigid body. Furthermore, we assume that $k_0 L < < 1$. The net pressure radiated in directions perpendicular to the rod, $\phi = \pi/2$, is $P_a = P_{a_u} + P_{a_h}$

$$P_a(x, \omega) = ik_0 \cos\phi \frac{e^{ik_0 r}}{4\pi r}\left(1 - \frac{2m_a}{m_c z}\right)\int_{-L/2}^{L/2} - F_{2h}(y_3, \omega)\, dy_3 \tag{5.114}$$

where $m_c = m_a + m_s$

$$z = 1 - \left(\frac{\omega_n}{\omega}\right)^2 + i\eta_T\left(\frac{\omega_n}{\omega}\right)$$

Eq. (5.114) applies to a highly idealized flow–structure interaction for which the cylinder radius is much less than an acoustic wavelength, the length of the cylinder is much shorter than the wavelength and the distance r is much greater than L; i.e., $k_0 d \ll 1$, $k_0 L \ll 1$, and $r \gg L$. Nonetheless, the result provides an elementary illustration of the importance of cylinder vibration. The most likely case for the radiation to be dominated by the

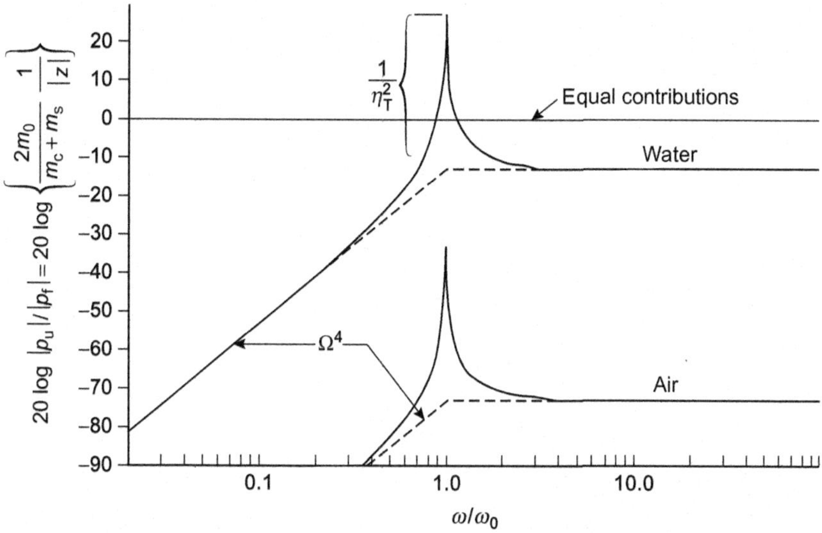

FIGURE 5.28 Radiated sound from flow-induced vibration relative to that from exciting force for 1/4-in. diameter steel cylinders in air or in water. Loss factor η_T assumed 0.01 in each case.

vibration-induced component occurs at coincidencewhen $\omega = \omega_s = \omega_n$. In this case the vibration-induced component dominates the force dipole by a factor $2m_a/(\eta_T m_c)$. This is illustrated in Fig. 5.28. In a heavy fluid, say, water, this factor may be much larger than unity. Note that η_T is typically between 10^{-3} and 10^{-2} for most practical structures unless some sort of damping treatment is applied. If the motion of the cylinder is mass controlled ($\omega \gg \omega_n$), then $z = 1$ and the sound pressure becomes proportional to $1 - 2m_a/m_c$. If the cylinder is neutrally buoyant, as it might be in water, i.e., $m_a = m_s$, then this factor is zero; i.e., the sound field from the mass-controlled surface motion identically cancels that from the forces on the fluid. Although this may be the case in a liquid medium; in air, however, the force dipole dominates since m_a is generally much less than m_s. Thus it is to be expected that the resultant sound from flow–body interactions may be substantially enhanced, or reduced, for structures of high ratios of added mass to dry mass. The effect of vibration will be dependent on frequency and damping in the structure. Only when the mass ratio (and fluid loading in general) is reduced can one view as separate the motion-induced and force-induced contributions to the sound field.

5.7.3 Self-Excited Vibration

It has been assumed in the above analysis that the flow-induced forces exciting the cylinder are not altered by the motion of the cylinder. In Section 4.3.4, Other Influences on Vortex Shedding, however, it was pointed out that motion of the cylinder across the flow can change the oscillating lift coefficient. The influence of the cylinder vibration on the unsteady flow forces will be dependent on the amplitude of the motion of the cylinder, the relationship between the frequency of vibration and the natural frequency of vortex shedding, and the Reynolds number of the flow. If we take, as a basis of comparison, the sound pressure radiated from the flow-induced surface motion given by Eq. (5.108) (for k_0L approaching zero), the part played by self-excitation may be easily illustrated. We assume that the excitation for $F_2(y_3, \omega)$ has a finite value only at the vortex-shedding frequency so that neglecting acoustic directivity factors the amplitude of the sound pressure is found from a combination of Eqs. (5.109), (5.110), and (5.111)

$$|P_{a_u}(x,\omega_s)| = \frac{\omega_s^2\pi}{4}\frac{\rho_0}{c_0}\frac{d^2}{r}\left|\frac{\omega_s F_h(\omega_s)}{m_c L_3\omega_n^2[(\omega_s/\omega_n)^2 - 1 + i\eta_T(\omega_s/\omega_n)]}\right| \tag{5.115}$$

where η_T is the sum of mechanical and radiation dampings, and where $F_h(\omega_s)$ is the net hydrodynamic lift force on the cylinder; i.e., it is $[F_h(\omega)]_n$ for $n = 0$. The natural vortex-shedding frequency is related to the flow velocity bywhere S is the Strouhal number,

$$\omega_s = 2\pi U_\infty S/d$$

Where S is the Strouhal number, as long as the motion of the cylinder does not alter the vortex-shedding process. Fig. 5.28 illustrates the dependence of radiated pressure amplitude normalized on the vortex-shedding force as a function of vortex-shedding frequency. We assume that without structure-to-fluid feedback the normalized force

$$\frac{F_h(\omega_s)}{\frac{1}{2}\rho_0 U_0^2 d L_3} = \text{constant}$$

is the same as a constant lift coefficient. Also without feedback, when $\omega_s = \omega_n$ the sound intensity is augmented by an amount η_T^{-1} due to resonant oscillation of the cylinder. The vortex-shedding frequency (as detected, say, by a velocity probe set on one side of the wake of the cylinder) would increase linearly with speed, as depicted in the lower part of the figure. Self-excitation feedback could occur as the vortex-shedding frequency approaches the resonance frequency. In such cases, the vibration of the cylinder and the resulting far-field sound pressure would both be larger than the mechanical resonance would provide. At the same time, the vortex-shedding frequency could be altered near the resonance frequency so that the ω_s versus U relationship would deviate from the normal straight line. The amount of deviation would depend on the over all coupling of the cylinder motion with the vortex formation. There may, in fact, be regions of speed over which the ω_s would be fixed or "locked-in" to a value near the resonance frequency.

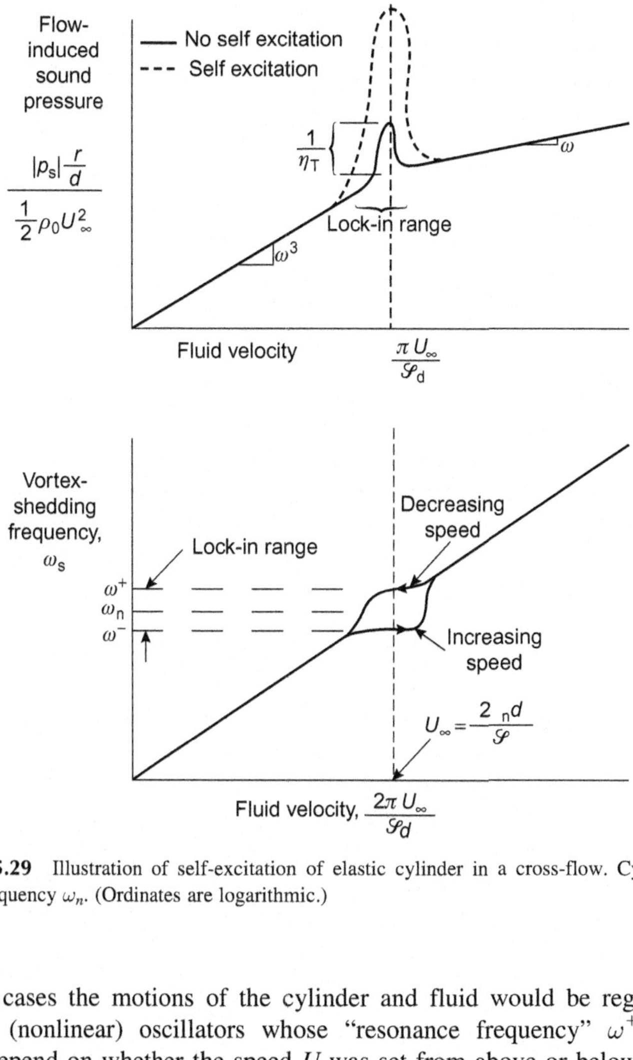

FIGURE 5.29 Illustration of self-excitation of elastic cylinder in a cross-flow. Cylinder has natural frequency ω_n. (Ordinates are logarithmic.)

In such cases the motions of the cylinder and fluid would be regarded as coupled (nonlinear) oscillators whose "resonance frequency" ω^+ or ω^- would depend on whether the speed U was set from above or below. This is indicated in the figures by the arrows. For lightly damped, high mass ratio (m_a/m_s) cylinders, the "resonance frequency" of the coupled cylinder–vortex system may be substantially different from that of the cylinder alone (Fig. 5.29).

Collectively, the measurements of Jones [69], Bishop and Hassan [70,71], Vickery and Watkins [72], and Hartlen et al. [73] and in the simulations of Lu [74] have shown that during lock-in, or self-excitation, the amplitude of the vortex-shedding force increases and its phase relative to the motion of the cylinder becomes dependent on the ratio ω_s/ω_n. Additional measurements have shown that the axial correlation length can be increased.

The increase in the axial integral length of the lift fluctuations Λ_3 with cylinder displacement amplitude $Y_m(\omega)$ has been shown by Burton and Blevins [75,76], using measurements of Toebs [77] to be roughly described as

$$\frac{2\Lambda_3}{d} = \frac{(2\Lambda_3)_0}{d} + \frac{100Y_m(\omega)}{A_m - Y_m(\omega)}, \quad A < A_m$$

$$= \frac{L}{d}, \quad A > A_m$$

where A_m is a threshold amplitude and $A_m = 0.5d$; $(2\Lambda_3)_0$ is the correlation length when there was not transverse motion. In the case of Toebs measurements $(2\Lambda_3)_0 \sim 5d$ (compare with Fig. 4.16). Measurements at low Reynolds numbers by Griffin and Ramberg [78] ($R_d \simeq 500$) disclosed complete correlation of vortices along the cylinder when $y_m(\omega) > 0.2d$. Therefore the threshold amplitude for complete correlation may decrease at low Reynolds number. The effective damping of the cylinder may be expressed simply as a combination of the naturally occurring damping force on the cylinder and the vortex-shedding force in phase with the cylinder velocity; i.e., this force per unit length is

$$F_D = m_s \omega_n \eta_T U_2 - \left(\frac{1}{2}\rho_0 U_\infty^2\right) dC_L(\omega) \sin(\phi + \pi/2)$$

where ϕ is the phase between the force and the cylinder displacement. If $\phi = 0$, then the flow-induced forces will oppose the natural damping in the cylinder so that F_D will be less than $m_s \omega_n \eta_T U_2$.

Since the unsteady lift is a function of the displacement of the cylinder, increasing the dissipation by increasing η_T will affect both the structural response and the fluid–structure feedback loop. Accordingly the dependence of the flow-induced cylinder vibration amplitude will depend on a damping factor

$$D_{cyl} = \frac{\eta_T m_c (2\pi S)^2}{\frac{1}{2}\rho_0 d^2}$$

as shown in Fig. 5.30. This figure is a compilation of data that were collected by Skop and Griffin [79] from various sources. The dependence on D_{cyl} for nonlinear excitation contrasts the reduction in vibration amplitude with increasing damping that would be dictated by Eq. (4.79) for linear excitation. The amplitude of linearly flow-excited force depends on D_{cyl} as in the notation of Fig. 4.23; since $-i\omega Y_{max} = u_{2_n}(\omega)$

$$\frac{Y_{max}}{d} = \frac{\sqrt{2}(\overline{C_L^2})^{1/2}}{D_{cyl}}$$

where $2\overline{C_L^2}^{-1/2}$ is the amplitude of the lift coefficient. As shown in the figure, the relationship of a linear flow–structure interaction given by this equation, especially at small values of D_{cyl}.

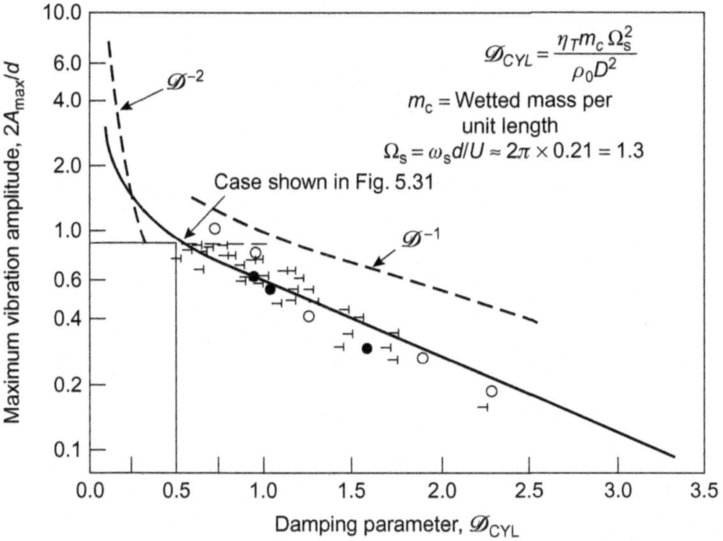

FIGURE 5.30 The maximum amplitude of vortex-excited oscillation, $2A_{max}$, for an elastically mounted, rigid circular cylinder, as a function of the response parameter, S_G. The density of points results from a composite of investigations done in air and in water. *Adapted from Skop RA, Griffin OM. On a theory for the vortex-excited oscillations of flexible cylindrical structures. J Sound Vib 1975;41:263–74.*

5.7.4 Semiempirical Modeling as a Nonlinear Oscillator

This behavior of the cylinder vibration with damping parameter has been described analytically by modeling the flow field as a nonlinear oscillator that is coupled to a linear harmonic oscillator representing the structure. This modeling was proposed by Hartlen and Currie [80] in order to describe the relationship between the lift coefficient and the cylinder velocity $U_2(y_3, \omega)$.

The nonlinear, limiting-cycle character of the oscillating lift resembles the behavior of both the van der Pol and the Rayleigh oscillators. The important characteristics of these oscillators is the nonlinear dependence of fluid damping on the amplitude of the fluid disturbance (Fig. 5.30). The disturbance function used by Hartlen and Currie [80] is the lift oscillation, which in the time domain is defined

$$C_L(t) = \text{Im}\left[\int_{-\infty}^{\infty} C_L(\omega)e^{-i\omega t + i\varphi(\omega)}\,d\omega\right] \tag{5.116a}$$

or

$$C_L(t) = C_L(\omega)\sin(\omega t - \phi) \tag{5.116b}$$

The lift was assumed to satisfy the nonlinear van der Pol equation

$$\ddot{C}_{\mathrm{L}}(t) - \alpha\omega_{\mathrm{s}}\dot{C}_{\mathrm{L}}(t) + \frac{\gamma}{\omega_{\mathrm{s}}}[\dot{C}_{\mathrm{L}}(t)]^3 + \omega_{\mathrm{s}}^2\dot{C}_{\mathrm{L}}(t) = b\frac{\dot{y}_m(0,t)}{d} \tag{5.117}$$

where $\dot{y}_m(0,t)$ is the flexural velocity of the cylinder, b is the coupling coefficient, α and γ are positive-valued empirical parameters. When $\dot{C}_{\mathrm{L}}(t)$ is small the fluid oscillation is simple harmonic, but negatively damped and therefore able to grow with time. This growth occurs until $\dot{C}_{\mathrm{L}}(t)$ is large enough to set the damping term positive. In the case of a rigid cylinder $b=0$, $\dot{y}_{\mathrm{T}}(0,t)=0$ and the lift coefficient is given its value on rigid surfaces which is written

$$C_{\mathrm{L}}(t) = C_{\mathrm{L}_0}e^{-i\omega_{\mathrm{s}}t} \tag{5.118}$$

where

$$C_{\mathrm{L}_0} = \sqrt{4\alpha/3\gamma} \tag{5.119}$$

is the lift coefficient for pressures induced by vortices shed from a cylinder.

Skop and Griffin [79,81] and Griffin et al. [82,83] modified the equation for the lift somewhat by adding another stiffness term to the equation. This term permitted action as a nonlinear spring.

The equation of motion for the m mode of the cylinder vibration in the time domain is from Eq. (5.27)

$$m_{\mathrm{c}}\ddot{y}_m(t) + \eta_{\mathrm{T}}m_{\mathrm{c}}\omega_m\dot{y}_m(t) + m_c\omega_m^2 y_m(t) = q_{\infty}dC_{\mathrm{L}}(t) \tag{5.120}$$

where $m_{\mathrm{c}} = m_{\mathrm{s}} + m_{\mathrm{a}}$ is the total wetted mass per unit length,

$$y_m(t) = Y_m(\omega)\sin\omega t$$

is the lateral displacement of the points of flow separation. In Eq. (5.120), ${k_n}^4D_{\mathrm{s}}$ is replaced by $m_{\mathrm{c}}{\omega_m}^2$ as the effective stiffness, C_d is replaced by $m_{\mathrm{c}}\eta_{\mathrm{T}}\omega_m$, and a_n is replaced by $Y_m(\omega)$. Eqs. (5.117) and (5.120) can be solved approximately and simultaneously without the use of a digital computer when Eq. (5.116b) is substituted into Eq. (5.117) and the coefficients of sines and cosines are grouped, with high-order terms that are of higher order than $\sin^2$ or $\cos^2$ ignored. This operation was followed by Hartlen and Currie [80], who found the amplitude, frequency, and phase:

$$C_{\mathrm{L}}(\omega)\cos\phi = \frac{(\omega_m^2 - \omega^2)}{a\omega_{\mathrm{s}}^2}\frac{Y_m(\omega)}{d}$$

$$C_{\mathrm{L}}(\omega)\sin\phi = -\frac{\eta_{\mathrm{T}}\omega\omega_m}{a\omega_{\mathrm{s}}^2}\frac{Y_m(\omega)}{d}$$

and solved the pair of equations

$$\left\{\left[1-\left(\frac{\omega}{\omega_m}\right)^2\right]\left[\left(\frac{\omega_s}{\omega_m}\right)^2-\left(\frac{\omega}{\omega_m}\right)^2+\alpha\eta_T\frac{\omega_s\omega^2}{\omega_m^3}\right]\right\}$$

$$-\frac{3}{4}\frac{\gamma}{a^2}\frac{\omega_m^2\omega^3}{\omega_s^5}\left(\frac{Y_m(\omega)}{d}\right)^2\left\{\eta_T^3\left(\frac{\omega}{\omega_m}\right)^3+\eta_T\frac{\omega}{\omega_m}\left[1-\left(\frac{\omega}{\omega_m}\right)^2\right]^2\right\}=0$$

$$\left[\eta_T\frac{\omega}{\omega_m}\left[\left(\frac{\omega_s}{\omega_m}\right)^2-\left(\frac{\omega}{\omega}\right)^2\right]-\alpha\frac{\omega_s}{\omega_m}\frac{\omega}{\omega_m}\left[1-\left(\frac{\omega}{\omega_m}\right)^2\right]-ab\frac{\omega\omega_s^2}{\omega_m^3}\right]$$

$$+\frac{3\gamma\omega^3\omega_m^2}{4a^2\omega_s^5}\left(\frac{Y_m(\omega)}{d}\right)^2\left[\left(1-\left(\frac{\omega}{\omega_m}\right)^2\right)^3+\dot{\eta}_T^2\frac{\omega^2}{\omega_n^2}\left(1-\left(\frac{\omega}{\omega_n}\right)^2\right)\right]=0$$

where

$$a=\frac{\rho_0}{2m_c}\left(\frac{U_\infty}{\omega_s}\right)^2$$

simultaneously for the system frequency ω and the amplitude $Y_m(\omega)/d$. A second method is to program Eqs. (5.117) and (5.120) on an analog computer and determine $Y_m(\omega)$, $C_L(\omega)$, and ϕ from observed oscillations in response to small initial conditions. This technique, although it preserves the complete nonlinearity of the functions, is difficult to program. The difficulty arises in the necessity of scaling the amplitudes of the functions and the frequency so that all terms in the equations remain within the dynamic range of the computer. The digital computer was used by Griffin and Ramberg [78] and Griffin et al. [83] to solve the linearized equations of Hartlen and Currie [80] in their modified form, which included the fluid stiffness term dependent on $C_L(t)$. An example is shown in Fig. 5.31.

5.7.5 Randomly Driven One-Dimensional Structures

If we relax the assumption that $k_0L \ll 1$ and indeed also assume that the mode shape function $\Psi_n(y_3)$ (in Eq. (5.110)) is now no longer replaceable by a constant equal to unity, then the expressions derived in Sections 5.7.1 and 5.7.2 take on a more generally meaningful connotation. Now it is not possible to replace the integrals appearing in Eqs. (5.108), (5.109), and (5.111) by simple averages because the values of the integrals depend on the details of how the velocity $U_2(y_3, \omega)$ and the force $F_2(y_3, \omega)$ interact with the phasing of $\exp[-ik_0y_3 \sin \beta]$. This is particularly true when the exciting force is a random variable with a zero spatial mean as discussed in Section 4.3.2, Oscillatory Lift and Drag Circular Cylinders. It is helpful to characterize the

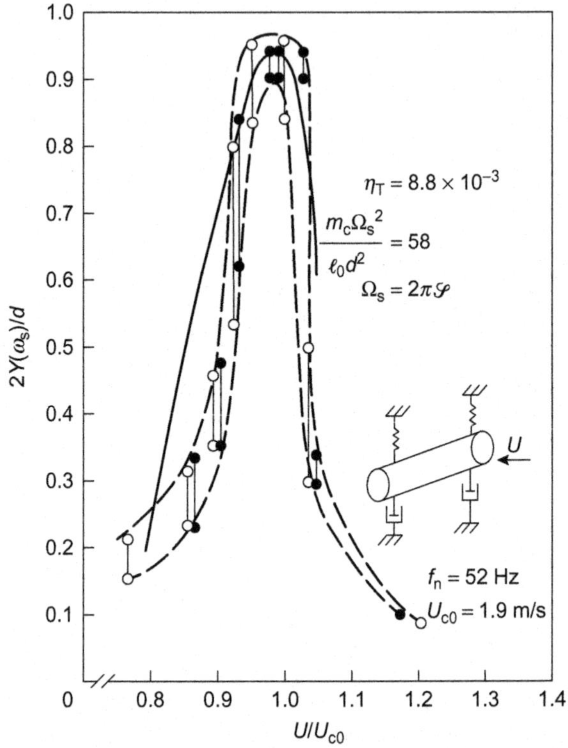

FIGURE 5.31 Measured amplitude of vortex-excited oscillation for cylinder as a function of free stream flow speed U. Points taken with the wind speed increasing (•), with wind speed decreasing (○). Wake oscillation mode $f_n = f_0 = 52$ Hz, $U_{c0} = 1.9$ m/s. *From Griffin et al. [83].*

force in terms of its statistics along the axis of the cylinder and to base the acoustic and vibration response on those statistics.

These concepts will be illustrated below for the case of flow-excited vibration of a circular cylinder. This will be done in the context of Sections 2.6.3 and 3.6.3, The Physical Significance of the Vorticity Source; Anisotropic Turbulence: Spectral Models Based on Stretched Coordinates and the example serves as a one-dimensional illustration, review, and application of the more general methods derived in Sections 5.3, 5.4, and 5.5. We define the spatial Fourier transform of the velocity function as the one-dimensional form of Eq. (5.39)

$$
\begin{aligned}
S_n(k_3) &= \int_{-\infty}^{\infty} e^{-ik_3y_3}\Psi_n(y_3)\,dy_3 \\
&= \int_{-L/2}^{L/2} e^{-ik_3y_3}\Psi_n(y_3)\,dy_3
\end{aligned}
\tag{5.121}
$$

with an inverse

$$\Psi_n(y_3) = \frac{1}{2\pi}\int_{-\infty}^{\infty} e^{ik_3 y_3} S_n(k_3)\, dk_3 \tag{5.122}$$

Eq. (5.109) then includes $S_n(k_3)$ evaluated at the wave number $k_0 \sin\beta$ and may be written in a form that represents a sum of modal contributions that represents a sound pressure resulting from the added mass force on the cylinder

$$P_{\mathrm{a}_u}(x,\ \omega) = ik_0 \cos\theta \sin\phi \frac{e^{ik_0 r}}{4\pi r}(-im_{\mathrm{a}}\omega)\sum_{n=0}^{\infty} U_{2n}(\omega) S_n(k_0 \cos\phi) \tag{5.123}$$

The values of $S_n(k_3)$ are known for a specified series of mode shapes of the structure. The modal amplitude $U_{2_n}(\omega)$ is given by Eq. (5.111) in terms of the stochastic force exciting the cylinder.

The cross-spectrum of $F_n(\omega)$ can be written in terms of the statistical properties of the unsteady lift. Using the notions of Eqs. (4.25) and (4.26) we have the one-dimensional analog of Eq. (5.40) for the stochastic line force

$$\Phi_{F_{mn}}(\omega) = \frac{1}{L^2}\int_{-L/2}^{L/2}\int_{-L/2}^{L/2} \Phi_{f_h}(\omega) R_{\mathrm{pp}}(y_3 - y_3')\Psi_m(y_3)\Psi_0(y_3')\, dy_3\, dy_3' \tag{5.124}$$

Assuming that the statistics of the unsteady forces are established independently of the length of the cylinder (i.e., they are spatially stationary along y_3), we can introduce a wave number spectrum of the forces,

$$R_{\mathrm{pp}}(r_3) = \int_{-\infty}^{\infty} e^{ik_3 r_3}\phi_{\mathrm{p}}(k_3)\, dk_3 \tag{5.124a}$$

with inverse

$$\phi_{\mathrm{p}}(k_3) = \frac{1}{2\pi}\int_{-\infty}^{\infty} e^{-ik_3 r_3} R_{\mathrm{pp}}(r_3)\, dr_3 \tag{5.124b}$$

so that using Eq. (5.121), Eq. (5.124) becomes

$$\Phi_{F_{mn}}(\omega) = \frac{1}{L^2}\int_{-\infty}^{\infty} \Phi_{f_h}(\omega)\phi_{\mathrm{p}}(k_3) S_m(k_3) S_n^*(k_3)\, dk_3 \tag{5.125}$$

The mean-square lift per unit length is

$$\overline{F_2^2} = \int_{-\infty}^{\infty} \Phi_{f_h}(\omega)\, d\omega \tag{5.126}$$

which is also

$$= \overline{C_L^2}\left[\frac{1}{2}\rho_0 U_\infty^2 d\right]^2$$

of Section 4.3, Measured Flow-Induced Forces and Their Frequencies. The wave number spectrum is defined to have a unity integral; i.e.,

$$\int_{-\infty}^{\infty} \phi_p(k_3)\, dk_3 = 1 \tag{5.127}$$

The far-field acoustic intensity spectrum, $I(\mathbf{x}, \omega)$ is given by combining Eqs. (5.106), (5.123) and (5.124) and carrying out the necessary integral operations to obtain

$$\begin{aligned} I(x, \omega) &= \frac{k_0^2 \cos^2\theta \sin^2 \phi}{\rho_0 c_0 16\pi^2 r^2} \left[\frac{m_a \omega^2}{m_c \omega^2}\right]^2 L_3^2 \sum_{n=0}^{\infty} \Phi_{f_n}(\omega) \\ &\cdots \times \frac{|S_n(k_0 \cos \phi)|^2}{|z|^2} \int_{-\infty}^{\infty} \phi_p(k_3) |S_n(k_3)|^2\, dk_3 \end{aligned} \tag{5.128}$$

where $m_a = \pi d^2 \rho_0/4$ and z is given after Eq. (5.114). Integrals of the cross products $S_m(k_3) \cdot S_n(k_3) \cdot S_m(k_3) \cdot S_n(k_3)$ are taken as zero, meaning that the modes are uncoupled. This is not always the case, but for purely sinusoidal mode shapes of a stretched wire modes are uncoupled, or orthogonal. Certain special forms of the excitation spectrum $\phi_p(k_3)$ could conceivably couple the mode shape functions of different order, but the major response would still be from principal modes $n = m$. Graphical representations of the various wave number functions are illustrated in Fig. 5.32. As described in Section 5.4 for spatially harmonic mode shapes $S_n(k_3)$ is peaked about a wave number k_n, in this case given by $k_n = n\pi/L$. Thus the string filters out components of $\phi_p(k_3)$; it acts as a spatial filter. If there happens to be concentration of lift at wave numbers near $k_3 = k_n$, then a substantial response of

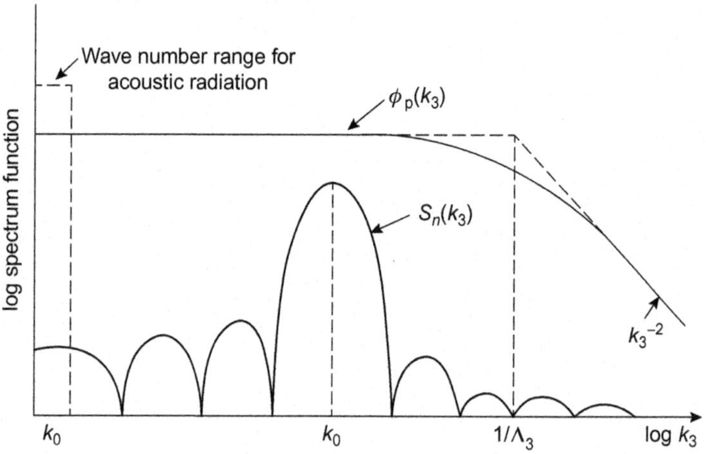

FIGURE 5.32 Illustration of wave number spectra and relative magnitudes of critical wave numbers in the derivations of Eqs. (5.125) and (5.128). The case depicted is for a wave number, $k_3 = 1/\Lambda_3$, which is much less than k_n occurring at low values of n.

that mode will occur. Indeed if, as described in the last section, lock-in occurs, the $\phi_p(k_3)$ may in fact have a component of lift at a k_3 equal to the k_n of the self-excited mode. This would occur because shed vortices will be in phase with the local cylinder vibration. The sound radiated will also depend on the coupling of the vibration with the acoustic medium. The strength of this acoustic coupling is represented by the presence of $|S_n(k_0 \cos \phi)|^2$. As in the case of the flow-excitation acceptance if $k_0 \cos \phi = k_n$, there will be the most intense sound in the far field at those angles $\phi = \cos^{-1}(k_n/k_0)$. These angles are those at which a trace wave speed of sound, $c_0/\cos \phi$, equals the phase speed of flexure in the cylinder.

The intensity of sound from the forces alone can be found in similar manner to be, instead of Eq. (4.27a),

$$I(x, \omega) = \frac{\omega^2 \cos^2 \theta \sin^2 \phi}{8\pi\rho_0 c^3 r^2} \Phi_{f_h}(\omega) L \phi_p(k_3 = k_0 \cos \phi) \tag{5.129}$$

Direct radiation from reaction to the unsteady forces is emitted by low wave number components at $k_3 = k_0 \cos \phi$. In deriving Eq. (5.129) it was assumed that the small centroid γ_c appearing in Eq. (4.27a) can be ignored in comparison to L. In the context of Section 4.5, Formulation of the Acoustic Problem for Compact Surfaces, the wave number spectrum for an exponential correlation function $\exp\{-|r_3|\Lambda_3\}$ is

$$\phi_p(k_3) = \frac{\Lambda_3/\pi}{(k_3\Lambda_3)^2 + 1} \tag{5.130}$$

Accordingly, for relatively long acoustic wavelengths such that $k_0\Lambda_3 \ll 1$, Eq. (3.72) may be invoked along with Eq. (5.126) to recover Eq. (4.27a) exactly (with $\gamma_c \ll L$). The form of the intensity indicated by the flow-induced vibration that is consistent with both $k_0\Lambda_3 \ll 1$ and $k_n\Lambda_3 = n\pi\Lambda_3/L \ll 1$ is (using the relationships derived in Section 5.4 to approximate the low wave number behavior of $S_n(k \ll k_n)$)

$$I(x, \omega) = \frac{\omega^2 \cos^2 \theta \sin^2 \phi}{16\pi^2 \rho_0 c_0^3 r^2} \left[\frac{m_a}{m_c}\right]^2 \sum_{n=1}^{\infty} \Phi_{fh}(\omega) \frac{8}{(k_n L)^2} \frac{2\Lambda_3 L}{|z|^2} \cos^2\left(\frac{1}{2} k_0 L \cos \phi\right) \tag{5.131}$$

which is a kind of edge mode radiation corresponding to $k_0 L > 1$ and $k_0/k_m < 1$ for those modes of vibration that have an *odd* number of half-wavelengths along L; i.e., for $k_n L = n\pi$, $n = 1, 3, 5, \ldots$.

The importance of the vibration-induced sound can again be seen for a long cylinder ($k_0 L > 1$) by examining the ratio of Eq. (5.131) to Eq. (5.129).

The ratio of vibration-induced Intensity to direct dipole component of intensity is

$$\frac{[I(x,\omega)]_{\text{VIB}}}{[I(x,\omega)]_{\text{DIP}}} = \left(\frac{m_a}{m_c}\right)^2 \frac{1}{|z|^2} \frac{8}{(k_n L)^2} \cos^2\left(\frac{1}{2} k_0 L \cos\phi\right) \tag{5.132}$$

which is diagramed in Fig. 5.33 for a moderately high frequency at the resonance condition, $\omega = \omega_n$ and $|z| = \eta_T$. The number of lobes is controlled by the magnitude of $k_0 L$. In the low-frequency limit for $k_0 L \ll \pi$ (or $L/\lambda_0 \ll \frac{1}{2}$) there will be two maxima in the plane (The $\theta = 0$ plane corresponds to the $y_1\ y_2$ plane illustrated in Figs. 5.25 and 5.26.) $\theta = 0°$ at $\phi = 90°$, $270°$ which correspond to broadside sound radiation above and below the cylinder. Any sound power radiated can be determined by integrating Eq. (5.131) over the r, θ, ϕ, surface to obtain the average intensity over a spherical surface at a far-field distance ($r \gg L$) from the cylinder for which $k_0 L \gg 1$

$$I(r) \simeq \frac{1}{6} I_{\max}(r)$$

and for which the behavior of $\cos^2(\frac{1}{2} k_0 L \cos\phi)$ has been approximated by $\frac{1}{2}$ for $k_0 L \gg 1$ and where $I_{\max}(r)$ corresponds to the maximum intensity occurring at $\theta = 0$, π and at $\phi = \pi/2$, $3\pi/2$. The sound power spectrum $\mathbb{P}_{\text{rad}}(\omega)$ from the vibration is found from Eq. (5.131)

$$\mathbb{P}_{\text{rad}}(\omega) = \frac{\pi}{3} \frac{k_n^2}{k_0} (k_0 d)^2 d^2 \rho_0 c_0 \frac{\Phi_{f_h}(\omega)}{m_c^2 \omega^2 |z|^2} \frac{2\Lambda_3}{L} \tag{5.133}$$

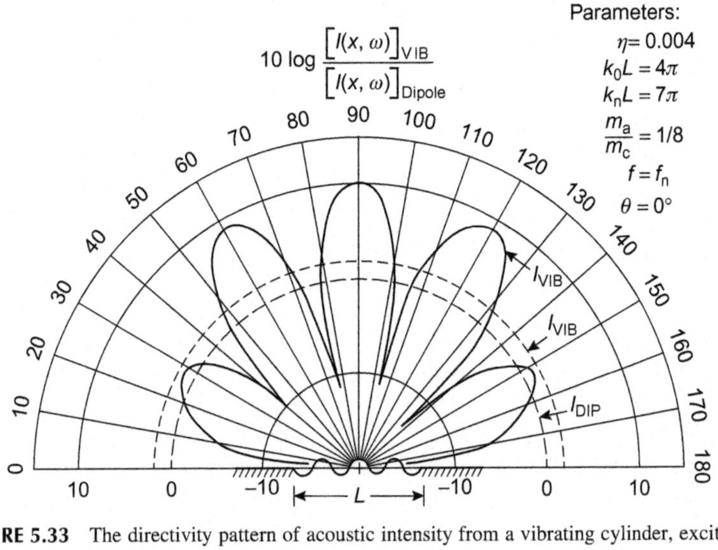

FIGURE 5.33 The directivity pattern of acoustic intensity from a vibrating cylinder, excited by a force at $f = f_s$, relative to dipole sound from same fluid force on a rigid cylinder.

It is useful to show how the sound power could have been derived using the general relationships of Sections 5.3 and 5.5. Eq. (5.133) could have been written down directly by combining Eq. (5.34) (with $m_c = m_a + m_s$) and Eq. (5.84). The modal force per unit length replaced the modal pressure and is found from Eq. (5.125) for functions as depicted in Fig. 5.32

$$\left[\Phi_f(\omega)\right]_n = \Phi_{f_h}(\omega)\frac{2\Lambda_3}{L}, \quad k_n\Lambda_3 < 1$$

The mass per unit length m_c replaces the mass per unit area in Eq. (5.34). The radiation efficiency is determined by integrating Eq. (5.122) to obtain

$$\frac{\mathbb{P}_{\text{rad}}(\omega)}{\rho_0 c_0 (dL)\Phi_n(\omega)} \equiv \sigma_{\text{rad}} = \frac{\pi}{12}(k_0 d)^2 \frac{k_0{}^2}{k_n}\frac{d}{L}$$

where $\Phi_n(\omega)$ is the spectral density of the modal velocity of the cylinder replacing the spectrum $\Phi_{mn}(\omega)$ in Eq. (5.84). This expression for the radiation efficiency is functionally similar to Eq. (5.90a) for the unbaffled beam.

Thus it can be seen in these alternative formulations that the sound pressure and cylinder vibration require special features of coupling. Both the flow field and the vibration field must match spatially; the structure sampling certain wave vector components from the excitation forces and the acoustic medium likewise sampling required wave vector components from the surface motion. Eqs. (3.43), (5.78), (5.128), and (5.129) are all of the same form in this regard, and so this feature of spatial acceptance of energy by both structures or fluid media is common to all areas of flow-induced sound and vibration.

5.8 SUMMARY AND PRINCIPLES OF NOISE CONTROL

The preceding analysis identifies several quantities which may be used to effect noise reductions in various applications. Putting together Eqs. (5.52e), (5.52d), and (5.91b), we can write sound power from a number of simultaneously excited structural modes

$$\mathbb{P}_{\text{rad}}(\omega, \Delta\omega) = \rho_0 c_0 A_p \overline{\sigma}_{\text{rad}} \frac{\mathbb{P}_{\text{in}}}{M_s \eta_T \omega} \tag{5.134}$$

with a similar one derivable for a single mode, e.g., Eq. (5.133). Although this relationship was derived here for the specific case of a flow-excited panel, its form is general. The radiated sound power depends on the area of the radiator, A_p, the structural radiation efficiency $\overline{\sigma}$, the input power or the power acceptance of the structure from the flow, $\mathbb{P}_{\text{in}}$, the mass of the structure, $M_s = \rho_p hA_p$, and the damping η_T. Any means of altering one or more of these variables provides a potential for noise control.

5.8.1 Reductions of Source of Excitation

The first, most obvious means of reducing radiated sound is by reducing the excitation forces or the input power into the structure. Eqs. (5.95) and (5.99) show a direct proportion between radiated sound pressure and the magnitude of a spatially localized exciting force. For more distributed force fields (say, in boundary layer or acoustic excitation of structures) Eqs. (5.17), (5.42), and (5.52d) show that by reducing the input power the mean-square velocity of the structure will be reduced in proportion. Noise control by this means certainly may be accomplished by reducing the flow-excited force levels (the value of $\hat{R}_{\mathrm{pp}}(\mathbf{r}, \tau)$ appearing in Eq. (5.37) or the magnitude of $\Phi_{p_n}(\omega)$ in Eq. (5.40)) or by simply mass loading the structure (Eqs. (5.41), (5.43), and (5.52d)). Alternative means of reducing $\mathbb{P}_{\mathrm{in}}$ involves modifications of spatial coupling characteristics of the flow or the structure. This requires a rather sophisticated knowledge of the fluid mechanical processes that lead to the particular wave number dependence of $\Phi_{\mathrm{p}}(\mathbf{k}, \omega)$, or of $\hat{R}_{\mathrm{pp}}(y_1 - y_2, \tau)$ appearing in Eq. (5.37). These means have been examined for the case of turbulent boundary layer excitation of aircraft fuselages (see Chapter 9: Response of Arrays and Structures to Turbulent Wall Flow and Random Sound). In flow-induced noise applications it may be possible to reduce either $\overline{p^2}$ or $\Phi_{ph}(\omega)$ simply by reducing relative velocities (effecting roughly a reduction in the magnitude of flow-induced pressures at a point) or by reducing the level of flow disturbances by minimizing turbulence, retarding boundary layer transition or separation, or minimizing discrete vortex production. These means are often accomplished by rather minimal changes of body shape or surface finish. Succeeding chapters will examine these measures in some detail.

5.8.2 Reductions in Noise by Structural Modifications

Any means by which the spatial scales of vibration may be altered to reduce either the acceptance of power ($S_n(\mathbf{k})$) in Eq. (5.40) or the radiation efficiency (Eq. (5.85)) offers potential for reducing noise. Structural modifications may be utilized to alter $\overline{\sigma}_{\mathrm{rad}}$ and possibly $\mathbb{P}_{\mathrm{in}}$ through changes in the wave number matching as quantified in the component functions of Eqs. (5.40) and (5.85). Such methods may be especially helpful in the reduction of the modal average radiation efficiency $\overline{\sigma}_{\mathrm{rad}}$ at a particular frequency. As an example we examine the $\overline{\sigma}$ for flat ribbed panels given by Eq. (5.92). Above acoustic coincidence, $k_0 > k_{\mathrm{p}}$, the radiation efficiency is nearly unity so no changes can be made. Below acoustic coincidence $\overline{\sigma}_{\mathrm{rad}}$ is generally dependent on the geometric factor

$$\frac{L_1 + L_3}{k_{\mathrm{p}} A_{\mathrm{p}}} = \sqrt{\frac{\pi\sqrt{3}f}{hc_\ell} \frac{2(L_1 + L_3)}{A_{\mathrm{p}}}}$$

The radiation efficiency below coincidence is therefore proportional to the ratio of the total perimeter of the plate to the area of the plate. As pointed out in Section 5.5.5, the addition of ribbing to the plates to provide stiffness would increase rather than decrease the sound by increasing the total perimeter of the panel. Another means of controlling $\overline{\sigma}_{\text{rad}}$ of plates below coincidence is to reduce the wave number ratio

$$\frac{k_0}{k_p} = 2\sqrt{\frac{\pi\sqrt{3}fhc_\ell}{c_0^2}}$$

This can be achieved at a given frequency by reducing both the panel thickness and the wave speed c_ℓ. This latter may be done by reducing the Young's modulus E (Eqs. (5.32) and (5.47)) or by increasing the density.

The power acceptance $\mathbb{P}_{\text{in}}$ can perhaps be reduced in some special cases by altering the modal function $S_{mn}(\mathbf{k})$. This function is generally peaked about wave numbers, $k_1 = k_n$ and $k_3 = k_n$, where k_m and k_n, defined in Section 5.4, are determined by the geometry of the structure. If the wave number spectrum of excitation $\Phi_p(\mathbf{k},\omega)$ has a peak in some restricted range of $\mathbf{k}$ and ω, then it may be possible to alter k_m and k_n of the structure so as to decouple $\Phi_{pp}(\mathbf{k}, \omega)$ and $S_{mn}(\mathbf{k})$ in Eq. (5.40). Unfortunately in most practical situations the structural mode density is large enough that even if the modal is decoupled from the excitation field for one combination of m and n, these parameters will be coupled for another combination of m and n. Therefore this method of noise control is expected to have only specialized application.

5.8.3 Damping and Mass Loading

Whereas the measures introduced above may have specialized or limited usefulness, damping and mass loading nearly always result in noise reductions when correctly applied. Mass increases can be accomplished by using structural materials of higher density. If mass loading can be accomplished without an increase in stiffness or Young's modulus, then so much the better because k_0/k_p and therefore $\overline{\sigma}_{\text{rad}}$ will be reduced also. The most obvious benefit of mass loading, however, is in its ability to increase the average impedance of the structure. This results in less vibration response (Eqs. (5.7), (5.42), and (5.52d)) and less power acceptance (Eqs. (5.43) and (5.52b)). Accordingly, by Eqs. (5.52b) and (5.99), sound power is inversely proportional to the mass per unit area of the structure.

By damping the structure using dissipating materials the sound power may be reduced. However, as inspection of Eqs. (5.99) and (5.100) will show, if radiation damping is already much larger than structural damping, then further modest additions in structural damping may not yield benefits. In particular, since η_{rad} is dependent on $\overline{\sigma}_{\text{rad}}$, and $\overline{\sigma}_{\text{rad}}$ is often dominated by particular subgroups of well-radiating modes, it is conceivable that the addition of certain amounts of structural damping will reduce the response of poorly radiating modes while not affecting the well-radiating modes. Therefore effectiveness of structural damping in all applications should not be taken for granted.

5.8.4 Estimation of Radiation Efficiency and Added Mass

Simplified forms of Eq. (5.92) can be written for the purpose of estimating the radiation efficiencies of flat baffled plates. Similar formulas can be set down for other geometries noting that

$$\left(\frac{k_0}{k_p}\right)^2 = \frac{f}{f_c}$$

where the acoustic coincidence frequency f_c is given in Fig. 5.11, and that for a plate of thickness h (see Eqs. (5.19), (5.20), and (5.32))

$$\frac{\lambda_c}{h} = \frac{\pi}{\sqrt{3}} \frac{c_b}{\sqrt{1-\nu_p^2}} \simeq 1.91 \frac{c_b}{c_0}$$

then Eqs. (5.86) and (5.92) can be written in approximate forms for easy computation. At very low frequencies that the panel behaves as a piston

$$\sigma_{00} \simeq \frac{32}{\pi^3} \frac{A_p}{\lambda_c^2} \left(\frac{f}{f_c}\right)^2 \quad \text{for} \quad \frac{f}{f_c} < \beta_0 \tag{5.135}$$

where β_0 is determined by the smaller of

$$\beta \simeq \frac{\lambda_c^2}{2\pi\sqrt{A_p}}$$

which is the condition $k_0^2 A_p = 1$ or

$$\beta \simeq \frac{\lambda_c^2}{2A_p} \sqrt{\frac{P^2}{8A_p} - 1}$$

where $P = 2(L_1 + L_3)$ is the perimeter of the panel, which is the value of β that corresponds to the frequency of the first ($m = 0$, $n = 0$) resonance in the panel and is determined by

$$k_p = \pi\sqrt{L_1^{-2} + L_3^{-2}}$$

At higher frequencies when the panel has corner, edge, and surface modes, then

$$\sigma \simeq \frac{1}{\pi^2} \frac{\lambda_c^2}{A_p} \left[\frac{2}{\pi}\sqrt{\frac{f_c}{f}} + \frac{P}{\lambda_c}\sqrt{\frac{f}{f_c}}\right] \tag{5.136}$$

for

$$\frac{f_e}{f_c} = \frac{\lambda_c}{\pi L_1} < \frac{f}{f_c} < 0.5, \quad L_1 > L_3$$

where the lower limit, f_e, corresponds to $k_0 L_1 > 2$ and the upper limit corresponds to $k_0/k_p < 0.7$. The two terms of Eq. (5.136) become equal at

$$\frac{f_1}{f_c} = \frac{2}{\pi}\frac{\lambda_c}{P}$$

Also, at coincidence we have the approximation

$$\overline{\sigma} \simeq \left(\frac{L_1}{\lambda_c}\right)^{1/2} + \left(\frac{L_3}{\lambda_c}\right)^{1/2} \simeq \frac{1}{\sqrt{2}}\sqrt{\frac{P}{\lambda_c}} \quad \text{for} \quad f = f_c \tag{5.137}$$

which is valid to within 2 dB. Above coincidence the approximation is

$$\overline{\sigma} = \frac{1}{\sqrt{1 - f_c/f}} \quad \text{for} \quad \frac{f}{f_c} > 1 \tag{5.138}$$

These relationships are diagramed in Fig. 5.34. The diagram shows critical intercepts that may be drawn once the values of P/λ_c, A/λ_c^2, and L_1/λ_c are known. In addition to the parameters already defined, also shown is $4f_1$. This frequency is selected for ease in computing Eq. (5.136) using only the

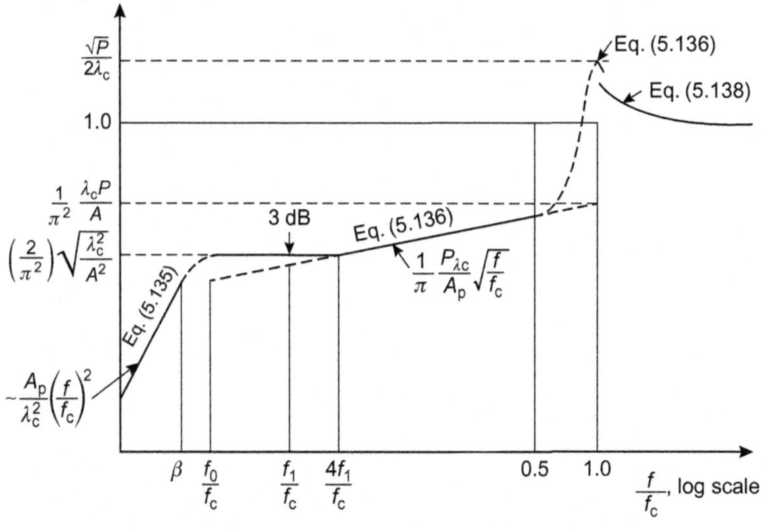

FIGURE 5.34 Chart for the computation of radiation efficiency for a flat baffled panel.

second term. Very often the first term is negligible. In the vicinity of $f = f_1$ the radiation efficiency is nearly constant and twice the value given by the second term at $f = f_1$. The intercept of the horizontal line with the $\sqrt{f}$ line occurs at $f = 4f_1$.

In the interval $0.5 < f/f_c < 1$ a more exact formula given by Maidanik is

$$\sigma \simeq \frac{1}{4\pi^2}\frac{P\lambda_c}{A_p}\left\{\frac{(1-\alpha^2)\ln[(1+\alpha)/(1-\alpha)]+2\alpha}{(1-\alpha^2)^{3/2}}\right\} \tag{5.139}$$

where $\alpha = \sqrt{f/f_c}$. Use of the above scheme for constructing a radiation efficiency curve for the arrangement shown in Fig. 5.15 is shown in Fig. 5.35. In this example there are subpanels within the main panel of over all dimensions 4 × 6 in. For frequencies less than approximately 125 Hz the subpanels are no longer separately resonant. Thus low-frequency sound is controlled by the motion of the main baffled panel. The effect of the ribs in modifying the mass and stiffness of the main panel was neglected for this rough estimate (Fig. 5.36).

Added masses at low frequencies for plates in water are found from Eq. (5.93). Fig. 5.31 shows relative added masses as a function of f/f_c; note when $f/f_c \geq 1$ the added mass is zero. As discussed in Section 5.5.6 m_a/m_s applies to both baffled and unbaffled plates. Eq. (5.93) should not be considered valid for $f/f_c > 0.5$.

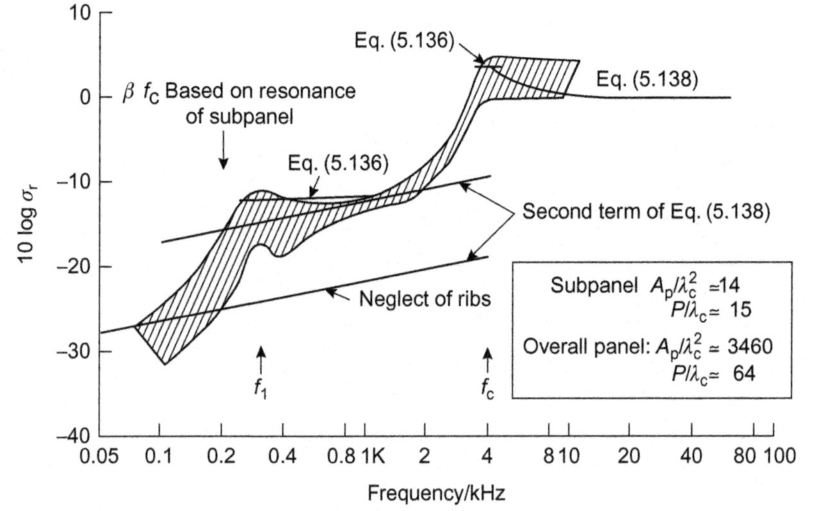

FIGURE 5.35 Estimates of radiation efficiencies for the ribbed aluminum panel shown in Fig. 5.16. Shaded region shows range of measured values; $h = \frac{1}{8}$ in.

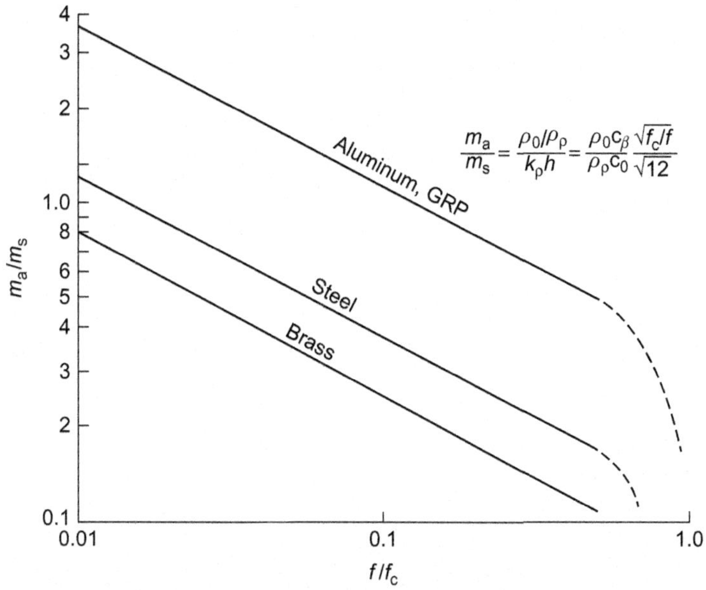

FIGURE 5.36 Approximate added masses of flat plates with water on one side.

REFERENCES

[1] Junger MC, Feit D. Sound structures and their interaction. 2nd ed. Melville, New York: American Institute of Physics; 1986.

[2] Skelton EA, James JH. Theoretical acoustics of underwater structures. London: Imperial College Press; 1997.

[3] Fahy F, Guardonio P. Sound and structural vibration. 2nd ed. Waltham, Mass: Academic Press; 2007.

[4] Lin YK. Probabilistic theory of structural dynamics. New York: McGraw-Hill; 1967.

[5] Crandall SH, editor. Random vibration, vol. I. Cambridge, MA: MIT Press; 1958.

[6] Crandall SH, editor. Random vibration, vol. 2. Cambridge, MA: MIT Press; 1963.

[7] Cremer L, Heckl M, Ungar EE. Structure-borne sound. Berlin and New York: Springer-Verlag; 1973.

[8] Skudrzyk E. Simple and complex vibratory systems. University Park, Pa: Pennsylvania State University Press; 1968.

[9] Maidanik G. Response of ribbed panels to reverberant acoustic fields. J Acoust Soc Am 1962;34:809–26.

[10] Maidanik G. Erratum response of ribbed panels to reverberant acoustic fields. J Acoust Soc Am, 57. 1975. p. 1552.

[11] Lyon RH, DeJong RG. Statistical energy analysis of dynamical systems: theory and applications. 2nd ed. Butterworth-Heinemann, Waltham, Ma; 1994.

[12] Lyon RH, Maidanik G. Power flow between linearly coupled oscillators. J Acoust Soc Am 1962;34:623–39.

[13] Newland DE. Mechanical vibration analysis and computation. Mineola, New York: Dover Publications; 2006.

[14] den Hartog JP. Mechanical vibrations. Mineola, New York: Dover Publications; 1985.

[15] Lyon RH. What good is statistical energy analyses anyway. Shock Vib Dig 1970;2:1–9.

[16] Smith Jr. PW, Lyon RH. Sound and structural vibration. NASA [Contract. Rep.] CR NASA-CR-160; 1965.

[17] Fahy FJ. Statistical energy analyses—a critical review. Shock Vib Dig 1974;6:1–20.

[18] Maidanik G. Some elements in statistical energy analysis. J Sound Vib 1977;52:171–91.

[19] Lyon RH. Response of strings to random noise fields. J Acoust Soc Am 1956;28:391–8.

[20] Kraichnan RH. Noise transmission from boundary layer pressure fluctuations. J Acoust Soc Am 1956;29:65–80.

[21] Powell A. On the fatigue failure of structures due to vibrations excited by random pressure fields. J Acoust Soc Am 1958;30:1130–5.

[22] Dyer I. Response of plates to a decaying and convecting random pressure field. J Acoust Soc Am 1959;31:922–8.

[23] Maidanik G, Lyon RH. Response of strings to moving noise fields. J Acoust Soc Am 1961;33:1606–9.

[24] Smith PW. Response and radiation of structural modes excited by sound. J Acoust Soc Am 1962;34:640–7.

[25] White PH, Powell A. Transmission of random sound and vibration through a rectangular double wall. J Acoust Soc Am 1966;40:821–32.

[26] Lyon RH. Boundary layer noise response simulation with a sound field. Acoustical fatigue in aerospace structures. Chap. 10. Syracuse, NY: Syracuse University Press; 1965.

[27] White PH. Transduction of boundary layer noise by a rectangular panel. J Acoust Soc Am 1966;40:1354–62.

[28] Davies HC. Sound from turbulent-boundary-layer excited panels. J Acoust Soc Am 1971;49:878–89.

[29] Jameson PW. Measurement of low wave number component of turbulent boundary layer pressure spectra density. Turbul Liq 1975;4:192–200.

[30] Leehey P. Trends in boundary layer noise research. Aerodyn. Noise, Proc. AFOSR-UTIAS Symp., Toronto 1968;273–98.

[31] Chandiramani KL. Vibration response of fluid-loaded structures to low-speed flow-noise. J Acoust Soc Am 1977;61:1460–70.

[32] Timoshenko S, Woinowsky-Krieger S. Theory of plates and shells. New York: McGraw-Hill; 1959.

[33] Kinsler LE, Frey AR, Coppens AB, Sanders JV. Fundamentals of acoustics. Hoboken, New Jersey: Wiley; 2000.

[34] Timoshenko S. Vibration problems in engineering. 3rd ed. New York: Van Nostrand; 1955.

[35] Strawderman WA. The acoustic field in a closed space behind a rectangular simply supported plate excited by boundary layer turbulence. Rep. 827. U.S. Navy Underwater Sound Lab. Rep. 827. New London, CT; 1967.

[36] Obermeier F. On the response of elastic plates backed by enclosed cavities to turbulent flow excitations. Acoust. Vib. Lab. Rep. 70208-6. Cambridge, MA: MIT; 1971.

[37] Arnold R. Vibration of a cavity backed panel. Acoust. Vib. Lab. Rep. 70208-7. Cambridge, MA: MIT; 1971.

[38] Leissa AW. Vibration of plates. NASA [Spec. Publ.] SP NASA SP-160; 1969.

[39] Aupperle FA, Lambert RF. On the utilization of a flexible beam as a spatial filter. J Sound Vib 1972;24:259–67.

[40] Martin NC. Wave number filtering by mechanical structures. PhD thesis. Cambridge, MA: MIT; 1976.
[41] Copson ET. Asymptotic expansions. London and New York: Cambridge University Press; 1965.
[42] Lamb H. Hydrodynamics. New York: Dover; 1945.
[43] Borisyuk AO, Grinchenko VT. Vibration and noise generation by elastic elements excited by a turbulent flow. J Sound Vib 1997;204:213–37.
[44] Maury C, Gardonio P, Elliott SJ. A wave number approach to modelling the response of a randomly excited panel, Part 1, General theory. J Sound Vib 2002;252:83–113.
[45] Hambric SA, Sung SH, Nefske DJ, editors. Engineering vibroacoustic analysis: methods and applications. Hoboken, New Jersey: John Wiley & Sons; 2016.
[47] Hambric, S.A., Jonson, M.L., Fahnline, J.B., Campbell, R.L. Simulating the vibro-acoustic power of fluid-loaded structures excited by randomly distributed fluctuating forces. Proc. NOVEM 2005. St Raphael, France, 18–21; 2005.
[46] Hambric SA, Hwang YF, Bonness WK. Vibrations of plates with clamped and free edges excited by low-speed turbulent boundary layer flow. J Fluids Struct 2004;19:93–110.
[48] Hambric SA, Boger DA, Fahnline JB, Campbell RL. Structure- and fluid-borne acoustic power sources induced by turbulent flow in 90° piping elbows. J Fluids Struct 2010;26:121–47.
[49] Ciappi E, DeRosa S, Franco F, Guyader J-L, Hambric SA. Flinovia—flow induced noise and vibration issues and aspects. New York: Springer; 2015.
[50] Esmialzadeh M, Lakis AA, Thomas M, Marcoullier L. Prediction of the response of a thin-structure subjected to a turbulent boundary layer induced random pressure-field. J Sound Vib 2009;328:109–28.
[51] Chase DM. Turbulent boundary layer pressure fluctuations and wave number filtering by non-uniform spatial averaging. J Acoust Soc Am 1969;46:1350–65.
[52] Hwang YF, Maidanik G. A wavenumber analysis of the coupling of a structural mode and flow turbulence. J Sound Vib 1990;142:135–52.
[53] Jones DS. Generalized functions. New York: McGraw-Hill; 1966.
[54] Blake WK, Maga LJ. On the flow-excited vibrations of cantilever struts in water. I. Flow-induced damping and vibration. J Acoust Soc Am 1975;57:610–25.
[55] Blake WK, Maga LJ. On the flow-excited vibrations of cantilever struts in water. II. Surface pressure fluctuations and analytical predictions. J Acoust Soc Am 1975;57:1448–64.
[56] Davies HG. Low frequency random excitation of water–loaded rectangular plates. J Sound Vib 1971;15:107–26.
[57] Davies HG. Excitation of fluid-loaded rectangular plates and membranes by turbulent boundary layer flow. Winter Annu Meet Am Soc Mech Eng Pap 70-WA/DE-15; 1970.
[58] Davies HG. Acoustic radiation by fluid loaded rectangular plates. Acoust. Vib. Lab. Rep. 71467-1. Cambridge, MA: MIT; 1969.
[59] Wallace CE. Radiation resistance of a baffled beam. J Acoust Soc Am 1972;51:936–45.
[60] Wallace CE. Radiation resistance of a rectangular panel. J Acoust Soc Am 1972;51:946–52.
[61] Blake WK. The acoustic radiation from unbaffled strips with application to a class of radiating panels. J Sound Vib 1975;39:77–103.
[62] Blake WK. The radiation of free-free beams in air and water. J Sound Vib 1974;33:427–50.
[63] Manning JE, Maidanik G. Radiation properties of cylindrical shells. J Acoust Soc Am 1964;36:1691–8.
[64] Chertock G. Sound radiation from prolate spheroids. J Acoust Soc Am 1961;33:871–80.

[65] Chertock G. Sound radiation from vibrating surfaces. J Acoust Soc Am 1964;36: 1305–13.

[66] Maidanik G, Kerwin EM. The influence of fluid loading in the radiation from infinite plates below the critical frequency. BBN Rep. 1320. Cambridge, MA: Bolt Beranek and Newman; 1965.

[67] Maidanik G. The influence of fluid loading on the radiation from orthotropic plates. J Sound Vib 1966;3:288–99.

[68] Feit D. Pressure radiation by a point-excited elastic plate. J Acoust Soc Am 1966;40: 1489–94.

[69] Jones GW. Unsteady lift forces generated by vortex shedding about large stationary, and oscillating cylinders at high Reynolds number. ASME Symp Unsteady Flow, Pap. 68–FE–36; 1968.

[70] Bishop RED, Hassan AY. The lift and drag forces on a circular cylinder in a flowing fluid. Proc R Soc London Ser A 1964;277:32–50.

[71] Bishop RED, Hassan AY. The lift and drag forces on a circular cylinder in a flowing fluid. Proc R Soc London Ser A 1964;277:51–75.

[72] Vickery BJ, Watkins RD. Flow induced vibrations of cylindrical structures. Proc Australasian Conf Hydraulics Fluid Mech 1st 1964;213–41.

[73] Hartlen RT, Baines WD, Currie IG. Vortex excited oscillating of a circular cylinder. Rep. UTME-TP 6809. Toronto: University of Toronto; 1968.

[74] Lu QS, To CWS, Jin ZS. Weak and strong interactions in vortex-induced resonant vibrations of cylindrical structures. J Sound Vib 1996;190:791–820.

[75] Blevins RD, Burton TE. Fluid forces induced by vortex shedding. J Fluids Eng 1976;98: 19–26.

[76] Burton TE, Blevins RD. Vortex shedding noise from oscillating cylinders. J Acoust Soc Am 1976;60:599–606.

[77] Toebs GH. The unsteady flow and wake near an oscillating cylinder. J Basic Eng 1969;91:493–505.

[78] Griffin OM, Ramberg SE. The effects of vortex coherence, spacing, and circulation on the flow-induced forces on vibrating cables and bluff structures. N.R.L. Rep. No. 7945; 1976.

[79] Skop RA, Griffin OM. A model for the vortex excited resonant response of bluff cylinders. J Sound Vib 1973;27:225–33.

[80] Hartlen RT, Currie IG. Lift-oscillator model of vortex induced vibration. J Eng Mech Div Am Soc Civ Eng 1970;96, No. EM5:577–91.

[81] Skop RA, Griffin OM. On a theory for the vortex-excited oscillations of flexible cylindrical structures. J Sound Vib 1975;41:263–74.

[82] Griffin OM, Skop RA, Ramberg SE. The resonant, vortex-excited vibrations of structures and cable systems. Proc Annu Offshore Technol Conf II, Pap. No. OTC 2319, 1975; 734–44.

[83] Griffin OM, Skop RA, Koopman GA. The vortex-excited resonant vibrations of circular cylinders. J Sound Vib 1973;31:235–49.

FURTHER READING

Ross D. Mechanics of underwater noise. Oxford: Pergamon; 1976.

Chapter 6

Introduction to Bubble Dynamics and Cavitation

In underwater acoustics, sound propagation may be influenced by the dynamics of suspended gas bubbles in the liquid and when liquids flow past bodies, cavitation may occur. Since an understanding of both propagation in bubbly fluids and cavitation requires some knowledge of the dynamics of bubbles; this chapter will examine the dynamics of bubbles; the linear perturbations of gas-filled bubbles by small amplitude pressures, the propagation through bubbly mixtures, the onset of nonlinear large-amplitude bubble motions, and implosive collapses. The subject of nonlinear bubble dynamics as covered here is introductory to our coverage of hydrodynamic cavitation inception and noise in the next chapter. Accordingly, we will emphasize the basic behavior of air bubbles in water and illustrate the various uses of the Raleigh–Plesset theory in estimating bubble motion and critical pressure for nonlinear growth. We will also show how a related linear modeling of the acoustics of bubble swarms is useful in describing the wave dispersion characteristics in water-filled hydrodynamic test environments. Readers desiring a deeper discussion of bubble dynamics and cavitation, theory and engineering, are referred to the texts by Knapp, Daily, and Hammitt [1], Brennan [2], Brennan [3], Leighton [4], Carlton [5] as well as extensive reviews by Plesset and Prosperetti [6], Feng and Leal [7], Arndt [8,9], Blake and Gibson [10], and Prosperetti [11].

6.1 BASIC EQUATIONS OF BUBBLE DYNAMICS

In this section we consider the conditions that are necessary for the maintenance of small-amplitude bubble variations, quasi-static bubble equilibrium, the necessary conditions for the nonlinear bubble motions that lead to cavitation, and the effects of bubble gas and liquid compressibility in the collapse of cavities.

6.1.1 Linear Bubble Motions

The dynamics of bubbles in a liquid responding to an imposed pressure fluctuation have been considered with varying degrees of complexity. The first

Mechanics of Flow-Induced Sound and Vibration, Volume 1.
DOI: http://dx.doi.org/10.1016/B978-0-12-809273-6.00006-3

and simplest analysis is that of Rayleigh [12], later elaborated on by Plesset [13], Neppiras and Noltingk [14,15], Plesset and Prosperetti [6], Commander and Prosperetti [16], Prosperetti [11] who considered the liquid surrounding the bubbles as essentially incompressible, except for providing acoustic radiation damping to the bubble oscillations. In Rayleigh's analysis the medium inside the bubble is liquid vapor so that the internal pressure is constant. Plesset [13] and Neppiras and Noltingk [14,15] allowed the pressure balance across the bubble wall to be determined by internal pressure and also by vapor pressure, pressure of insoluble gas as well as by surface tension and viscosity, introduced by Houghton [17]. In related work, Blue [18] calculated and Howkins [19] measured resonance frequencies of a bubble attached to a wall, Shima [20] examined effects of liquid-phase compressibility finding them small in water, and Strasberg [21] examined photographically the resonance of nonspherical bubbles. Lauterborn and Bolle [22] and Lauterborn and Ohl [23] have observed collapses near wall boundaries, and Chang and Ceccio [24] observed growth and collapses of bubbles in vortex cores, relating the sound levels to various stages in the bubble history. Ref. [24] is the most recent example to show that the sound is controlled by the volume history and that deformation of the bubble contributes little because the wave length of the sound is large compared with the bubble radius. Accordingly, bubble deformation represents higher order acoustic multipoles the sound from which is swamped by the bubble's fundamental volume monopole.

Fig. 6.1 shows the relevant geometry, and we follow the gist of the various theoretical developments in the just-quoted references to derive the well-known Rayleigh–Plesset equation. The pressure differences across

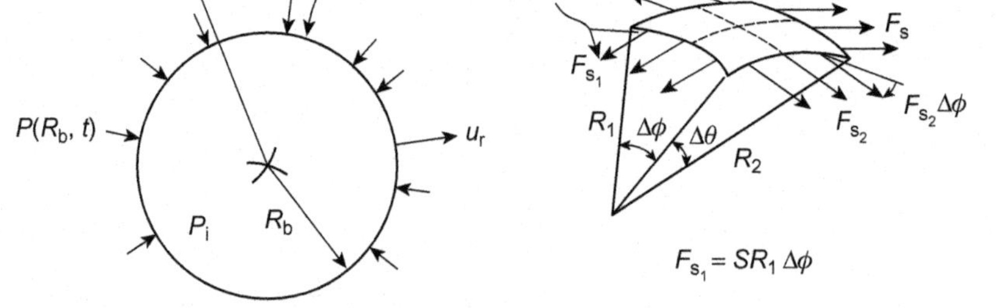

FIGURE 6.1 Force diagrams for a spherical bubble. (A) Spherical bubble in an unbounded liquid; (B) Force diagram for a thin film.

a segment of the bubble wall are balanced by the surface tension forces. Using the notation of Fig. 6.1B, this instantaneous force balance is (e.g., Commander and Prosperetti [16])

$$(P_i - P(R,\ t))R_2R_1\,\Delta\phi\,\Delta\theta = (SR_1\,\Delta\phi)\Delta\theta + (SR_2\Delta\theta)\Delta\phi$$

where R_1 and R_2 are the (possibly different) radii of curvature in the orthogonal directions. For the spherical bubble $R_1 = R_2 = R$ so that the condition necessary for force equilibrium is

$$P_i - P(R,t) = 2S/R + 4\mu_0\dot{R}/R \tag{6.1}$$

In the general case, the internal pressure in the bubble is the sum of partial pressures of the vapor pressure P_v and the total equilibrium partial pressure of gas P_g, which is dissolved in the liquid. Thus

$$P_i = P_g + P_v$$

During the bubble motion, this gas is compressed or expanded so that the partial pressure varies with the bubble radius. Typically, the ideal gas law is assumed in order to simplify the thermodynamic properties of the enclosed gas and so to provide a simple equation of state. The dependence of pressure on volume is then given by the simple relationships

$$P_g = P_{g_0} \times (V_0/V)^{\gamma}$$
$$P_g = P_{g_0} \times (R_0/R)^{3\gamma}$$

where the variables subscripted zero apply to the initial state. For adiabatic motions (no heat transfer from the gas to the liquid) γ is the ratio of the specific heat at constant pressure to that at constant volume ($\gamma \approx 1.4$ for a diatomic gas), for isothermal contraction $\gamma = 1$. Plesset and Hsieh [25] have subsequently analyzed periodic forced linear oscillations of bubbles, finding that the motions are isothermal for oscillation frequencies less than the resonance frequency and adiabatic for oscillation frequencies above resonance.

The equilibrium pressure at the bubble wall is then given by

$$P(R,t) = P_v + P_{g_0} \times (R_0/R)^{3\gamma} - 2S/R - 4\mu_0\dot{R}/R \tag{6.2}$$

Spherically symmetric motions in the liquid are governed by Eq. (2.2),

$$\rho_0\frac{\partial u_r}{\partial t} + \rho_0 u_r\frac{\partial u_r}{\partial r} = -\frac{\partial p}{\partial r}$$

For the *incompressible* motion in the liquid we let the radial velocity be the gradient of a potential,

$$u_r = \nabla_r\phi$$

so that

$$\nabla_{\mathrm{r}}\left[\frac{\partial\phi}{\partial t}+\frac{1}{2}(\nabla\phi)^2\right]=-\frac{1}{\rho_0}\nabla_{\mathrm{r}}p$$

and integrating along a stream tube from $r=R$ to a distant point $r(r>R)$,

$$\frac{\partial[\phi(r)-\phi(R)]}{\partial t}+\frac{1}{2}\left([\nabla_{\mathrm{r}}\phi(r)]^2\right)-\left[\nabla_{\mathrm{r}}\phi(R)^2\right]=\frac{1}{\rho_0}[P(R,t)-P(r,t)]$$

For incompressible motions, the methods of Section 2.1.3.1 give the spherically symmetric potential as a solution of Laplace's equation

$$\frac{\partial}{\partial r}\left(r^2\frac{\partial\phi}{\partial r}\right)=0$$

subject to the boundary condition on the sphere $r=R$ that $U_r=\dot{R}$. Thus the potential is

$$\phi(r)=-\frac{\dot{R}R^2}{r}$$

so that an equation for the bubble wall velocity is

$$\ddot{R}R+\frac{3}{2}(\dot{R})^2=\frac{P(R,t)-P(r,t)}{\rho_0} \tag{6.3}$$

where r is now selected so that $r\gg R$ and $P(r, t)$ is the ambient pressure at the bubble's infinity. Eqs. (6.2) and (6.3) may be combined since the pressure balance (6.2) across the wall may be taken to apply for any value of R. The pressure, $P(r, t)$ may be considered as a time-varying hydrodynamic driving pressure.

Eq. (6.3) is the basic equation for incompressible liquid motion adjacent to a bubble, and it is accurate to within an order of $\dot{R}/c_0$, where c_0 is the speed of sound in the liquid. When the local hydrodynamic pressure $P(r, t)$ decreases, the bubble wall accelerates outward. For small oscillations the term quadratic in $\dot{R}$ is small, but for a critically small value of $P(r, t)$ this term, which is always positive, will dominate the linear acceleration term and control the bubble growth.

Several useful alternative forms of Eq. (6.3) may be derived to describe the bubble dynamics, for example, Keller and Miksis [26], Prosperetti and Lezzi [27], Commander and Prosperetti [16]. However, keeping with the line of development begun above we present the oscillations of bubble volume in terms of an applied perturbation pressure that is superimposed on a static equilibrium pressure. The response of the bubble volume can be found by substitution of

$$\dot{R}R^2=\frac{\dot{V}}{4\pi}$$

where $\dot{V}$ is the volume velocity of the bubble into Eq. (6.3), which gives the alternative relationship

$$\frac{\rho_0}{4\pi R}\ddot{V} - \frac{\rho_0}{2}\left(\frac{\dot{V}}{4\pi R^2}\right)^2 = P(R) - P(r,t) \tag{6.4}$$

The instantaneous volume will oscillate about its equilibrium value V_0 under the influence of the driving pressure $P(r,\ t)$ which oscillates about a static value that determines the equilibrium state of the bubble. Thus let this static pressure be P_0 so that

$$P(r,t) = P_0 + p(r,t) \tag{6.5}$$

We replace the pressure on the liquid side of the bubble wall by Eq. (6.2) and note that the equilibrium pressure of gas in the bubble of equilibrium radius R_0 is, by Eq. (6.2),

$$P_{g_0} = P_0 + 2S/R_0 - P_v \tag{6.6}$$

The static conditions are $\dot{R} = 0$ and $R = R_0$, and the static ambient pressure is $P_0(R_0) = P_0$.

Therefore we find by substitution

$$\begin{aligned}\frac{\rho_0}{4\pi R}\ddot{V} - \frac{\rho_0}{2}\left(\frac{\dot{V}}{4\pi R^2}\right)^2 + [-P_{g_0}] \times \left(\frac{V_0}{V}\right)^\gamma - \left(P_v - \frac{2S}{R} - \frac{4\mu_0\dot{R}}{R} - P_0\right) \\ = P_0 - P(r,t) \\ = -p(r,t)\end{aligned} \tag{6.7}$$

which is an alternative form to both Eqs. (6.3) and (6.4). For small oscillations, we use the first term of a Taylor series

$$P_{g_0}(R)\left(\frac{V_0}{V}\right)^\gamma - P_{g_0}(R) \simeq -\frac{\gamma}{V_0}P_{g_0}(R)(V - V_0) \tag{6.8}$$

and assume that the bubbles are large enough that the changes in surface-tension contribution can be ignored. Eq. (6.3) now reduces to the linearized form derived by Strasberg [28]

$$\frac{\rho_0}{4\pi R_0}\ddot{V} + \frac{\mu}{\pi R^3}\dot{V} + \frac{\gamma P_{g_0}}{V_0}(V - V_0) = -p(r,t) \tag{6.9}$$

This equation describes a simple harmonic oscillation that we will pursue below.

To determine the resonance frequency of the bubble, assume free simple harmonic motion at a frequency ω_0 such that the volume fluctuations are given by

$$V - V_0 = v e^{-i\omega_0 t}$$

The frequency of free motion (neglecting the damping term) satisfies

$$-\omega_0^2\left(\frac{\rho_0}{4\pi R_0}\right)+\left(\frac{\gamma P_{g0}}{V_0}\right)=0 \tag{6.10}$$

Utilizing Eq. (6.6) the equilibrium gas pressure can be replaced by the components involving the equilibrium static pressure and the surface tension so that the resonance frequency is

$$(\omega_0 R_0)^2=(3\gamma/\rho_{0^2})(P_0+2S/R_0-P_v) \tag{6.11}$$

At atmospheric pressure with large enough bubbles and with $\gamma=1.4$ for adiabatic (also isentropic) harmonic bubble motion, Eq. (6.11) yields

$$f_0 R_0=330\ \mathrm{cm/s} \tag{6.12}$$

This result was first derived by Minnaert [29]; it is now found in nearly all subsequent analyses of linear spherical bubble motion. Resonance frequencies of the fundamental breathing modes of nonspherical bubbles also closely follow Eq. (6.11) as shown by Strasberg [21].

In Eq. (6.9) the term multiplying the volumetric acceleration represents the added mass of the contiguous liquid so that the first term is the inertially controlled motion. The second term represents damping on the bubble motion. The third term represents the compressibility of gas inside the bubble, which dominates the motion for pressure oscillations that have a frequency much less than the resonance frequency. At resonance, the bubble motions are controlled by dissipation, which may be included in its more general context by introducing an ad hoc loss factor, call it η_T, into Eq. (6.9), so that the linear single degree of freedom oscillations are given by

$$\left(\frac{\rho_0}{4\pi R_0}\right)\left[\ddot{v}+\eta_T\omega_0\dot{v}+\omega_0^2 v\right]=-p(r,t) \tag{6.13}$$

where $v=V-V_0$ is the volumetric fluctuation. This equation has been derived by Devin [30] and used by Strasberg [28] in this form to describe transients of bubble formation and by Whitfield and Howe [31] in description of a bubble moving through the pressure field of a nozzle as well as others interested in the acoustic attenuation in bubble warms, e.g., Ref. [16] and Chapter 3. The damping of bubbles at high frequencies (10 kHz or greater) has been given substantial attention; an extensive review of that work has been given by Flynn [32].

The total loss factor η_T is the sum of the contributions,

$$\eta_T=\eta_{rad}+\eta_{th}+\eta_{vis}$$

where η_{rad} is the radiation loss factor, η_{th} is the thermal loss factor, and η_{vis} is the viscous loss factor. These are derived in various forms in the references cited above; two of the most comprehensive being Commander and

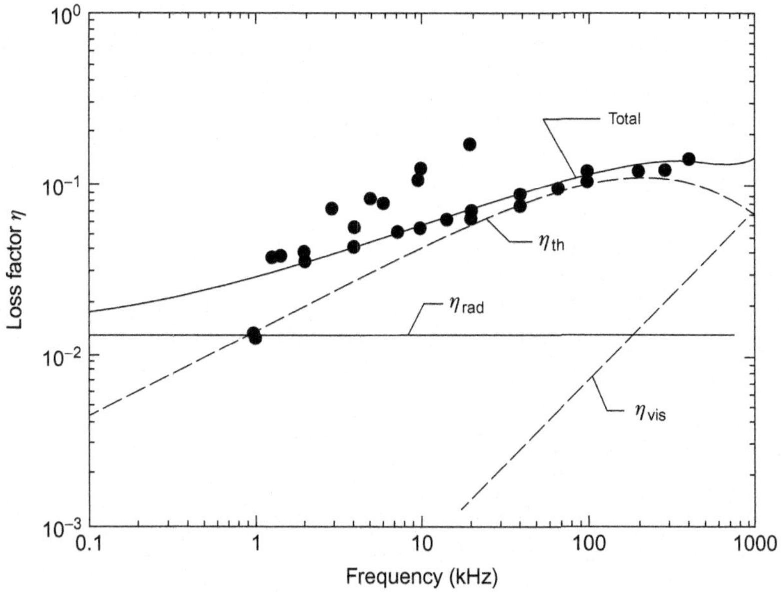

FIGURE 6.2 Loss factors of air bubbles in water at 1 atm. *From Devin C. Survey of thermal, radiation, and viscous damping of pulsating air bubbles in water. J Acoust Soc Am 1959;31:1654–67; points represent measurements from various sources.*

Prosperetti [16] and Prosperetti [33]. Fig. 6.2 shows some measured loss factors together with the individual damping contributions. The radiation, thermal, and viscous losses that determine the total damping will be summarized below.

The thermal loss factor is given by [16,33]

$$\eta_{th} = \frac{P_0}{\rho_0 (\omega R_0)^2} \operatorname{Im} \mathcal{F}\left(\frac{i}{D}\right)$$

where

$$\mathcal{F}\left(\frac{i}{D}\right) = \frac{3\gamma\zeta^2}{\zeta[\zeta + 3(\gamma - 1)A_-] - 3i(\gamma - 1)(\zeta A_+ - 2)}$$

$$A_\pm = \frac{\sinh\zeta \pm \sin\zeta}{\cos\zeta - \cosh\zeta}$$

$$\zeta = \sqrt{2/D} = R_0\sqrt{\left(\frac{2\omega}{\chi}\right)}$$

$$\chi = \frac{K_g}{\rho_{g_0} C_p}$$

and K_g is the thermal conductivity of the gas (5.6×10^{-5} cal/(cm sec deg C)), C_{p_g} is the specific heat at constant pressure of the gas ($\simeq 0.24$ cal/(g deg C)), and $\gamma = 1.4$ is the adiabatic constant.

Prosperetti [33] provides asymptotes

$$\operatorname{Im} \mathcal{F}\left(\frac{i}{D}\right) = \frac{\gamma - 1}{10\gamma}\zeta^2, \quad \zeta < 4$$

$$\operatorname{Im} \mathcal{F}\left(\frac{i}{D}\right) = 9\gamma\frac{\gamma - 1}{\zeta}1 - 2/\zeta + O\left[\zeta^{-3}\right], \quad \zeta > 10$$

Between these two asymptotic extremes, for $4 < \zeta < 10$, there is a smooth transition. When D is large (i.e., at lower frequencies) Prosperetti [34,35] finds

$$\eta_{\text{th}} = \left(\frac{\gamma - 1}{\gamma}\right)\frac{P_0 \rho_{g_0} C_p}{5\omega K_g \rho_0}$$

At atmospheric pressure, Devin finds

$$\eta_{\text{th}} \simeq 4.4 \times 10^{-4} (\omega_0 / 2\pi)^{1/2} \tag{6.14}$$

Qin et al. [36] have demonstrated dependence of calculated rebounding of nonlinear bubble rebounds after collapse on damping, finding that thermal damping has the largest control over the rebounding magnitude.

The radiation loss factor can be found by expanding Eqs. (2.21) and (2.23) to find the pressure on the bubble surface for small $k_0 R$:

$$p(R_0, t) \simeq \frac{-i\omega\rho_0 \dot{v}}{4\pi R_0}(1 - ik_0 R_0)e^{-i\omega t} \tag{6.15}$$

The first term is the familiar inertial pressure; the second term is dissipative, and it represents an acoustic resistance to the bubble wall motion by the liquid. Now the pressure given by Eq. (6.3) may be compared with Eq. (6.13) to disclose a radiation loss factor, which is

$$\eta_{\text{rad}} = k_0 R_0 = \frac{\omega_0 R_0}{c_0} \tag{6.16}$$

It is notable that the first terms of Eqs. (6.13) and (6.15) are identical, representing the inertial liquid-phase load on the bubble. The second term in Eq. (6.15) is a second-order acoustic correction to the liquid-phase loading that is valid as $k_0 R_0 \ll 1$. The reader can thus appreciate the gradual transition from the first-order liquid-phase dynamics as an incompressible medium to the more complete first- and second-order dynamics that include the acoustic importance of slight liquid compressibility embodied in finite $k_0 R_0$. Utilizing Eq. (6.11), Eq. (6.16) has the alternative forms

$$\eta_{\text{rad}} = \sqrt{3\gamma P_0 / \rho_0 c_0^2} \tag{6.17}$$

$$= \sqrt{3\rho_g c_g^2 / \rho_0 c_0^2} \tag{6.18}$$

showing that the radiation loss factor is independent of frequency as illustrated in Fig. 6.2. Using the ideal gas law for adiabatic volume changes $p_g \sim p^{\gamma}{}_g(\text{const})$, leads to $dP/d\rho = c_g{}^2 = \gamma P_0/\rho_g$. Note that reactive loading by the ambient fluid can affect the resonance frequency as derived by many authors, e.g., Brennan [2] or Prosperetti [11]

$$(\omega_0 R_0)^2 = (3\gamma/\rho_{0^2})(P_0 + 2S/R_0 - P_v - 2S/(\rho_0 R_0) \tag{6.19}$$

The frequency spectrum of linear bubble motions can be determined from Eq. (6.13) by Fourier transformation as in Chapter 5, Fundamentals of Flow-Induced Vibration and Noise

$$\frac{\rho_0 V(\omega)}{4\pi R_0}(\omega_0^2 - \omega^2 - i\eta_T \omega\omega_0) = -p(r, \omega)$$

where $V(\omega)$ is the Fourier transform of the volume pulsation. The spectral density of volume fluctuation can be found in terms of the spectral density of the excess pressure

$$\Phi_{vv}(\omega) = \left(\frac{4\pi R_0}{\rho_0}\right)^2 \frac{\Phi_{pp}(r, \omega)}{|\omega_0^2 - \omega^2 - i\eta_T \omega\omega_0|^2} \tag{6.20}$$

The linearized bubble response for low-amplitude excitation pressure therefore resembles a single-degree-of-freedom simple harmonic oscillator.

6.1.2 Sound Propagation in Bubbly Liquids

This theory of linear bubble motions has been used to describe the steady-state propagation and absorption of sound waves in bubbly mixtures. The bubbles increase both the compressibility and the acoustic absorption of the two-phase fluid. We shall pursue a simplified mixture theory to describe these properties. In so doing we will assume that the sizes of the bubbles are much smaller than an acoustic wavelength, that they do not interact, and that the bubbles are homogeneously dispersed throughout the liquid phase. The concentration of gas, in terms of the volume of gas per volume of liquid, will be designated as β so that the density of the mixture is given by

$$\rho_m = \rho_g \beta + \rho_0(1 - \beta) \tag{6.21}$$

where ρ_g is the density of the gas phase and ρ_0 is that of the liquid phase. Since for air and water mixtures $\rho_\ell/\rho_g \simeq 800$ (at standard temperature and pressure) the density of the mixture is nearly identically ρ_0.

The speed of sound in the mixture at an angular frequency ω is related to a complex wave number k_m through a complex wave speed, c_m;

$$k_m = \omega/c_m = (k_m)_r + i(k_m)_i \tag{6.22}$$

The propagation characteristics of acoustic pressure are given essentially by

$$\begin{aligned} p &= p_0 e^{i(k_\mathrm{m} r - \omega t)} \\ &= p_0 e^{i((k_\mathrm{m})_r r - \omega t)} e^{-(k_\mathrm{m})_i r} \end{aligned} \tag{6.23}$$

where r is a distance referring to some origin inside the mixture and $k_{\mathrm{m}i}$ gives rise to attenuation of the sound pressure. In order to determine the wave speed in the mixture, we calculate a resultant compressibility of the liquid–gas mixture. To this end note that the total volume reduction, δV, to a region of the mixture resulting from a pressure disturbance δp is the sum of the individual compressions of the liquid and gas phases, δV_ℓ and δV_g, respectively,

$$\delta V = \delta V_\mathrm{l} + \delta V_\mathrm{g} \tag{6.24}$$

The size of this mixture need only include a uniform distribution of bubbles. In turn, δV_g is the total gas compression, which for the ith component bubble in the mixture can be written down by using Eq. (6.13),

$$(\delta V_\mathrm{g})_i = \frac{(\delta P) 4\pi R_i / \rho_0}{\omega^2 + i\eta_\mathrm{T} \omega\omega_0 - \omega_0^2}$$

where ω_0 is the resonance frequency of the bubble of radius R_i so that the fractional volume change is an integral over the entire distribution of bubble radii,

$$\frac{\delta V_\mathrm{g}}{V} = \frac{-1}{\rho_0} \frac{\delta P}{\omega^2} \int_0^\infty \frac{4\pi R n(R)\, dR}{(\omega_0/\omega)^2 - 1 - i\eta_\mathrm{T}(\omega_0/\omega)} \tag{6.25}$$

The integrand contains the distribution of radii $n(R)$ in the form of the number of the bubbles of radius R per unit volume of liquid in an incremental range of radii. The total volumetric concentration of gas suspended in bubble form (not dissolved) in the liquid is just

$$\beta = \int_0^\infty \frac{4\pi}{3} R^3 n(R)\, dR = \int_0^\infty \frac{d\beta}{dR} dR \tag{6.26}$$

The compressibility of the mixture is given by

$$\frac{\delta V}{V} = \frac{-\delta P}{\rho_\mathrm{m} c_\mathrm{m}^2} \tag{6.27}$$

From Eq. (6.24),

$$\frac{\delta V}{V} = \frac{-\delta P}{\rho_0 c_0^2} + \frac{\delta V_\mathrm{g}}{V}$$

where $1/\rho_m c_m^2$ is the "compressibility" of the mixture, and c_m is the associated speed of sound in the mixture. Combination of Eqs. (6.24) through (6.27) gives

$$k_m^2 = \left(\frac{\rho_m}{\rho_0}\right)\left\{k_0^2 + \int_0^\infty \frac{4\pi Rn(R)\,dR}{(\omega_0^2/\omega^2) - 1 - i\eta_T(\omega_0/\omega)}\right\} \tag{6.28}$$

as the complex acoustic wave number in the mixture. With knowledge of a bubble size distribution this equation is easily evaluated using modern mathematics software packages. In the absence of a model for $n(R)$ several approximations can be made as shown below.

This relationship has been derived and compared with measurements by Cartensen and Foldy [37] (who also derived reflection and transmission coefficients for bubble screens) Meyer and Skudrzyk [38], Hsieh and Plesset [39], who showed that c_m is an isothermal sound speed for values of β of practical interest, and Commander and Prosperetti [16] who show many additional comparisons with measurement. Additional experimental confirmation of Eq. (6.28) can be found in Fox et al. [40] using measurements of acoustic transmission through bubble screens and by Silberman [41] using acoustic transmissions down a wave tube. The measurements are generally difficult to interpret in terms of the theory because of uncertainties in knowing the thickness of the bubble screen, bubble size distribution, and bubble damping, as the early measurements of Cartensen and Foldy [37] will attest. Commander and Prosperetti [16] provide a broad set of comparisons with many data sources finding generally good agreement with measurement for frequencies for which resonant response occurs or for bubble volume concentrations in excess of about 2%. Fig. 6.3, from the measurement program of Fox et al. [40], shows phase velocity and attenuation measurements in a bubble cloud field that was narrowly distributed between radii of 0.06 and 0.24 mm with an average of 0.12 mm. The transmission loss (TL) over a distance r is determined from the ratio of pressures at distances x and $x + r$ using Eq. (6.23); i.e.,

$$\begin{aligned} \mathrm{TL} &= 20\log|p(x+r)|/|p(x)| \\ &= 8.69(k_m)_i r \end{aligned} \tag{6.29}$$

The lines of Fig. 6.3 represent alternative theoretical estimates derived from Eq. (6.28) for a narrow distribution of bubble sizes and for a large distribution of bubble radii. We shall derive these relationships below. For a narrow-radius distribution, such that the range of radii ΔR satisfies $\Delta R/R < \eta_T$, Eq. (6.28) becomes

$$\frac{\rho_0}{\rho_m}k_m^2 = k_0^2\left[1 + \frac{c_0^2\rho_0}{c_g^2\rho_g}\beta\frac{(\omega_0/\omega)^2}{(\omega_0/\omega)^2 - 1 - in_T(\omega_0/\omega)}\right] \tag{6.30}$$

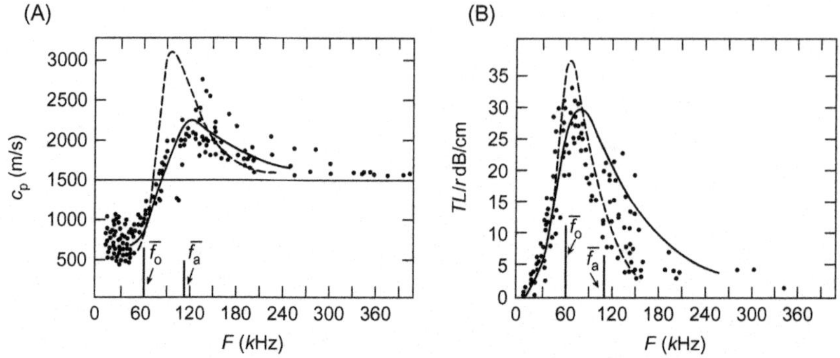

FIGURE 6.3 Phase speed and absorption in a bubbly liquid containing radii 0.06–0.25 mm diameter ($\overline{R}\simeq 0.12$ mm, $\sigma_R/\overline{R}\simeq\frac{1}{3}$). Points are result of measurements; lines computed form Eq. (6.33) using measured distribution (————) and a narrow distribution (- - - - -) with $R=0.11$ mm, $n=0.5$. $\beta=2\times10^{-4}$ cm^3/cm^3 for both. (A) Phase speed; (B) Absorption. *From Fox FE, Curley SR, Larson GS. Phase velocity and absorption measurements in water containing air bubbles. J Acoust Soc Am 1955;27:534–9.*

or

$$\frac{\rho_0}{\rho_m}\frac{c_0^2}{c_m^2}=\left[1+\frac{c_0^2\rho_0}{c_g^2\rho_g}\beta\frac{(\omega_0/\omega)^2}{(\omega_0/\omega)^2-1-i\eta_T(\omega_0/\omega)}\right] \tag{6.31}$$

where we have replaced the static pressure in the bubble with $\gamma P_0=\rho_{g_0}c_g^2$. This replacement follows the use of the definition of the speed of sound in the gas phase, Eq. (2.5), with the equivalence between ωR_0 and $\gamma P_0/\rho_0$ given by Eq. (6.11) and continuing to assume negligible effects of surface tension. For such a distribution, all bubbles are resonant at frequency ω_0, and so all participate equally in the dynamics of the medium. This situation is illustrated schematically in Fig. 6.4A. We can see that an assumption of a narrow range of bubble radii does not provide a good match with the data.

However, an alternative relationship may be derived when the bubble radii extend over a broad range, ΔR, such that there is at any frequency of excitation a broad population of vibrating bubbles of which only some motions are resonant but others are stiffness- or mass-controlled. If we let the bubble sizes be distributed about an average radius $\overline{R}$ with an associated resonance frequency $\overline{\omega}_0$, we can denote the bubble distribution as a function of the differential radius

$$n(R)=n(R-\overline{R})$$

In Eq. (6.28) the integration over R includes the variable resonance frequency that is a function of radius through

$$\omega_0 R=\sqrt{3\rho_g c_g^2/\rho_0}\equiv c$$

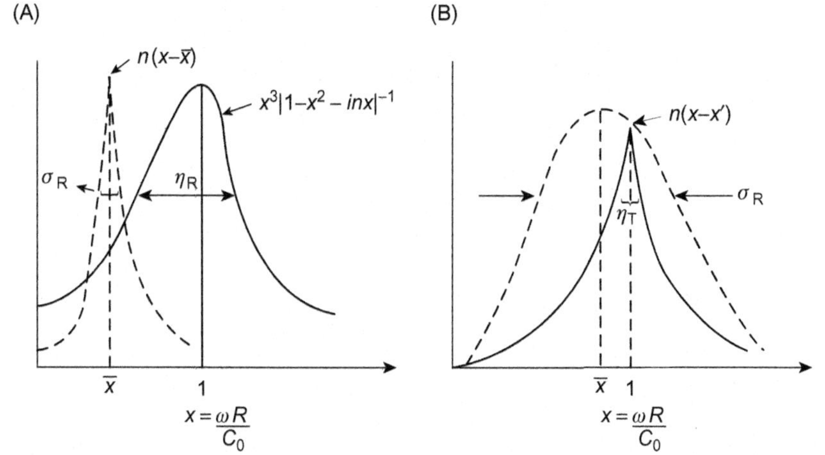

FIGURE 6.4 Illustrations of two extremes in the relationships between bubble distribution and admittance functions. (A) Narrow distribution of bubbles—see Eq. (6.30) $\sigma_R/R < \eta_T$; (B) Broad distribution of bubbles, Eqs. (6.33) and (6.34) $\sigma_R/R < \eta_T$.

For convenience therefore we normalize the radius on the fixed frequency ω and the equivalent speed c so that we may write (with $x = \omega R/c$ and $\overline{x} = \omega\overline{R}/c = \omega/\overline{\omega}_0$)

$$n(R) = n\left(\frac{\omega R}{c} - \frac{\omega\overline{R}}{c}\right)\frac{\omega}{c} = n(x - \overline{x})\frac{\omega}{c}$$

so that

$$n(R)\,dR = n(x - \overline{x})\,dx$$

Eq. (6.28) becomes accordingly

$$\frac{\rho_0}{\rho_m}\frac{k_m^2}{k_0^2} = 1 + \left\{\frac{\rho_0 c_0^2}{\rho_g c_g^2}\frac{4}{3}\pi\overline{R}^3\left(\frac{\overline{\omega}_0}{\omega}\right)^3\int_0^\infty \frac{x^3 n(x - \overline{x})\,dx}{(1 - x^2) - i\eta_T x}\right\} \tag{6.32}$$

When the bubble radii are widely distributed so that the width of the peak of $n(x - \overline{x})$ is broader than the admittance peak near $x = 1$ as illustrated in Fig. 6.4B, it is still possible to "decouple" the function $n(R)$ from the admittance function of resonant bubble motion in the mixture. A measure of the breadth of such a bubble distribution is a standard deviation of radii that is much larger than the resonance band expressed in terms of the damping and the mean radius, i.e., it is required that $\sigma_R \gg \overline{R}\eta_T$. A statistical approximation that is really the same as that used in Section 5.3.2 for modeling multimode vibration of structures is obtained from Eq. (6.31) as

$$\frac{\rho_0}{\rho_m}\frac{k_m^2}{k_0^2} \simeq 1 + \left\{\left(\frac{\rho_0 c_0^2}{\rho_g c_g^2}\right)\frac{4}{3}\pi\overline{R}^3\left(\frac{\overline{\omega}_0}{\omega}\right)^3 n(1 - \overline{x})\int_0^\infty \frac{x^3[(1 - x^2) + i\eta_T x]}{(1-x^2)^2 + \eta_T^2 x^2}\,dx\right\} \tag{6.33}$$

for $\omega \approx \overline{\omega}_0$ where the bubble distribution function is regarded as smooth enough to be separated from the resonance characteristics of the bubble admittance. In the integral, the real part of the admittance function involving $1-\omega/\overline{\omega}_0$ passes through zero at $\omega=\overline{\omega}_0(x=1)$, and it also has oppositely signed peaks of equal magnitude slightly above and below ω_0. For such a broadband distribution $n(x-\overline{x})$ the real part of this equation gives a negligible contribution. The imaginary part, however, has a single peak at $x=1$ with an effective value $\pi/2$. Accordingly Eq. (6.33) simplifies to

$$\frac{1}{1-\beta}\frac{\rho_0}{\rho_{\mathrm{m}}}\frac{k_{\mathrm{m}}^2}{k_0^2} \simeq 1 + i\frac{\pi}{2}\left(\frac{\rho_0 c_0^2}{\rho_{\mathrm{g}} c_{\mathrm{g}}^2}\right)\frac{d\beta}{dR}\frac{\overline{R}\beta}{1-\beta} \tag{6.34}$$

This expression is physically valid whenever sound in bandwidth $\Delta\omega$ is transmitted into a bubbly medium that contains bubbles resonant in the frequency band of the sound, i.e., whenever $\Delta\omega$ contains resonances ω_0.

These equations, see also Refs. [42–46], accordingly provide a useful means of determining $d\beta/dR$ from propagation characteristics. In the simplest case a bubble swarm distribution of radii there is a corresponding resonance frequency and bubble size for which the sound is absorbed, i.e., the TL is high. The behavior of the mixtures at these and other frequencies is summarized in Table 6.1 for both narrow and broad distribution of bubble radii and without any reference to a particular bubble distribution.

Below the critical frequencies $\omega=\overline{\omega}_0$, and $\omega=\omega_{\mathrm{a}}$, all mixtures behave similarly since bubble resonances are not relevant here, and the TL is reduced. For $\omega<\overline{\omega}_0$ medium is controlled by the total stiffness of the suspended gas, while at well above resonances at very high frequencies, such that $\omega>\overline{\omega}_0$, the bubbles are dynamically stiff and oscillate as small rigid spheres in response to the sound. Since their radii are much less than an acoustic wavelength, the propagation approaches that of the liquid medium. At the resonance frequency, where absorption is large, the propagation velocity is relatively low, and in bubble distributions the standard deviation of radius replaces the damping. At the frequency ω_{a} (called "anti-resonance" and discussed in greater detail by Junger and Cole [47]) the wave speed increases. For narrowly distributed radii, the speed of sound increases as damping decreases, but the increase would be somewhat reduced as the distribution is made broader. The relationships in Table 6.1 agree with measured propagation characteristics closely. The calculations made by Fox et al. [40] using a value for the loss factor of 0.5, which is now considered excessive in light of the more recent measurements of damping factor (as illustrated in Fig. 6.2), are shown in Fig. 6.2. Eq. (6.34) cannot be used to calculate the propagation at $\omega_{\mathrm{a}}/\overline{\omega}_0$ because at this frequency the approximations leading to Eq. (6.34) are not valid. Note also that near $\overline{\omega}_0$ and ω_{a} the analog of the loss factor in the distributed medium replaces $1/\eta_{\mathrm{T}}$ by $(\pi/2)(R/\beta)(d\beta/dR)$, whence the expression for c_{m} at $\omega=\omega_{\mathrm{a}}$ in the second column of Table 6.1.

TABLE 6.1 Asymptotic Ranges for Propagation in Bubbly Mixtures[a]

	Speed of Sound, $c_m/c_0 = k_0/(k_m)_r$		Transmission Loss/$(8.69k_0r) = (k_m)_i/k_0$	
Frequency	Narrow-radius distribution ($\eta_T \gg \sigma_R/\overline{R}$)	Broad-radius distribution ($\eta_T \ll \sigma_R/\overline{R}$)	$\eta_T < \sigma_R/\overline{R}$	$\eta_T < \sigma_R/\overline{R}$
$\omega < \omega_0$	$\left[1+\left(\frac{\beta\rho_0 c_0^2}{\rho_g c_g^2}\right)\right]^{-1/2}$	Same	$\frac{\overline{\eta}}{2}\frac{\omega}{\omega_0}\left(\frac{\beta\rho_0 c_0^2}{\rho_g c_g^2}\right)$	Same
$\omega = \omega_0$	$\left[\frac{2\overline{\eta}_T}{\beta}\frac{\rho_g c_g^2}{\rho_0 c_0^2}\right]^{1/2}$	$\left[\frac{\pi}{4}\overline{R}\frac{d\beta}{dR}\frac{\rho_0 c_0^2}{\rho_g c_g^2}\right]^{-1/2}$	$\frac{(k_m)_i}{k_0} = \frac{c_0}{c_m}$	$\frac{(k_m)_i}{k_0} = \frac{c_0}{c_m}$
$\omega = \omega_a > \omega_0$	$\left[\frac{2}{\overline{\eta}_T}\left(\beta\frac{\rho_0 c_0^2}{\rho_g c_g^2}\right)^{1/2}\right]^{1/2}$	$\simeq\left[\pi\frac{\overline{R}}{\beta}\frac{d\beta}{dR}\left(\beta\frac{\rho_0 c_0^2}{\rho_g c_g^2}\right)^{1/2}\right]^{1/2}$	$\frac{(k_m)_i}{k_0} = \frac{c_0}{c_m}$	$\frac{(k_m)_i}{k_0} = \frac{c_0}{c_m}$
$\omega > \omega = \omega_a > \omega_0$	$\left[1-\beta\frac{\rho_0 c_0^2}{\rho_g c_g^2}\left(\frac{\overline{\omega}_0}{\omega}\right)^2\right]^{-1/2}$	Same	$\left[\frac{\overline{\eta}_T}{2}\left(\frac{\overline{\omega}_0}{\omega}\right)^3\left(\frac{\beta\rho_0 c_0^2}{\rho_g c_g^2}\right)\right]$	Same

[a] *Values at resonance* $\omega = \omega_0$ *are predicated on* $\beta > \eta_T\rho_g c_g^2/(\rho_0 c_0^2)$ *or* $R\, d\beta/dR > \rho_g c_g^2/(\rho_0 c_0^2)$ *and represent reasonably significant concentrations.* σ_R = *standard deviation of radius population,* $\omega_a/\omega_0 \approx [\beta/(1-\beta)\cdot\rho_0 c_0^2/\rho_g c_g^2]^{1/2}$ *when* $\sigma_R/\overline{R} \le 1$, $\overline{R}$ = *average radius, and* β = *volumetric concentration of gas.*

TLs in bubbly media of relatively low gas concentration have been used with good success to determine bubble populations [16,42–46,48]. In these cases either of two expressions may be used. For a narrow range of bubble sizes, such that $\sigma_R/\overline{R} < \eta_T$ the TL depends on the value of the bubble damping: for $\beta/\eta_T < (\rho_g c_g^2)/\rho_0 c_0^2$

$$\frac{\mathrm{TL}}{r} = 8.69 \frac{\beta}{2} \frac{\omega \rho_0 c_0}{\eta_T \rho_g c_g^2} \tag{6.35}$$

with $\eta_T \simeq 0.1$. For a broad bubble distribution, the transmission only depends on the concentration gradient $d\beta/dR$ rather than on the loss factor explicitly so that for $R\, d\beta/dR < (\rho_g C_g^2)/(\rho_0 c_0^2)$

$$\frac{\mathrm{TL}}{r} = 8.69 \frac{\sqrt{3}\pi}{4} \frac{d\beta}{dR} \sqrt{\frac{\rho_0 c_0^2}{\rho_g c_g^2}} \tag{6.36}$$

for $\sigma_R/\overline{R} > \eta_T$.

The acoustics of bubbly mixtures is potentially important in modifying sound propagation near the ocean surface, e.g., Refs. [43,45,46]. It is also important in much underwater acoustic testing and experimentation such as is done in many test facilities. In water tunnels without resorbers, for example, after continued operation with a cavitating body, the free gas content can increase appreciably. This class of application will be discussed in Chapter 7, Hydrodynamically Induced Cavitation and Bubble Noise. At lower frequencies than the bubble resonance, relatively small volumetric concentrations of gas can appreciably reduce the sound speed in the fluid. One consequence of this reduction will be to increase the local Mach number of the flow around moving bodies that are immersed in the two-phase fluid. It is quite possible that surface motions that are subsonic in pure water could accordingly become supersonic so that the wave-bearing qualities of the fluid will be altered relative to the characteristic length and velocity of the test body. Shock waves could also be formed, which would make it necessary to bring into analysis thermodynamic properties of the fluid.

These relationships have been used by Zhang et al. [49] to calculate the complex propagation mode numbers in a two-phase channel flow with suspended carbon dioxide bubbles. The channel of dimensions $0.15 \times 0.15 \times 2$ m long had a bubbly mixture flow with a controllable swarm of CO_2 bubbles with radii ranging from about 0.15 to 0.55 mm and volumetric void content, β, ranging from about 0.006 to 0.024. Figs. 6.5 and 6.6 show comparisons of measured wave numbers compared with calculations made with Eq. (6.30). Given the uncertainties in knowing the precise values of bubble distributions the calculations assumed either a range of concentrations at fixed radius or a range of radii at fixed concentration. The attenuation coefficient in these figures is the TL/z defined with Eq. (6.29), where z is the

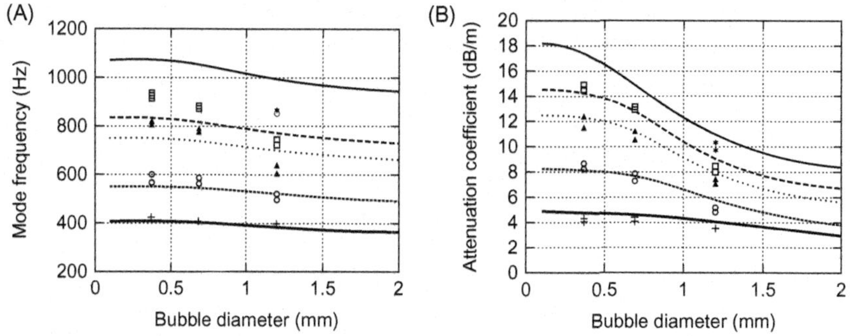

FIGURE 6.5 Mode frequency and attenuation coefficient for duct cross modes calculated assuming a volumetric void fraction, $\beta = 0.0115$ and a range of bubble radii. The table in Fig. 6.6 shows the legend for both Figures 6.5 A and B and 6.6 A and B.

Mode	1.0	1.1	2.0	1.2	2.2	3.0	3.2
Model	▬▬▬	--------	· · · ·	▬ ▬ ▬	— —	—	··········
Data	+	O	▲	□	*	×	▼

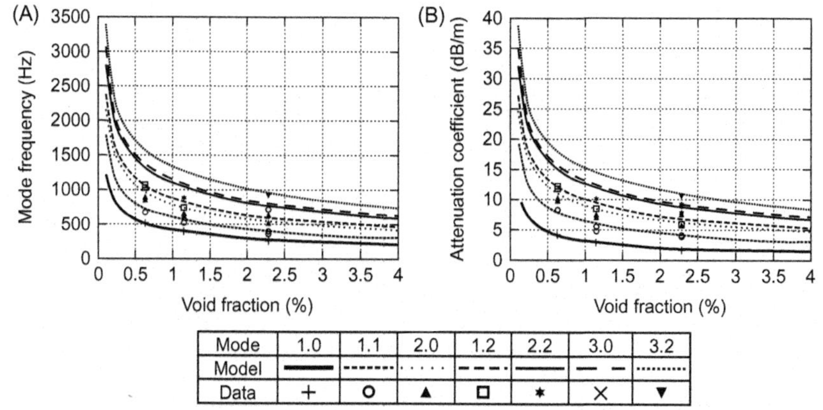

FIGURE 6.6 Mode frequency and attenuation coefficient for duct cross modes calculated assuming a bubble radius of 1.2 mm and a range of volumetric void fractions, β. The table in Fig. 6.6 shows the legend for both Figures 6.5 A and B and 6.6 A and B.

acoustic path between hydrophones in the channel, and the real part of the $(k_3)_i$ propagation wave number, in Eq. (2.167);

$$(k_3)_i = \operatorname{Im}\left(\sqrt{k_m^2 - k_{mn}^2} \right)$$

or

$$TL/z = 8.69k_i$$

where in Fig. 6.6 TL is in dB and z is in m.

and with cross-mode resonance frequencies given by

$$\omega_{m,n} = \operatorname{Re}\left(\sqrt{c_m^2} \right) k_{mn}$$

The propagation wave numbers for the acoustic cross modes of the channel, k_{mn}, were calculated as continuous functions of β or R_0 for each m,n mode assuming rigid channel walls with propagation boundary conditions, Eq. (2.159) as defined with Eq. (2.167). Measurements were made for each mode using either vibration of a channel wall or wall pressure. Note that the behavior of the duct mode propagation characteristics is more sensitive to bubble concentration than to bubble size.

Some other treatments of two-phase media along these lines have been published by Plesset [50] in connection with the stability and thermodynamics of single spherical bubbles; propagation of shock waves in such liquids has been considered by van Wijngaarden [51], Whitam [52], and Benjamin [53]. Extensive analytical treatments of the continuum mechanics of bubbly fluids have been given by Zwick [54–56] and Isay and Roestel [57,58] (who consider the effects of the compressibility on the lifting characteristics of hydrofoils); a monograph on the wave motions in bubbly mixtures has been provided by Wallis [59].

6.2 THEORETICAL CAVITATION THRESHOLDS AND NONLINEAR OSCILLATIONS OF SPHERICAL BUBBLES

6.2.1 The Onset of Nonlinear Oscillations

When an harmonic perturbation pressure is of such a magnitude that the velocity-squared term in the bubble equations (Eqs. (6.3) and (6.4)) is important, the bubble motions cease to be sinuous and take on a more complicated time history. Fig. 6.7 illustrates this behavior for a variety of amplitudes of

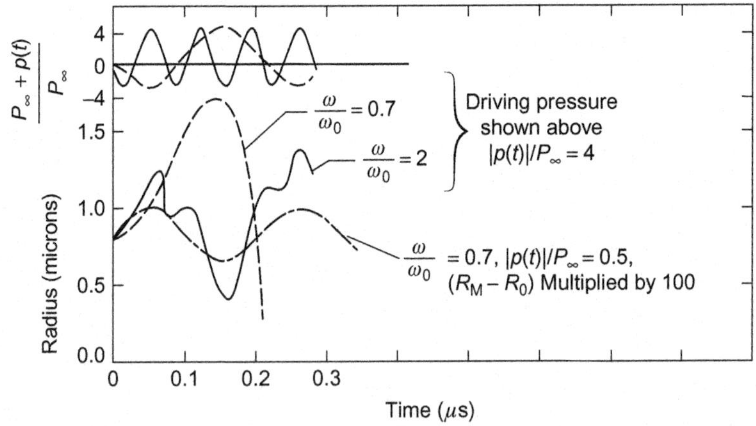

FIGURE 6.7 Radius–time curves for forced oscillations of a gas bubble, $R_0 = 0.8 \times 10^{-4}$ cm, $\omega_{res} = 4.3 \times 10^7$/s ($6.8 \times 10^6$ Hz). Curves (— — —) and (– – –) from Neppiras and Noltingk [14], curve (———) from Solomon and Plesset [60] ($\eta_T = 0$).

driving pressure and a variety of resonance frequencies of the bubbles. The figure illustrates the theoretical behavior that the linear or nonlinear motions of a bubble depend both on the amplitude of the pressure oscillation and on its frequency relative to the linear resonance frequency of the bubble. For rather large pressure oscillations, e.g., four times the local ambient equilibrium pressure P_0, simple harmonic motion of the bubble will not exist. If the frequency of oscillation is below the resonance frequency, the bubble will grow and then rapidly collapse, this behavior is characteristic of cavitation. For larger frequencies the motion will consist of a superposition of two harmonics, one at the resonance frequency and the other at the driving frequency. For small-pressure amplitudes the motion is nearly simple harmonic, as shown. The amplitude of the pressure fluctuations responsible for the cavitation is such that the pressure applied to the bubble actually becomes negative so that a tension is placed on the bubble causing the large rate of expansion necessary for cavitation. There is in fact a critical value of P_0 (say, P_{crit} that will be determined below) for which cavitation is to be expected for small reductions of pressure below this value. The implication made by the solid and dashed lines in the lower part of Fig. 6.7 for the example of $P(r, t)/P_0 = 4$ is that the exciting pressure must be applied for a time long enough to permit the necessary bubble growth and that this time must be measured in terms of the characteristic period of resonant oscillation. At frequencies that are small enough relative to the resonance frequency, an adequate criterion for determining the critical pressure can be determined by considering the static equilibrium of the bubble.

6.2.2 The Critical Pressure for Vaporous Cavitation

The critical pressure for cavitation, based on a theory of static equilibrium was first determined by Blake [61] and later extended by Strasberg [62], Akulichev [63], van der Walle [64], Prosperetti et al. [65], Boguslavskii and Korets [66], and many others reviewed by Feng and Leal [7]. Fundamentally, this condition can be written in terms of the difference in liquid static pressure at the bubble wall, $P(R)$ and the field pressure $P(r)$ which is an ambient value and a differential as expressed by Eq. (6.5). Then, using Eq. (6.2)

$$P(R) - P(r) = P_v - 2S/R - P(r) + P_{g0}(R_0/R)^{3\gamma} = \Delta P \tag{6.37}$$

Note that since static conditions ($\omega \ll \omega_0$) are being examined, the volume velocity ($\dot{V}$) and volume acceleration ($\ddot{V}$) are neglected. The static equilibrium will exist at a critical radius when for a further increase in radius the pressure difference will decrease. This condition for equilibrium is (see also van der Walle [64])

$$\left.\frac{d(\Delta P)}{dR}\right|_{R=R_{crit}} = 0$$

or

$$\frac{3\gamma P_{g_0}}{R_{crit}}\left(\frac{R_0}{R_{crit}}\right)^{3\gamma}=\frac{2S}{R_{crit}^2} \tag{6.38}$$

where R_{crit} is the critical bubble radius. To relate this condition to a corresponding ambient critical pressure, we rewrite Eq. (6.2) to find the partial pressure of gas in the bubble when it reaches its critical radius

$$P_{g_{crit}}=P_{crit}(r)+\frac{2S}{R_{crit}}-P_v \tag{6.39a}$$

Also rewriting Eq. (6.38) to replace P_{g_0},

$$P_{g_{crit}}=P_{g_0}(R_0/R_{crit})^{3\gamma}=2S/3\gamma R_{crit}$$

so the critical value of the ambient hydrodynamic pressure at $r > R$ is

$$P_{crit}(r)-P_v=-\left(\frac{3\gamma-1}{2\gamma}\right)\frac{4S}{3R_{crit}} \tag{6.39b}$$

The equilibrium field pressure for which $\Delta P=0$ is $P_0(r)=P_0$ and from either Eq. (6.2) or (6.37) we can find the corresponding P_{g_0} in terms of P_v and R_0. Substitution into Eq. (6.38) eliminates P_{g_0}. Further substitution for R_{crit} in Eq. (6.39b) gives the final result of the critical pressure required for the cavitation of a bubble with radius R_0 in an initial ambient pressure of $P_0=P_0(r)$ thus

$$\begin{aligned}P_{crit}(r)-P_v&=\frac{2S}{R_{crit}}\left[\frac{1}{3\gamma}-1\right]\\&=-\left(\frac{2S}{R_0}\right)^{3\gamma/(3\gamma-1)}\left[\frac{3\gamma-1}{3\gamma}\right]\frac{1}{(3\gamma)^{1/(3\gamma-1)}}\\&\quad\times\left[P_0(r)-P_v+\frac{2S}{R_0}\right]^{1/(1-3\gamma)}\end{aligned} \tag{6.40}$$

Eq. (6.40) shows that for cavitation to occur, i.e., for instability to exist, the critical pressure outside the bubble must become less than the vapor pressure. The relationship also shows that as R_0 decreases, this critical pressure must be more negative. In other words, the tensile strength of the liquid increases as the size of suspended bubbles decreases. The limiting tensile strength is large and negative approaching $P_{crit}=-280$ atm [32,63] and is known to be influenced by many chemical and thermodynamic factors. Furthermore when cavitation occurs, as in the example shown by the dashed line in the lower part of Fig. 6.7, any gas in the original bubble will be expanded into a volume many times larger than in the original bubble. In such cases since the partial pressure of trapped gas is then greatly exceeded

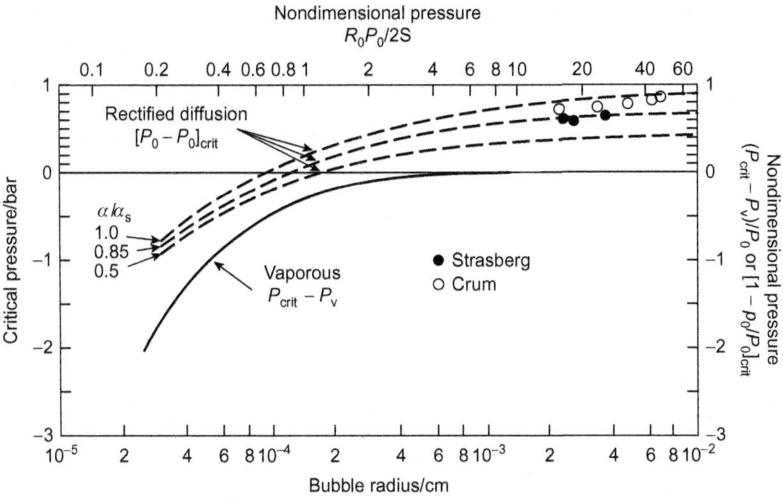

FIGURE 6.8 Critical pressure for static equilibrium at $P_0 = 1$ atm for vaporous cavitation is shown with the solid line. Critical pressures for rectified gaseous diffusion at $\alpha/\alpha_s = 0.5$, 0.85, 1.0 are shown with the dashed lines. Measurements for rectified diffusion are from Strasberg [67] and Crum [68].

by the vapor pressure of liquid, the cavitation is called vaporous. Eq. (6.40) is shown in Fig. 6.8 (taken from Strasberg [48]) for the case of isothermal expansions ($\gamma = 1$) at an initial pressure of $P_0(r) = 1$ atm; the expression becomes

$$P_{crit}(r) - P_v = -\left(\frac{2S}{R_0}\right)^{3/2} \frac{2/3}{\sqrt{3}} \left[P_0(r) - P_v + \frac{2S}{R_0}\right]^{-1/2} \tag{6.41}$$

For bubble radii of 10^{-3} cm and smaller the cavitation threshold will be at increasingly negative pressures when $P_0 = P_0(r) = 1$ atm.

A more comprehensive time-domain theory has been developed for the dynamic equilibrium of bubbles; however. It gives the thresholds of nonlinear motions and cavitation for periodic excitation pressures evaluated numerically by Solomon and Plesset [60], using Eqs. (6.3) and (6.7). Assuming a steady-state sinusoidal driving pressure of the form given by Eq. (6.5) radius–time curves similar to those in Fig. 6.4 were calculated for $\omega/\omega_r = 0.011$, 0.04, and 0.069. The forced oscillations persist after the initial resonant disturbance decays. The curve in Fig. 6.4 for $\omega = 2\omega_0$ is misleading in this regard because bubble damping had not been included in either Neppiras and Noltingk's or Solomon and Plesset's analyses. If bubble damping had been included, the resonant motion would have persisted over roughly $1/(\pi\eta_T)$ natural periods. The ambient pressure is given by

$$P(r, t) = P_0 + p_0 \sin(\omega t + \pi) \tag{6.42}$$

As the amplitude of the excitation pressure increases so that the critical pressure is reached, the damped transient motion is replaced by the unstable transient of cavitation.

In the instances of large-amplitude motion such that the $\dot{r}$ term dominates the $\ddot{r}$ term in Eq. (6.3), the bubble radius linearly increases with time, as shown by the dashed line in Fig. 6.7 for t less than 0.1 μs. In this case Eq. (6.3) suggests that

$$R \simeq t\sqrt{\frac{2}{3}}\sqrt{\frac{P_v - (P_0 - p_0)}{\rho_0}}, \quad \omega t < \pi \tag{6.43}$$

i.e., the bubble radius will be proportional to the square root of the difference between the vapor and external pressure when the latter is smaller. The limit $\omega t < \pi$ determines the length of the time that the pressure fluctuation is negative, i.e., that the pressure difference $P_v - P_0 - p_0 \sin(\omega t + \pi)$ is positive. The maximum radius will then be determined by the time $t = \pi\omega^{-1}$ so that

$$R_M \simeq 2.6\omega^{-1}\sqrt{\frac{P_v - (P_0 - p_0)}{\rho_0}} \tag{6.44}$$

Eq. (6.44) demonstrates that the maximum bubble radius is independent of the initial radius, a result first determined analytically by Neppiras and Noltingk [15]. Linear dependence of the bubble radius with time has been observed for motions in hydrodynamic cavitation (see, e.g., Arndt and Ippen [69]).

Fig. 6.9 summarizes the results of calculations using the complete bubble equation and Eq. (6.42) made by Neppiras and Noltingk [15] and Solomon and Plesset [60] in a form that is consistent with the above analyses. The solid lines represent the steady-state first-order linear bubble amplitude given by Eq. (6.20), rewritten

$$\frac{(R_M - R_0)}{\sqrt{p_0/\rho_0}} = \frac{\omega}{\omega_0}\left[\frac{\sqrt{p_0/3P_0}}{1 - (\omega/\omega_0)^2 - i\eta_T\omega/\omega_0}\right] \tag{6.45}$$

and the contrasting transient nonlinear cavitation amplitude, which is (see Eq. (6.44))

$$\frac{\omega(R_M - R_0)}{\sqrt{p_0/\rho_0}} \simeq 2.6\sqrt{\frac{p_0 - P_0}{p_0}} \tag{6.46}$$

Recall that the condition of static equilibrium, $\omega/\omega_0 = 0$, would be given by Fig. 6.7 and neglecting surface tension would indicate that for $A \geq 1$ explosive growth would occur on the negative portions of the sinusoidal pressure oscillation of Eq. (6.42). Then a critical oscillation pressure, of amplitude $(p_0)_{crit}$ at ambient P_0, is given by Eq. (6.41), letting

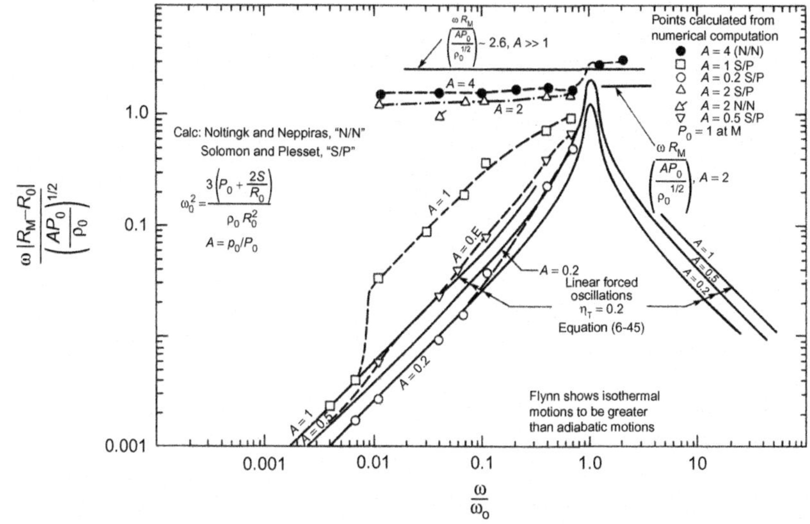

FIGURE 6.9 Amplification factor for sinusoidally excited bubbles shown as a function of frequency, ω/ω_0, for different pressure amplitudes, $AP_0 = |p| = p_0$.

$P_{crit}(r) = P_0 - (p_0)_{crit}$. The calculated points, in Fig. 6.9 connected by dashed lines, are normalized on the ambient P_0 and indicate conditions of dynamical stability. For fractional values of p_0/P_0 at $P_0 = 1$ atm the minimum pressure is greater than the vapor pressure of water at any instant in the pressure cycle. For pressure amplitudes $A > 1$ such that instantaneously $P_v - P_0 + p_0 > 0$ the limiting values of Eq. (6.46) roughly apply. On the other hand for small ratios of $p_0/P_0 = A \ll 1$, so that $P_v - P_0 + p_0 < 0$, the admittance Eq. (6.45) closely agrees with the more exact numerical calculations. For $p_0/P_0 = A \leq 1$ it is seen that the threshold between linear and nonlinear motions and the degree of departure from Eq. (6.44) depends on frequency: the smaller the value of $P_0/P_0 = A$, the higher the frequency. For oscillations above the resonance frequency, only one point has been calculated, and although the motion appears not to be cavitation-like (Fig. 6.7), the computed amplitude exceeds that which would be estimated using the linear theory by about a factor of 3. Cavitation thresholds in oscillating pressure fields will be independent of frequency and given by Eqs. (6.40) and (6.41) as long as the frequency of oscillation is much less than the bubble resonance [66]. At high frequencies, $\omega > \omega_0$, the inertial terms of Eqs. (6.3) and (6.7) become dominant, and the threshold pressure is given by the requirement that the pressure around the bubble must be less than the vapor pressure. However, as shown by Eq. (6.2), as the frequency increases the pressure difference that is necessary to generate a bubble of a given size will also increase. An analysis by Guth [70] has indicated that large-amplitude nonlinear growth does not occur for negative relative pressures, $A > 1$, when $\omega > 1.6\omega_0$.

6.2.3 The Importance of Diffusion

Diffusion can be important in both ultrasonically and hydrodynamically induced cavitation. When the growth of a bubble is determined by the diffusion of dissolved gas into the bubble from the surrounding liquid, the cavitation is called gaseous *cavitation*, and it is not explosive as in the case of the vaporous cavitation discussed in the preceding section. Diffusion can play a part in vaporous cavitation; a bubble nucleus that is too small to grow explosively may slowly grow because of diffusion of gas into the bubble until the radius increases to the critical radius given by Eq. (6.39b) for the ambient pressure in question. Once having grown to the critical radius, it will explosively cavitate. Also, quite often some types of hydrodynamic cavitation, for example, sheets which are bound to the surface, leave a cloud of microbubbles that disappear downstream slowly because of the amount of gas captured in the collapsed bubbles. In this cavitation the maintenance of a steady sheet cavity provides a continuous liquid–gas interface across which vapor may continuously pass and downstream of which (the end of the cavity) the vapor-filled bubbles may be continuously ejected.

This is another area of bubble dynamics with a rich literature; Prosperetti [11] provides an extensive review of the most recent work on the subject of all mechanisms of diffusive growth. The essentials are well known and have dealt with four phenomena that in some way are relevant to the subject of hydrodynamic cavitation inception and collapse: a steady bubble growth or solution in a quiescent liquid that is either supersaturated (so the bubble will grow) or undersaturated (so the bubble will dissolve) [39,64,71]; a bubble that is fixed (as bonded to a surface) while being scrubbed by a moving liquid in which case it grows by "convective diffusion" [72,73]; a bubble in an otherwise quiescent liquid that is excited with an oscillating pressure (as for example generated by an acoustic transponder) in which case it may grow by a process called *rectified diffusion* [48,61,74–79]; and finally a bubble that expands in a liquid as by vaporous cavitation during which a small quantity of gas is diffused into the bubble [80]. This diffused gas may later limit the collapse radius of the bubble. These processes all rely on the fact that gas will come out of solution whenever the partial pressure of gas in the bubble, p_g, is less than the equilibrium pressure of dissolved gas in the solution.

What is relevant to hydrodynamic cavitation and cavitation noise is that the critical pressure (as we shall see, the cavitation inception pressure) is dependent on the bubble sizes in suspension in the flowing liquid. This becomes important in the generation of sound will be shown to be governed by bubble collapse and bubble collapse and rebound are known to be dependent on the partial pressure of undissolved gas in the bubble when it has expanded. As with the onset of vaporous cavitation, the onset of diffusion of gas out of solution into the bubble to generate gaseous cavitation or the growth of gas-filled bubbles involves a critical pressure. This pressure depends on both the sizes of the bubbles in suspension and the concentration of dissolved gas. For example, in steady diffusion for a fully saturated solution of gas in water at a pressure of 1 atm and relatively large bubbles in suspension, any reduction of the pressure below 1 atm will cause gas to diffuse out of solution and the bubble to grow. Very small bubbles, on the other hand will require a reduction in ambient pressure to somewhat less than 1 atm both because of the smaller surface area across which diffusion may occur and because of layer surface tension in smaller bubbles. In their essentials these effects may be estimated in a simple calculation (see Strasberg [48]) using the relationship below. From Eq. (6.2), noting that

$$p_g = H\alpha$$

where α is the concentration and H is Henry's law constant, the critical pressure for steady diffusion may be written

$$P_{\text{crit}}(r) = P_0\left(\frac{\alpha}{\alpha_s}\right) - \frac{2S}{R}$$

where P_0 is the saturation pressure and α_s is the saturation concentration of the gas in the liquid.

In rectified diffusion, free bubbles which would normally dissolve are made to grow under the stimulation of an undulating pressure. This is partly because of unequal mass transfer across the bubble wall during oscillation of the bubble. For a particular ambient pressure, P_0, there is a critical value of undulating pressure amplitude, p_0, as in Eq. (6.42), such that a combined critical minimum pressure is $[P_0 - p_0]_{\text{crit}}$. This critical value is, approximately [67], for a bubble of radius R_0

$$\left(1 - \frac{p_0}{P_0}\right)_{\text{crit}} = 1 - \sqrt{\frac{3}{2}}\left[1 + \frac{2S}{R_0\, P_0} - \frac{\alpha}{\alpha_s}\right]^{1/2}$$

This approximation is formally valid for small values of S/R_0P_0, and $\alpha/\alpha_s \simeq 1$, and for pressure oscillation frequencies which are much less than the resonance frequency of the bubble. The relationship assumes that the bubble gas dynamics are isothermal. Fig. 6.8 illustrates this approximate relationship and compares it with the measured values of Strasberg [67] and Crum [68]. Further relationships which are valid for a broader range of α/α_s, temperature, and S/R_0P_0 have been derived by, e.g., Safar [76], Eller [77,78], Crum [68,81], and Lee and Merte [82]. These concepts will be revisited in Section 7.2.2.2 when we examine the effects of gas diffusion on cavitation inception.

6.3 THE COLLAPSE OF CAVITATION BUBBLES

6.3.1 Spherical Vapor-Filled Bubbles

It will be shown in the next section that the central issues in determining the noise from single cavitation bubbles are determined during the collapse phase. Therefore we shall now examine the physical processes that determine volume velocities in the final stages of collapse. The time histories of bubble radii shown in Fig. 6.7 show that when nonlinear bubble growth occurs (dashed line) a second state of motion, the collapse, occurs when the rarefaction is replaced by compression. Motions of this type, generated hydrodynamically by Plesset [13], are shown in Fig. 6.10 and the observed pattern of bubble history shows that the collapse stage has a short time scale. Since the wall acceleration is large during collapse, it is reasonable to conclude that the collapsing motion will contribute heavily to sound production. Therefore we shall examine this aspect of the dynamics closely to establish what the important controlling variables are at various frequency ranges in the cavitation-noise spectrum. We shall see that at the termination of the collapse phase, the motion will be influenced by the presence of any gas in the bubble and the properties of that gas. This gas becomes important when the radius of the bubble becomes very small because the compressed gas fills the bubble. Also, the compressibility of the liquid (or two-phase fluid) surrounding the bubble will become important if the wall velocity of the bubble wall becomes comparable to the speed of sound in that fluid.

The first theoretical treatment of the collapse of spherical bubbles on which much modern thinking is fundamentally based is due to Rayleigh [12]. The pressure inside the bubble was considered to be constant; therefore Rayleigh's problem would apply to the physical circumstance of vapor-filled bubbles only. The pressure difference in Eq. (6.3), letting $P_{g_0}=0$ and neglecting the surface tension pressure to give $P(R,\ t)=P_v$, is $P_v-P(r)$, where, now, $P(r)$ is supposed to be much larger than the vapor pressure so that the bubble will collapse. Rewriting the left-hand side of Eq. (6.3), we find equivalently

$$\frac{1}{2\dot{R}R^2}\frac{d}{dt}(R^3\dot{R}^2)=\frac{P_v-P(r)}{\rho_0} \tag{6.47}$$

which may be rearranged to

$$\frac{d}{dt}\left(R^3\dot{R}^2\right)=[P_v-P(r)]\frac{d}{dt}\left(\frac{2}{3}R^3\right) \tag{6.48}$$

under the assumption that $P_v-P(r)$ is invariant over the time scale of collapse. Further, assume the initial condition

$$\dot{R}=0 \quad \text{and} \quad R=R_M \quad \text{at} \quad t=0$$

to find the wall velocity squared

$$(\dot{R})^2=\frac{2}{3}\frac{\Delta P}{\rho_0}\left[1-\frac{R_M^3}{R^3}\right] \tag{6.49}$$

where $\Delta P=P_v-P(r)$ is the differential pressure across the wall of the bubble. Note that as the bubble radius becomes small Eq. (6.49) says that

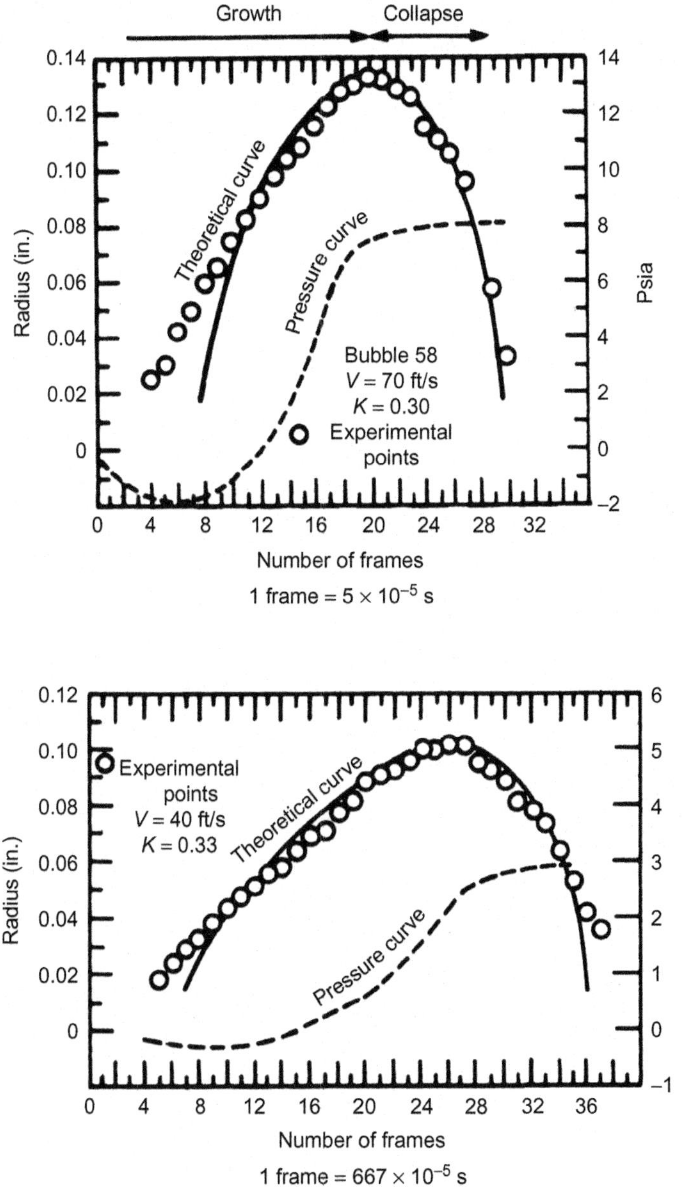

FIGURE 6.10 Measured and theoretical bubble histories for the indicated pressure distributions on a headform from Plesset [13]. © *American Society of Mechanical Engineers.*

the magnitude of $\dot{R}$ will become infinite. The velocity, however, becomes physically limited by the compression of any small amount of gas trapped in the collapsing cavity, the presence of which was ignored in this simple analysis.

Even though the wall velocity becomes infinite in the Rayleigh model of constant pressure difference, the time it takes for the bubble to collapse can be determined. The reciprocal of Eq. (6.49) gives the radius–time relationship

$$t = \sqrt{\frac{3\rho_0}{2(-\Delta P)}} \int_R^{R_M} \frac{R^{3/2}\, dR}{(R_M^3 - R^3)^{1/2}}$$

which is integrated over the interval $0 \leq R \leq R_M$ giving the time for complete collapse:

$$\tau_c = 0.915 R_M \sqrt{\rho_0/(P(r) - P_v)} \tag{6.50}$$

In spite of its simplicity, Rayleigh's equation gives an excellent representation of the gross characteristics of single-bubble cavitation dynamics. Plesset's [13] measurements of the cavitation characteristics at the nose of a body of revolution in a high-speed water tunnel, using high-speed bubble show that the presence of the internal gas is a more important motion pictures (more than 20,000 frames per second), trace the trajectory of a cavitation bubble as it passed through the region of minimum pressure on the body. Fig. 6.10 shows representative bubble histories together with the matching local hydrodynamic pressure. The solid lines represent the bubble history computed from the simple Rayleigh–Plesset equation, Eq. (6.47). To do this, $P(r)$ is considered to be the local hydrodynamic pressure in the reference frame of the bubble.

A similar investigation had been conducted earlier by Knapp and Hollander [83], who measured five rebounds of the bubble following the initial collapse, as illustrated in Fig. 6.11. Alternate rebounds are shown above and below the $R = 0$ datum in order that they may be readily examined. It is suspected that the multiple rebounds are strongly influenced by the stored energy in the bubble as result of the compression of gas in the collapse phases. The collapse radius versus time was well approximated by the Rayleigh bubble equation, as will be discussed later in connection with Fig. 6.15. For further reading, see Brennan [2].

6.3.2 Spherical Bubbles With Internal Gas

Rayleigh considered the limiting effects of noncondensable gas using a calculation of the compression work on the gas by considering the change in potential energy from the initiation of collapse to radius R is converted into the total kinetic energy of the entrained water plus the work done in compressing the gas in the bubble. Accordingly he found that a limiting radius did exist for which the velocity of the bubble wall could be retarded to $\dot{R} = 0$.

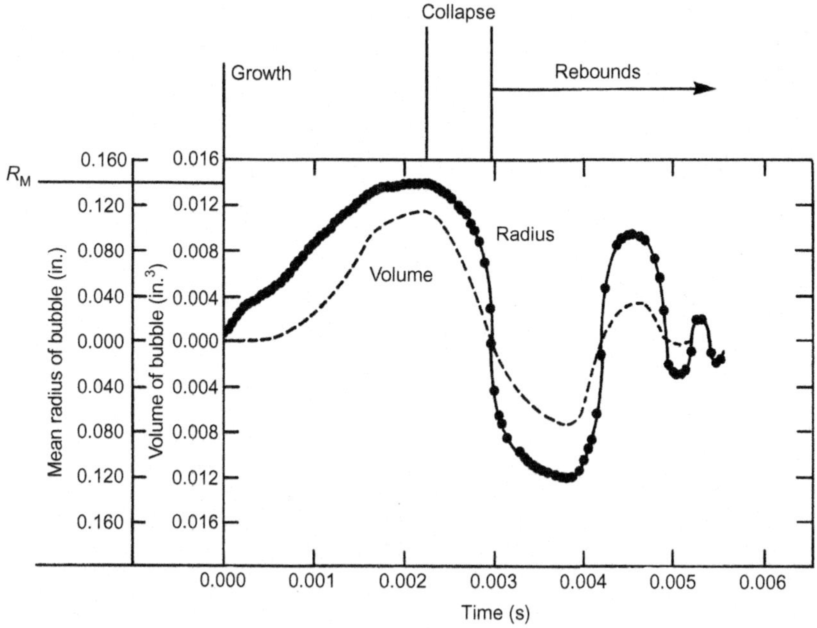

FIGURE 6.11 Observed radius history for a cavitation bubble on a 1.5-caliber ogive nose. (Note rapid collapse compared to slower growth, and multiple rebounds due to compressible gas in the bubble.) Also shown is the corresponding volume history. *From Knapp RT, Hollander A. Laboratory investigations of the mechanism of cavitation. Trans. ASME 1948;70:419–35. © American Society of Mechanical Engineers.*

Subsequent early refinements of the theory accounted for the compressibility of the liquid [84–97] to various degrees of approximation as we shall outline in the following; these efforts have been surveyed in some detail elsewhere [1,98,99]. The first most detailed calculations of bubble collapse and rebound for cases involving gas-filled and empty bubbles in compressible and incompressible liquids are probably those of Hickling [92,93], samples of which are shown in Figs. 6.12 and 6.13. The effect of included compressible gas is to reduce the wall Mach number, but all empty bubbles (i.e., those for which the internal pressure remains constant and effectively equal to the vapor pressure) have unlimited $\dot{R}$ at zero radius. The introduction of gas, even in small quantities, limits the collapse so that the thermodynamic characteristics of the gas controls the final limiting radius more than does hydrostatic pressure $P(r)$. Furthermore, calculations for an incompressible or compressible liquid surrounding the bubble show that the presence of the internal gas is a more important limitation on $\dot{R}$ than is the compressibility of the liquid phase. Fig. 6.13 shows computed values of the instantaneous liquid pressures as a function of distance from collapsing and rebounding bubbles at sequences shortly before and after collapse. The positive values of time denote times after the minimum radius occurs. In the collapse phase, maximum

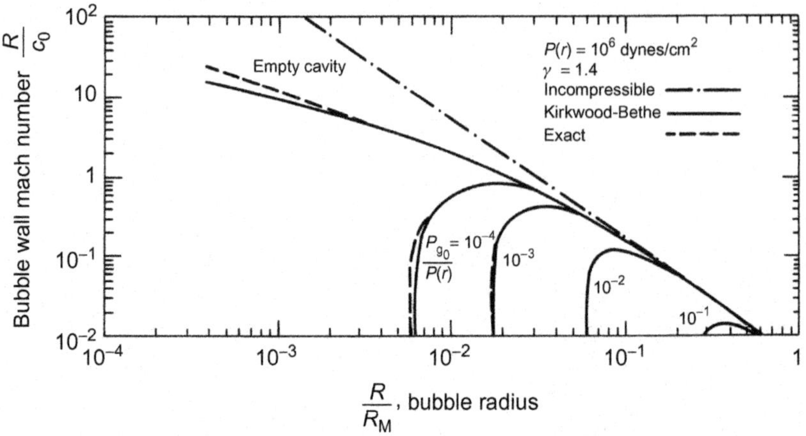

FIGURE 6.12 The bubble wall Mach number as a function of the bubble radius for decreasing gas content. The gas content is determined by its initial pressure P_0 in atmospheres. The index γ has the value 1.4 and the ambient pressure P_∞ is 1 atm. *From Hickling R. Some physical effects of cavity collapse in liquids. J Basic Eng 1966;88:229–35. © American Society of Mechanical Engineers.*

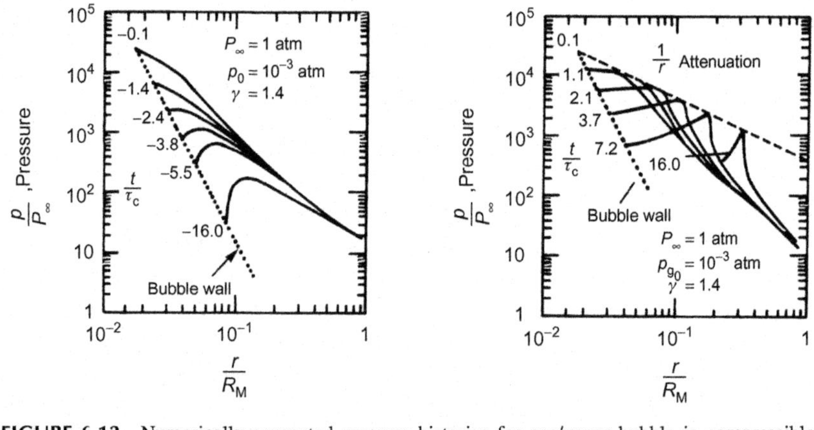

FIGURE 6.13 Numerically computed pressure histories for gas/vapor bubble in compressible water. Times t/τ_c are referred to the total collapse time τ_c, and pressures to the hydrostatic pressure $P(r)$. *From Hickling R, Plesset MA. Collapse and rebound of a spherical bubble in water. Phys Fluids 1964;7:7–14; Hickling R. Some physical effects of cavity collapse in liquids. J Basic Eng 1966;88:229–35.*

pressures occur at a distance $r \sim R$ from the bubble wall, in the rebound phase, a compression wave develops and propagates outward with $1/r$ attenuation.

The effect of compressible gas in bubbles collapsing in an incompressible liquid was examined theoretically by Neppiras and Noltingk [15] and Guth [96] and Khoroshev [100] leading to simple closed-form approximations.

The influence of included gas on the minimum radius of the collapsing bubble can be examined by modifying Eq. (6.47) to represent the gas compression as thermodynamically adiabatic as depicted in Eq. (6.37). Making the necessary substitutions and reinstating the expression for the partial pressure of included gas yet still ignoring surface tension effects, we find the modified form of Eq. (6.47) to be

$$\rho_0 \frac{d}{dt}\left(R^3\dot{R}^2\right) = [P_v - P(r)]\frac{d}{dt}\left(\frac{2}{3}R^3\right) + \frac{P_{g_0}}{2\pi}\left(\frac{V_0}{V}\right)^{\gamma}\frac{dV}{dt}$$

where the radius function on the right has been replaced by the bubble volume to simplify notation and when $\gamma \neq 1$

$$V^{-\gamma}\frac{dV}{dt} = \frac{d}{dt} = \frac{d}{dt}\left\{\frac{V^{1-\gamma}}{1-\gamma}\right\}$$

Introducing the initial conditions as before, and holding $P(r)$ constant with time, we find for the wall velocity

$$(\dot{R})^2 = \left(\frac{2}{3}\right)\left(\frac{P(r)-P_v}{\rho_0}\right)\left\{\left[\left(\frac{R_M}{R}\right)^3 - 1\right] + \frac{P_{g_0}}{(P(r)-P_v)(\gamma-1)}\left(\frac{R_M}{R}\right)^3\left[1-\left(\frac{R_M}{R}\right)^{3(\gamma-1)}\right]\right\} \tag{6.51}$$

The formal behavior of $(\dot{R})^2$ with R shows negative values when $R_M/R > \{[P(r)-P_v](\gamma-1)/P_{g_0}\}^{1/3(\gamma-1)}$ which obviously have no physical significance. A measure of a limiting minimum radius R_m/R_M may, however, be defined by the condition $\dot{R}=0$ in order to obtain a measure of dependence of R_m on P_g. In the limit as R_m/R_M also approaches zero, Eq. (6.51) reduces to the asymptotic result for the minimum radius in an incompressible liquid,

$$\frac{R_m}{R_M} \simeq \left[\frac{1}{\gamma-1}\frac{P_{g_0}}{P(r)}\right]^{1/3(\gamma-1)} \tag{6.52}$$

Finally, in the contrasting limit of isothermal gas compression, $\gamma = 1$, the equivalent form of Eq. (6.51) is identical to that of Rayleigh

$$(\dot{R})^2 \simeq \left(\frac{2}{3}\right)\left(\frac{(P(r)-P_v)}{\rho_0}\right)\left\{\left[\left(\frac{R_M}{R}\right)^3 - 1\right] + \frac{3P_{g_0}}{P(r)-P_v}\left[\left(\frac{R_M}{R}\right)^3 \ln\left(\frac{R_M}{R}\right)\right]\right\} \tag{6.53}$$

The minimum radius corresponding to $\dot{R}=0$ is, for small R_m/R_M,

$$\frac{R_m}{R_M} \simeq \exp\left(-\frac{P(r)-P_v}{3P_{g_0}}\right) \tag{6.54}$$

Fig. 6.14 summarizes the variation of the minimum radius with the gas pressure P_{g_0}, using these equations, as well as the general trends given by Hickling's analysis. Pairs of curves (1 and 2) and (3 and 4) illustrate the difference between the assumption of incompressible and compressible liquids respectively for either the adiabatic or the isothermal gas compressions. The asymptotic dependence shown by curve 4, given by Eq. (6.54), is not physically realizable since by Eq. (6.53) $(\dot{R})^2$ is singular in limit as R approaches zero. This singularity is removed by allowing less heat transfer ($\gamma \neq 1$) in which case Eq. (6.51) applies. Fig. 6.14 shows the effect of gas content and liquid compressibility on the collapse of cavitation bubbles. It is to be noted that at large values of hydrostatic pressure, fluid compressibility influences collapse only slightly more than 1 atm. Although the minimum bubble radius is dramatically influenced by the presence of gas, Khoroshev [100] has shown less than 10% increase in the collapse time compared with the incompressible value (6.51) for values of $P_{g_0}/P(r)$ less than 0.1. The minimum bubble radius is furthermore dependent on both the presence of gas and the liquid compressibility, especially for small gas pressures.

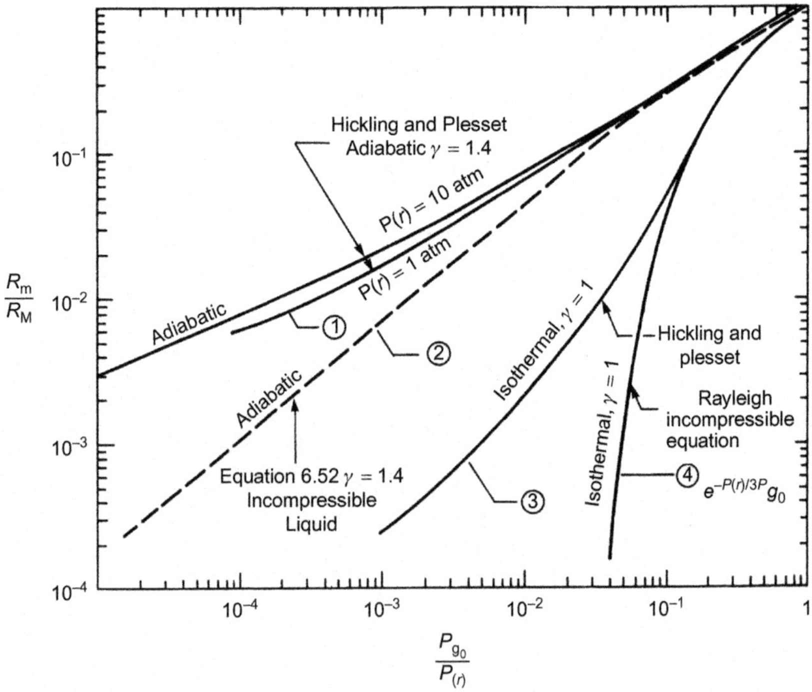

FIGURE 6.14 Numerically computed minimum bubble radius for collapsing bubbles for which $\dot{R} = 0$. (Rayleigh model assumes incompressible liquid; Hickling and Plesset assume compressible liquid.)

The Rayleigh–Plesset theory provides a radius–time relationship for empty ($P_{g_0} = 0$) bubbles for times somewhat earlier than $t = \tau_c$, but still for which $R_M/R \gg 1$, yet $(\dot{R})^2 \gg \ddot{R}R$. Eq. (6.50), reduces to

$$(\dot{R})^2 \simeq -\frac{2}{3}\frac{\Delta P}{\rho_0}\left(\frac{R_M}{R}\right)^3 \tag{6.55}$$

This equation is integrated from some R and t to $R = 0$ at $t = \tau_c$ to give

$$\frac{R}{R_M} \simeq a^{2/5}(\tau_c - t)^{0.4}, \quad t < \tau_c \tag{6.56}$$

where

$$\begin{aligned} a &= \frac{5}{2}\sqrt{\frac{2}{3}\frac{P(r)}{\rho_0 R_M^2}} \\ &= \frac{5}{2}\sqrt{\frac{2}{3}\left(\frac{0.915}{\tau_c}\right)^2} \end{aligned} \tag{6.57}$$

therefore

$$\frac{R}{R_M} \simeq 1.3\left(\frac{\tau_c - t}{\tau_c}\right)^{0.4} \tag{6.58}$$

This approximation closely approximates the complete Rayleigh solution near collapse; see Fig. 6.15. The inset of Fig. 6.15 on the upper right shows

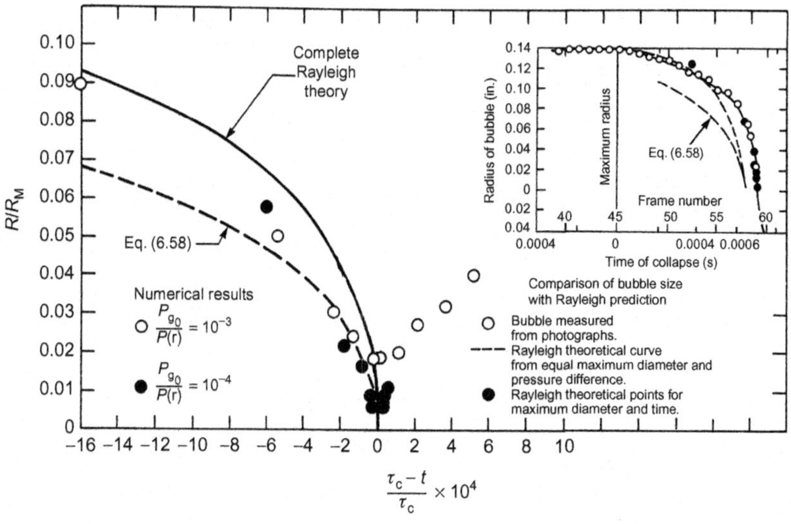

FIGURE 6.15 Time history models of bubble radius in final stage of collapse.

the example of bubble collapse in hydrodynamic cavitation observed photographically by Knapp and Hollander [83] and Knapp [101] compared with both the numerically calculated radius history and the approximation given by Eq. (6.58). The measurement shown here is simply an expanded view of the initial collapse shown in Fig. 6.11. Unfortunately the details of collapse are not easily observed experimentally because of the short time scales involved during which the radial velocity must decrease from nearly sonic (see Fig. 6.12) to zero at the final stage of collapse unless the content of entrapped gas is large. Therefore these results yield only rough approximations with which estimations of the effect of gas on the sound will be estimated in Section 7.2.

Further observations of bubble collapses have shown that a spherical geometry is only maintained for single bubbles in virtually unbounded media. Harrison [102] and Lauterborn [103,104], Lauterborn et al. [23], and Tomita et al. [121] have shown such spherical symmetry in the initial collapse phase, but with some distortion in rebound. Rebound dynamics appears strongly influenced, if not controlled, by thermal conduction and, to a lesser extent radiation. This is illustrated by the simulations shown in Fig. 6.16. Lauterborn et al.'s [23] physically-generated bubble history appears well modeled in an analytical Rayleigh–Plesset model of Qin et al. [36] with heat conduction and thermal radiation included, yet not so well in a simulation that did not include effects of liquid condensation and heat transfer that was done, by Popinet et al. [105]. Hydrodynamically induced bubbles collapsing

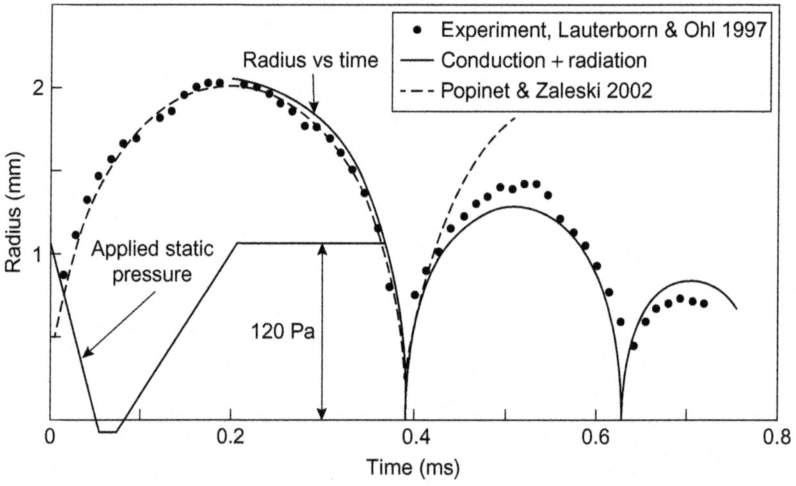

FIGURE 6.16 Qin et al.'s [36] calculated time history of radius for a bubble of initial radius 30 mm exposed to a pressure time history shown in the lower left in set. Comparison is made with the measurement of Lauterborn and Ohl [23]; both axes are linear. *Revised from Qin Z, Bremhorst K, Alehossein H, Meyer T. Simulation of cavitation bubbles in a convergent–divergent nozzle water jet. J Fluid Mech 2007;573:1–25.*

on a hydrofoil [106,107] or in a Venturi [108,109] depart from the spherical geometry because of the pressure gradients imposed on the bubbles, yet the qualitative behavior of bubble volume does resemble that of spherical bubbles. Similarly, pressure gradients associated with pressure wave reflections from boundaries or from other bubbles have contributed to similar effects in spark-induced bubbles collapsing near walls [103,110,111] or within a radius of another bubble [103,104,112,113]. Departures from spherical symmetry often involve the formation of a small jet toward the boundary that is often regarded as a major contributor to erosion. In hydrodynamically induced cavitation, nonspherical bubbles appear with prolate spheroidal shapes and as irregularly shaped clouds of microbubbles. A notable common result of analyses by Levkovskii [114], Shima and Nakajima [115], and Shima [116] is that as long as a collapsing bubble is no closer to the boundary than about one radius (i.e., nearly touching the boundary), the collapse time differs from the Rayleigh time (Eq. (6.50)) by no more than 20%. Furthermore, during collapse, the distortion of the bubble is such that the far surface approaches the wall and the near surface remains essentially fixed with respect to the wall. Plesset and Mitchell [117] showed analytically that while collapsing cavity shapes are unstable as noted, the shapes of expanding cavities near walls remain essentially spherical. This is in agreement with observations on hydrofoils. Thus the Rayleigh collapse time forms an important basis for defining a cavitation time scale for the spectrum of hydrodynamically induced cavitation noise.

6.4 THEORY OF SINGLE-BUBBLE CAVITATION NOISE

In this section we shall deal with the classical theory of sound from cavitation of single bubbles. Although cavitation noise rarely follows the theoretical behavior closely enough to be exactly predicted from fundamentals, the classical theory provides a basis for scaling sounds measured experimentally. It also accounts qualitatively and parametrically for the spectral shape of radiated sound as often observed and it shows the abrupt onset of cavitation noise as ambient pressure is reduced. The single-bubble noise theory is also the basis of much modern theoretical modeling of hydrodynamic traveling bubble cavitation noise [118]. Therefore, we shall consider in detail the sounds emitted as a bubble grows and collapses under the influence of a pulsation of ambient pressure in an unbounded liquid and examine the theoretical effects on the noise of compressible gas in the bubble and compressibility of the liquid.

6.4.1 Dependence of Sound on Stages of Bubble History

In dealing with cavitation noise, we are really concerned with noise resulting from the time variation of voids in the liquid. Since there are volume changes the noise resulting is monopole governed, equivalent to Eq. (2.24b)

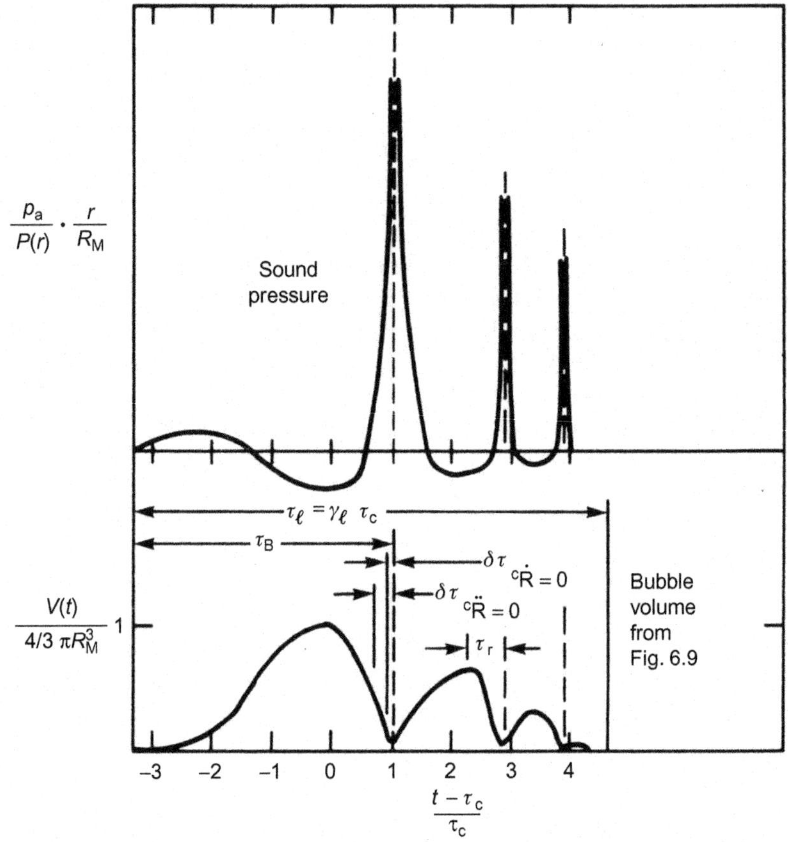

FIGURE 6.17 Sound pressure and volume of a single cavitating bubble with two rebounds. Also shown are characteristic times appropriate for Eqs. (6.50), (6.57), and (6.72). *After Fitzpatrick HM, Strasberg M. Hydrodynamic sources of sound. Proc Symp Nav Hydrodyn 1st. Washington, DC; 1956. p. 241–80.*

in the linear acoustic approximation requiring that $\dot{R}/c_0 < 1$. Fig. 6.17 illustrates the volume history of a hydrodynamically induced cavitation bubble. It can be deduced from this that the volume acceleration will be largest at times when the radial velocity changes direction, i.e., at the times of the minimum radius. The figure is constructed using as a guide the volume history observed by Knapp and Hollander [83] and shown in Fig. 6.11. The maximum sound pressure is attained in a time interval of less than $\tau_c/2$ that is centered on the instant of collapse. The peak sound pressure would be expected to depend on the time scale τ_c, which, by Eq. (6.50), depends on the relative hydrostatic pressure $P(r) - P_v$ and the maximum bubble radius. Eq. (2.24b) can then be made dimensionless on the maximum radius R_M and the collapse time τ_c, i.e.,

$$\frac{p_\mathrm{a}(r/R_\mathrm{M}, t/\tau_\mathrm{c})}{P(r)}\frac{r}{R_\mathrm{M}} = \frac{\partial^2(v/R_\mathrm{M}^3)}{\partial(t/\tau_\mathrm{c})^2}$$

In the following $P(r) = P_0$ is taken as a constant local static pressure where the bubble collapses.

In the frequency domain the Fourier transform (Eq. (2.100)) of the acoustic pressure is given by

$$p_\mathrm{a}(r,\omega) = \frac{\rho_0}{4\pi r}\ddot{V}(\omega)e^{ik_0(r-a)} \tag{6.59}$$

where (using the conditions $\lim_{T\to\pm\infty}\dot{V}(T) = \lim_{T\to\pm\infty}V(T) = 0$)

$$\ddot{V}(\omega) = \frac{-\omega^2}{2\pi}\int_{-\infty}^{\infty} e^{i\omega t}v(t)\,dt \tag{6.60a}$$

with a spectral density of radiated sound pressure Φ_{prad}

$$\tau_\ell \Phi p_\mathrm{rad}(r,\omega) = S_p(r,\omega) \tag{6.60b}$$

where τ_ℓ, illustrated in Fig. 6.17, is the total life time of the bubble including rebounds: $\tau_\ell \simeq 4-7\ \tau_\mathrm{c}$ and where $S_p(r,\,\omega)$ is a spectrum function that may best be defined by relating it to the integrated pressure-squared which is proportional to the mean square averaged over the lifetime as τ_ℓ, i.e.,

$$\int p_\mathrm{a}^2(t)\,dt = \tau_\ell\,\overline{p^2} = \int_{-\infty}^{\infty} S_p(r,\omega)\,d\omega$$

The function $S_p(r,\,\omega)$ may be formed from the Fourier coefficient $\ddot{V}(\omega)$ by the methods of Section 2.6.2

$$S_p(r,\omega) = 2\pi|p_\mathrm{a}(\omega)|^2 = \frac{\rho_0^2}{8\pi r^2}|\ddot{V}(\omega)|^2$$

The distinction between $S_p(r,\,\omega)$ and $\Phi_p(\omega)$ is made because each cavitation event emits a pressure pulse to which may be attributed a level of emitted acoustic energy that is proportional to $\tau_\ell\,p_\mathrm{a}^2$. If a sequence of these events occurs at a temporal rate say, $\dot{N}$, then it makes sense to consider a power function and an associated mean square pressure which have the relationships

$$\overline{p_\mathrm{a}^2} = \lim_{T\to\infty}\frac{1}{2T}\int_{-T}^{T} p_\mathrm{a}^2(t)\,dt = \dot{N}\int_{-\infty}^{\infty} S_p(r,\omega)\,d\omega = \int_{-\infty}^{\infty}\Phi_{p_\mathrm{rad}}(r,\omega)\,d\omega$$

For a single event the reciprocal τ_ℓ^{-1} replaces $\dot{N}$ giving the equivalent definition in Eq. (6.60b). Following the methods of Section 1.5.2 these spectrum functions can be rewritten in a pair of dimensionless forms that will be used interchangeably. Letting

$$\tau_\mathrm{c} \simeq R_\mathrm{M}\sqrt{\rho_0/P_0}, \tag{6.61}$$

then

$$\frac{S_p(r,\omega)r^2}{P_0\,\rho_0 R_M^4} = \tilde{S}_p\left(\frac{r}{R_M}, \omega\tau_c\right) \tag{6.62}$$

and

$$\frac{\Phi_{p_{rad}}(r,\omega)r^2}{P_0^{3/2}\rho_0^{1/2}R_M^3} = \tilde{\Phi}\left(\frac{r}{R_M}, \omega\tau_c\right) \tag{6.63}$$

and

$$\tilde{S}_p(r/R_M, \omega\tau_c) = \tau_\ell R_M \sqrt{\rho_0/P_0}\,\tilde{\Phi}p_{rad}(r/R_M, \omega\tau_c)$$

Fitzpatrick and Strasberg [119] were the first to determine the frequency spectrum of the radiated sound from a cavitating bubble. Using the illustrative pressure–time history shown in Fig. 6.17, they were able to perform the necessary Fourier transforms. The resulting sound spectrum is shown in Fig. 6.18. There are three distinct frequency regions that are linked to corresponding time zones in the collapse history.

We shall, in the following discussion, use the asymptotic approximations of the preceding section to identify certain portions of the bubble history that contribute to corresponding frequency ranges. To perform the calculation it is useful to note that some elements of the time history of the bubble volume may be segmented into a sequence of time intervals of duration $\Delta\tau_n$, i.e.,

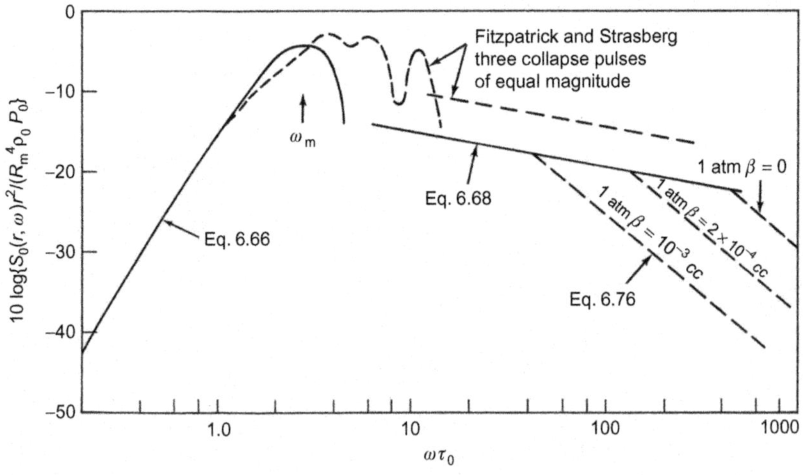

FIGURE 6.18 Ideal spectrum function for the sound pressure generated by a single cavitation bubble with a volume history such as shown in Figs. 6.9 and 6.14. *After Fitzpatrick HM, Strasberg M. Hydrodynamic sources of sound. Proc Symp Nav Hydrodyn 1st. Washington, DC; 1956. p. 241–80.*

$$v(t) = \sum_{n=0}^{N} v_n(t)u(t, \Delta\tau_n) \tag{6.64}$$

where the unit function $u(t, \Delta\tau_n)$ is unity only within the interval $|t-t_n|<\tau_n$ and zero otherwise. The function $v_n(t)$ is a power-law expression used to approximate the bubble motion in the successive intervals $\Delta\tau_n$. The appendix at the end of this chapter illustrates this calculation. Basically, if $v_n(t)$ is of the form

$$\begin{aligned} v_n(t) &= a_n(t-t_n)^m, \quad t>t_n \\ &= 0, \qquad\qquad\quad\ \ t<t_n \end{aligned} \tag{6.65a}$$

then as long as $\omega\Delta\tau_n>1$ as derived in the appendix

$$V_n(\omega) \simeq \frac{a_n\Gamma(m+1)}{2\pi(i\omega)^{m+1}}, \quad \frac{\Delta\tau_n\omega}{2}>1 \tag{6.65b}$$

We shall now consider the specific frequency ranges in Fig. 6.18 that are commensurate with the various events in the bubble history highlighted in Fig. 6.17 and derived in Section 6.3.2.

a. *Low Frequencies*: $\omega\tau_c<1$, $\omega\tau_\ell<1$
In this case Eq. (6.60a) gives only

$$\ddot{V}(\omega) \simeq -\frac{\omega^2}{2\pi}\int_0^\infty V(t)\,dt = -\frac{\omega^2\tau_\ell}{2\pi}\overline{V}(t)$$

Therefore

$$S_p(r,\omega) \simeq \frac{\tau_\ell^2\omega^4\rho_0^2\overline{V}^2}{32\pi^3 r^2} \tag{6.66}$$

where $\overline{V} = \int_0^\infty V(t)\,dt \simeq -1.3V_M\tau_c$.

b. *Frequency of Maximum Intensity*: $\omega\tau_\ell>\omega\tau_c>1$
In this case, the details of collapse are unimportant, and the initial bubble life can be approximated by

$$V(t) \simeq V_M\cos\left(\frac{t-\tau_c}{\tau_c}\frac{\pi}{2}\right) \quad \text{for} \quad -\tau_c<t<-\tau_c<\tau_c$$

Thus

$$V(\omega) \simeq \frac{V_M\tau_c}{2}\frac{\cos\omega\tau_c}{(\pi/2)^2-(\omega\tau_c)^2}$$

so that

$$\frac{S_p(r,\omega)r^2}{R_M^4\rho_0P_0} \simeq \frac{\pi}{18}(\omega\tau_c)^4\left[\frac{\cos\omega\tau_c}{(\pi/2)^2-(\omega\tau_c)^2}\right]^2 \tag{6.67}$$

There is also a contribution from rebounds, which add to $V(\omega)$, of the form

$$V_{r_n}\tau_{r_n}\frac{\cos\omega\tau_{r_n}/2}{(\pi/2)^2-(\omega\tau_{r_n}/2)^2}e^{-i\omega[2\tau_c-(n-1)\tau_{n-1}]}$$

where τ_{r_n} are collapse times of successive rebounds. Because of the phase factor in the rebound contribution, there can be certain interferences at some frequencies in the vicinity of the maximum spectrum level at $\tau_c\omega\simeq\pi$. Fig. 6.18 shows an example of the influence of three equal rebounds on the spectrum, as calculated by Fitzpatrick and Strasberg [119].

c. *Moderate Frequencies*: $\omega\tau_\ell>\omega(\delta\tau_c)_{\dot{R}=0}>1$
This frequency range is limited by the segment of the collapse phase controlled by constant bubble wall velocity as considered in Section 6.3.2. In this range we use the approximate function, Eq. (6.65b) with Eq. (6.58)

$$|\ddot{V}(\omega)|\simeq\frac{2V_M\tau_c}{(2\pi)(\omega\tau_c)^{1/5}}$$

and in the form of Eq. (6.63)

$$\frac{S_p(r,\omega)r^2}{R_M^4\rho_0P_0}\simeq\frac{2.4}{9\pi}(\omega\tau_c)^{-2/5} \tag{6.68}$$

d. *High Frequencies Controlled by Noncondensible Gas*: $\omega\tau_c>\omega(\delta\tau_c)_{\dot{R}=0}\gg 1$

In this case, Eqs. (6.55) and (6.68) yield

$$|V_n(\omega)|\simeq\frac{V_M\tau_c}{16\pi}\left(\frac{R_m}{R_M}\right)^3\left(\frac{\tau_c}{(\delta\,\tau_c)_{\dot{R}=0}}\right)^2\frac{720}{(\omega\tau_c)^7}$$

and

$$\frac{S_p(r,\omega)r^2}{\rho_0P_0R_M^4}\simeq 3.6\times10^5\left(\frac{R_m}{R_M}\right)^6\left(\frac{\tau_c}{(\delta\,\tau_c)_{\dot{R}=0}}\right)^4(\omega\tau_c)^{-10}\quad\text{for}\quad\omega(\delta\tau_c)_{\dot{R}=0}\gg 1 \tag{6.69}$$

This time constant is given by Eq. (6.57). Eq. (6.69) shows a rapid decrease of the spectrum level with frequency and represents the existence of an upper frequency limit on the spectrum. This band-limiting is an indicator that an upper frequency limit ensures that the spectrum represents an acoustic pulse of finite energy. A rough numerical estimate of $(\delta\tau_c)_{\dot{R}=0}$ for a partial pressure of gas in the bubble at maximum radius of $P_{g_0}/P(r)\simeq10^{-3}$, using the values of R_m/R_M from Fig. 6.13 ($\gamma=1.4$), is

$$\tau_c/(\delta\tau_c)_{\dot{R}=0}\sim6\times10^3$$

This means that when $\omega\tau_c>6000$, the sound spectrum will be band limited.

6.4.2 Spherical Collapses in Compressible Liquids

The compressibility of the surrounding liquid limits the high frequency sound emitted from a collapsing bubble and the effect becomes more important when the concentration of suspended small gas bubbles in the surrounding liquid increases. These suspended bubbles, even though they do not cavitate, alter the propagation of sound so that as the bubble wall velocity becomes supersonic in the final stages of collapse, the resulting sound pulses take on the theoretical forms shown in Fig. 6.19. The "N" waves are formed because some of the acoustic energy is initially propagated at a speed U that is greater than the ambient value of speed of sound. This sound overtakes the sound propagating at speed c_0 until, because of geometrical spreading, the high-speed particle velocities slow down to c_0. The "N" wave then propagates as a pulse waveform resembling a ramp function of amplitude p_0 and temporal width δt_s as shown in the figure.

The conditions of bubble dynamics that control the wall velocity in the final stages of collapse in an incompressible liquid can be deduced from the equations of bubble motion that were described earlier. By substituting

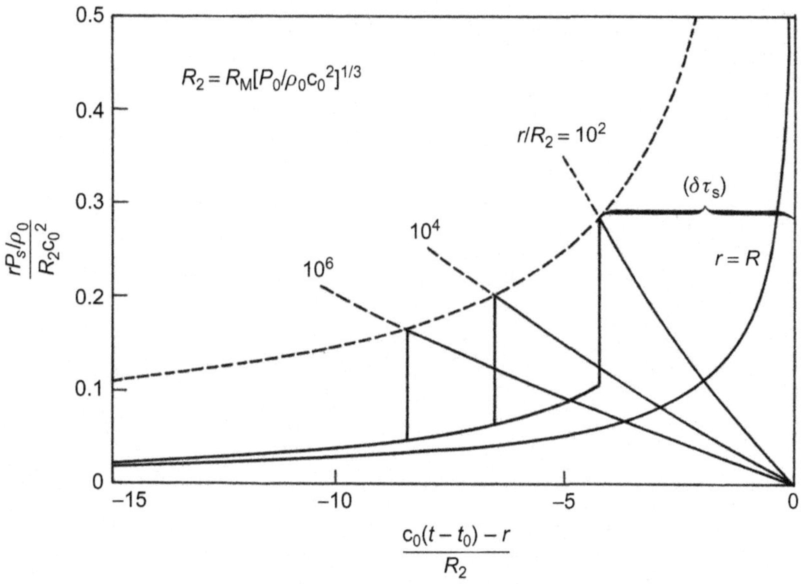

FIGURE 6.19 Pressure pulses of developing shock waves emitted from a vapor cavity collapsing in a compressible liquid. Pulses are shown as a function of the retarded time variable at successive distances from the bubble wall. *Values computed by Mellen RH. Spherical pressure waves of finite amplitude from collapsing cavities. Rep. No. 326. U.S. Navy Underwater Sound Lab. New London, CT; 1956; curves are adapted from Fitzpatrick HM, Strasberg M. Hydrodynamic sources of sound. Proc Symp Nav Hydrodyn 1st. Washington, DC; 1956. p. 241–80.*

Eqs. (6.2) and (6.52) into Eq. (6.3) an expression for the acceleration for the bubble wall, $\ddot{R}$, is derived. The maximum velocity $\dot{R}$ is determined by the condition that $\ddot{R}=0$. The instantaneous value of the bubble radius at this point in time is

$$\left(\frac{R_M}{R}\right)_{\ddot{R}=0}=\left[\frac{\gamma-1}{\gamma}\frac{P(r)-P_v}{P_{g_0}}\right]^{1/3(\gamma-1)} \tag{6.70}$$

where the term $P_{g_0}/P(r)$ has been neglected compared to unity. The maximum wall velocity is found by substituting into Eq. (6.51) to obtain

$$\left(\frac{\dot{R}}{c_0}\right)_{\max}\simeq\left(\frac{P(r)-P_v}{\rho_0 c_0^2}\right)^{1/2}\left\{\frac{2}{3}\left(\frac{\gamma-1}{\gamma}\right)\left[\left(\frac{\gamma-1}{\gamma}\right)\left(\frac{P(r)-P_v}{P_{g_0}}\right)\right]^{1/(\gamma-1)}\right\}^{1/2} \tag{6.71}$$

Eq. 6.71 is graphed in Fig. 6.20 for two values of ambient pressure and this figure should be examined in conjunction with Fig. 6.12. The relationship shows that the limiting velocity is influenced both by the amount of heat transfer that may occur and by the partial pressure of gas in the cavity. As P_{g_0} diminishes to zero, the wall Mach number increases indefinitely. The rate of increase is less pronounced, however, than this simple theory shows. Compressibility of the liquid limits this process; numerically computed

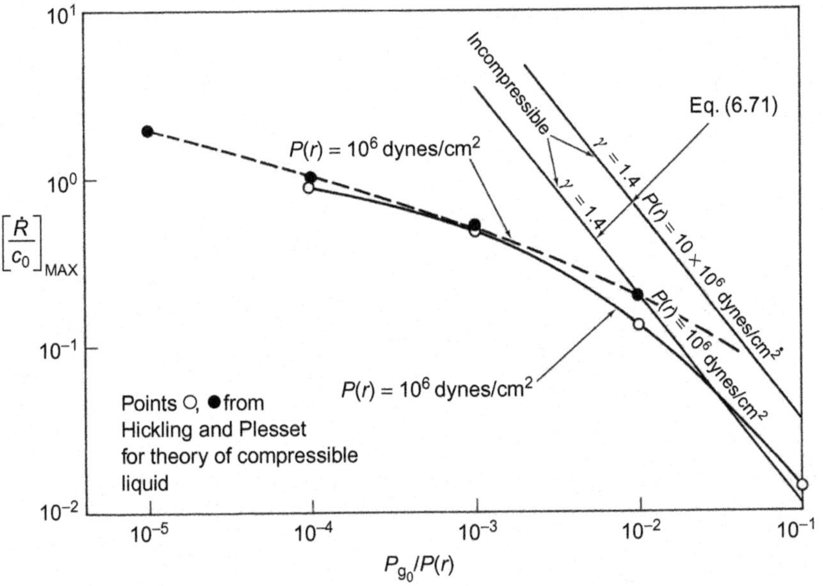

FIGURE 6.20 Maximum bubble-wall Mach number as a function of initial inside gas pressure and hydrostatic pressure; adiabatic collapse.

values of $\dot{R}$ given by Hickling and Plesset [92] show that $(\dot{R}/c_0)$ theoretically exceeds unity when $(P_{g_0}/P(r))$ is less than 10^{-4} for ambient pressures of 1–10 atm. Compressibility of the water becomes important, however, for bubble wall Mach numbers in excess of only 0.1 and at $(P_{g_0}/P(r))$ as high as 10^{-2} atm. This result can be generalized somewhat by noting that when $\dot{R}/c_0 > 0.1$ Eq. (6.55) for the simple theory gives a corresponding approximate radius

$$\frac{R}{R_M} \simeq 4\left(\frac{P(r)}{\rho_0 c_0^2}\right)^{1/2}$$

at an approximate time before collapse of

$$\frac{\delta\tau_s}{\tau_c} \simeq 40\left[\frac{P(r)}{\rho_0 c_0^2}\right]^{5/6} \tag{6.72}$$

given by Eq. (6.58). Therefore when $t - \tau_c < \delta\tau_s$, the waveform of the acoustic pressure pulse will begin to deform and progressively develop the "N" wave shape. For bubbles in water, the radius at this point is $0.15R_M$ and $0.31R_M$ for $P(r) = 1$ and 10 atm, respectively. Examination of Fig. 6.15 shows that these bubble sizes are small enough that the degree of heat transfer from the bubble will also be influential as indicated by the additional dependence on γ. For such short time scales as for events preceding complete collapse, the processes should be adiabatic.

Fitzpatrick [120] proposes a similar criterion for $\delta\tau_s$, but based on an analytical function for radius, $R(t)$ different than that used here. Largely because of this his time $\delta\tau_s$ is shorter than that found by a factor of 2. Thus a good rule is for the constant in Eq. (6.72) to be taken between 20 and 40.

According to Fig. 6.21, $\delta\tau_s$ depends on the distance from the bubble; at larger distances it appears that

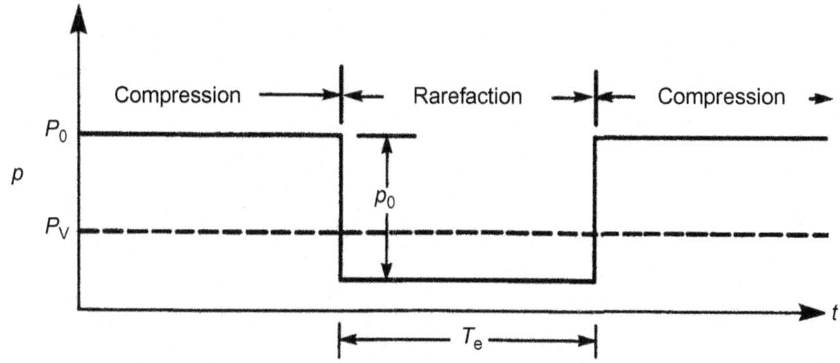

FIGURE 6.21 Pressure–time history used in a hypothetical model of cavitation sound. Bubble cavitation will occur when $P_v > p = P_0 - p_0$, as illustrated.

$$\frac{\delta \tau_s c_0}{R_M [P(r)/\rho_0 c_0^2]^{1/3}} \simeq 9 \tag{6.73}$$

which corresponds approximately to

$$\frac{\delta \tau_s}{\tau_c} \simeq 9 \left[\frac{P(r)}{\rho_0 c_0^2} \right]^{5/6} \tag{6.74}$$

Comparison with the alternative result, Eq. (6.72), suggests that the numerical coefficient (9 or 40) is really just a "ball-park" value. The ramp-like shock-induced waveforms shown in Fig. 6.19 have the (dashed line) envelope

$$p(t) = p_0 e^{-|t|/\alpha},$$

and these have the temporal correlation

$$\overline{p(t)p(t+\tau)} = p_0^2 e^{-|\tau|/\alpha} \tag{6.75}$$

which gives the auto-spectral density

$$\Phi_{pa}(\omega) = \frac{1}{2\pi} \frac{p_0^2 \alpha}{1 + (\omega \alpha)^2} \tag{6.76}$$

where $\alpha = \delta t_s$ is given by Eq. (6.72) or (6.74). In Fig. 6.18 several lines are drawn which correspond to $\rho_0 c_0 = 20{,}000$ atm for pure water and to other values of speeds of sound corresponding to bubbly mixtures of the indicated

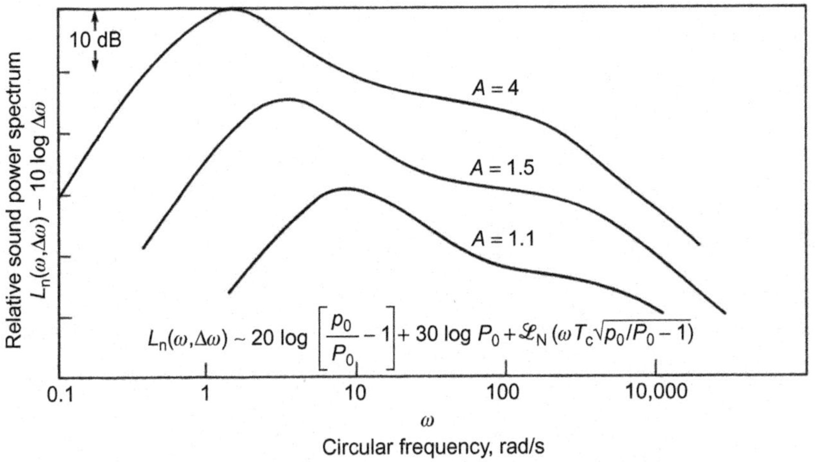

FIGURE 6.22 Sound pressure spectrum (relative to arbitrary reference) for ideal cavitation noise induced by a hydrodynamic pressure pulsation of duration T_e and amplitude $p_0 = Ap_0$. T_e and P_0 are fixed for the variable rarefaction pressure $-p_0$ shown in Fig. 6.21, and the cavitation rate is fixed. $p_v/P_0 \ll p_0/P_0 - 1$ for the examples shown.

volumetric concentrations β. Eqs. (6.22) and (6.30) and Table 6.1 were used to calculate the speed of sound. It can be seen that moderate concentrations of free bubbles can influence dramatically the high-frequency spectrum levels. In water-tunnel facilities [98,121] volumetric concentrations of $\beta < 10^{-7}$ may be expected in undersaturated water; as far as the cavitation noise spectrum is concerned, such fluid is essentially pure water.

Experimental evidence of the existence of shock waves has been reviewed by Fitzpatrick and Strasberg [119] Ellis [122], and Brennan [2]. Generally the shocks had been observed and associated with spherical collapses of spark-induced bubbles; however, Ellis [122] has also observed shock formation from nonsymmetrical bubbles collapsing on a surface, and from groups of bubbles; see also the review by Brennan [2]. Such observations have not been made in hydrodynamically induced cavitation, however. Evidence of the ω^{-2} behavior in the acoustic spectrum from single bubble collapses was deduced early-on by Harrison [102], who observed on an oscilloscope wave traces of the time history of pressure pulses from hydrodynamically induced cavitation in a Venturi tube. The limiting frequency above which the ω^{-2} dependence should be expected is expressed in the dimensionless form

$$\frac{\omega_s}{\omega_m} \simeq \frac{1}{27}\left[\frac{\rho_0 c_m^2}{P(r)}\right]^{5/6} \tag{6.77}$$

where c_m is the speed of sound in the surrounding fluid, or in terms of the cavitation index defined as (see also Chapter 7: Hydrodynamically Induced Cavitation and Bubble Noise)

$$K - (P(r) - P_v) \Big/ \frac{1}{2}\rho_0 U_\infty^2, \quad P_v \ll P_\infty$$

$$\frac{\omega_s}{\omega_m} \simeq \frac{1}{15}[KM^2]^{-5/6} \tag{6.78}$$

where for both equations it has been assumed that $\omega_m \tau_c \simeq 3$ (see Fig. 6.18 and Eq. (6.61)) defines the frequency of the maximum spectrum level and $M = U_\infty / c_m$, where U_∞ is the velocity of the fluid relative to the body.

6.4.3 Features of Noise in Developing Cavitation

As cavitation develops, the sizes of the participating bubbles increase, moving the spectrum of the sound to higher levels and to lower frequencies. This behavior is typical of all types of cavitation observed in practical situations, although the details of spectral behavior may vary widely from case to case. The features of cavitation noise that distinguish it from other forms of flow-induced noise are its abrupt onset, rapid increase with early cavitation development, moderate increase with further development of fully developed cavitation, and a shift in sound spectrum to lower frequencies as the cavitation develops. These features can be exemplified by examining the theoretical behavior of single-bubble cavitation noise as the bubbles develop according to the theory of Section 6.2.2. The sound can be described in terms of the parameters R_M, ρ_0, P_0, and a time scale for the pressure pulsation that causes bubble growth. We hypothesize that a bubble is subjected to a pressure rarefaction of magnitude—p_0 for a duration T_e. During this period, illustrated in Fig. 6.21 the bubble grows to radius R_M; when the rarefaction pressure is increased to zero, returning the pressure around the bubble to the ambient P_0, the bubble collapses implosively. As long as T_e is longer than the natural period of oscillation of the bubble $2\pi/\omega_0$, the initial bubble growth will occur as described in Section 6.2.2.

As shown by Eq. (6.44) for $P_0 - p_0 < P_v$ the maximum bubble radius attained will scale as

$$R_M \simeq T_e\left(\frac{P_v-(P_0-p_0)}{\rho_0}\right)^{1/2}$$

where $P_0 - p_0$ represents the absolute value of pressure at the bubble wall. The growth of cavitation is presumed to occur by a reduction in the ambient pressure P_0 or by an increase of the magnitude of the negative pressure pulse $-p_0$. The time T_e is presumed arbitrary and constant.

The collapse time of the bubble of maximum radius R_M is given by Eq. (6.61). If as in Section 6.4.1 we assume that the bubbles cavitate at a rate $\dot{N}$ that does not change with either P_0 or p_0, then the time-averaged far-field sound power from the ensemble of bubbles will be $\dot{N}$ times the total acoustic energy radiated from the ensemble as

$$\mathbb{P}_{rad}(\omega, \Delta\omega) = \dot{N}\frac{4\pi r^2}{\rho_0 c_0}S_p(r,\omega)$$

so that parametrically, with $\tau_c \sim T_e\sqrt{(P_v - P_0 + p_0)/P_0}$,

$$\mathbb{P}_{rad}(\omega, \Delta\omega) \sim \frac{\dot{N}P_0^3(T_e)^4}{\rho_0^2 c_0}\left[\frac{P_v-(P_0-p_0)}{P_0}\right]^2 \Delta\omega\tilde{S}_p\left(r, \omega T_e\sqrt{\frac{P_v-(P_0-p_0)}{P_0}}\right)$$

The sound power level L_n can then be written in a nondimensional form to depend on P_0 and p_0 as

$$L_n(\omega, \Delta\omega) \sim 10\log\Delta\omega + 20\log\left(\frac{P_v - P_0 + p_0}{P_0}\right) + 30\log P_0 + S_n\left(\omega T_e\sqrt{\frac{P_v - P_0 + p_0}{P_0}}\right) \quad (6.79)$$

where $S_n(\Omega)$ is a dimensionless spectrum function that is proportional to $10 \log \tilde{S}_p(r, \omega)$ of Eq. (6.62) put in terms of the dimensionless frequency $\Omega = \omega T_e \sqrt{(P_v - P_0 + p_0)/P_0}$. In this form it can be seen that the *overall* sound power level depends on the parameters P_0, P_v, and p_0 as

$$L_n \sim 30 \log P_0 + 15 \log\left(\frac{P_v - P_0 + p_0}{P_0}\right) \tag{6.80}$$

where it has been assumed for simplicity that $P_0 > P_v$. According to Fig. 6.9 when the scaled rarefaction amplitude is large, i.e., $A = p_0/P_0$, nonlinear motion as described by Eqs. (6.43) and (6.44) occurs. We shall assume that Eq. (6.44), and therefore the above relationships for $L_n(\omega, \Delta\omega)$, apply for all large scale amplitudes, i.e., for $A > 1$. In deriving Eqs. (6.79) and (6.80) the event rate $\dot{N}$ was assumed to be constant. However in a more complete theory, it may also be dependent on the parameters. There is no generally-accepted relationship.

Fig. 6.22 shows how the power spectral density of radiated sound increases with p_0 for a fixed P_0. The marked increase in sound power at low frequencies as the cavitation develops is typical of all types of ultrasonic and hydrodynamic cavitation. The inception state, or cavitation threshold, is taken in this model as $A = 1$, and the rapid rise of cavitation noise with increasing A is a result of the function $20 \log(A - 1)$. The overall sound power level also shows this behavior, but in Fig. 6.23 the rate of increase of L_n with A lends to flatten as $A \gg 1$. This may be regarded as a fully developed cavitation state. When P_0 is increased, the threshold value of p_0 also increases as shown but the curves of L_n versus p_0 cross each other due to the larger value of $30 \log P_0$.

A parallel with hydrodynamically induced cavitation may be easily drawn. Since rarefaction is related to a reduced Bernoulli pressure, $\frac{1}{2}\rho_0 U^2$, the p_0 in Eqs. (6.79) and (6.80) may be replaced by a pressure proportional to $\frac{1}{2}\rho_0 U^2$. Fig. 6.23 shows that as p_0 increases by a factor of 2 over the value of inception the sound power increases 20 dB. Such an increase in pressure could be induced by only a factor $\sqrt{2} \simeq 1.4$ (or a 40%) increase in velocity. Thus the speed dependence of L_n for cavitation noise will be much more pronounced than for any noncavitation noise. To illustrate this point we make the substitution $p_0 = \frac{1}{2}\rho_0 U^2$ so that

$$\frac{U^2}{U_i^2} = \frac{\frac{1}{2}\rho_0 U^2}{P_0}$$

where U_i is the velocity for which $p_0 = P_0$ (or $A = 1$), which gives Eq. (6.80) in the form

$$L_n \sim 30 \log P_0 + 15 \log\left[(U/U_i)^2 - 1\right] \tag{6.81}$$

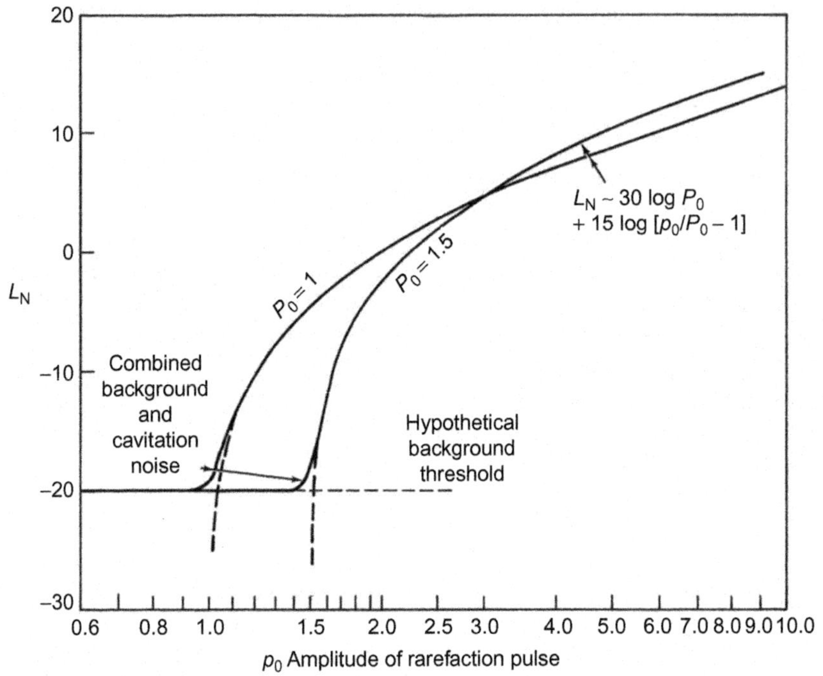

FIGURE 6.23 Overall sound power level (arbitrary reference) for the same source type as in Fig. 6.22. Calculations values of L_n as function of p for two values of P_0 and hypothetically fixed arbitrary T_e.

In this example U_i is the inception velocity for which the onset of cavitation just occurs. The implicit dependence on speed is on the order of at least 130 log U for $U/U_i < 1.3$ with a less-pronounced dependence at greater velocities as implied by Fig. 6.23.

These simple notions are put to work in the collection of illustrations in Fig. 6.24 which depict the general behavior of cavitation onset and sound radiation from hydrodynamic bubble cavitation on hydrofoils. Such a cavitation occurs as a sequence of "events" associated with the convection of ambient (micro)bubbles through the low pressure zone on a hydrofoil. The occurrence of cavitation as the local pressure is reduced below the critical pressure (Eq. (6.41)) is presented as a nondimensional cavitation index K which is defined

$$K = \frac{P_\infty - P_v}{\frac{1}{2}\rho_0 U_\infty^2} \tag{6.82}$$

where P_∞ is the ambient static pressure in the vicinity of the hydrofoil, U_∞ is the relative water velocity, and boundary layer and other real fluid effects

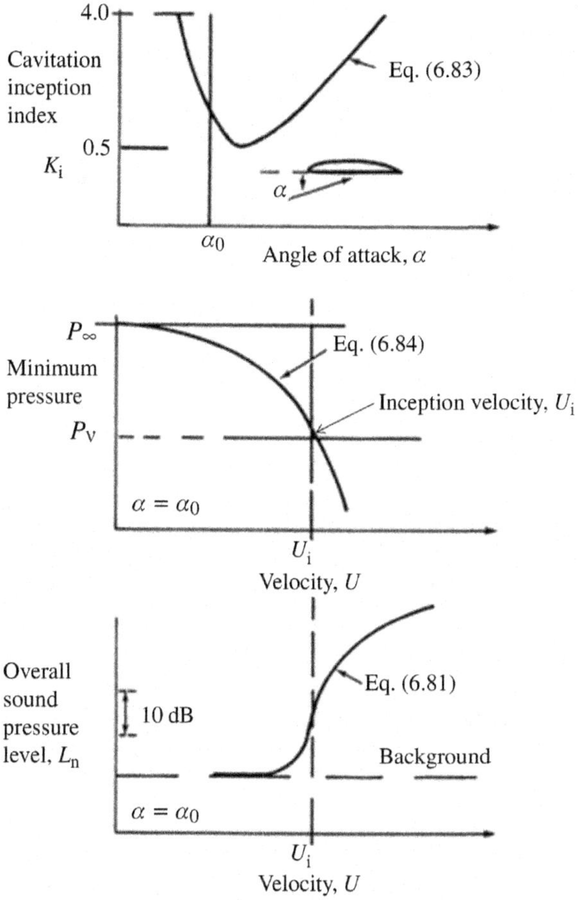

FIGURE 6.24 Illustration of a typical behavior of cavitation inception. (A) Cavitation inception index as function of angle of attack; (B) Minimum pressure at fixed angle of attack $\alpha = \alpha_o$ as a function of velocity; (C) Sound pressure level radiated as a function of velocity at fixed angle of attack $\alpha = \alpha_o$.

(see Chapter 7: Hydrodynamically Induced Cavitation and Bubble Noise) are neglected. The cavitation inception occurs when the minimum pressure equals the critical pressure; i.e., wherever

$$P_{min} = P_{crit}$$

or, expressed in dimensionless terms,

$$\frac{P_{min} - P_\infty}{\frac{1}{2}\rho_0 U_\infty^2} = \frac{P_{crit} - P_\infty}{\frac{1}{2}\rho_0 U_\infty^2}$$

Adding and subtracting the vapor pressure P_v we have

$$\frac{P_{min} - P_v}{\frac{1}{2}\rho_0 U_\infty^2} - \frac{P_\infty - P_v}{\frac{1}{2}\rho_0 U_\infty^2} = \frac{P_{crit} - P_\infty}{\frac{1}{2}\rho_0 U_\infty^2}$$

or since as Fig. 6.8 shows, $P_v \approx P_{crit}$ for large enough bubble nuclei then cavitation occurs whenever P_∞ reaches a critical (inception) value $(P_\infty)_i$ and $P_{min} = (P_{min})_i = P_v$. Thus

$$\frac{(P_\infty)_i - P_v}{\frac{1}{2}\rho_0 U_\infty^2} = \frac{[P_\infty - P_{min}]_i}{\frac{1}{2}\rho_0 U_\infty^2}$$

or introducing the incipient cavitation index and the pressure coefficient

$$K_i = [-C_p]_{min} \tag{6.83}$$

where $[C_p]_{min}$ is the minimum pressure coefficient. Thus cavitation will occur in this simple motion whenever the cavitation index becomes equal to the $[-C_p]_{min}$ on the hydrofoil. The inception index as shown in Fig. 6.24A is dependent on the angle of attack of the hydrofoil because the minimum pressure will increase or decrease according to the angle of attack for a given thickness distribution and camber.

As long as the minimum pressure coefficient on the surface is negative, the magnitude of the minimum will decrease with an increase in velocity for a fixed ambient pressure P_∞ according to this notion because

$$P_{min} = \frac{1}{2}\rho_0 U^2 [C_p]_{min} + P_\infty \tag{6.84}$$

and $[C_p]_{min}$ is less than zero. This behavior is depicted in Fig. 6.24B for some particular angle of attack. Cavitation is incipient at the critical, or inception, velocity U_i such that $P_{min} = [P_{min}]_i = P_v$ as shown in the figure. The sound power level L_n resulting in the onset of cavitation will follow a trend much as given by Eq. (6.81) but more generally expressed as

$$L_n \approx 30 \log P_\infty + S(U/U_i) \quad \text{for} \quad U/U_i > 1.0$$

where the function $S(U/U_i)$ will be dependent on many factors that are associated with real cavitation flow. It will, however, generally resemble qualitatively the behavior shown in Fig. 6.23 or 6.24C.

The behavior of real hydrodynamic cavitation and cavitation noise is more complex than that illustrated here. In spite of these complexities, however, the general behavior is qualitatively very much like that illustrated in Figs. 6.22, 6.23, and 6.24. Real fluid motions involve viscous effects, surface roughness, variation in nucleation physics, and these all modify the values of inception index K_i, the function $S(U/U_i)$, and the spectral content of cavitation noise. These factors shall all be discussed in Chapter 7, Hydrodynamically Induced Cavitation and Bubble Noise.

APPENDIX: DERIVATION OF APPROXIMATE SPECTRAL FUNCTIONS

In Section 6.4.1, various time intervals were highlighted to emphasize different aspects of the bubble collapse. These times will have corresponding frequency intervals that determine the spectral form of $S_p(r, \omega)$. To isolate each of these events in the life of a bubble, the instantaneous volume will be approximated by a sum of N functions that combine to approximate the original $\upsilon(t)$

$$\upsilon(t) = \sum_{n=0}^{N} \upsilon_n(t)u(t, \Delta\tau_n) \tag{6.A1}$$

where $\upsilon_n(t)$ has the functional form given by Eq. (6.65a)

$$\begin{aligned}\upsilon_n(t) &= a_n(t-t_n)^m,\\ &= 0,\end{aligned} \tag{6.A2}$$

and as illustrated in Fig. 6.A1. The unit function $u(t, \Delta\tau_n)$ is defined so that

$$\begin{aligned} u(t, \Delta\tau_n) &= 1, \quad t_n < t < t_n + \Delta\tau_n \\ &= 0 \quad \text{t outside the interval} \end{aligned}$$

Fig. 6.A1 illustrates the use of functions that generally describe the maximum volume, collapse, and rebound phases of the bubble that were highlighted in Figs. 6.15 and 6.17.

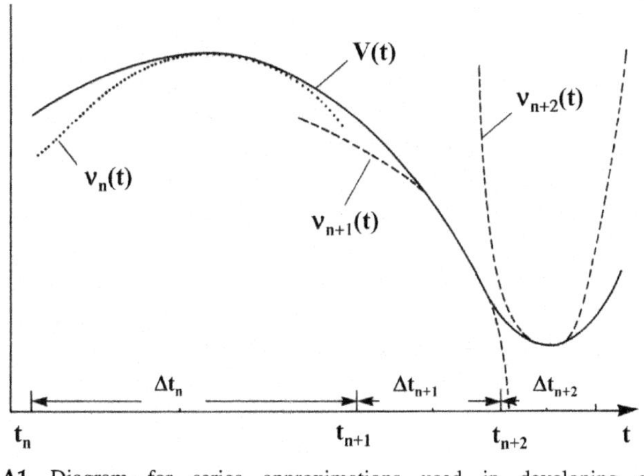

FIGURE 6.A1 Diagram for series approximations used in developing equations of Section 6.4.1, beginning with Eq. (6.64).

The Fourier transform of the nth contribution to Eq. (6.A1) gives

$$\begin{aligned} V_n(\omega) &= \frac{1}{2\pi}\int_0^\infty v_n(t)u(t,\Delta t)e^{i\omega t}dt \\ &= \int_{-\infty}^\infty \tilde{V}_n(\Omega)u_n(\omega-\Omega)\, d\Omega \end{aligned} \tag{6.A3}$$

where

$$\tilde{v}_n(\Omega) = \frac{1}{2\pi}\int_0^\infty e^{i\omega t}v_n(t)\, dt$$

and

$$\begin{aligned} U_n(\omega) &= \frac{1}{2\pi}\int_{-\Delta\tau_{n/2}}^{1/2\Delta\tau_{n/2}} e^{i\omega t}\, dt \\ &= \frac{\Delta\tau_n}{2\pi}\left(\frac{\sin \Delta\tau_n\, \omega/2}{\Delta\tau_n\, \omega/2}\right) \end{aligned} \tag{6.A4}$$

The Fourier transform $\tilde{V}_n(\omega)$ is formally convergent for arbitrary noninteger values of m only in a limit for which $\omega = \omega + i\delta$ for which δ must be positive and it can be taken arbitrarily small, i.e., $\delta \to 0^+$. Thus in this limit

$$\tilde{V}_n(\omega) = \frac{a_n\Gamma(m+1)}{2\pi(i\omega)^{m+1}} \quad m > -1 \tag{6.A5}$$

where $\Gamma(m+1)$ is the gamma function. It has the following approximate values needed for numerical evaluations:

$$\begin{aligned} &\Gamma(1+3(0.4)) = \Gamma(2.2) \sim 1 && \text{for Eq. (6.71)} \\ &\Gamma(1+3(2)) = \Gamma(7) = 6 = 720 && \text{for Eq. (6.72)} \end{aligned}$$

Therefore for any interval Eqs. (6.A3)–(6.A5) give

$$V_n(\omega) \sim \int_{-\infty}^\infty \frac{a_n\Gamma(m+1)}{2\pi(i\Omega)^{m+1}}\frac{\Delta\tau_n}{2\pi}\frac{\sin \Delta\tau_n(\omega-\Omega)/2}{\Delta\tau_n(\omega-\Omega)/2}d\Omega \tag{6.A6}$$

In the limit as $\omega\Delta\tau_n > 1$ we make the approximation given by Eq. (2.128) so that Eq. (6.A6) may be formally evaluated in the approximate form

$$V_n(\omega) \sim -\frac{a_n\Gamma(m+1)}{2\pi(i\omega)^{m+1}}, \quad \frac{\Delta\tau_n\omega}{2} > 1 \tag{6.A7}$$

REFERENCES

[1] Knapp RT, Daily J, Hammitt FG. Cavitation. New York: McGraw-Hill; 1970.

[2] Brennan CE. Cavitation and bubble dynamics. Oxford; 1995.

[3] Brennan CE. Hydrodynamics of pumps. Oxford; 1994.

[4] Leighton TG. The acoustic bubble. Academic Press; 1994.

[5] Carlton J. Marine propellers and propulsion. Elsevier; 2012.

[6] Plesset MS, Prosperetti A. Bubble dynamics and cavitation. Annu Rev Fluid Mech 1977;9:145–85.

[7] Feng ZC, Leal LG. Nonlinear bubble dynamics. Annu Rev Fluid Mech 1997;29:201–43.

[8] Arndt REA. Cavitation in fluid machinery and hydraulic structures. Annu Rev Fluid Mech 1981;13:273–328.

[9] Arndt REA. Cavitation in vortical flows. Annu Rev Fluid Mech 2002;34:143–75.

[10] Blake JR, Gibson DC. Cavitation bubbles near boundaries. Annu Rev Fluid Mech 1987;19:99–123.

[11] Prosperetti A. Vapor bubbles. Annu Rev Fluid Mech 2017;49:221–48.

[12] Rayleigh L. On the pressure developed in a liquid during the collapse of a spherical cavity. Philos Mag 1917;34:94–8.

[13] Plesset MS. The dynamics of cavitation bubbles. J Appl Mech 1949;16:277–82.

[14] Neppiras EA, Noltingk BE. Cavitation produced by ultrasonics: theoretical conditions for the onset of cavitation. Proc Phys Soc London Sect B 1951;64:1032–8.

[15] Neppiras EA, Noltingk BE. Cavitation produced by ultrasonics. Proc Phys Soc London Sect B 1950;63:674–85.

[16] Commander KW, Prosperetti A. Linear pressure waves in bubbly liquids: comparison between theory and experiments. J Acoust Soc Am 1989;85:732–46.

[17] Houghton G. Theory of bubble pulsation and cavitation. J Acoust Soc Am 1963;35:1387–93.

[18] Blue JE. Resonance of a bubble on an infinite rigid boundary. J Acoust Soc Am 1967;41:369–72.

[19] Howkins SD. Measurements of the resonant frequency of a bubble near a rigid boundary. J Acoust Soc Am 1965;37:504–8.

[20] Shima A. The natural frequency of a bubble oscillating in a viscous compressible liquid. J Basic Eng 1970;92:555–62.

[21] Strasberg M. The pulsation frequency of non-spherical gas bubbles in liquids. J Acoust Soc Am 1953;25:536–7.

[22] Lauterborn W, Bolle H. Experimental investigations of cavitation bubble collapse in the neighbourhood of a solid boundary. J Fluid Mech 1975;72:391–9.

[23] Lauterborn W, Ohl C-D. Cavitation bubble dynamics. Ultrason Sonochem 1997;4:65–75.

[24] Chang NA, Ceccio SL. The acoustic emissions of cavitation bubbles in stretched vortices. J Acoust Soc A 2011;130:3209–19.

[25] Plesset MS, Hsieh D. Theory of gas bubble dynamics in oscillating pressure fields. Phys Fluids 1960;3:882–92.

[26] Keller JB, Miksis MJ. Bubble oscillations of large amplitude. J Acoust Soc Am 1980;68:628–33.

[27] Prosperetti A, Lezzi A. Bubble dynamics in a compressible fluid, Part 1, First-order theory. J Fluid Mech 1986;168:457–78.

[28] Strasberg M. Gas bubbles as sources of sound in liquids. J Acoust Soc Am 1956;28:20–6.

[29] Minnaert M. On musical air bubbles and the sounds of running water. Philos Mag 1933;16:235–48.

[30] Devin Jr. C. Survey of thermal, radiation, and viscous damping of pulsating air bubbles in water. J Acoust Soc Am 1959;31:1654–67.

[31] Whitfield OJ, Howe MS. The generation of sound by two-phase nozzle flows and its relevance to excess noise of jet engines. J Fluid Mech 1976;75:553–76.

[32] Flynn HG. Physics of acoustic cavitation in liquids. Phys Acoust (W. P. Mason, ed.) 1964;16.

[33] Prosperetti A. The thermal behavior of oscillating gas bubbles. J Fluid Mech 1991;222:587–616.

[34] Prosperetti A. Bubble phenomena in sound fields: part one. Ultrasonics 1984;22:6S77.

[35] Prosperetti A. Bubble phenomena in sound fields: part two. Ultrasonics 1984;22:115–24.

[36] Qin Z, Bremhorst K, Alehossein H, Meyer T. Simulation of cavitation bubbles in a convergent–divergent nozzle water jet. J Fluid Mech 2007;573:1–25.

[37] Cartensen EL, Foldy LL. Propagation of sound through a liquid containing bubbles. J Acoust Soc Am 1947;19:481–501.

[38] Meyer E, Skudrzyk E. Uber die Akustischen Eigenschaften von Gasblasenschleiern in Wasser. Acoustica 1953;3:434–40; transl. by Devin C. On the acoustical properties of gas bubble screens in water. DTBM Transl. No. 285. David Taylor Naval Ship R & D Center. Washington, DC; 1958.

[39] Hsieh D, Plesset MS. On the propagation sound in a liquid containing gas bubbles. Phys Fluids 1961;4:970–5.

[40] Fox FE, Curley SR, Larson GS. Phase velocity and absorption measurements in water containing air bubbles. J Acoust Soc Am 1955;27:534–9.

[41] Silberman E. Sound velocity and attenuation in bubbly mixtures measured in standing wave tubes. J Acoust Soc Am 1957;29:925–33.

[42] Schiebe FR, Killen JM. An evaluation of acoustic techniques for measuring gas bubble size distributions in cavitation research. Rep. No. 120. Minneapolis, MN: St. Anthony Falls Hydraulic Laboratory, University of Minnesota; 1971.

[43] Commander KW, McDonald RJ. Finite-element solution of the inverse problem in bubble swarm acoustics. J Acoust Soc Am 1991;89:592–7.

[44] Commander K, Moritz E. Off-resonance contributions to acoustical bubble spectra. J Acoust Soc Am 1989;85:2865–8.

[45] Medwin H. In situ acoustic measurements of bubble populations in coastal ocean waters. J Geophys Res 1970;75:599.

[46] Medwin H. Acoustical determinations of bubble-size spectra. J Acoust Soc Am 1977;62:1041.

[47] Junger MC, Cole JE. Bubble swarm acoustics: Insertion loss of a layer on a plate. J Acoust Soc Am 1980;68:241–7.

[48] Strasberg M. The influence of air-filled nuclei on cavitation inception. DTMB Rep. No. 1078. Washington, DC: David Taylor Naval Ship R & D Center; 1956.

[49] Zhang MM, Katz J, Prospretti A. Enhancement of channel wall vibration due to acoustic excitation of an internal bubbly flow. J Fluids Struct 2010;26:994–1017.

[50] Plesset MS. Bubble dynamics. In: Davies R, editor. Cavitation in real-liquids. Amsterdam: Elsevier; 1964. p. 1–18.

[51] van Wijngaarden L. One-dimensional flow of liquids containing small gas bubbles. Annu Rev Fluid Mech 1972;4:369–96.

[52] Whitam GB. On the propagation of weak shock waves. J Fluid Mech 1956;1:290–318.

[53] Benjamin TB, Feir JE. Nonlinear processes in long-crested wave trains. Proc Symp Nav Hydrodyn, 6th. 1966. p. 497–8.

[54] Zwick SA. Behavior of small permanent gas bubbles in a liquid, Part I, Isolated bubbles. J Math Phys 1958;37:246–68.

[55] Zwick SA. Behavior of small permanent gas bubbles in a liquid, Part II, Bubble clouds. J Math Phys 1959;37:339–53.

[56] Zwick SA. Behavior of small permanent gas bubbles in a liquid, Part III, a forced vibration problem. J Math Phys 1959;37:354–70.

[57] Isay WH, Roestel T. Berechnung der Druck-vert eilung und Flugelprofilen in gashaltiger Wasserstromung. Z Angew Math Mech 1974;54:571–88.

[58] Isay WH, Roestel T. Die niederfrequent instationaire Druckverteilung an Flugelprofilen in gashaltiger Wasserstromung. Rep. No. 318. Inst. Schifflau, Univ. Hamburg; 1975.

[59] Wallis GB. One-dimensional two-phase flow. New York: McGraw-Hill; 1969.

[60] Solomon LP, Plesset MS. Non-linear bubble oscillations. Int Shipbuild Prog 1967;14:98–103.

[61] Blake FG. The onset of cavitation in liquids. Tech. Memo No. 12. Cambridge, MA: Acoustics Research Laboratory, Harvard University; 1949.

[62] Strasberg M. Onset of ultrasonic cavitation in tap water. J Acoust Soc Am 1959;31:163–76.

[63] Akulichev VA. The calculation of the cavitation strength of real liquids. Sov Phys Acoust (Engl Transl) 1965;11:15–18.

[64] van der Walle F. On the growth of nuclei and the related scaling factors in cavitation inception. *Proc Symp Nav Hydrodyn, 4th.* 1962.

[65] Prosperetti A, Crum LA, Commander KW. Nonlinear bubble dynamics. J Acoust Soc Am 1988;83:502–14.

[66] Boguslavskii YY, Korets VL. Cavitation threshold and its frequency dependence. Sov Phys Acoust (Engl Transl) 1967;12:364–8.

[67] Strasberg M. Rectified diffusion: comments on a paper of Hsieh and Plesset. J Acoust Soc Am 1961;33:161.

[68] Crum LA. Measurements of the growth of air bubbles by rectified diffusion. J Acoust Soc Am 1980;68:203–11.

[69] Arndt REA, Ippen A. Rough surface effects on cavitation inception. J Basic Eng 1968;90:249–61.

[70] Guth W. Nichtlineare Schwingungen Von Luftblasen in Wasser. Acustica 1956;6:532–8.

[71] Epstein PS, Plesset MS. On the stability of gas bubbles in liquid-gas solutions. J Chem Phys 1950;18:1505–9.

[72] Parkin BR, Kermeen RN. The roles of convective air diffusion and liquid tensile stresses during cavitation inception. Proc IAHR Symp Cavitation Hydraul Mach. Sendai, Japan. 1963.

[73] van Wijngaarden L. On the growth of small cavitation bubbles by convective diffusion. Int J Heat Mass Transf 1967;10:127–34.

[74] Hsieh D, Plesset MS. Theory of rectified diffusion of mass into gas bubbles. J Acoust Soc Am 1961;33:206–15.

[75] Eller A, Flynn HG. Rectified diffusion during non-linear pulsations of cavitation bubbles. J Acoust Soc Am 1965;37:493–503.

[76] Safar MH. Comment on papers concerning rectified diffusion of cavitation bubbles. J Acoust Soc Am 1968;43:1188–9.

[77] Eller AI. Growth of bubbles by rectified diffusion. J Acoust Soc Am 1969;46:1246–50.

[78] Eller AI. Bubble growth by diffusion in an 11-kHz sound field. J Acoust Soc Am 1972;52:1447–9.

[79] Pode L. The deaeration of water by a sound beam. DTMB Rep. No. 854. Washington, DC: David Taylor Naval Ship R & D Center; 1954.
[80] Boguslavskii YY. Diffusion of a gas into a cavitation void. Sov Phys Acoust (Engl Transl) 1967;13:18–21.
[81] Crum LA. Rectified diffusion. Ultrasonics 1984;22:215–23.
[82] Lee HS, Merte H. Spherical bubble growth in uniformly superheated liquids. Int J Heat Mass Transf 1996;39:2427–47.
[83] Knapp RT, Hollander A. Laboratory investigations of the mechanism of cavitation. Trans ASME 1948;70:419–35.
[84] Trilling L. The collapse and rebound of a gas bubble. J Appl Phys 1952;23:14–17.
[85] Gilmore FR. The growth or collapse of a spherical bubble in a viscous compressible fluid. Rep. No. 26–4. Pasadena: Hydromech. Lab., California Inst. Technol.; 1952.
[86] Kirkwood JG, Bethe HA. The pressure wave produced by an underwater explosion. OSRD Rep. No. 588. Office of Scientific Research and Development, National Defense Research Committee; 1942.
[87] Mellen RH. An experimental study of the collapse of a spherical cavity in water. J Acoust Soc Am 1956;28:447–54.
[88] Mellen RH. Spherical pressure waves of finite amplitude from collapsing cavities. Rep. No. 326. New London, CT: U.S. Navy Underwater Sound Laboratory; 1956.
[89] Hunter C. On the collapse of an empty cavity in water. J Fluid Mech 1960;8:241–63.
[90] Johsman WE. Collapse of a gas-filled spherical cavity. J Appl Mech 1968;90:579–87.
[91] Esipov IB, Naugol'nykh KA. Collapse of a bubble in a compressible liquid. Sov Phys Acoust (Engl Transl) 1973;19:187–8.
[92] Hickling R, Plesset MA. Collapse and rebound of a spherical bubble in water. Phys Fluids 1964;7:7–14.
[93] Hickling R. Some physical effects of cavity collapse in liquids. J Basic Eng 1966;88:229–35.
[94] Ivany RD, Hammitt FG. Cavitation bubble collapse in viscous, compressible liquids—numerical analysis. J Basic Eng 1965;87:977–85.
[95] Esipov IB, Naugol'nykh KA. Expansion of a spherical cavity in a liquid. Sov Phys Acoust (Engl Transl) 1972;18:194–7.
[96] Guth W. Zur Enstehung der Stosswellen bei der cavitation. Acustica 1956;6:526–31.
[97] Lofstedt R, Barber BP, Putterman SJ. Toward a hydrodynamic theory of sonoluminescence. Phys Fluids A 1993;5(11):2911–28.
[98] Beyer RT. Nonlinear acoustics. Washington, DC: Nav. Ship Syst. Command Publication; 1974.
[99] Pernik AD. Problems of cavitation. Probl. Kavitatsii, IZD-VO Sudostroenie, Leningrad; 1966 (in Russ.).
[100] Khoroshev GA. Collapse of vapor-air cavitation bubbles. Sov Phys Acoust (Engl Transl) 1964;9:275–9.
[101] Knapp RT. Cavitation mechanics and its relation to the design of hydraulic equipment. Proc Inst Mech Eng Part A 1952;166:150–63.
[102] Harrison M. An experimental study of single bubble cavitation noise. DTMB No. 815. Washington, DC: David Taylor Naval Ship R & D Center; 1952.
[103] Lauterborn W. Kavitation durch Laserlight. Acoustica 1974;31:51–78.
[104] Lauterborn W. General and basic aspects of cavitation. In: Bjorno L, editor. Proc Symp Finite Amplitude Eff Fluids. Guildford, Surrey, England: IPC Sci. Technol. Press; 1974. p. 195–202.

[105] Popinet S, Zaleski S. Bubble collapse near a solid boundary: a numerical study of the influence of viscosity. J Fluid Mech 2002;464:137–63.
[106] Parkin BR. Scale effects in cavitating flow. Rep. No. 21-7. Pasadena: Hydrodyn. Lab., California Inst. Technol.; 1951.
[107] Blake WK, Wolpert MJ, Geib FE. Cavitation noise and inception as influenced by boundary layer development on a hydrofoil. J Fluid Mech 1977;80:617–40.
[108] Ivany RD, Hammitt FG, Mitchell TM. Cavitation bubble collapse observations in a Venturi. J Basic Eng 1966;88:649–57.
[109] Ill'in VP, Morozov VP. Experimental determination of the ratio of cavitation noise energy to the initial bubble energy. Sov Phys Acoust (Engl Transl) 1974;20:250–2.
[110] Blake WK, Wolpert MJ, Geib FE, Wang HT. Effects of boundary layer development on cavitation noise and inception on a hydrofoil. D. W. Taylor, NSRDC Rep. No. 76–0051. Washington, DC: Naval Ship R & D Center; 1976.
[111] Schutler ND, Messier RB. A photographic study of the dynamics and damage capabilities of bubbles collapsing near solid boundaries. J Basic Eng 1965;87:511–17.
[112] Kozirev SP. On cumulative collapse of cavitation cavities. J Basic Eng 1968;90:116–24.
[113] Hammitt FG. Discussion to "On cumulative collapse of cavitation cavities". J Basic Eng 1969;91:857–8.
[114] Levkovskii YL. Collapse of a spherical gas-filled bubble near boundaries. Sov Phys Acoust (Engl Transl) 1974;20:36–8.
[115] Shima A, Nakajima K. The collapse of a non-hemispherical bubble attached to a solid wall. J Fluid Mech 1977;80:369–91.
[116] Shima A. The behavior of a spherical bubble in the vicinity of a solid wall. J Basic Eng 1968;90:75–89.
[117] Plesset MS, Mitchell TP. On the stability of the spherical shape of a vapor cavity in a liquid. Rep. No. 26-9. Pasadena: Hydrodyn. Lab., California Inst. Technol.; 1954.
[118] Blake WK. Aero-hydroacoustics for ships. U.S. Gov. Publ. DTNSRDC Rep. 84-010, 2 vols. 1984.
[119] Fitzpatrick HM, Strasberg M. Hydrodynamic sources of sound. Proc Symp Nav Hydrodyn, 1st. Washington, DC; 1956. p. 241–80.
[120] Fitzpatrick HM. Cavitation noise. *Proc Symp Nav Hydrodyn 2nd*. Washington, DC; 1958. p. 201–5.
[121] Tomita Y, Shima A. High speed photographic observations of laser-induced cavitation bubbles in water. Acustica 1990;71(3):161–71.
[122] Ellis AT. On jets and shock waves from cavitation. Proc Symp Nav Hydrodyn 6th. 1966.

Index

Note: Page numbers followed by "*f*" and "*t*" refer to figures and tables, respectively.

E

F

G

H

I

J

K

L

M

T

U

V

Printed in the United States
By Bookmasters